Bird Atlas of Botswana

Bird Atlas of Botswana

Huw Penry

With cartography by
Huw Penry and Andre Jahns

UNIVERSITY OF NATAL PRESS
PIETERMARTIZBURG
1994

University of Natal Press, Pietermaritzburg
Private Bag X01, Scottsville 3209

ISBN 0 86980 895 8 (paperback)
0 86980 894 X (hardback)

Cover: Crested Francolin
Original painting: Penny Meakin
Design: David Moon

Frontispiece: Redcrested Korhaan
Photo: John Carlyon

Typeset in the University of Natal Press
Printed by Kohler Carton and Print
Box 955, Pinetown 3600

Contents

Figures and tables

Figures *Page*

Tables

Foreword

The growing number of bird atlases is a sign of their value in biological research, conservation and birdwatching. Atlases are a reference for ornithologists, a baseline for ecologists and conservationists, a measure of wildlife resources and quality of life, a tool for land-use planners, and an information source for residents and tourists.

The cost of data collection for a bird atlas cannot be met from corporate funds. Bird atlas projects can be completed only by harnessing the enthusiasm of amateurs who devote time, money and expertise to the task. That such willingness exists demonstrates the popularity of birds. In the case of Botswana – a developing nation with a large land area – fieldwork for the *Bird Atlas of Botswana* was conducted by a remarkably small group of dedicated individuals.

Botswana's avifauna has been treated comprehensively only twice before. Smithers' 1964 checklist, using mostly specimen evidence, provided a sound basis for fieldwork for the *Bird Atlas of Botswana*, while Newman's 1989 field guide incorporates advances in field identification that make the handling of most birds unnecessary. The sheer amount of information in the *Bird Atlas of Botswana* adds greatly to our knowledge of Botswana's birds. In particular, the systematic coverage of the whole country, not just the more popular areas, provides a more detailed ornithological assessment of the Kalahari than was previously possible.

I got to know Dr Huw Penry over twenty years ago when he worked on the Zambian Copperbelt. When fieldwork for the Zambian bird atlas was launched in 1976, we were enthusiastic members of the team and spent much time together. Huw is cautious in his judgements, rigorous in his methods and has a record of seeing projects through. Having moved to Jwaneng in 1980, he initiated the Botswana Bird Atlas. Publication of the *Bird Atlas of Botswana* completes the project, makes an important contribution to Afrotropical ornithology and provides a firm foundation for the next generation of students of Botswana's birds.

September 1994

DYLAN R. ASPINWALL
Chairman
Zambian Ornithological Society

Acknowledgement to sponsors

The publishers have pleasure in acknowledging the generous donations given by the following organisations towards the costs of the book:

The Anglo American and De Beers Chairman's Fund

 Sheraton Gaborone

Chapter 1

Introduction

This book is the culmination of ten years of fieldwork in Botswana from July 1980 to June 1990 by resident and visiting birdwatchers. Its purpose is to produce a distributional statement in the form of a map for each regularly occurring bird species. It is the first such statement for Botswana and is intended as a blueprint for future more detailed assessments for individual species, families or the total avifauna. Primarily it is concerned with the common species but information on the distribution of uncommon or rare species was also accumulated and is included.

Background

In 1980 there was only one publication in existence dealing with the avifauna of Botswana as a national entity. This was *A Checklist of the Birds of the Bechuanaland Protectorate and Caprivi Strip* by R.H.N. Smithers published in 1964. That this work was nearly 20 years old was an important reason for undertaking the *Botswana Bird Atlas* as a project. Subsequently, and partly because some of the information contained in this atlas project was made available, a fieldguide, *The Birds of Botswana*, was published by Kenneth Newman in 1989.

A reasonably accurate knowledge of the distribution and composition of the avifauna of a country is a desirable baseline for more detailed study. Standard works containing extrapolations of distribution for Botswana in a southern African context have been misleading in two senses (a) they tended to give the impression that areas had been studied when in fact they had not even been visited, and (b) some were inaccurate and showed that in certain instances authors had insufficient knowledge of Botswana to extrapolate the distribution.

In 1980 there was no ornithological infrastructure in Botswana, that is no resident ornithologist, no ornithological collection in its National Museum, no bird club and no communication medium dealing specifically with bird information. Unless this was created by birdwatching enthusiasts, such an infrastructure was unlikely to materialize in the foreseeable future when more important priorities existed for Botswana at that stage of its national development. This was a common feeling among the small, temporarily resident group of birdwatching expatriates and no different from circumstances which had led to the formation of ornithological societies and clubs in other African countries, such as Zambia and Malawi. As a result of the efforts of several individuals the Botswana Bird Club was formed in 1980, as a branch of the Botswana Society. The Society had been founded in 1968 to encourage research and publication in Botswana on a wide range of subjects such as history, economics, archaeology, anthropology, sociology, botany, the conservation of wildlife, hydrology, biology, law and arts. The first edition of the *Babbler*, the magazine of the Botswana Bird Club, appeared in 1980 and has subsequently been published bianually.

It was in this climate, but prior to the formation of the Botswana Bird Club, that I considered a basic distributional study of the avifauna to be a priority which could be achieved with dedication and perseverance by amateur birdwatchers over 10 years. This optimism was based on nine years of birdwatching experience in neighbouring Zambia of which four years had been spent on fieldwork for the *Bird Atlas of Zambia* by Dowsett & Aspinwall (in prep).

Bird-atlas studies were becoming popular in many countries on several continents in the early 1980s. Such projects had several advantages: they did away with the unnecessary slaughter of birds as proof that species existed in a particular area; they ensured that all geographical areas of the same size were studied and thus eliminated inaccuracies from extrapolations; they could be compared with and, therefore, assessed against similar projects in other areas of the world; they allowed non-professional ornithologists to contribute meaningfully to the scientific study of birds; and, above all, most participants found that the work involved was enjoyable and richly rewarded their efforts.

With a nucleus of enthusiastic birdwatchers Botswana had the potential to make its own assessment of its avifauna. Most of what little was known about birds in Botswana had originated from visiting ornithologists from South Africa and Zimbabwe. This contribution to bird knowledge in Botswana had in some respects discouraged ornithological endeavour in Botswana itself, although recognizing and acknowledging at the same time that those pioneers had laid some good foundations. There had been, for example, a heavy bias towards studying the Chobe, Okavango and Makgadikgadi regions and, with the major exception of Smithers (1964), an almost total neglect of the rest of the country. Statements made about the role of Botswana in the ornithology of southern Africa related almost entirely to the contribution of the northern wetlands—the importance of the vast expanse of the Kalahari as a vulnerable and interesting semidesert and its contribution to migration patterns in southern Africa seemed to have been overlooked. These statements are by no means criticisms of the contribution of South African and Zimbabwean ornithologists or of their institutions; they are to show that ornithological contributions from outside the country were not able to address some basic issues and the time was ripe for Botswana to do the groundwork itself, otherwise these issues would not be resolved.

Figure 1 Basic map of the project

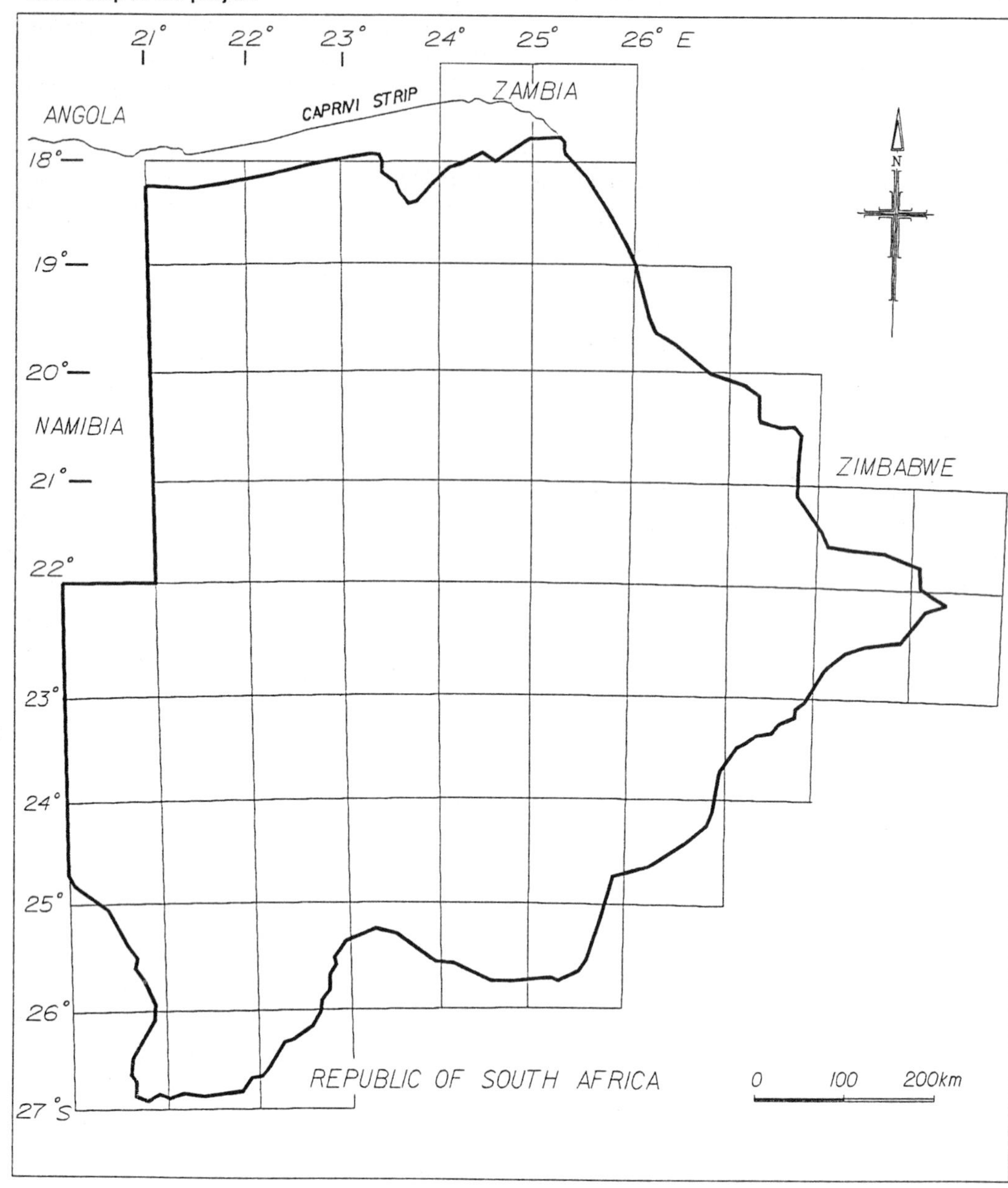

The first such issue was to study in greater detail the composition of the avifauna of the Kalahari and its relationship to adjacent areas. This a bird atlas could present. Other issues were: How well had areas been studied and who had the information? Who should control ornithological information collected in Botswana and decide on direction and priorities? Where and how should new and historical information be stored? Botswana as an independent country ought to make a statement on bird distribution as a national entity and not only as part of a subcontinent.

The above reasons are outlined so that people unfamiliar with birdwatching in Africa can appreciate the background to ornithological endeavour in many parts of the continent where birdwatching is not a well known hobby nor a national priority and the birdwatcher is often treated with suspicion.

The project, therefore, was governed by certain realities which may surprise people who have not experienced the circumstances in which it was undertaken. As a conscious decision it was not widely advertised. Previous experience had indicated that the most manageable system was to utilise a group of enthusiastic birdwatchers whose field skills were known to the Coordinator (in this case, the author), who would contribute on a regular basis and from several localities with credible and consistent bird lists. Since it was important to reduce as much as possible the number of records which would need verification, people new to African birds were not actively encouraged. Most of the active participants were likely to be amateur birdwatchers with full-time jobs and little spare time. Access to many places in Botswana requires the correct vehicle and equipment. It would not be justified to encourage people without the necessary experience to venture into such areas. Another limitation was the turnover of expatriates, usually on two- or three-year contracts, some of whom were reaching competence in field identification after one or two years, and thus with only a short period available for contributing. Efforts were made to encourage residents to participate but the outcome was disappointing, with a handful of notable exceptions. At the end of the project only two people had been involved throughout the 10 years. Fortunately there were major contributors in the intervening period—their contribution is acknowledged below.

The project foundered in 1982 when I was transferred by my employers to South Africa. At that time there was still only tentative support for it, even from the Botswana Bird Club, and it became necessary to set up structures in Botswana to facilitate the transfer of information and to find persons willing to commit themselves to the required tasks. This did not materialise until early 1984 when a sufficiently interested Bird Recorder was available. Because of the small number of people with the necessary time and dedication and with the regular turnover of recruits—for example, there were six different Bird Recorders for the Botswana Bird Club during the 10 years—both the amount of fieldwork and the transfer of information fluctuated. In addition, I was handicapped by distance in making personal contact with potential contributors but with the help of a few major participants the project survived.

About the book

The book is designed primarily as an atlas, i.e. a book of maps. There are 496 main species maps which show distribution symbols in 30-Minute Squares on a grid size of One-Degree Square (these units are explained in Chapter 2). Figure 1 shows the base map used for illustrating species distribution. A simplified version of this is the base map used for illustrating the distribution of the main species. Species are discussed and illustrated in systematic order. The systematic order, English and scientific names and numbering follows those of *Roberts' Birds of Southern Africa*, 5th edition 1985 by G.L. Maclean which itself is based on the Southern African Ornithological Society's *Checklist of Southern African Birds* 1980 by P.A. Clancey. The text which accompanies each species is a short summary of the following factors:

Status

A statement is made with respect to each species. This includes: whether the species is resident or a migrant; a synopsis of the distribution shown in the main map; an assessment of how common or rare it is in different regions of Botswana; some comments on movements and seasonality; the relationship to its distribution in neighbouring countries; breeding dates; a comment on the numbers of birds usually seen at one encounter. The term resident is used in the context that the species is not a regular migrant and is expected to breed in Botswana. If there is no further comment on breeding it means that no breeding has been recorded. Breeding details are given when these are available and the months quoted are for egglaying dates based on known records. The months mentioned are not necessarily the only months in which the species may breed. A project on birds breeding in Botswana is in progress; records on the main maps in the current work, therefore, do not attempt to define breeding localities. Place names are mentioned to define limits of distribution and sometimes to indicate good localities for observing the species.

Habitat

The main habitat in which the species is found in Botswana is given first, followed by additional or less commonly used habitats. Subdivisions of habitats are included when they may assist in differentiating between species. Habitat classification is defined in the section on Vegetation (page 25).

Analysis

Under this section, the first figure denotes the number of squares in which the species has been recorded and the ensuing percentage is that total expressed as a percentage of the 230 squares in Botswana. Total Count is the total of all records for that species on the database of the atlas project. The ensuing percentage in this case is the portion of the total database which that species' records occupy. These statistics allow a quick and simple comparison of one species with another and add a dimension of overall assessment to the frequency analysis of each species in each square. In statistical terms this is a crude measurement although the figures themselves are accurate and are given only as a guide. The objective of the study is to produce a descriptive statement and not a full quantitative assessment.

Example: **43 squares (19%)** **Total count 138 (0,075%)**

Continental distribution

A map of Africa south of the Tropic of Cancer is inserted in the top right corner of each main map so that, at a glance, readers can relate the distribution of the species in Botswana to the species' distribution on the continent. As discussed later, the Kalahari represents a barrier to species of tropical woodland and savanna affinities and the way in which this affects distribution is best illustrated by including the continental distribution visually. These maps have some limitations in representing information. It is difficult to show the true distribution of some Palaearctic and intra-African migrants which may vary from year to year. The separate ranges of the Palaearctic and African races of such species as Little Bittern and Peregrine, for example, are not delineated. Where clearly defined off-season or wintering ranges are known these ranges are drawn. Where the ranges are unclear or unknown but the species is known to travel through a country, the migration route is included in the total African range. This caters also for species which, for example, spend the nonbreeding season at higher latitudes in some years but migrate further south in others. The information on which these maps are primarily based are the first three volumes of *The Birds of Africa* (1982 to 1988) (References, page 305) and the *Atlas of Speciation in African Passerine Birds* (Hall and Moreau, 1970). In the southern African region they include information from *Birds of the Transvaal* (Tarboton *et al.*, 1987), *First Atlas of Bird Distribution in the Orange Free State* (Earle & Grobler, 1987), *The Complete Book of Southern African Birds* (Ginn *et al.*, 1989) as well as *Roberts' Birds of Southern Africa* (Maclean, 1985).

Analysis by month

The histogram in the bottom right corner of each main map shows the months in which the species has been recorded. Each bar represents the sum of all the records in the month shown on the x-axis. A discrepancy often exists between the sum of all the bars and the Total Count reported in the text. The missing records are those for which no month date was

Figure 2 **Areas visited by the author**

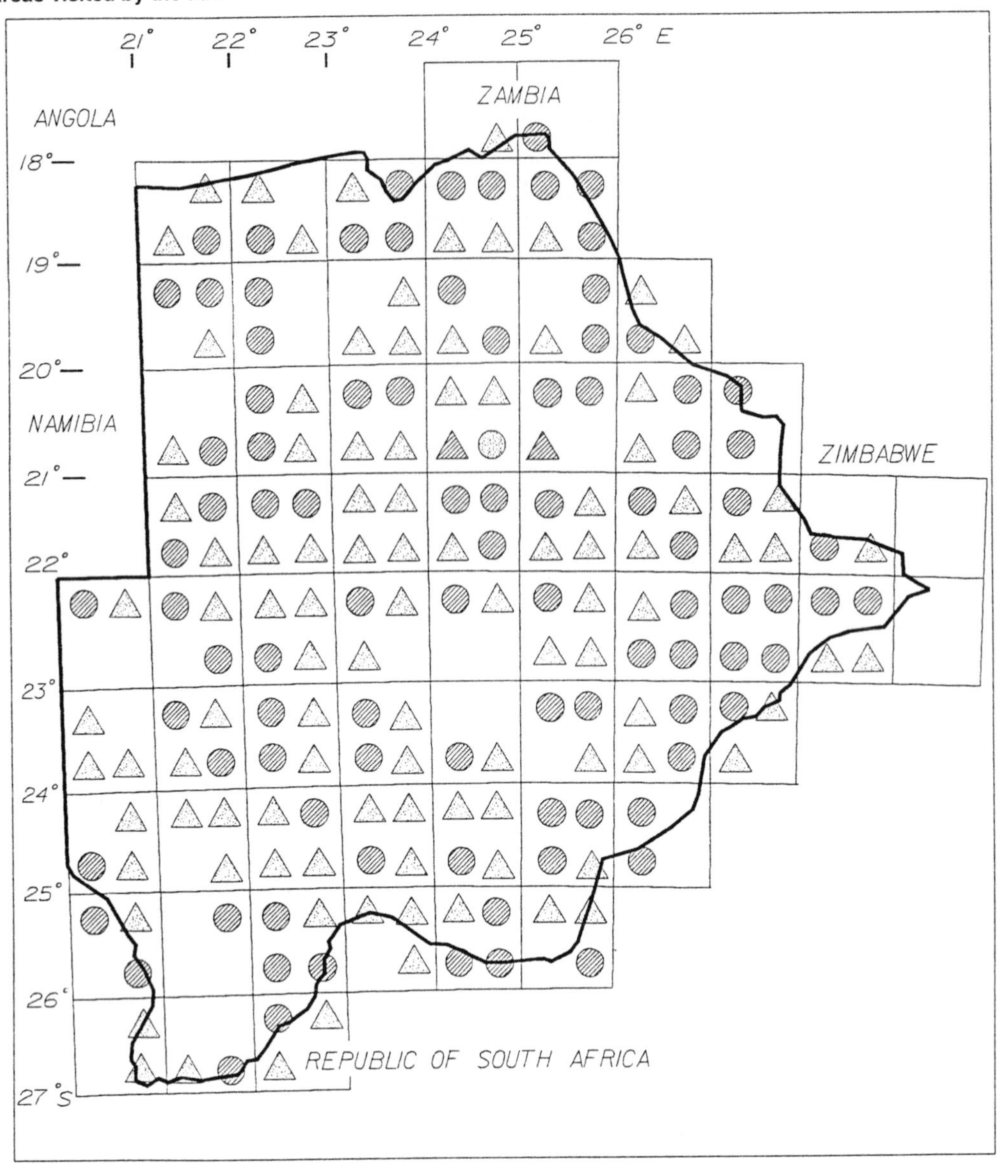

△ Squares camped in 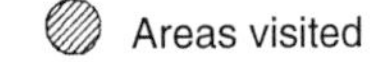 Areas visited

submitted and this is more fully explained under Methods (page 7). Because very wide fluctuations occur in the number of records for different species, the graphs have **different scales.** Whenever possible the same scale is used for species in the same family or those which might make up an acceptable subdivision of a family. The purpose of these histograms is to highlight genuine seasonal fluctuations and to hint at possible movements without exaggerating changes which may look significant because the sample size is too small. However, some graphs have exaggerated changes because the scale could not eliminate this artefact. The histograms are at their most effective for migrants which are totally absent or greatly reduced at certain times of the year. For local species there may or may not be significant monthly or seasonal movement or, in other cases, significant evidence of no movement. These histograms are included for all the main species. Even if it is not possible to draw accurate conclusions on the current evidence there may be pointers for future study or the information may be found to support or refute other analyses now or in the future.

The future

One objective of this work is to smooth the path for the future study of birds in Botswana. The distributional information included in it is probably sufficient to exclude distributional research as a priority for future effort, but this information is by no means complete. The publication may induce a response from people who discover omissions in the distributional information for several species in certain areas and birdwatchers are urged to be bring these gaps to the attention of the Botswana Bird Club. Such updating is needed and should be continued—even after the publication is revised at some future date. All such records should ideally be in comparable format—namely square, observer, date; even if the square size changes through changed circumstances, it is likely to be a denominator of the current square size. All submissions should be regarded with the same value as museum specimens and thus properly and accurately labelled. It must not be construed that distributional information has now ended. Circumstances which affect distribution are constantly changing as discussed elsewhere. This work covers mainly a drought period and its conclusions will differ from those in a wet climatic cycle. Moreover, knowledge of many common species is still inadequate and there is a large pool of information which needs to be accumulated.

Whenever possible the species text mentions the need for future study. Newcomers to Botswana may require such guidance in the limited period which they have available for bird research. There is ample scope for study in other areas not highlighted in this work and careful scrutiny of the contents should provide some guidance in this direction. A project on breeding is currently in progress and in this field there is much to be done. Basic information on common species is again lacking so that all people of all levels of competence in birdwatching can contribute usefully by reporting birds breeding in their gardens. Information is required on almost any aspect of avian biology—for example, behaviour, food items, movements, moult, colour variations, vocalisation and unusual numbers. It is important to be aware that a central organisation now exists in Botswana for the collection of such information and everyone is encouraged to contribute any information which they consider to be interesting. Visitors should note that it is an expected courtesy in any country to communicate bird information to the resident Bird Society or Club.

The Botswana Bird Club is active in the fields of conservation and education. These are fundamental and pressing issues of our time and active helpers are constantly required. In the text of this work mention is made of some of the species which are in danger of extinction in Botswana. The issues are dealt with more fully in two articles on the threatened birds of Botswana (Penry, 1986) which discuss the Botswana species mentioned in the Red Data Books by Collar & Stuart (1985) and Brooke (1984). Detailed mention of conservation issues are outside the scope of this work, but it should be appreciated that this bird atlas project is intended as a contribution to the conservation effort in Botswana.

Names and pronunciation

Setswana and English are the official languages of Botswana. In this work, Setswana names for different bird species are not given and criticism of this fact may arise. In defence, the work is primarily a scientific report of a 10-year field study and is unlikely to appeal to people who speak only Setswana as English is the main language used in the teaching of applied science. Beyond this Setswana names for some birds vary from one area of the country to another and often are generic (e.g. the same word is used for all species of canary in some areas but may be more specific in others). The same Setswana word may refer to different species of the same family in different parts of the country. Some generic Setswana names cross taxonomic family lines. It is beyond the authority of this work to prescribe Setswana names where none currently exist or where all may not agree. It is sufficient cause for thanks that there is reasonable, but not total, consensus on the English names of birds without confusing the issue further. The pronunciation and different spellings of Setswana names also requires some clarification. Wherever possible, the Setswana spelling is used for place names and this is the most recent version available. Anglicised versions found in old publications have been eliminated but some old forms are included for reference in the Gazetteer. Some examples are Xau not Dow, Khakhea not Kakia, Lobatse not Lobatsi. In this work Kalahari is spelt in the English version and Makgadikgadi in the Setswana.

Atlas comparisons

Table 1 enables comparisons to be made with other Atlas projects in the southern African region.

Table 1 **Dynamics of southern African atlas projects**

Area	Scale	No. of squares	30MS equiv.	Contributors
Transvaal	15MS	456	126	344
Natal	15MS	426	103	299
O.F.S.	15MS	234	70	180
Botswana	30MS	230	230	187

15MS = 15-Minute Square and 30MS = 30-Minute Square (which is four times larger than 15MS)

Knowledge of the avifauna in each area at the conclusion of these atlas projects is illustrated in Table 2.

Table 2 **Outcome of southern African atlas projects**

Area	Main species	Rare species	Need confirming
Transvaal	639	Included	36
Natal	530	+ 28	14
O.F.S.	383	+ 53	12
Botswana	496	+ 59	36

Contributors

Contributors to the fieldwork are listed below and their contribution is acknowledged with gratitude. Contributions from two or more travelling companions are credited to each name. The solo efforts of the main contributors are shown in brackets after their total contribution.

Over 100 cards

Penry E.H.	501	(208)
Hunter N.D.	464	(252)
Aldiss D.T.	211	(126)
Borello W.D. & R.M.	196	(138)
Brewster C.A.	159	(159)
McGowan G.	155	(150)
Barnes J.E.	135	(89)
Skinner N.J.	134	(119)
Aspinwall D.R.	116	(87)

40–100 cards

Botswana Bird Club	92
Bushell B. & D.	86
Tarboton W.R.	79
Pickles R.	58
Bird M. & E.	56
Newman K.B.	56
Grobler N.J.	55
Oake K.	48
Nelson R.	43
Randall R.D.	42

20–39 cards

Culverwell J., Dowsett R.J., Greenwood-Penny P., Jacobsen N., Kvist A., Lindsay K., Longden K., Marokane W., Muller M., Pryce E., Randall R.M., Sheldon D., Soroczynski M., Teuten R., White R.H., Wilson H.E.

10–19 cards

Balden J., Brodie C. & R., Butchart D., Cooper-Poole B., Dowsett-Lemaire F., Fabian D.T., Graham B. & J., Hepburn J., Hodgson M.C., Longden T., Maclean G.L., Meadows B., Pollock C. & J., Rea M., Schroeder H. & L., Spawls S., Woollard E. & J.

5–9 cards

Bishop D., Boshoff A.P., Colebrook-Robjent J.F.R., Comley P., Davidson I., Dwyer S., Halliday T.A., Hancock P., Hartmann M., Heery P., Herholdt J.J., Hopson E., Jones C. & D., Marsland D., Nielsen M.E., Osborne T.O., Reed H. & M., Rockingham-Gill D., Searle R.F., Shaw W.C., Springer P., Start J.M., Steyn P., Syversten P.O., Thompson N.F., Tindall R., Williams R., Wilson J., Wragge D., Youthed S.

1–4 cards

Allen B., Badubi D., Beasley A., Bell C., Bendsen H., Bentzen C., Bissett R., Blundell A., Breen K., Brooke R.K., Brown C.J., Bulley B.G., Burgess J., Calder D.R., Carter C.C., Carr-Hartley K.P., Carter J., Chouler J., Collias E. & N., Corlett J., Creek R., Crouse R., De Greyling G., Dobbs D. & J., Douse M., Drynan J., Fisher J.H., Flatt J., Fowkes J. & S., Frere T., Gerhardt J., Gillard L., Hall D., Hall R.S., Hanmer D., Harrison M., Hester A., Hughes C., Hurwitz E., Irvine K., Ives M.A.P., Jones C., Keast J., Kerton A., Kurtz D. & J., Lockwood G., Loon R., Lorantz M., Louw G. & R., Maret P., Marshall B., Martens A., B. & G., McAllister J., McGowan J., McKenzie P., McLuskie J., Monahan P., Moore P.C.L., Muller B., Mundy P., Ntshogotha J., O'Connel B., Owens D. & M., Palmer N.G., Peake N., Pearson D., Pearson J., Peckover B., Penney M., Petame Mr, Richardson D.M., Rickets J.A., Robertson A., Roussou J., Sassoon S., Sethoko Mr, Skarpe C., Smith C., Smith K.M., Spriggs J., Stowe B., Stuart C. & T., Sullivan E., Sussman Q., Sydes C.L., Tarrant I., Trott F., Vernon C.J., van de Reep J. & S., von Plato A., Wall H., Whitehead P., Whittlesey R.W., Wilkinson D. & V., Williams A.J., Williams N., Williams R.G., Williamson J.E., Wilson M., Wolf D., Wright B., Wright R., Yellen J.

Photographs

A limited number of photographs has been included to show different habitats throughout Botswana. The Okavango Delta has been omitted from the illustrations because there are many books with excellent photographs of this famous wetland. Most serious ornithologists and wildlife enthusiasts with an interest in Botswana will possess such books. The vegetation of the Okavango is discussed at length in the text of Chapter 3. The photographs concentrate mainly on the varieties of vegetation in the Kalahari. Sixteen of the twenty-four illustrations depict the Kalahari savannas ranging from Deception Valley in the northern part of the Central Kalahari Game Reserve near the Makgadikgadi Depression to the western Molopo Valley in the southwest. These areas are not well-known and misconceptions exist amongst birdwatchers who have not visited these dry savannas concerning their appearance and variety. Also missing from the habitat illustrations are the Makgadikgadi Pans which are also well illustrated in books and journals.

Acknowledgements

In addition to the many people who contributed to the fieldwork, there are many who gave assistance and support at various stages of the project. Base maps: Roy Anderson, Wendy Borello, Robert White, Hilton Williams, Clem Vernon. Bird recorders: Janet Barnes, Di Bushell, Neville Skinner, Sue Walker (nee Dwyer). Vetting and follow-up of unusual records: Records Subcommittee of the Botswana Bird Club and Bird Recorders. Free checklists: Botswana Bird Club. Computer programme: Brian Brink. Computing: Andre Jahns, Cobus Rogers, Dick Starling. Printing: Stuart Armstrong, Jo Demelenne, Basil Brown, Francois Retief, Bill Reville. Safety: Keith Beaver and Chloride SA (Pty) Ltd. Exceptional hospitality: Julia and Nigel Hunter, Green Tabangwe of Sepako, Basil Erasmus of Medenham. Geology: Don Aldiss. Safaris: Don Aldiss, Nick Grobler, Nigel Hunter, Warwick Tarboton. Climatic data: Michael Main and Meteorological Office, Gaborone. Permission to use staff and equipment on Vaal Reefs mine: Bob Williams and Nap Mayer.

Unless there had been sufficient moral support and enthusiasm, through some dark hours, against the odds and at crucial times over twelve years, this work would not have been concluded. For sustenance or timely encouragement special thanks are due in varying measure to Don Aldiss, Dylan Aspinwall, Molly Benson, Chris Brewster, Bob Dowsett, Len Gillard, Nick Grobler, Nigel and Julia Hunter, Gordon Maclean, Graham Madge, James Monk, Kenneth Newman, Patrick Niven, David Rockingham-Gill, Humphrey and Jackie Sitters, Simon Stuart, Warwick Tarboton, Emil Urban, and Hardy Wilson. However, nobody has been more long-suffering and supportive than my wife Sue and my two neglected children.

Chapter 2

Objectives, methods and results

Objectives

The primary objective of a bird-distribution atlas is to confirm the presence of all bird species which occur in each geographical unit of the study area and to report the results in the form of a map for each species. Such an objective cannot be reached *in toto* because bird distribution varies with time, climatic cycles, changes in vegetation, etc. The aim is to get as close to the total as possible so that only rare species may be missing from the list. A guideline as to how many species can be expected to occur in each unit is necessary, otherwise the study could stop short of a reasonable result or continue for an unnecessary length of time. As there were no previous atlas studies and virtually no locality bird lists in existence in Botswana in 1980, a crude estimate of the number of species expected to occur in each 30-Minute Square had to be made (see Methods). This calculation was considered to be the **target** for the Botswana Bird Atlas Project.

In order to achieve the target and to be able to measure the outcome, certain goals were set as follows:

1. To visit every square in Botswana.
2. To record at least a basic 50 species for each square.
3. To obtain a species-per-square count commensurate with the calculated potential number of species in each square.
4. To visit each square at least five times in the 10-year period so that some assessment of the frequency of occurrence of the commonest species could be made.

An ideal would have been to visit every square in Botswana every month for ten years. This was not practical and not set as an objective. However, information on migration, seasonal variations and movements was desirable in a country where virtually no such information existed. The amount of information collected in the project—even at the very low 4% sample rate of five visits per square—should allow an estimate of frequency sufficient to make a preliminary statement. This goal was set as a secondary objective which would allow the distributional symbols to be graded according to frequency of occurrence. The grading of symbols would illustrate species frequency on a geographical basis and thus allow assessments against other biological factors such as vegetation and rainfall. This would also redress false impressions of distribution which are given in maps which illustrate distribution as a continuous shaded area.

In order to show species movements, it was agreed that the parameters of the project would include the date, i.e. month and year. Seasonal variations and migrations were to be analysed only on a monthly basis as it was anticipated that insufficient data would be collected in every year from which to make a meaningful assessment of annual variations. Seasonal variations and movements would be shown as histograms accompanying each map. These histograms would relate to the whole country and thus occasionally obliterate regional variations in movements which, therefore, would need to be mentioned in the text.

Methods

The measurements, terminology, strategies, data recording and data presentation of the project are described below.

Measurements

The primary measurement of the project is an expression of the simple formula: what? where? when?

The study unit of the atlas project therefore concerned **the presence of a species (what?) in a square (where?) in a month (when?)** The different parameters of **presence** were: accepted sight record, accepted sound record, confirmed taperecording, identifiable bird part, e.g. carcass, wing, tail or feather, recognizable photograph, specimen.

A **square** is the geographical area defined by the boundaries of the degree and half-degree lines of latitude and longitude.

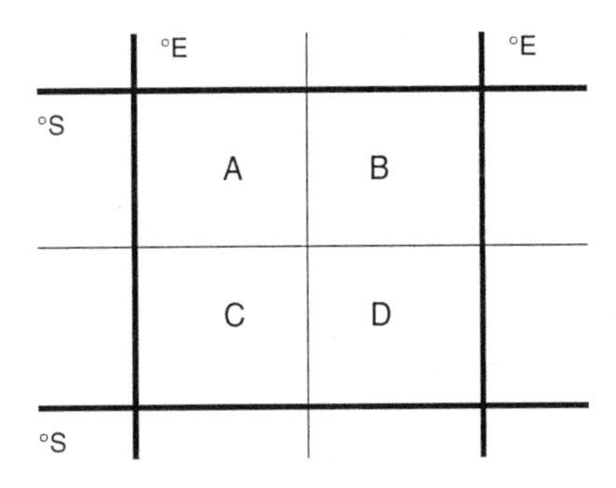

In Botswana the sides of each square are approximately 50 km in length thus giving an area of about 2 500 km^2 in each. As the side of each square is 30 minutes of latitude and longitude, the square was called a 30-Minute Square to avoid confusion by different interpretations of terms such as Quarter-Degree Squares or Half-Degree Squares. Each Degree Square occupies the area to the south and east of the degree lines which describe it—thus the Degree Square 2124 lies immediately to the south of 21°S latitude and east of 24°E longitude. This Degree Square is subdivided into four 30-Minute Squares by allocating the suffixes A in the west and B in the east to the two northern components, and C in the west and D in the east to the southern components. Thus

Figure 3 Species per square: numerical

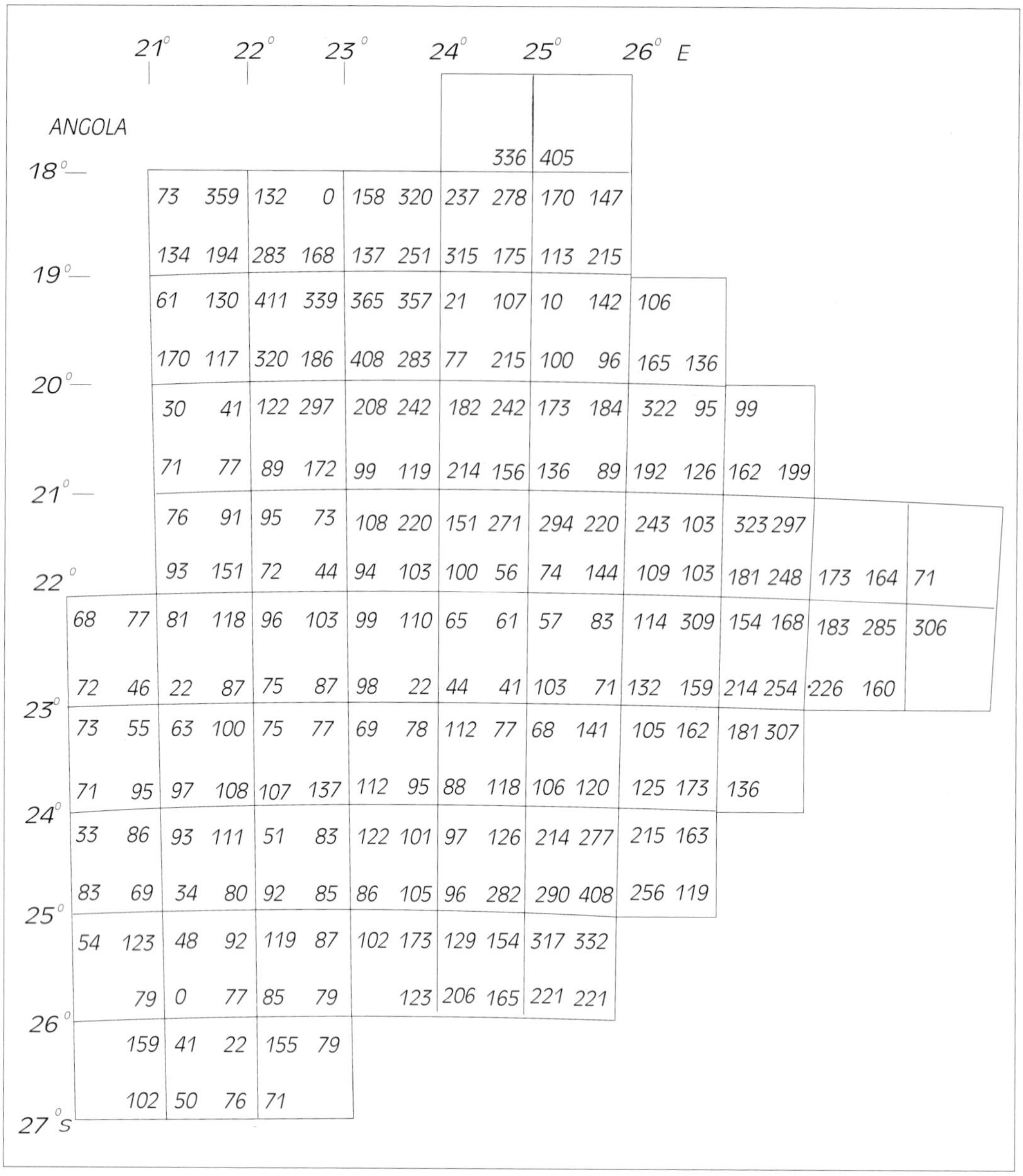

21° 22° 23° 24° 25° 26° E

ANGOLA

18°

336 405
73 359 132 0 158 320 237 278 170 147
134 194 283 168 137 251 315 175 113 215

19°

61 130 411 339 365 357 21 107 10 142 106
170 117 320 186 408 283 77 215 100 96 165 136

20°

30 41 122 297 208 242 182 242 173 184 322 95 99
71 77 89 172 99 119 214 156 136 89 192 126 162 199

21°

76 91 95 73 108 220 151 271 294 220 243 103 323 297
93 151 72 44 94 103 100 56 74 144 109 103 181 248 173 164 71

22°

68 77 81 118 96 103 99 110 65 61 57 83 114 309 154 168 183 285 306
72 46 22 87 75 87 98 22 44 41 103 71 132 159 214 254 226 160

23°

73 55 63 100 75 77 69 78 112 77 68 141 105 162 181 307
71 95 97 108 107 137 112 95 88 118 106 120 125 173 136

24°

33 86 93 111 51 83 122 101 97 126 214 277 215 163
83 69 34 80 92 85 86 105 96 282 290 408 256 119

25°

54 123 48 92 119 87 102 173 129 154 317 332
79 0 77 85 79 123 206 165 221 221

26°

159 41 22 155 79
102 50 76 71

27° S

Gaborone, lying in the southeast quarter of Degree Square 2425, is in 30-Minute Square 2425D and Gomare, in the northwest quarter of 1922, is in 1922A. Observers were encouraged to use 1 : 250 000 scale maps whenever possible to accurately define the boundaries of squares. The main grid for each species map is in Degree Squares but the species presence detail is in 30-Minute Squares.

The unit of measurement of time is expressed in months calculated sequentially over the 120 months of the project to give an unit called a **square month** (defined below). **Month analysis**, which is shown graphically with each map, is based on a 12-month cycle, for example by adding all the square months for March for each species. Each bar represents the sum of all square month records occurring in that month over the 10-year period of the project. A discrepancy between the sums of all the bars and the total records given in the text is as a result of records which have been submitted without the month being specified. These undated records are not included in the bar charts. This discrepancy can vary from about 1% in some species to over 30% in others. This presentation is also only a guideline and is satisfactory for some species but less meaningful for those with few records.

The bias introduced by the different number of visits in different months can be eliminated by creating a factor from the numbers of visits per month given in Table 3:

Table 3 Visits per month to atlas squares

Month		Month	
July	338	January	260
August	260	February	264
September	227	March	361
October	244	April	392
November	267	May	236
December	319	June	181
Undated lists	198		

This formula has not been applied in the calculations and results documented in this work.

The **target** for the number of species which should occur in each square was established by considering the vegetation, rainfall and diversity of habitats in each region of Botswana and by reference to four ornithological publications (see Strategies). Allowances were made for access routes, remoteness and how much of the whole square was in Botswana. The potential number of species in each square (see Table 4) was rounded off to the nearest 50 species (Penry 1984).

Table 4 Potential number of species per square

Square type	No. of species
Wet northern	300
Lush eastern	250
Mixed habitat	150
Dry Kalahari	100

These general regions are defined as follows:

Wet northern squares—Chobe, Linyanti, the Okavango region, the Makgadikgadi region, Boteti River and Lake Ngami.

Lush eastern squares—the Francistown region, the Tuli Block, Limpopo, Marico and Ngotwane drainage, Kanye and Lobatse to Ramatlabama.

Mixed-habitat squares—mixed woodland areas in the south, west, northwest and northeast.

Dry Kalahari squares—the central Kalahari and the southwest.

On the above basis the **target** for the project was set at **40 000** species. The unit of measurement is species per square (described below) and the target is the sum of the number of species occurring uniquely in each of the 230 squares in Botswana.

By 1986 the effect of the drought, particularly on the Makgadikgadi, Lake Ngami and the Kalahari squares necessitated a revision of the target. The Makgadikgadi squares were reduced to 200 species each and many Kalahari squares to 50. The **revised target** because of drought became **35 000**.

Terminology

Record

The unit which records each species reported by an observer in a square. It is composed of the names of the species, the square, the observer, the month and the year. Undated records are those which cannot be labelled with a specific month and year. A species recorded in the same square in the same month by a different observer is a separate record, but the same species recorded by the same observer in the same square on another day in the same month is discounted.

Species per square

The unit which measures each species uniquely in each square, i.e. it discounts subsequent records of the same species in the same square. It is the unit for measuring diversity, not frequency, and is summed to give the total number of species in each square. (Figure 3)

Square month

A unit of time and place combined, which is allocated to each square for the duration of the project. There are 120 sequential square months for each square and 230 potential square months for the whole of Botswana for each month of the project. The unit is activated by a visit but is lost to the project if no species is recorded in that square in the specified month. This is not the unit used in the month analysis which is calculated on a 12-month cycle. The square month is related to visits and is used as a denominator in frequency analysis.

Card

The list of birds submitted by each observer for each square in each month. A piece of paper with records from the same square in two different months by the same observer generates two cards and similarly two cards are generated by the observer reporting from two different squares in the same month.

Visit

A visit records the presence of an observer in a square in a month. Only one such occurrence (or no occurrence) can take place in each of the 120 months in each square, irrespective of the number of observers or cards. All of the different species recorded by all observers are included but the impersonalised square month is the denominator. Figure 4 lists the visits diagramatically.

Strategies

The **philosophy** of the project was to make a fresh statement on the distribution of birds in Botswana, prospectively and without inhibition from previous statements on any species. Research into past literature was avoided and all records were accepted without vetting for the first 5 years. Selected records from Smithers (1964) and Ginn (1976), i.e. only those records where place names were specified, were entered into the database in 1983—836 records from 66 squares and 1 124 from 15 squares respectively. Together with information from Irwin *et al.* (1969) from the Caprivi and 1724D, and Maclean (1970) from the Nossob Valley, these four works provided a loose framework on which to check the likelihood of records. By 1986 sufficient data had been accumulated to give a guideline on distribution patterns for many species. At this time also, the Records Subcommittee of the Botswana Bird Club commenced a scrutiny of all records of all species and produced a rarities list of 122 species for which there were fewer than 10 confirmed records in Botswana. Subsequently this subcommittee began work on a detailed review of the avifauna in taxonomic order which is not yet complete.

The **fieldwork** and coverage were achieved by using different practical strategies governed by the availability, ability, mobility and enthusiasm of contributors. Monthly lists were received regularly in some years from certain individuals resident in widely scattered localities, for example Shakawe, Gomare, Moremi, Linyanti, Kasane, Shashe, Francistown, Orapa, Serowe, Martin's Drift Molepolole, Gaborone, Jwaneng, Lobatse, Bray, Tshabong, Nossob Camp, Ghanzi and eastern Okwa. This was important for

Figure 4 Visits per square

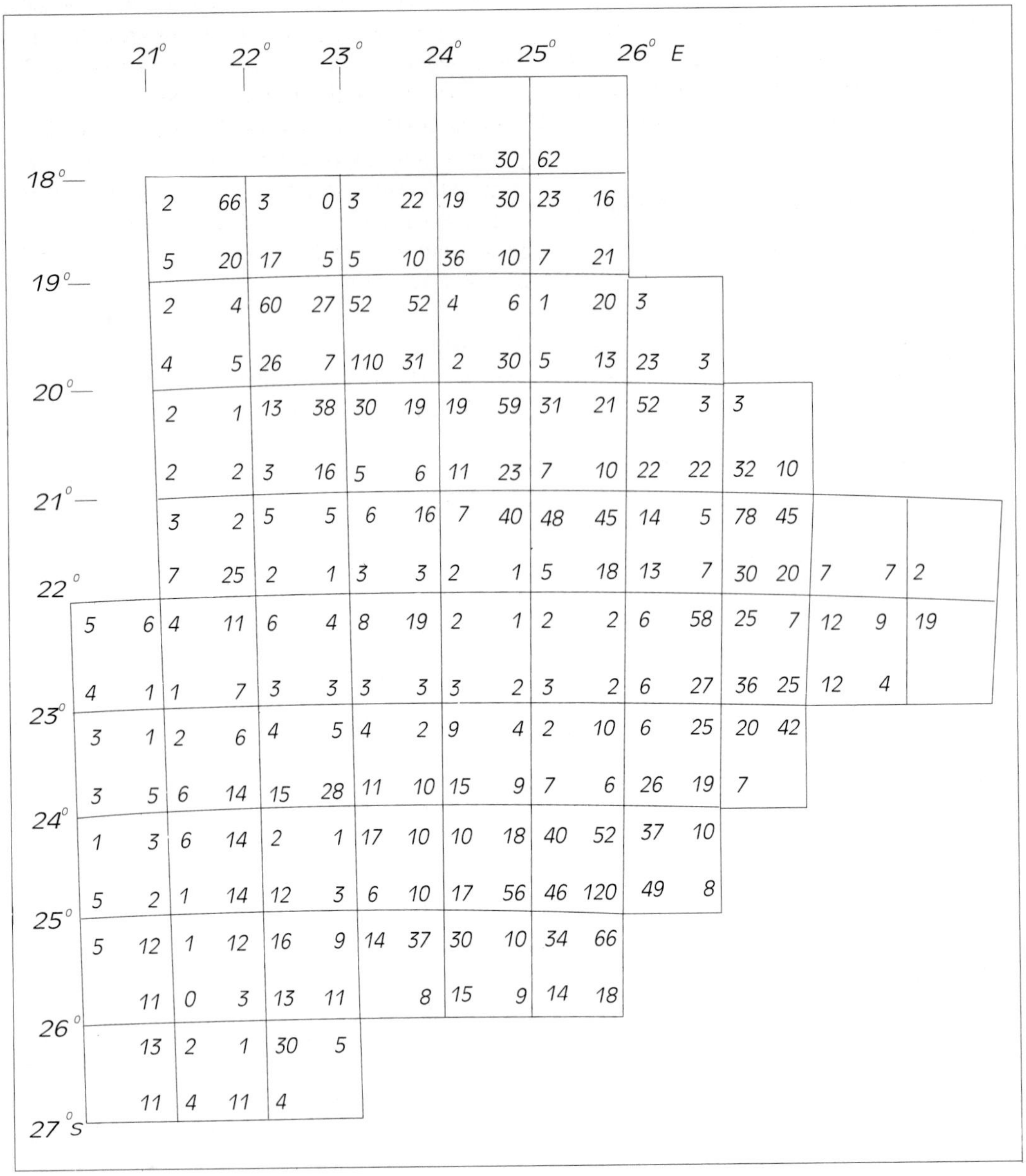

21° 22° 23° 24° 25° 26° E

30 62
18°
2 66 3 0 3 22 19 30 23 16
5 20 17 5 5 10 36 10 7 21
19°
2 4 60 27 52 52 4 6 1 20 3
4 5 26 7 110 31 2 30 5 13 23 3
20°
2 1 13 38 30 19 19 59 31 21 52 3 3
2 2 3 16 5 6 11 23 7 10 22 22 32 10
21°
3 2 5 5 6 16 7 40 48 45 14 5 78 45
7 25 2 1 3 3 2 1 5 18 13 7 30 20 7 7 2
22°
5 6 4 11 6 4 8 19 2 1 2 2 6 58 25 7 12 9 19
4 1 1 7 3 3 3 3 3 2 3 2 6 27 36 25 12 4
23°
3 1 2 6 4 5 4 2 9 4 2 10 6 25 20 42
3 5 6 14 15 28 11 10 15 9 7 6 26 19 7
24°
1 3 6 14 2 1 17 10 10 18 40 52 37 10
5 2 1 14 12 3 6 10 17 56 46 120 49 8
25°
5 12 1 12 16 9 14 37 30 10 34 66
11 0 3 13 11 8 15 9 14 18
26°
13 2 1 30 5
11 4 11 4
27° S

consolidating and building confidence in bird knowledge for those areas. Some of these people visited adjacent squares on a regular basis and sent in lists while visiting other areas of Botswana. This type of contribution amounted to 44% of the numerical input. Reliable contributions came from regularly visiting birdwatchers from other countries, while those from casual visitors required more careful scrutiny. Short lists were also received from keen participants travelling through squares on a long journey (see Biases).

Very important sources of distributional records were the planned safaris to infrequently visited regions by the most active and skilled birdwatchers in Botswana. The adoption of this tactic was the main factor in achieving the coverage of 99% of Botswana squares. By going from square to square in sequence over a period of 10–14 days, visiting different habitats whenever possible (some squares had only one access route), a good representation of the avifauna of the region was more likely. An intimate knowledge of the species in minor variations of habitat, the diversity of microhabitats, subtle changes of the avifauna and margins of distribution of some species was achieved by this method. The input from this process constitutes 36% of the database. Without this effort most of the central and western areas of Botswana would still be poorly known ornithologically and knowledge of many other areas would be inadequate.

The fieldwork was announced as closed on 31 December 1989. A few observers who submitted records direct to the author continued to send records until the end of June 1990. This strategy was adopted because records sent via the Bird Recorder usually took 3 months to process and as little delay as possible was needed before the analysis commenced in July 1990.

The **validity and verification** of records was a matter of concern in the early stages of the project. By 1987 most areas of the country had been visited, patterns of distribution had emerged for most species, much field experience had been built up by a nucleus of birdwatchers and confirmation of records had been received from different observers. As a result, from this time the validity of records for most species was not in doubt.

Unusual records were discussed and accepted or rejected by the Records Subcommittee on the basis of submitted evidence. The rarities list had increased to 124 species by the end of 1987. Any record for which the Subcommittee had insufficient experience was referred to referees in other countries. In spite of this protective process it is inevitable that some uncertainty will remain about some of the records.

For this reason the author takes responsibility for the records included in this work—a decision on the inclusion or exclusion of a very few records has had to be made pending decisions from the sources mentioned above. It is not a primary objective of this work to make accurate assessments of rare species but care has been taken not to mislead. The purpose of the Atlas is to make accurate distribution maps for the common species so that future effort can concentrate on accumulating more information on the scarcer species.

Biases and deficiencies occur in this work which will be obvious to statistical analysts. These are discussed below to illustrate some techniques and strategies which need to be used in bird atlas work to reduce these influences:

Observer bias

There was no attempt to limit this bias by excluding people who had less than the required competence and experience in the field. The required competence is not easily measurable —even so-called 'experts' have weaknesses and can make mistakes of identification in their own specialist field let alone when reporting on other species. Most people have gaps in their knowledge and tend to record birds in which they have a special interest more than they do other birds. People fail to record common and conspicuous species because they are concentrating on something else at the time. As much as possible, observer bias was reduced by the vetting process but inevitably there will be some incorrect acceptances and rejections. Some species will be under-recorded because only some observers know them well (there are many examples of fairly common species in this category, e.g. Tinkling Cisticola, Rufouseared Warbler).

Purpose bias

It was accepted that lists submitted specifically for Atlas purposes as a comprehensive sample of all the species occurring within that area, and considered within the observer's competence, could be relied upon as representative samples. Lists submitted by observers who went to an area to find specific birds with which they were unfamiliar are heavily biased towards those species to the detriment of the common species. An attempt was made to eliminate this bias by encouraging participation only from those persons known by me to understand the principles and requirements of Atlas contributions. However lists from casual visitors and 'twitchers' were not excluded on this principle, even though the bias would be introduced. A bias in the same style came from persons who submitted only additions to lists previously submitted to the Atlas.

Travelling bias

A certain amount of time is required to record a representative sample from a visit to a square. One or two stops in order to get out of the vehicle, listen and observe are essential and a stop at each different habitat is considered desirable. Birds recorded from a vehicle without stopping tend to be biased towards species which are conspicuous. Thus lists received from several consecutive squares during a long journey without stopping may be biased in this way. In the frequency-of-occurrence analysis the denominator (the number of visits) is high in relation to the number of times inconspicuous species are recorded. This phenomenon is particularly seen in squares along the Gaborone-to-Francistown road in the east, and on the Nata-to-Maun road and the Nata-to-Kasane road. This bias was reduced on atlassing safaris by setting a target of 50 species, or approximately 50% of the expected species, before moving on to the next square, and by planning the itinerary to sample no more than three or four squares per 24 hours.

Access bias

Squares in which birdwatchers reside had more visits and a more complete monthly list than other squares. The Gaborone square was the only square which submitted a monthly list for every month of the project and this factor biases the analysis of squares in the southeast sector which do not have the same coverage nor diversity of habitats. Furthermore the input volume from the Gaborone square is 17 965 records, 10% of the total database, and this also introduces a bias. Squares which required special planning and equipment were visited least frequently. Advertising the need to visit such areas in the 6-monthly reports by the author attempted to reduce this bias, but it was a deliberate philosophy not to put inexperienced people at risk through undertaking such visits. Squares with easy access from main roads were visited more often than those without such access. There was also a bias from visitors and short-term expatriates towards tourist routes and favourite areas and a considerable effort was made by myself and a nucleus of birdwatchers to concentrate on less attractive areas in order to counteract this bias. For this reason the tourist areas are not considered to have been adequately assessed except where regular contributors resided.

Sampling bias

Some squares had only one access route. This route usually allowed access to only some of the different habitats in that square thus producing a sampling bias. It was not practical to visit all the remote squares in every season, let alone in every month, and a sampling bias by season was thus introduced in some squares. Very few squares with five visits or fewer will be free from this bias. Winter visits to Kalahari squares can result in a poor return in species numbers for the effort, and such visits were made less frequently than desirable. Conversely there was a bias towards visiting Kalahari squares only in summer months. Many lists from the tourist areas in the northern wetlands included mainly waterbirds and species from the woodlands and savannas were not adequately represented.

Presentation bias

The frequency analysis is divided into three analysed categories (symbol sizes) because a more detailed analysis of the available data was not statistically justifiable. As discussed below, the squares with fewer than 10 visits are subject to very crude analysis and should be studied in this light. Month-analysis histograms have different scales and thus views of the data are from different distances. This bias needs to be taken into consideration when looking at the charts.

Recording of data

Data for the project were collected from any person willing to contribute. This was mainly from the most active amateur birdwatchers residing in Botswana but included amateur and professional ornithologists known to me from South Africa, Zimbabwe and Zambia. No attempt was made to coerce persons to contribute nor to advertise the project aggressively. In this way the intention was to receive reliable and high-volume inputs from well-known observers without the need to question much of the data and to receive quick and acceptable responses when queries were made. As a result of this tactic the number of record cards received may not be as high as those of other atlas projects. At the same time it was anticipated that the volume of information received would be manageable. All data were processed and entered onto Master Cards, and later onto computer, by the Coordinator.

After 1982 contributors were requested to submit their records via the office of the Bird Recorder of the Botswana Bird Club. Several benefits accrued from this.

1. The Botswana Bird Club was able to keep up to date with distributional information and rarity submissions.
2. The Bird Recorder could take over the task of requesting additional detail from observers regarding queries and pass the information direct to the Records Subcommittee.
3. Unusual records were more likely not to be missed by the two scrutinies of the Bird Recorder and the Coordinator.
4. Contributors in Botswana would be spared additional postage charges (the Coordinator had moved to live in South Africa from 1982).

In practice some sent contributions in accordance with the above, others direct to the Coordinator and some to both. Although the Bird Recorder accepted records in any format, (e.g. on food, nesting, visit lists, behaviour, etc.), the bulk of information received by the Recorder during the project period was for the distribution atlas. Contributors were requested to submit their records on Field Cards or printed Checklists as the accepted standard for atlas input. However submissions in non-standard format were accepted unless the records contained only species already reported for that square in that month although there were very few such occurrences, mainly from the well visited Okavango squares.

All **records** were entered into the Atlas database by the Coordinator throughout the 10 years. Initially a manual system of Master Cards for each square with a code symbol for each visit was used. A check on this system in 1986 produced an input error figure of only 0,36%. This system is useful for quick assessment of the status of squares but cumbersome for an assessment of species in the country. In July 1987 the information was transferred to a personal computer using dBase 111 Plus software. Over a three-month period, 1 791 record cards were entered containing 66 225 entries. Subsequently all new records were entered into the computer though the manual system was kept up to date as a means of checking errors until the end of 1988. The dBase input file contained the following fields: Square; Observer; Date; Species (62 three-character species numbers); Adspecies (an additional 51 three-character species numbers). This was a simple input file system making it easy to list birds consecutively in the Species and Adspecies fields compatible with the list of birds received from field cards. However, such a file does not allow analysis of each individual species in the list. A computer program was designed in April 1988 to convert each entry in the Species and Adspecies fields into unique records each labelled with the Square, Observer, Date information. The Atlas dBase file created by the program permitted **analysis** (sorting and indexing) of every species entry by Square, by Observer, by Month(date) or by Species. In order to exclude duplicates for the purpose of counting the true number of cards and visits, the unique facility of dBase was used.

Species were entered as three-character numbers, e.g. Ostrich = 001, etc., except those few with suffixes which were four-character, e.g. 645X = Chirping Cisticola (using 1978 Roberts' numbering). Checks were performed regularly to correct input typing errors.

The database used in excess of 26 megabytes. The main components were; input file (raw data) 1.93 Mb, base file (labelled records) 4.69 Mb, square-per-species index 3.94 Mb and species-per-square index 3.94 Mb. Other dBase files included Observer, Cards, Visits, Mapsize (symbol size), Breeding, Gazetteer, References, Square (square count) and various indexes for each. Multimate Advantage 11 software was used for wordprocessing the text and imported into Ventura Desktop-publishing software for trial layout and style. Graphics were prepared on Intergraph and plotted on Intergraph 7585B plotter.

Presentation of data

The presence of a species in a 30-Minute Square is represented on the distribution maps by symbols. The **frequency of occurrence** of each species in each square is shown by different-sized symbols and calculated from the formula below.

Total number of records for the species in the square divided by the total number of visits to the square.

The resulting frequency is grouped into four categories (Table 5).

Table 5 Categories of frequency

Category	Symbol	Meaning
10,0% or less	Open	Sparse or rare
10,1 – 49,9%	Hatched	Uncommon to common
50,0% or more	Solid	Very common
Pre-1980 only	Star	(Not analysed)

Further subdivisions were not considered reliable because many of the squares did not have enough visits to make a more accurate assessment of frequency—51% of the squares had fewer than 10 visits. The current system should be accepted as a helpful guideline only. In the species text accompanying each map the middle category is subdivided as follows—uncommon (10,1–16,0%), fairly common (16,1 –30,0 %) and common (30,1–49,9%). The term 'common' relates to the likelihood of recording the species and does not refer to the number of birds of a species.

On the above basis, frequency data in squares with more than 10 visits could be calculated with an acceptable degree of confidence. Species occurring in squares with fewer than 10 visits could not have an analysis below 10%, (i.e. in the sparse category), and have been given a hatched symbol unless the species was recorded on more than 50% of the visits, in which case a solid symbol applies. This places the species in the uncommon category at worst for squares with six or more visits and this is reflected in the text where necessary when it applies to a significant area. Squares with five visits were allocated a hatched symbol when a species had been recorded once or twice, and a solid symbol for three or more recordings. Squares with four visits were given the hatched symbol for species recorded once or twice, and a solid symbol for three or four recordings. Squares with three visits were given a hatched symbol for species recorded once or twice, and a solid symbol only when the species had been recorded on every visit. Squares with two visits were given a hatched symbol for one or two recordings of the species but a solid symbol when there were two records and an adjacent square had a solid symbol. Squares with one visit were given hatched symbols for species records except when two adjacent squares had solid symbols, in which case a solid symbol was allocated.

Results

The primary purpose of the project is illustrated by the species-distribution maps which occupy the main part of the book. Maps have been drawn for 496 species which have been recorded in four or more squares or have more than 10 confirmed records in Botswana .

The rare species which have been accepted as occurring in Botswana total 59 and are listed separately (on pages 291–295). They are not shown in map form.

The number of species recorded in each square is shown in Figure 3 which is further illustrated graphically in Chapter 4 to show the diversity of species in different regions of the country. The numbers are categorized in units of 50 in Table 6 which shows that the target of 50 species in every square was not achieved in 19 squares (8%)—a 92% success in this objective. Table 6 also allows comparison with the set milestones for the species per square and shows inadequate consolidation of squares with a potential of 100 to 300 species per square, i.e. about 50% of Botswana. This is attributed to many factors of which the most significant appear to have been an insufficient number of observers for the size of Botswana, not enough residents able to travel off the main roads and the discouraging effects of the drought. The influence of these factors is particularly noticeable in the east and southeast where most of the birdwatchers in Botswana live. However, sufficient numbers (about a dozen) of keen birdwatchers visited or lived in some of the high-diversity squares so that the number of squares with over 300 recorded species exceeded expectation. The small number of people with the ability to travel widely were fully occupied achieving overall coverage of the country and had insufficient time to consolidate the better-worked squares. Two of the squares, 1822B and 2521C, were not visited because of difficult access and security risk. In total 99% of the objective was reached in the coverage of all squares in Botswana.

Table 6 Analysis of species per square

Number of species	Objective (squares)	Outcome (squares)	% of Botswana	% of objective
0	0	2	1%	Failed
1 – 49	0	17	7%	Failed
50 – 99	16	74	32%	Exceeded
100 – 149	90	53	23%	59%
150 – 199	43	36	16%	84%
200 – 249	46	15	6%	33%
250 – 299	22	14	6%	64%
300 – 349	14	12 }	5% }	}
350 – 399	N/A	3 }	1% }	136% }
400 +	N/A	4 }	1% }	}

The number of visits to each square, i.e. the coverage of the country, is shown in Figure 4 and an analysis is given in Table 7. The figures include 238 sources before the start of the project. Sources of old information are listed on page 299. The significant finding is that 31% of squares received fewer than five visits, and a further 21% only between five and nine visits. Seventy squares had fewer than five visits, and therefore only 69% of the objective for this parameter was achieved. These results are probably due to the inaccessability of large areas of Botswana to people without suitable transport. The two squares in which 90% coverage (110 visits) was achieved were Gaborone and Maun, though the quality of the records from Maun (mainly from casual visitors) was inferior to that of Gaborone where most resident birdwatchers live. However, the two squares recorded the same number of species which is an indication of the beneficial influence of manmade habitats to species diversity in Gaborone. Francistown West/Shashe received over 70 visits; Gomare, Shakawe, Kasane and Lobatse over 60 and Jwaneng, Molepolole, Serowe, Kudiakam Pan (2024B), Moremi Central (1923A) and Moremi East(1923B) over 50 each. At the end of the project these were the best recorded bird localities in Botswana.

Table 7 Analysis of visits per square

Visits	No. of squares	% of Botswana	Visits	No. of squares	% of Botswana
0	2	1	35 – 39	5	2
1 – 4	68	30	40 – 44	3	1
5 – 9	49	21	45 – 49	5	2
10 – 14	35	15	50 – 59	7	3
15 – 19	19	8	60 – 69	4	2
20 – 24	12	5	70 – 79	1	0,4
25 – 29	7	3	80 – 109	0	0
30 – 34	11	5	110 +	2	1

A breakdown of visits per month for the whole of Botswana is given in Table 3 on page 8. A breakdown of visits by months in each square is not given.

The total **number of records** in the database is **180 202** of which 8 044 (4,45%) are pre-1980 and 797 (0,44%) are from January to June 1990. Part of the objective was to update bird information in Botswana, thus the pre-1980 figures are to fill the gap between Smithers (1964) and the beginning of this work.

The total **number of cards** received was **3 547** of which 238 (6,7%) were from sources pre-1980.

Figure 5 Records per annum

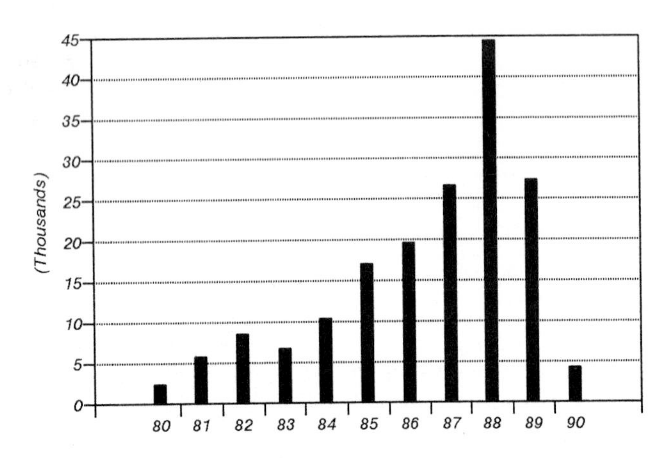

Figure 6 Cards per annum

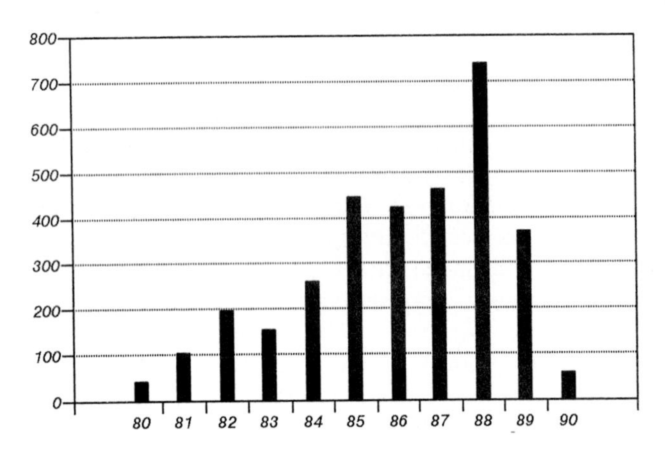

The accumulation of records over the 10 year period is shown in Figures 5 & 6 and the progress towards target in Figure 7.

From Figure 7 it can be seen that to achieve the revised target it was necessary to accumulate at the rate of 3 500 new records (species-per-square) per annum. This rate was only achieved between 1985 and 1989. The progress to target shows that gains were made from the substandard effort of 1981 to 1984, that the target was obtained in 1985 to 1987, and that after this the success tailed off again as the addition of new species to a worked square became more difficult. The sum of the species occurring uniquely in each of the 230 squares in Botswana at the close of the project was **33 238**. This is the sum of the numbers shown in Figure 3 and is 83% of the original target of 40 000 and 95% of the revised target of 35 000.

Figure 7 Progress towards target

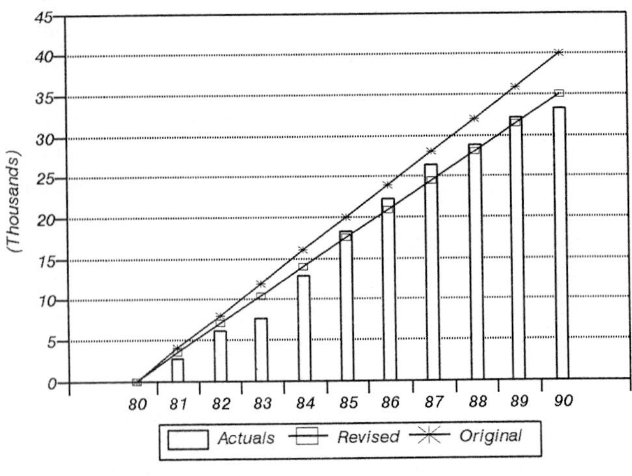

Had all squares been visited once in every month during the 120 months of the project, the total visits would have been 27 600.

The **sample size** of this project with 3 085 visits is thus an **11% sample.** This calculation excludes duplicate visits to any square in any month and pre-1980 records.

The average number of species per visit was 50,8 and the average number of species occurring in each 30-Minute Square is 144. These are divided per degree of latitude in Table 8.

Table 8 Species diversity by degrees of latitude

Degree	Number of squares	Average number of species per square
17°S	2	370
18°S	19	203
19°S	23	188
20°S	27	156
21°S	31	147
22°S	37	120
23°S	31	111
24°S	27	143
25°S	22	139
26°S	9	84

A general conclusion from Table 8 is that, as expected, the diversity of species decreases southward representing a higher proportion of squares which are located in the Kalahari. Western squares did not have enough coverage but the effects of drought are still noticeable in the Kalahari squares when compared with those southeastern squares where manmade water environments exist (this effect is also seen in the southwest). There is a bias in the southeastern squares where many birdwatchers reside which affects the number of species seen in latitudes 24° and 25°S.

The sum of all records from north of 23°S is 113 964 (63%) and south of this latitude 66 414 (37%). This corrects a bias existing before this project when 3 861 of 4 262 traceable records, i.e. 90,5% were from north of 23°S.

The analysis of the longitudinal diversity (Table 9) shows higher species numbers in the moist east than in the dry west, even though the Okavango Delta contributes to the high diversity in the west for 22° and 23°E.

drawn and its limits are based on current knowledge of the movements of many tropical bird species into the Kalahari under the most suitable conditions of vegetation and climate. The interzone represents an area where Kalahari species intrinsically form part of the avifauna and are the major components under most conditions. The extent of the tropical invasion into this area depends on the species as well as the prevailing climate, with much fewer tropical species during drought.

In addition, some species extend from the south to southern areas of Botswana either as seasonal visitors (e.g. Fiscal Flycatcher and Fairy Flycatcher), occur at the northern limit of their South African range (e.g. Jackal Buzzard and Orangethroated Longclaw) or are recent extensions of ranges (e.g. Black Sparrowhawk and Speckled Mousebird). Similarly some appear in some years or seasonally in the north at the southern limit of their range in Zambia (e.g. Purple-banded Sunbird) or southern limit of migration (e.g. Great Snipe). Other species occur throughout Botswana because they are adapted to the climatic condititions and vegetation within their endemic range or are migrants which occur at certain times of year throughout parts of or the whole subcontinent. There are inevitably some species whose pattern of distribution is still not easily explained because current knowledge of the species is scant.

As the experienced birdwatcher knows, it is the structure —the resulting shape and character of the vegetation—which determines in which area particular species are likely to be found. This structure (physiognomy) may alter from year to year from natural and unnatural influences such as plant growth, drought, flood, rainfall or destruction by fire, animals or humans. Botswana is a semidesert country with a wide range of daily and seasonal temperature and rainfall. Factors which might have less influence in temperate countries, such as local precipitation or the building of a farm dam, can have a major impact on local birdlife in Botswana especially in the Kalahari. Unleached Kalahari soils are rich in soluble chemicals (nutrients) and, as a result, many of its hardy plants respond dramatically to water.

Physical geography

Botswana is a landlocked country 581 730 km^2 in extent in the centre of the southern third of the African continent. It extends between latitudes 17°S and 26°S and longitudes 20°E and 29°E. The Tropic of Capricorn bisects the country so that the northern 67% (155 atlas squares) is in the tropics. Except in the west and northeast, its borders are formed by rivers.

Rivers have a significant impact on bird distribution. The northern rivers of the Okavango, Linyanti and Chobe are perennial, the former forming the huge Okavango Delta occupying 16 000 km^2 and producing one of the most important and famous wetlands in the world. The Okavango River rises on the Benguela Plateau in Angola where the annual rainfall is 1200–1500 mm. Owing to massive evaporation while meandering through the delta and its sand substrate, the outflow of the Okavango Delta is only about 5% of the inflow and these important dregs flow irregularly down the Nhabe River to Lake Ngami or down the Boteti River to the Makgadikgadi Pans. On route near Orapa some of the water from the Boteti is channelled into the Mopipi Dam. As a result of the drought, neither Lake Xau nor the western main pan (Ntwetwe) of the Makgadikgadi received water from the Boteti River during the duration of this project, though both are natural terminuses of this river in the Kalahari sands.

The eastern rivers of Botswana all drain into the Limpopo River which forms the southeast boundary of the country. The watershed of these eastern rivers is very important for bird distribution. The watershed divides Botswana into its two main geological regions—the **sandveld** (i.e the Kalahari) to the west, and the **hardveld** to the east. Rivers which may have flowed west of this watershed in ancient times to the Great Kalahari Lake are now represented only by **fossil valleys** as was the fate of other ancient rivers flowing from the western regions of Botswana (see Figure 11). The only fossil valley which crosses the watershed is the Serorome Valley whose headwaters are in the sandveld. All the eastern rivers, including the Limpopo itself, are seasonal. Water flowed in them for only a few weeks of the year during most years of the past decade. In high-rainfall cycles the major tributaries flow from December to September in some years but more usually for four or five months of the year. The variety of soils in the hardveld (see page 21) results in a wide diversity of plant life. Many of the plants in this region grow in profusion in response to the seasonal flow in the rivers. In contrast, there is a smaller diversity and poorer response of plant growth in the sandveld to the seasonal flows of the southern and southwestern rivers.

The Limpopo River runs northeast from its formation at the junction of the Crocodile and Marico rivers which arise in the Transvaal. The Ngotwane River is the western headwater of the Limpopo and arises in Botswana. Birds which funnel from the tropics along the Limpopo Valley extend their range southward along this tributary. At Gaborone, the Ngotwane River is dammed and this major expanse of water, the Gaborone Dam, affects positively the diversity of bird species in this southeast region.

The whole of the southern border of Botswana is formed by the Molopo River, approximately 550 km in length. In essence, it is a seasonal river but its annual flow is erratic and of short duration—often only days or weeks. Its main influence as a river is in the east, and water-dependent birds are mainly limited to this sector. West of Bray the Molopo is a dry, sandy river bed and west of 22°E it is mostly a wide arid canyon. Except in the east it has not flowed during the past 10 years.

The Nossob River forms the southwestern border. It has flowed about once in every 30 years in the last 100 years. It runs through duneland and in such an arid environment that the presence of the riverine vegetation, particularly large trees which are absent on the dunes, influences the diversity of bird species.

The Makgadikgadi Pans occupy about 8 000 km^2 in the northeast. They are relics of the Great Kalahari Lake, probably the deepest point of what was known as the Kalahari Basin. The alkaline waters of these huge expanses of flat salt pans attract large numbers of waterbirds. During most of this survey they have been wholly dry except at the Nata Delta (the inflow of the Nata River). Thousands of flamingos and pelicans breed on the pans under suitable conditions. The two large main pans are Ntwetwe, in the west,

Figure 11 Rivers and fossil valleys

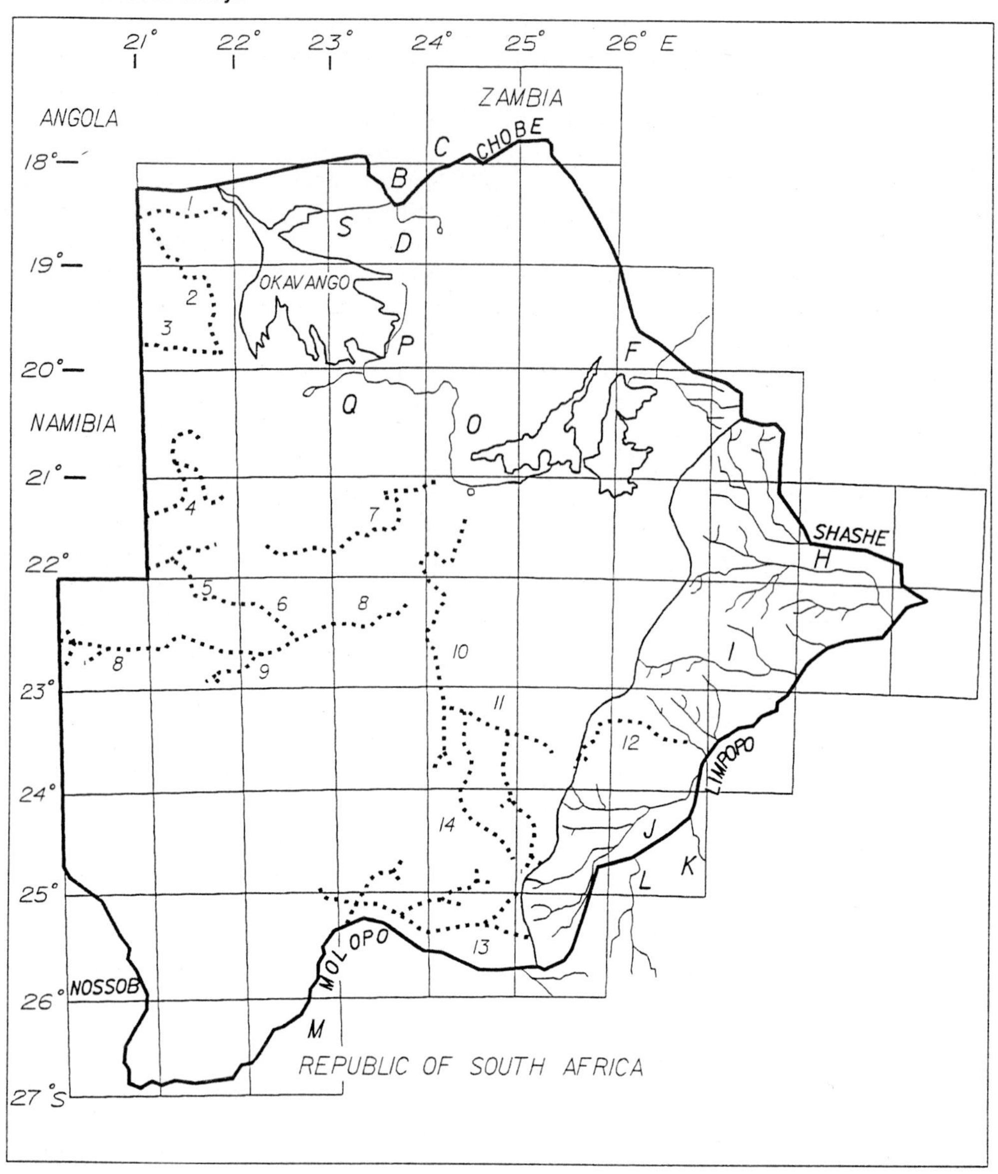

Rivers

B Kwando	F Nata	J Ngotwane	M Molopo	Q Nhabe
C Linyanti	H Matloutse	K Crocodile	O Boteti	S Selinda Spillway
D Savuti	I Lotsane	L Marico	P Thamalakane	

Fossil valleys

1 Nxamaseri	4 Groot Laagte	7 Deception	10 Quoxo	13 Moselebe
2 Xaudum	5 Buitsivango	8 Okwa	11 Metatswe	14 Naledi
3 Qangwadum	6 Hanahai	9 Takatshwane	12 Serorome	

and Sua, in the east. The inflow from the west into Ntwetwe is from the Boteti River but barely any water has flowed that far during the past 10 years. The inflow from the east into Sua is from the Nata River and to a lesser extent from the Semowane, Mosetse and Lepashe rivers. The Nata River arises in Zimbabwe and also drains the Tutume and Nkange rivers to the northwest from the Shashe River watershed.

The Shashe River itself drains southeast into the Limpopo and forms part of the northeastern border with Zimbabwe. Before its confluence it is joined by numerous rivers flowing south from the Plumtree-Bulawayo watershed in southwestern Zimbabwe and is thus a very large river at its confluence with the Limpopo. Southwest of Francistown it flows through the Shashe Dam from whence many of the important waterbird records arise in this region of Botswana.

Two further rivers need mention as they may influence bird distribution. The Zambezi River touches Botswana at Kazungula where the Chobe River joins it. This major river of Africa may act as a funnel for some birds, such as the Rock Pratincole and the Olive Bee-eater, and may have an influence on the presence of Caspian Terns and African Skimmers in northern Botswana. The Kwando River passes through southwestern Zambia as the Mashi River to discharge into the Linyanti swamps. Its valley, too may act as a funnel for some of the sparse species in Botswana such as the Green Lourie, Böhm's Spinetail and Sharptailed Starling, although the Okavango and Zambezi rivers may also contribute in the same way.

Figure 12 Topography

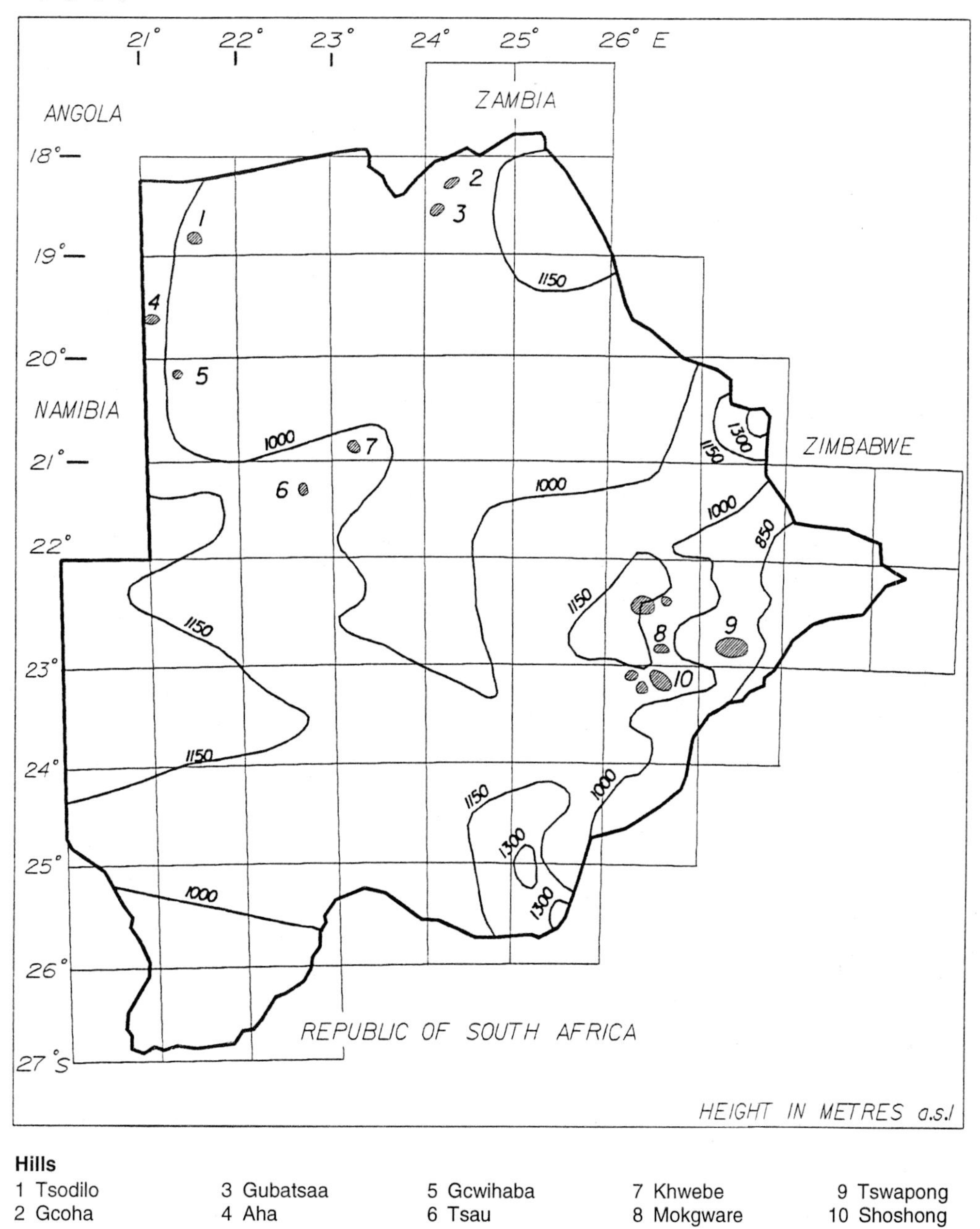

Hills

1 Tsodilo
2 Gcoha
3 Gubatsaa
4 Aha
5 Gcwihaba
6 Tsau
7 Khwebe
8 Mokgware
9 Tswapong
10 Shoshong

Botswana lies on the Great African Plateau and its **altitude** fluctuates very little between about 1 000 m to 1 150 m a.s.l. It is only in the southeast and east that wider altitudinal variations occur. A range of granitic hills in the southeast rises to just above 1 300 m and extends from Kanye to Ootse and Gaborone although few have sheer faces more than 100 m high. Further north within a 50 km radius of Mahalapye and Palapye are several separate ranges of which the best known to birdwatchers are the Shoshong and Tswapong hills. The Limpopo Valley starts at about 900 m and drops down to about 550 m at the Shashe confluence.

Changes in altitude have an insignificant effect on the avifauna of Botswana but the physical nature of the hills and the composition of the rocks and hillside vegetation influence the status of species such as Cape Vulture, Rock Kestrel, Black Eagle, Mocking Chat and Striped Pipit. From Sefophe and Selebi Phikwe north and east there are rounded granite boulders similar to those found in the Matobo Hills in Zimbabwe and on which the Boulder Chat is found. Maps of Botswana mark several other hills but in the main these are small bumps in the otherwise flat landscape of the Kalahari Basin and there is no change in bird composition from the surrounding savanna. The Tsodilo and Aha hills in the northwest are more renowned for their archaeological and anthropological interest than for their birds, though the Black Eagle and Freckled Nightjar are known from the former.

National **Parks**, Game, Wildlife and Forestry **Reserves** occupy about 15% of the country. Because of the necessary restrictions on leaving the confines of one's vehicle where predatory animals roam, they are not the best places for birdwatching but increasingly special arrangements are being made for birdwatching tourists. The largest single area is the central Kalahari Game Reserve, for entry into which a special permit is required. With its southern appendage, the Kutse Game Reserve, it occupies about 43 000 km^2 (about half the size of Natal). These reserves protect the Kalahari savanna at its most vulnerable point in the centre of the country where there is no surface water except after rain. The next largest

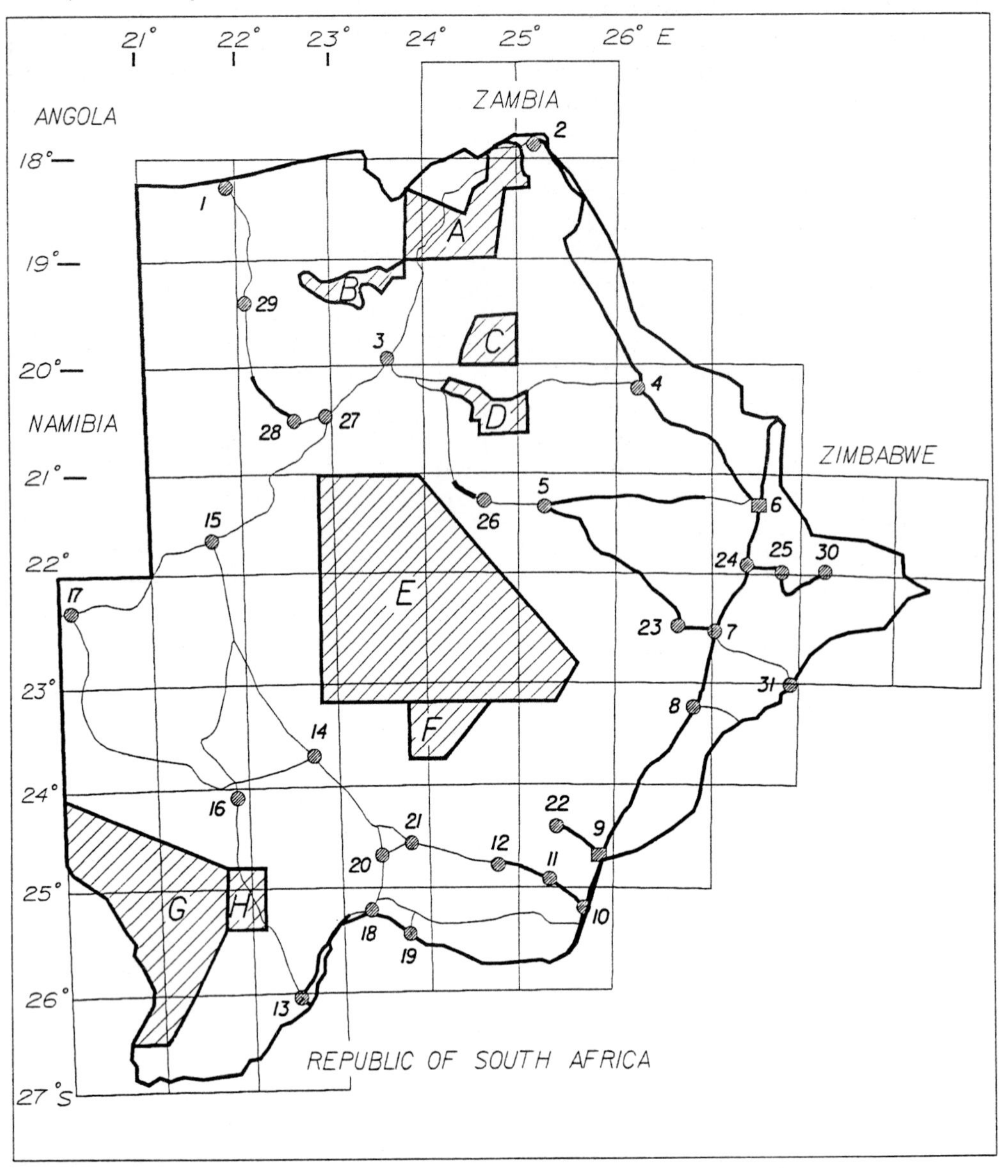

Figure 13 Towns, roads, and protected areas

Towns

1 Shakawe
2 Kasane
3 Maun
4 Nata
5 Orapa
6 Francistown
7 Palapye
8 Mahalapye
9 Gaborone
10 Lobatse
11 Kanye
12 Jwaneng
13 Tshabong
14 Kang
15 Ghanzi
16 Tshane
17 Mamuno
18 Werda
19 Bray
20 Khakhea
21 Sekhoma
22 Molepolole
23 Serowe
24 Serule
25 Selebi Phikwe
26 Mopipi
27 Toteng
28 Sehitwa
29 Gomare
30 Bobonong
31 Martin's Drift

National Parks and Reserves

A Chobe N.P.
B Moremi W.R.
C Nxai Pan N.P.
D Makgadikgadi Pans N.P.
E Central Kalahari G.R.
F Kutse G.R.
G Gemsbok N.P.
H Mabuasehube G.R.

reserve is the Gemsbok National Park which, with its western appendage, the Mabuasehube Game Reserve, occupies about 19 000 km^2 of the southwestern duneland and wooded savanna. Surface water is rarely available in this area. It is bounded on the west by the Nossob River and is contiguous with the Kalahari Gemsbok National Park in South Africa. The remaining protected areas are all in the north. The Chobe National Park with its nearby Forest Reserves (Sibuyu, Maikelelo, Chobe, Kasane and Kazuma) occupies about 15 000 km^2. In the northwest an area in the centre and eastern part of the Okavango Delta is protected as the Moremi Wildlife Reserve. Further east is the Nxai Pan National Park and south of it, in the northwest corner of the Makgadikgadi Pans, is the game reserve of that name. The distribution of these protected areas is illustrated in Figure 13.

Geology

The central landmass of Africa has been warping and rifting over millions of years. The oldest rocks which underlie Botswana are granitoid gneisses of the Archaean Basement Complex which are about two-thirds of the age of the earth itself. A major warp had caused the tilting of the Plateau initially to form the Kalahari Basin into which major rivers flowed to create the Great Kalahari Lake. Over millions of years large-scale rifts on the continent caused the formation of the Great Rift Valley running down the centre of the landmass. Near the southern end of this great rift are the Victoria Falls Gorges, just north of Botswana and about 100 km east of Kasane. Over time, the rifting process diverted some of the rivers flowing into the Great Kalahari Lake from the north—in particular the rift which caused the formation of the Middle Zambezi Valley diverted the Zambezi and Kwando rivers eastwards so that they no longer flowed into the Great Kalahari Lake which consequently began to dry out. To appreciate the significance of this rifting process one has to imagine the effect on Botswana if the huge Zambezi River still flowed into the Magkadikgadi region today—there would be no Kalahari desert in most of Botswana. The outlet of the Lake still remains a mystery but two theories exist—that it crossed the southern Kalahari to the Molopo River or that it flowed east to the Limpopo. Over subsequent millions of years and arid climatic conditions the drying lake bed known as the Kalahari Basin became filled with windblown sand on a late-Cretaceous crust. These events provide the underlying geology of the whole of Botswana except for the Hardveld.

The **Kalahari** sands are Tertiary and Quaternary deposits and in some places lie up to 150 m deep but usually much less. They extend for 3 200 km from the Orange River, to 1°N in Congo (Brazzaville), and 1 500 km from the Etosha basin in Namibia to central Zimbabwe in the east. The Kalahari sand mantle forms the world's largest continuous sand surface covering 1,2 million km^2. These sands are made up of soft sandstones, some clays, silts and gravel and collectively are known as the Kalahari Beds. In many places the beds have been leached over aeons to form a hard white rock-type called **calcrete** and in other areas a flinty grey **silcrete**. Areas where calcrete lies near the surface are often indicated by the woody shrub *Catophractes alexandrii* which is abundant in some areas of the Kalahari. Calcrete and silcrete are often seen in pans. Sand does not always cover the older rocks which emerge on the surface in areas such as the Ghanzi Ridge, isolated hills, sectors of the larger valleys and the margins of some pans. The windblown Kalahari sands also caused the silting up of the ancient rivers running to and from the Great Kalahari Lake which today are visible only as fossil valleys.

The **pans** of the Kalahari are important features for the birdwatcher because of their capacity to hold water which can result in major transformations between extremely dry and extremely wet conditions. Pans are not confined to the Kalahari, but, as most of eastern Botswana is drained by well defined channels in the hardveld, the type of localized drainage system which terminates in pans is less numerous there. The typical pans of the Kalahari are rounded or elongate, flat-bottomed depressions ranging from a few kilometres to less than 100 m in diameter. Different processes have caused their formation. Some form chains along the fossil valleys or occupy interdune hollows in old dunefields. Others are deflation hollows unrelated to other topographic features. Pans of the latter type usually have one or more low hills on their southern or southwestern margin. These are relict sand dunes formed from material blown off the surface of the pan by the prevailing winds in more arid times past. Small pans are particularly numerous on the continental divide—an ill-defined zone extending from Ncojane in the west, to near Kanye in the east (the 'Bakalahari Schwelle' of Passarge).

The lack of a clear definition of the eastern edge of the Kalahari can be a cause of frustration to the birdwatcher in the field. In theory it is the western reaches of all the tributaries of the Limpopo River but these are widely spaced and rarely flowing and, in the field, their influence on the vegetation is difficult to discern from the true Kalahari. In some places north of the continental divide, headward erosion by the tributaries of the Limpopo River has cut back into the sediments of the Kalahari and its edge is clearly discernible as a low escarpment. However, most of the eastern edge of the Kalahari is poorly defined as a result of various complex geological processes. For example, the lower layers of the Karoo Supergroup (see below) are made up of several sedimentary units including water-deposited sandstones and mudstones. These lower Karoo strata are overlain by massive aeolian sandstones which weather to thick sandy soils similar in appearance to those sands found in the Kalahari (but which the geologist can differentiate by analysis from the more earthy soils and calcrete patches of the older strata). Where these younger Karoo sediments abut the Kalahari, such as between Mochudi and Mahalapye, it is difficult to diferentiate visually the two types of sand and therefore define the eastern limits of the Kalahari.

Today the most conspicuous remnants of the Great Kalahari Lake are the Makgadikgadi Pans, Nxai Pan, Lake Xau, Lake Ngami and the Mababe depression, although rivers flow into them irregularly or not at all. Climatic changes, the continued deposition of aeolian sands into the Kalahari Basin, the formation of the Magikwe Sand Ridge and further faulting causing the Thamalakane, Gomare and Khunyere Faults have blocked most of the waters of the Okavango River from flowing to those lowlying remnant areas of the Great Kalahari Lake—the Zambezi and Kwando Rivers had already been diverted by another geological process (see above). Instead, the Okavango waters formed swamps as they terminated in the sands and, by the continued deposition of the sands and sediments which they carried, the delta as it exists today has been formed. The Okavango Delta is therefore much younger than the Makgadikgadi Pans. Even over the past century the flows, channels and configuration of the delta have changed and continue to change as blockages and new channels are formed. Slowly the delta is being choked by sand and vegetation.

The **dune** sands of the southwest are thought to have taken on their present form about 10 000 years ago. Dunes which may have existed in the Kalahari Basin have mostly lost their characteristic shape because of climatic changes, but were probably formed at the end of the Cretaceous and early Tertiary periods.

The rock formations underlying the **hardveld** can be divided into five broad age divisions: the Archaean Basement; Early Proterozoic sequences; mid-Proterozoic sequences; the Karoo Supergroup; and Quaternary sedments. Their varied and numerous forms have led to the puzzling complexity of rock formations and vegetation in the east and southeast. The Archaean Basement includes various schists and gneisses as well as numerous igneous intrusions, e.g. Gaborone Granite. These rock types underlie flat or gently undulating country which is punctuated in some areas by boulder koppies or larger rocky hills formed of the more potassium-rich gneisses and granitoids, rarely by basic or ultrabasic igneous rocks. Kgale Mountain, which overlooks Gaborone, is mainly Gaborone Granite with part of a dolerite sheet, while Modipe Hill east of Gaborone is part of the Modipe Gabbro. More extensive granitic hills, similar to the Matobo Hills of Zimbabwe, occur north and east of Bobonong.

Figure 14 Geology of Botswana

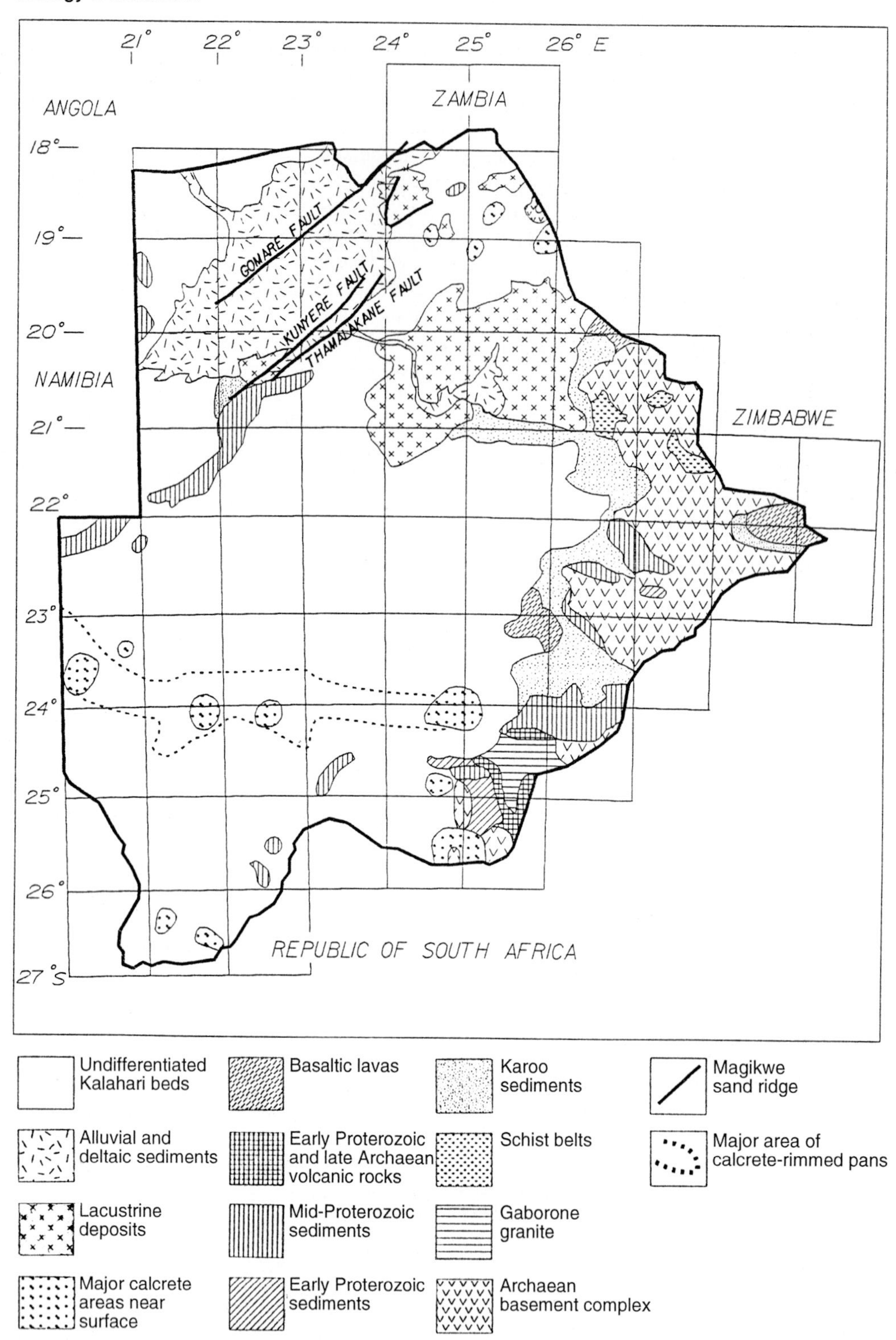

Schists and allied rock types are concentrated in three areas of the northeast (Figure 14). In some places the relatively clay-rich soils over these 'schist belts' give rise to scrubby vegetation dominated by *Acacia*, whereas broad-leafed woodland (including mopane) occurs on the more sandy soils overlying granites and gneisses.

The early Proterozoic sequences which occur in southeast Botswana include thick layers of sedimentary and volcanic rock. The sediments include quartzites, mudstones, cherts and dolomites which underlie rather subdued topography with sporadic ridges formed of the harder rock types (as at Mogobane), but the more resistant volcanic deposits form the rounded low hills ranging from Gaborone to Lobatse and Kanye. In contrast the prominent rocky plateaux between Ootse, Gabane and Moshaneng are formed by thick red sandstones and conglomerates in the mid-Proterozoic sequence, accompanied in places by sheets of dolerite. The tops of these hills are usually well wooded and their steep sides often form cliffs. Similar formations occur near Mochudi and Molepolole and in the Tswapong and Mokgware Hills near Palapye and Mahalapye. These mid-Proterozoic sandstone cliffs are used by Cape Vulture, Black Eagle, Black Stork and Mocking Chat. The Boulder Chat is found only on granitic koppies extending from Zimbabwe. The southern limit of the species appears to be blocked by the distribution of the Karoo lavas near Bobonong, the sandstone hills near Palapye and the Kalahari to the west. This species occurs 10 km north of the Tswapong Hills but has not been recorded south of these hills. In contrast the Longtailed Starling is common on Karoo basalts and not in adjacent areas.

A knowledge of the diversity of the geology in the east and southeast of Botswana (on the hardveld) helps to explain the diversity of the vegetation in this region. Here the bird-watcher needs to use his or her experience of vegetation structure. In contrast to the relatively uniform soil types of the Kalahari where changes in vegetation are more closely related to climate, the vegetation of the hardveld changes rapidly with differences in soil types because of changes in the underlying geology.

Climate

Between Botswana and the sea to the east and west are higher plateaux and mountains which precipitate moist air from the oceans before they reach far inland. **Rainfall** over landlocked

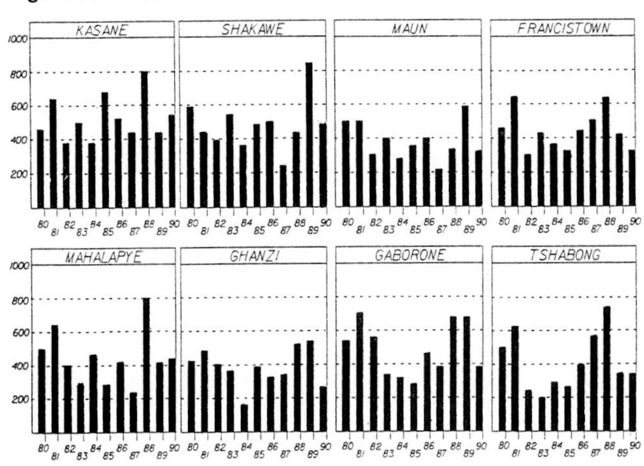

Figure 15 **Annual rainfall 1980–90**

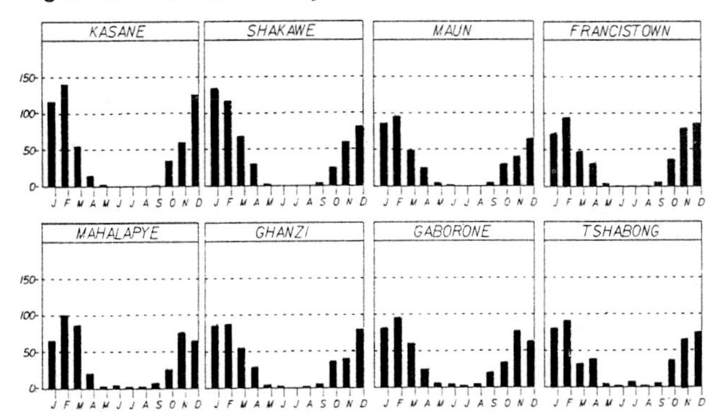

Figure 16 **Mean monthly rainfall 1980–90**

Botswana is therefore generally low, but fluctuates erratically and unpredictably from year to year and from place to place. Most of the rainfall occurs in the hot summer months and there is usually little or no rain in other months. The Intertropical Convergent Zone (ITCZ) supplements the annual rainfall in the northern third which historically has an annual average rainfall (AAR) of 650 mm with a variability of about 25%. In the east the AAR is about 500 mm with a variability of about 30% and this decreases gradually towards the south west corner where the AAR is about 200 mm with a

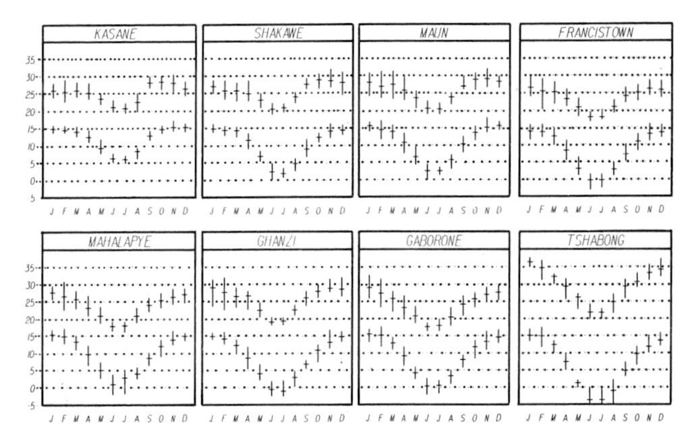

Figure 17 **Mean monthly temperatures and range of mean 1980–90**

variability of 80%. These historic figures (Figure 19) can be compared with the rainfall which actually occurred during the ten years of the project (Figures 15 & 20). Rainfall generally decreases in quantity and predictability from north to south and from east to west. The rainy season usually lasts from late September to April in the east, and from October to March in the west. Much of the rain falls as storms; gentle

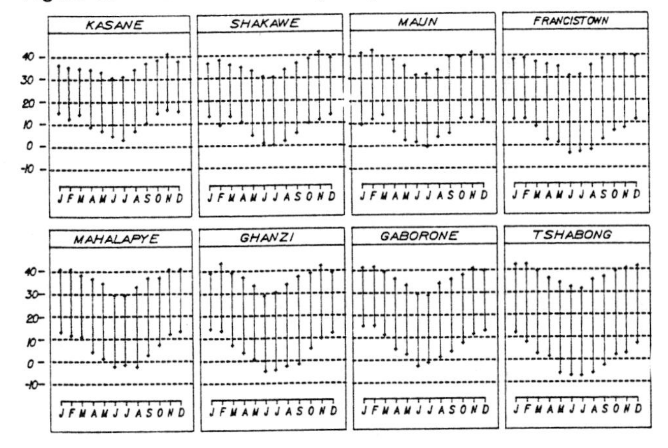

Figure 18 **Extremes of daily temp 1980–90**

rain or drizzle is uncommon. This means that a proportion of each downfall can have little benefit on the land as it rushes off into gullies and washaways or waterlogs the surface and evaporates in the intense heat of the following day. Storms usually deposit between 15 mm and 90 mm of rain in a few hours but a heavy tropical storm can produce up to 150–200 mm in a similar period. The effect of the rainfall has a major seasonal influence on the vegetation and consequently on the avifauna.

Over most of the country the climate and vegetation are semi-arid because of long winter periods of little or no rain with moderate or high daytime temperatures in plateau conditions of low humidity. There are no rivers in the flat Kalahari landscape to sustain vegetation during this winter period and water evaporates or drains readily from the sandy soils. Figures 15 and 20 show the annual rainfall in eight towns in different regions of the country over the 10-year period of this survey.

During the decade of the project, February was the wettest month in seven of the eight towns studied (Figure 16)—January was the wettest month at Shakawe. Only Kasane (AAR 554 mm) recorded an AAR in excess of 500 mm and Gaborone (AAR 478 mm) had the third highest rainfall.

Figure 19 Rainfall variability (derived from J.G. Pike, 1971)

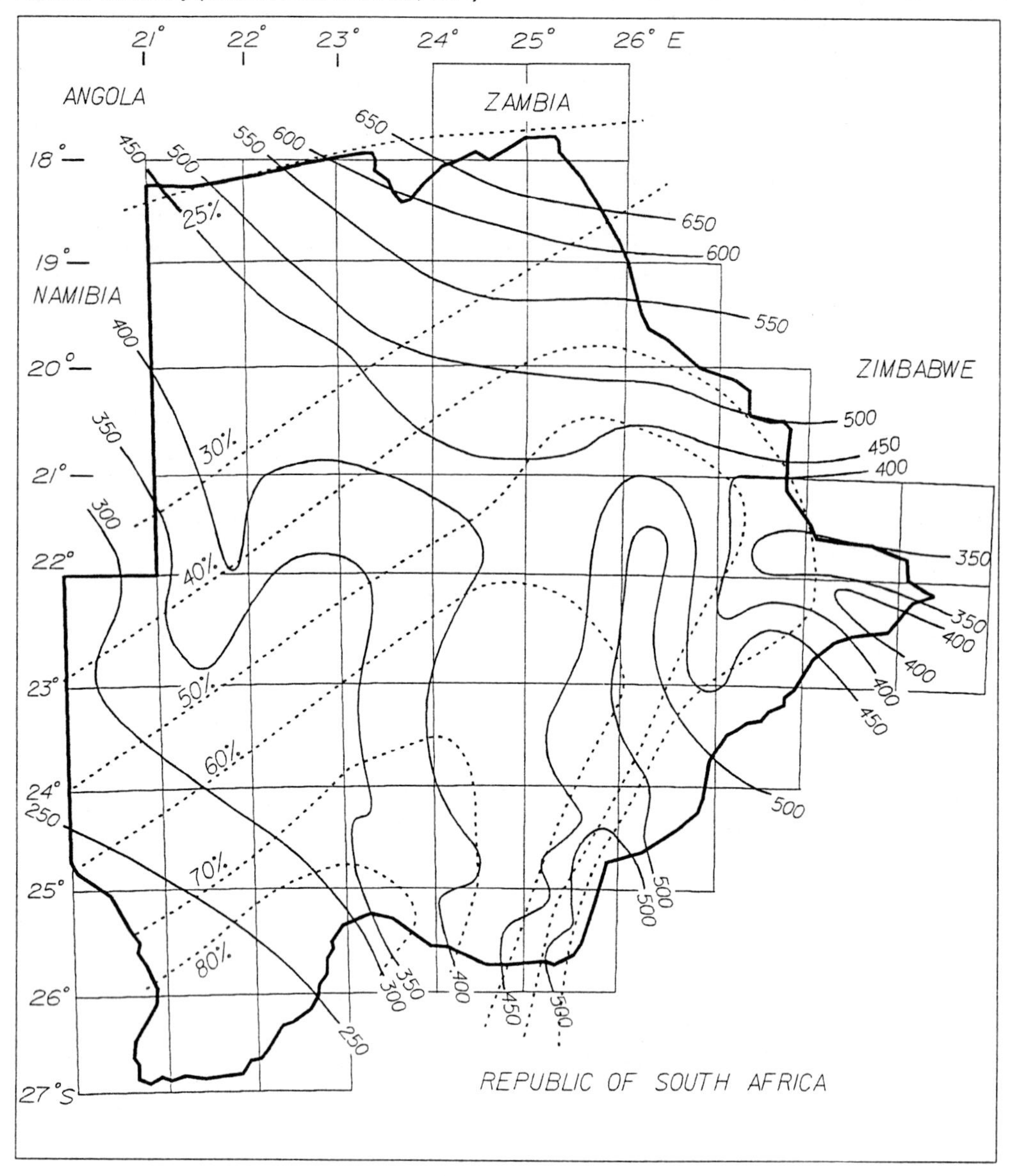

— Average annual rainfall in mm - - - Percentage variability

Ghanzi (AAR 380 mm) received the least rain and the next driest places were Maun (AAR 381 mm) and Tshabong (AAR 412 mm). Half the annual rainfall fell during one month of the year in several of the towns in several years. Kasane had no rain at all for 6-month periods in 1980 and 1981 and for 5-month periods in 1985 and 1986 but a deluge of 441 mm fell in February 1988. Nearly all the towns studied had their highest annual rainfall in 1988 which was the year when the drought was broken. Throughout Botswana there are usually three months or more of no rain at all and only in the southwest is a small amount of winter rainfall a regular feature.

There are two main **seasons**—winter, from June to August, and summer, from November to March. The months of April and May and September and October are short transition periods representing autumn and spring in temperate climes. The shortest daylight hours occur on June 22 (midwinter) and the longest on December 21 (midsummer). Day lengths only vary by about 2 hours from midsummer to midwinter in the tropical parts with sunset at about 18:30 in midwinter and 19:30 in midsummer and sunrise at 07:00 and 06:00 respectively. Below the Tropic of Capricorn the day lengths vary by up to 3½ hours increasing gradually southward. Winter is a long dry period of sunshine and mostly cloudless days with midday temperatures up to 17°C dropping to −8°C at night in the coldest areas. Frost is rare north of the Tropic but regular in the southern Kalahari and most severe in the arid southwest (see Tshabong minimum temperatures—Figures 17 & 18). Summer is characterised by very hot midday temperatures of about 32–35°C, although temperatures as high as 40°C occur. Summer nights are warm (15–20°C) and dry because of the relatively low humidity of the plateau except when storms are developing. Hot humid nights occur when rain is imminent.

Atmospheric pressure drops significantly at ground level during and after a storm, sometimes causing up to 10°C drop in temperature. The complexity of the changes in atmospheric pressure over the flat landmass can result in several storms occurring from different directions during a 12-hour period, each presaged by winds from equally varied direc-

Okavango River at Shakawe Fishing Camp in the northern part of the Okavango Panhandle. Papyrus beds on the left of the picture. 1821B, March 1985.

Northern woodland, north of the Okavango Delta – 1823A, March 1985. Here is one of the richest examples of Zambezian dry deciduous woodland.

Baikiaea woodland showing well developed understorey (mutemwa). Sibuyu Forest Reserve, 1825D, April 1987.

Grass plain on Kalahari sand north of Nata. 1926C, April 1987. The effect of drought is clearly shown by the colour and height of grass for that time of year.

Granite outcrop north of Francistown. 2027C, April 1987. Typical habitat for Boulder Chat. The vegetation shows the effect of drought.

Boteti River north of Xhumaga. 2024A, February 1989.

A mature and undisturbed stand of mopane woodland on Kalahari sand. Maikelelo Forest Reserve, 1824D, April 1987.

Limpopo River at Stockport showing no water flow in midsummer. 2327A, December 1989.

Central Kalahari savanna, looking due west along the Kalahari Traverse into Deception Valley. 2123B, February 1989. Good grass growth after rain.

Central Kalahari savanna after six months without rain. 2122D, November 1988. The cutline can be followed to the horizon.

Fossil valley, looking due south directly across the Okwa Valley on the western boundary of the Central Kalahari Game Reserve. 2222B, March 1988.

Typical central Kalahari (*Acacia*) woodland. 2223A, March 1988.

A scrub savanna type seen in widely scattered sectors of the eastern Kalahari. 2225A, January 1986.

Grassed pan in the Central Kalahari Game Reserve. The apron of woody shrubs shows clearly in the foreground. 2223C, March 1988.

Cataphractes alexandrii on calcrete. A regular vegetation type in the Kalahari which occurs in extensive patches such as this. 2323A, March 1988.

Western pan near Morwamosu after rain. Attractive to cattle and Abdim's Storks. 2423A, April 1988.

Western woodland on the cutline from Khakhea to Mabuasehube. 2422D, March 1985.

Ukwi pan with its surface of bare calcrete is the largest pan in the southern and western Kalahari (12 km in circumference). 2320D, March 1987.

Western woodland with very little grass cover, showing an area with a parklike arrangement of trees. 2321C, March 1987.

Western woodland with a sea of grasses for comparison with the drought conditions of the previous year (illustrated above). 2321C, March 1988.

Acacia erioloba savanna near Good Hope pan in the southeast. 2525A, April 1988.

The middle Molopo Valley from the dry bed of the Molopo River at Medenham. Botswana is on the right (north) of the fence. 2523D, January 1990.

The western Molopo Valley from the floor of the river bed showing a section of the escarpment on the southern side. 2621D, April 1990.

Acacia erioloba in the dry river bed of the Nossob Valley near Twee Rivieren. The road is on the international boundary with Botswana on the right (east). 2620B, April 1990.

Figure 20 **Average annual rainfall 1980–90**

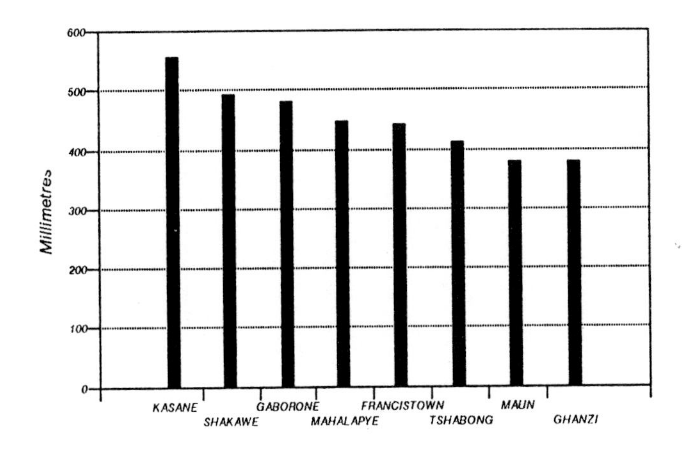

tions. In years of good rain, February and March often have days of high humidity due to ground evaporation. Bird activity is significantly highest in the cooler hours of the morning and an hour or two before sunset, and markedly reduced in mid-afternoon.

Vegetation

Of all the factors affecting bird distribution, changes in vegetation, whether due to soil, climate, season, altitude or human influence are the most immediate and this is the factor to which most birdwatchers relate first. For this reason bird distribution in relation to vegetation is discussed at length.

Charles Darwin stated (according to Haviland in Polunin, 1960), 'A traveller should be a botanist, for in all views plants form the main embellishment. But when the zoologist, forsaking botanical terms, tries to classify environments in the language of his own science, he cannot construct a workable scheme . . . he finds he must fall back on the language of the botanist or geologist.' Thus the botanist's classification of vegetation for the African continent given by White (1983) forms the basis of the discourse that follows. Because this is not a sufficiently detailed view of Botswana for the birdwatcher, further subdivisions for Botswana's vegetation given by Weare & Yalala (1971) and by Acocks (1988) for South Africa are also used. Finally, a birdwatcher's map based on experience in the field over the 10 years of this project is given. It takes a major account of physical geography and geology (as discussed earlier) and its zones are simplistic as this is the first time Botswana has been divided into ornithological zones. The birdwatcher's interpretation given here is on an atlas-mapping-unit basis (square) which is 2 500 km^2 and does not show the variety of different habitats in that square, only the main vegetation type to which a birdwatcher should go to see certain species characteristic of that type.

Phytochoria

Within the classification of the vegetation of the continent by White (1983), Botswana falls into two **phytochoria** (biotic zones)—the Zambezian regional centre of endemism in the north and northeast, and the Kalahari-Highveld regional transition zone in the southern and western two-thirds (Figure 21).

ZAMBEZIAN

The first, which is also discussed as the tropical woodland savanna component elswhere, includes the following subdivisions (Figure 22):

22a – Mosaic of Zambezian dry deciduous forest and secondary grassland;
28 – *Colophospermum mopane* woodland and scrub woodland.
29d – South Zambezian undifferentiated woodland and scrub woodland.
75 – Herbaceous swamp and aquatic vegetation.
76 – Halophytic vegetation.

KALAHARI

The second phytochorium, which characterises most of the true Kalahari component in Botswana, has the following subdivisions:

35a – Transition from undifferentiated Zambezian woodland to Kalahari *Acacia* deciduous bushland and wooded grassland.
44 – Kalahari *Acacia* wooded grassland and deciduous bushland.
56 – Kalahari/Karoo–Namib transition.

These two phytochoria form a good basis on which to explain bird distribution in the southern third of Africa. In relation to Botswana they divide the Kalahari component from the tropical savanna and bushveld component and align well with species distribution for this and other regions shown on the continental maps for each species.

SUMMARY OF WHITE'S DESCRIPTION

22a – **Dry deciduous forest** is typified by *Baikiaea* woodland on Kalahari Sand where *Baikiaea plurijuga* is the dominant tree. It has a better developed understorey than Miombo (*Brachystegia*) woodland. Other trees in this woodland include *Acacia erioloba, Pterocarpus antunesii, Combretum collinum, Ricinodendron rautanenii, Commiphora angolensis, Boscia albitrunca, Adansonia digitata* and some *Brachystegia* species. The shrub layer, the mutemwa, often forms thickets.

28 – ***Colophospermum mopane* communities** vary in height and density but have a remarkable structural uniformity because of the almost complete dominance of the mopane to the exclusion of other vegetation. It is easily reduced to shrub condition by fire and elephant browsing. The most conspicuous associated trees are *Acacia nigrescens, Adansonia digitata, Combretum imberbe, Sclerocarya birrea* and *Kirkia acuminata.*

29d – **South Zambezian undifferentiated woodland** occurs between the floor of the Limpopo Valley and the northern limits of the Highveld. Trees may reach 9 m in height locally but it is mostly scrub woodland. Its composition includes *Acacia gerrardii, A. nigrescens, A. tortilis Adansonia digitata, Albizia antunesiana, Burkea africana, Combretum apiculatum, C. collinum,*

Figure 21 **Phytochoria (after White, 1983)**

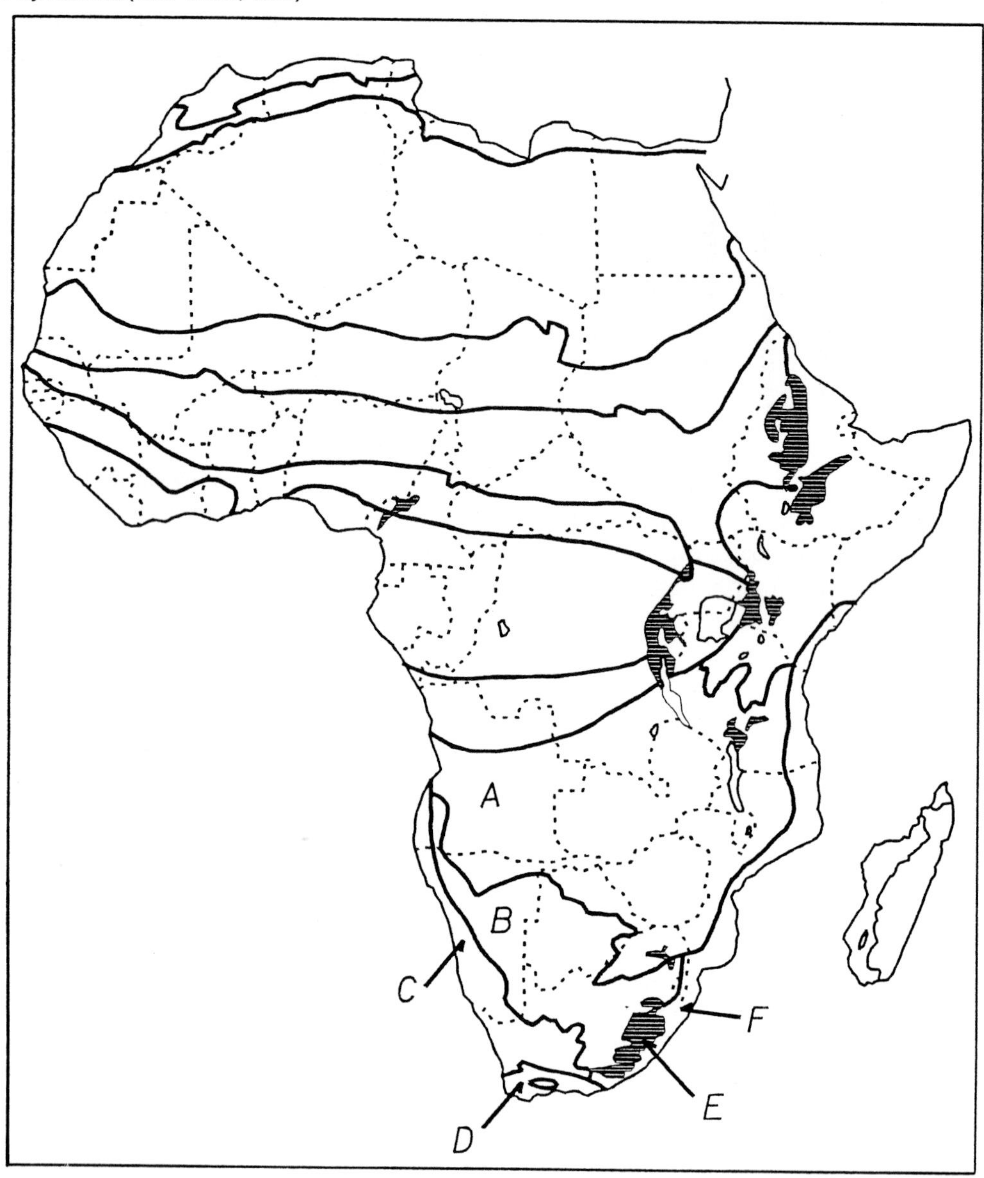

A Zambezian
B Kalahari–Highveld transition
C Karoo–Namib
D Cape

C. imberbe, *C. molle*, *C. zeyheri*, *Commiphora mollis*, *Dichrostachys cinerea*, *Diospyros mespiliformis*, *Euphorbia ingens*, *Peltophorum africanum*, *Sclerocarya caffra*, *Terminalia prunioides*, *T. sericea* and *Ziziphus mucronata*.

35a – **Wooded grassland** is the characteristic vegetation of the thick mantle of Kalahari sand. In Botswana it is a more or less continuous sward less than 1 m high and includes the grasses *Anthephora argentea*, *A. pubescens*, *Digitaria pentzii*, *Eragrostis biflora*, *E. ciliaris*, *E. lehmanniana*, *E. pallens*, *Panicum kalaharense*, *Schmidtia kalahariensis* and *Stipagrostis uniplumis*. The dominant trees are *Acacia* species but some broadleafed trees are abundant such as *Combretum collinum*, *Commiphora africana*, *C. angolensis*, *Ochna pulchra* and *Ziziphus mucronata*. The trees are usually less than 7 m tall and widely spaced. On the Ghanzi Ridge some Karoo woody plants and shrubs are plentiful such as *Rhigozum brevispinosum*, *Leucosphera bainesii*, *Phaeoptilum spinosum* and *Montinia caryophyllacea*.

44 – **Kalahari *Acacia* wooded grassland** has an underlying composition similar to the previous unit. *Acacia* trees dominate to the virtual exclusion of broadleafed trees. The principal trees are *Acacia erioloba*, *A. fleckii*, *A. hebeclada*, *A. luederitzii*, *A. mellifera*, *A, tortilis*, *Boscia albitrunca*, *Dichrostachys cinerea* and *Terminalia sericea*.

56 – **Kalahari/Karoo–Namib transition** is an area of wind-blown sand which occurs as fixed dunes in the form of long parallel ridges. Sand covers 90% of the surface. The lower slopes of the dunes are largely consolidated by vegetation but the upper slopes and crests are subjected to wind erosion and cover is sparser. It is a mosaic of lightly wooded grassland on the crests, pure grassland in the depressions between the dunes and *Rhigozum trichotomum* shrubby grassland in deeper hollows. In undisturbed areas the

Figure 22 **Vegetation of Botswana (after White)**

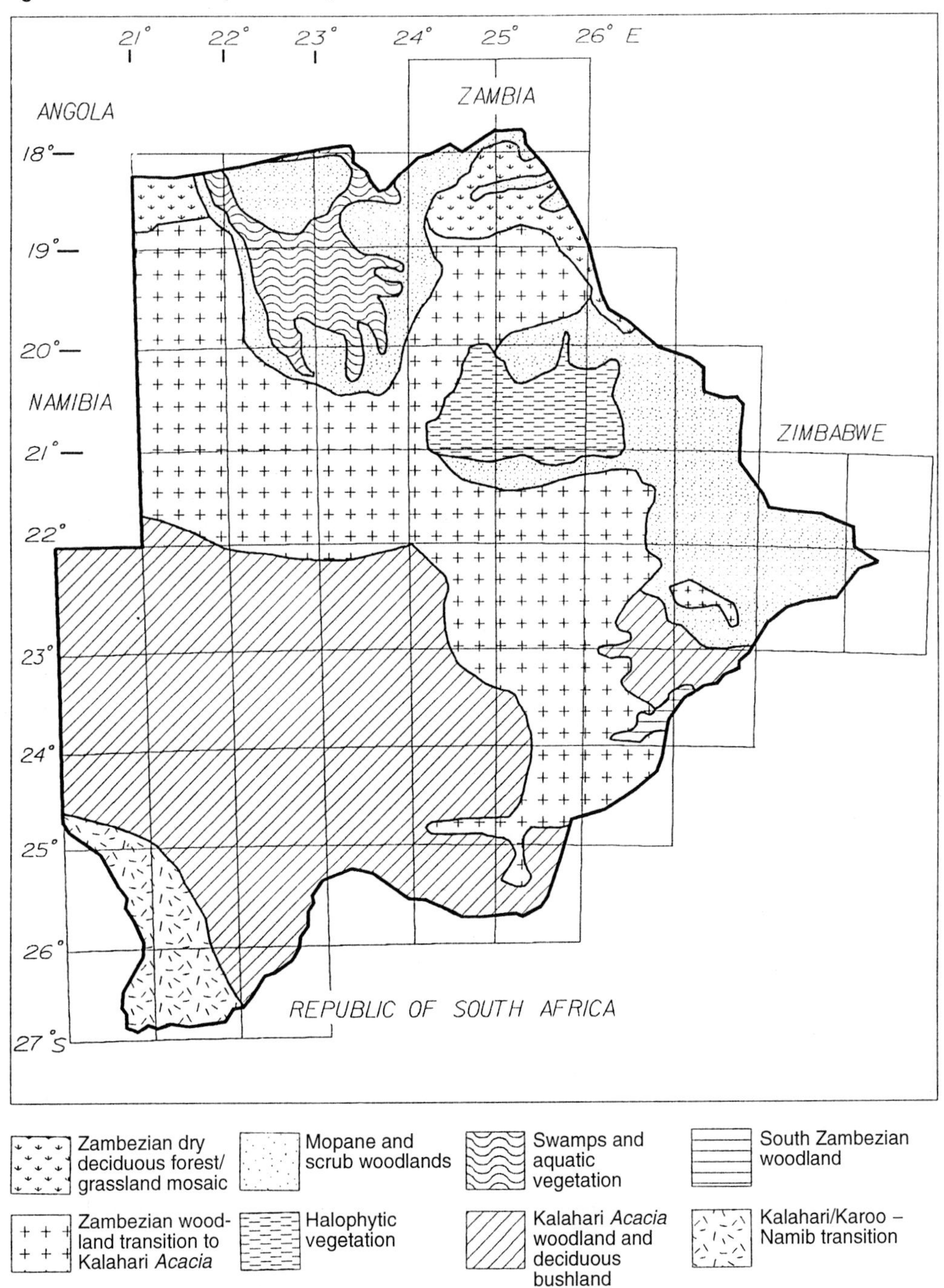

grasses are mostly *Asthenatherum glaucum*, *Stipagrostis uniplumis*, *Eragrostis lehmanniana*, *Stipagrostis ciliata* and *S. amabilis* whereas *Schmidtia kalahariensis* dominates in disturbed areas. The commonest trees are *Acacia erioloba*, *Boscia albitrunca*, *Acacia reficiens*, *A. haematoxylon* (mainly as a shrub), *Albizia anthelmintica* and *Terminalia sericea*.

75 – **Herbaceous** freshwater swamp and aquatic vegetation is found in the Okavango Delta. The main constituent is *Cyperus papyrus* but many species of aquatic plant such as water hyacinth (*Eichhornia crassipes*), water lettuce (*Pistia stratiotes*), water fern (*Salvinia molesta*), sedges, reeds and grasses occur.

76 – The **Makgadikgadi Pans** are examples of alkaline flats. They are surrounded by a narrow fringe of grassland dominated by *Sporobolus spicatus* and *Odyssea paucinervis*.

Zonal effects of vegetation on bird distribution

From the birdwatcher's viewpoint it helps to classify the country into smaller subdivisions based on the vegetation as well as other influences such as geology and physical geography. Starting in the northeast, these subdivisions are discussed together with the birds that occur in them.

ZAMBEZIA

From Kasane west through the Linyanti swamps to the areas north of the Okavango Delta south through Savuti to Tsebanana in the east. It includes the Chobe and Linyanti wetlands and the Kwando River. Away from the river valleys

Figure 23 Vegetation zones: ornithological

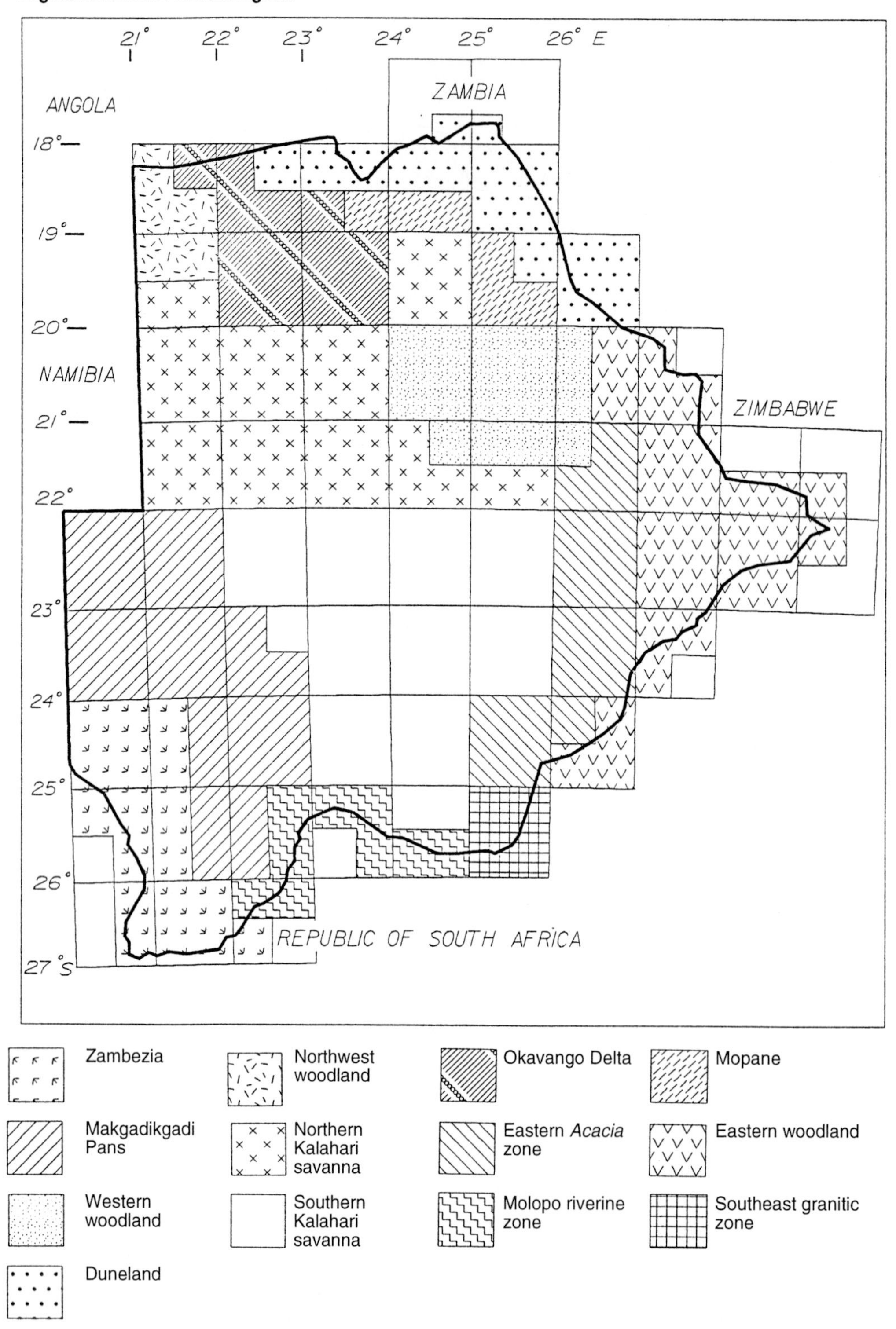

and the wetlands there are large areas of dry deciduous woodland with trees 10–15 m in height. The *Baikiaea* composition has been discussed above and the botanical term refers hereafter to those areas of dry deciduous woodland with a well developed understorey of bushes, creepers and grass. Some areas with very little understorey, but often good grass cover, are more typical of Miombo woodland and occur in some parts of the Sibuyu and Maikelelo Forest Reserve. These areas contain trees such as *Brachystegia spiciformis*, *B. boehmii*, *Julbernardia globiflora*, *Pterocarpus angolensis*, *Afzelia quanzensis*, *Sterculia africana*. The avifaunal composition of *Baikiaea* and Miombo are similar in these areas. This zone also has areas of *Acacia* in river valleys and mopane, but such areas do not predominate and it is the teak (*Baikiaea*) woodlands that are of special interest to the birdwatcher.

This is a zone of rich and diverse habitats which supports over 400 species of birds. Noteworthy species of the wetlands are Woollynecked Stork, Wattled Crane, Great Snipe, Longtoed and Whitecrowned Plovers, Black Coucal, Natal Nightjar, Halfcollared Kingfisher, Chirping Cisticola, Greater Swamp Warbler (in papyrus on the Kwando River),

Pinkthroated Longclaw, Thickbilled Weaver and Redshouldered Widow. Interesting birds of the woodland and wooded savanna include Dickinson's Kestrel, Dark Chanting Goshawk, Lizard Buzzard, Cape Parrot, Blackbellied Korhaan, Green Lourie, Pennantwinged Nightjar, Racket-tailed Roller, Broadbilled Roller, Grey, Redbilled, Yellowbilled, Bradfield's and Ground Hornbills, Whitebreasted Cuckooshrike, Blackheaded Oriole, Blackeyed Bulbul, Miombo Rock Thrush, Greencapped Eremomela, Stierling's Barred Warbler, Neddicky, Pallid Flycatcher, Purplebanded Sunbird, Goldenbacked Pytilia, Broadtailed Paradise Whydah, Blackeared and Streakyheaded Canaries. In forest and thickets are Black Sparrowhawk, African Goshawk, Wood, Barred and Giant Eagle Owls, Narina Trogon, Terrestrial and Yellowbellied Bulbuls and Bearded Robin. In this zone there are also dambos (near Jolley's Pan and around Mpandamatenga), seasonally flooded grassland on drainage lines, which are inhabited by Dickinson's Kestrel and Flappet Lark and rarely the Croaking Cisticola and Cuckoo Finch.

OKAVANGO DELTA

The Okavango Delta is characterized by huge tracts of *Papyrus* and reed beds with a dense mat of underlying floating aquatic vegetation. There are channels, lagoons (madiba), floodplains, floodplain woodland, marshes and forest-clad islands. In the north the Okavango River enters near Mohembo and continues southeast as a widening river forming an area known as the Panhandle until it splits up into many different channels and is slowly choked by the mass of vegetation. The water in the delta is crystal clear and the substrate is mainly sand. Surrounding the delta are mopane and mixed woodlands. Riparian forest lines the banks of the Okavango River and most of its subdivisions including patches along the Thamalakane and Nhabe rivers. The diversity of the vegetation is enormous and is presented in summary in Table 10.

Table 10 Diversity of vegetation in the Okavango Delta (after P. A. Smith, 1976)

Trees	
SEMIAQUATIC	18 species: *Acacia galpinii, A. karroo, A. nigrescens, Albizia versicolor, A. harveyi, Antidesma venosum, Berchemia discolor, Carissa edulis, Cassine transvaalensis, Croton megalobotrys, Diospyros mespiliformis, Ficus burkei, F. sycomorus, Garcinia livingstonei, Hyphaene benguellensis, Kigelia pinnata, Lonchocarpus capassa, Sclerocarya caffra*
FLOODPLAIN EDGES AND MAINLAND	10 species *Acacia erioloba, A. fleckii, A. hebeclada, A. tortilis, A. sieberana, Colophospermum mopane, Combretum imberbe, Gardenia spatulifolia, Terminalia sericea, Ziziphus mucronata*
MISCELLANEOUS SITES	10 species: *Acacia luederitzii, Adansonia digitata, Boscia albitrunca, Combretum hereroense, Commiphora* spp, *Euclea divinorum, Lonchocarpus nelsii, Peltophorum africanum*
RARE	12 species including: *Burkea africana, Combretum apiculatum, C. zeyheri, Euphorbia ingens, Kirkia acuminata, Ochna pulchra, Pterocarpus angolensis, Schinziophyton rautanenii*
UPPER PANHANDLE ONLY	5 species including: *Acacia albida, Syzygium guineense, Rhus quartiniana*
Shrubs and woody plants	41 species including: *Bauhinia, Boscia, Combretum, Commiphora, Cardaria, Dichrostachys, Ehretia, Ficus, Grewia, Hibiscus, Lantana, Markhamia, Plumbago, Rhus*
Creepers and lianes	10 species:
Aquatic herbs and ferns (106 spp)	
FREE-FLOATING	4 species
FLOATING SUBMERGED	11 species
ROOTED SUBMERGED	15 species
ROOTED FLOATING	18 species
ROOTED EMERGENT	57 species
Grasses and sedges	
PERENNIAL AND SEASONAL SWAMPS	17 grasses; 19 sedges
SEASONAL SWAMPS AND FLOODPLAINS	34 grasses; 36 sedges
ISLANDS AND PANS	41 grasses including: *Andropogon, Aristida, Chloris, Cymbopogon, Digitaria, Eragrostis, Panicum, Pennisetum, Phragmites, Setaria, Sporobolus* 8 sedges including: *Cyperus, Eleocharis, Fuirena, Pycreus, Scirpus*
Herbs	128 genera

In this luxuriant environment are found 16 species of heron including the Slaty Egret and Whitebacked Night Heron, 10 duck species including the Pygmy Goose, and representatives of most of the waterbird families occurring in Botswana. Birds of special interest are African Hobby, Longcrested Eagle, Western Banded Snake Eagle, Redchested Flufftail, African Skimmer, African Mourning Dove, Copperytailed and Black Coucals, Pel's Fishing Owl, Natal Nightjar, Böhm's Spinetail, Wiretailed and Mosque Swallows, Whiterumped Babbler, Terrestrial and Yellowbellied Bulbuls, Greater Swamp and African Sedge Warblers, Chirping, Blackbacked and Redfaced Cisticolas, Swamp Boubou, Sharptailed Starling, Yellowbilled Oxpecker, Collared Sunbird, Brownthroated, Thickbilled and Spectacled Weavers, Brown Firefinch and Orangebreasted Waxbill. Common birds typical of the area include Saddlebilled and Openbilled Storks, Hadeda, Pygmy Goose, African Fish Eagle, African Marsh Harrier, Purple Gallinule, African and Lesser Jacanas, Blacksmith, Crowned, Wattled and Longtoed Plovers, Water Dikkop, Cape Turtle and Laughing Doves, Green Pigeon, Palm Swift, Woodland Kingfisher, Carmine Bee-eater, White Helmetshrike, Longtailed, Burchell's and Glossy Starlings, Spottedbacked, Golden and Redheaded Weavers, Jameson's and Redbilled Firefinches, Common Waxbill and Pintailed Whydah.

NORTHWEST WOODLAND

The Northwest Woodland, west of Mohembo and the road to Nokaneng and along the Namibian border to as far south as the Aha Hills, is botanically a transition of *Baikiaea* woodland into *Acacia* woodland. Deep, soft Kalahari sand in the western parts makes exploration difficult. It is a drier deciduous woodland than that found in the northeast (AAR 400 mm compared with 600 mm) and dry thickets and woodland with dense understorey and creepers occur in patches. In this latter habitat the Blackfaced Babbler occurs as an extension of its range from Namibia and Angola. The dry dense thicket at Nokaneng looks ideal for the Angola Pitta and Buffspotted Flufftail which have hitherto not been recorded in Botswana. Rare species found in this zone include the Rufousbellied Tit and the Chestnut Weaver.

For the birdwatcher the region has the potential that birds from the north and west hitherto unknown in Botswana may occur. Whereas the northeastern woodlands are mainly closed, this zone has a more open pattern of *Baikiaea*, *Acacia*, broadleafed (*Combretum* and *Terminalia* especially) and mopane woodlands and savanna. Birds typical of the woodlands north of 19°S include Dark Chanting Goshawk, Bradfield's Hornbill, Whitebreasted Cuckooshrike, African Golden Oriole, Black and Pallid Flycatchers, Blackfaced Babbler, Southern Black Tit, Yellowbreasted Apalis, White Helmetshrike and Orange-breasted Bush Shrike, Plumcoloured, Burchell's and Longtailed Starlings, Yellowthroated Sparrow and Yelloweyed Canary. In tree and bush savanna typical birds include Redbilled Hornbill, Sabota Lark, Ashy and Cape Penduline Tits, Kalahari Robin, Burntnecked Eremomela, Melba Finch and Redbilled Buffalo Weaver. In *Acacia* savanna typical Kalahari species such as Pale Chanting Goshawk, Kori Bustard, Fawncoloured Lark and Tinkling Cisticola occur.

COLOPHOSPERMUM MOPANE

Colophospermum mopane occurs almost throughout the north and east of Botswana. It is not illustrated to a great extent on a birdwatcher's map because, in pure form, it is a rather dull birdwatching habitat, structurally uniform as previously mentioned and thus with a limited variety of species. However, it is an important habitat for certain species such as Redbilled Hornbill and Arnot's Chat and when 'mopane worms' (caterpillars of *Gonimbrasia belina*) are prolific, it attracts large numbers of different birds, particularly Yellowbilled Kites and other migrant raptors. Areas from west of Savuti through the southern part of the Chobe National Park southeastwards to Francistown are dominated by this woodland. Large areas can be in bush or scrub form and very few birds occur in this type. A drawback of the area for birdwatchers is that it is well populated with elephants migrating from the Okavango and Chobe regions to the Hwange National Park in Zimbabwe. At Tsebanana, which abuts the southwestern boundary of the Hwange National Park in Zimbabwe, there are some tall stands of mopane woodland on Kalahari sand. Birds commonly found in mopane, but not confined to this habitat, include Threebanded Courser, Doublebanded Sandgrouse, Cape Turtle Dove, Pearlspotted Owl, Mozambique Nightjar, Striped Kingfisher, Crested Barbet, Southern Black Tit, Crombec, Longtailed Starling, Whitebellied Sunbird and Whitebrowed Sparrowweaver. South of a line approximately joining Francistown to Orapa mopane woodland continues to the Tropic of Capricorn but pure stands are uncommon and the diversity of the avifauna is determined by the more predominant *Acacia* and other broadleafed trees. For this reason, the squares south of Francistown are placed in the Eastern Woodland zone for the birdwatcher.

MAKGADIKGADI PANS

The Makgadikgadi Pans form a unique environment within Botswana. The two large main pans are alkaline flats similar to those of the Rift Valley Lakes of Kenya. When dry, the main areas are almost devoid of birds except for species such as Chestnutbanded and Kittlitz's Plovers. The Nata Delta sector of the Sua Pan is almost always wet and has a constant population of waterbirds such as grebes, cormorants, ducks, plovers, terns and gulls and usually a few if not larger numbers of both species of flamingos and pelicans. When the main pans become wet, thousands of waterbirds move in and most do so to breed. Noteworthy breeding birds include Blacknecked Grebe, pelicans, flamingos and Caspian Terns.

Flat grassland surrounds the pans from Nata in the east to near Orapa, Rakops and Xhumaga in the west. This area is the flat bottom of the old Kalahari Lake (about 900 m a.s.l.) and has very few trees except near the Boteti River and in patches in the Gweta and Odiakwe sectors. *Hyphaene* palms fringe many drainage lines and extend north to Kudiakam and Nxai Pans to an extent that some authors define this as a separate zone (Palm savanna) but there appears to be no ornithological necessity to do so. Typical birds of the flat Makgadikgadi grassland are Secretarybird, Lanner, Rednecked Falcon, Blackshouldered Kite, Blackbreasted Snake Eagle, Pallid and Montagu's Harriers, Harlequin Quail, Black Korhaan, Spotted Dikkop, Doublebanded Courser,

Blackwinged Pratincole, Burchell's Sandgrouse, Marsh Owl, Palm Swift, Rufousnaped, Spikeheeled and Redcapped Larks, Banded Martin, Capped Wheatear, Anteating Chat, Fantailed and Desert Cisticolas, Richard's Pipit, Lesser Grey Shrike and Scalyfeathered Finch.

In the west there is riparian forest along the Boteti River. Many species funnel down this narrow fertile strip from the northern wetlands. These include Purple, Squacco and Greenbacked Herons, Hamerkop, Spurwinged and Pygmy Geese, Black Crake, African Jacana, Water Dikkop, Redeyed and African Mourning Doves, Green Pigeon, Meyer's Parrot, Grey Lourie, Barred Owl, Pied and Giant Kingfishers, Bluecheeked and Little Bee-eaters, Greater and Lesser Honeyguides, Greyrumped Swallow, African Golden Oriole, Whiterumped Babbler, Terrestrial Bulbul, Cape Reed Warbler, Tawnyflanked Prinia, Burchell's Starling, Lesser Masked Weaver, Red Bishop and Redbilled Firefinch.

EASTERN WOODLAND

The Eastern Woodland is a mosaic of *Acacia* and broadleafed woodlands mainly on the hardveld east of the Limpopo watershed but extending north of Francistown onto Kalahari sand west of the Shashe watershed until it meets the dry deciduous woodland of the northwest and the Makgadikgadi Pan zone. The classification of this zone by White (1983) into *Colophospermum mopane* does not reflect the true diversity of the zone south of Francistown which ornithologically is not dominated by mopane woodland. Its diversity is better illustrated in Weare & Yalala (1971). In its northern part there are significant areas of mopane which become less dominant as one moves south towards the Tropic of Capricorn. A feature of the northeastern sector south of Francistown, and around Bobonong, Sefophe, and the Palapye region south to Seleka Hill near Martin's Drift, are the hills, rocky outcrops and boulders under woodland canopy. To as far south as Selebi Phikwe this includes the habitat of the Boulder Chat and other species include Black and Saddlebilled Storks, Threebanded Courser, Mocking Chat and Striped Pipit.

A major habitat for birdwatchers in this zone is the riparian forest and rich riverine vegetation of the Limpopo River and its tributaries, particularly the Shashe and Motloutse Rivers along which this habitat stretches westwards towards the Kalahari. Characteristic species of these forests and the Limpopo River are African Black Duck, Greenbacked Heron, Hamerkop, Hadeda Ibis, African Fish Eagle, Natal Francolin, African Finfoot, Redeyed Dove, Redchested, Striped and Klaas's Cuckoos, Giant Eagle Owl, Woodland Kingfisher, Whitefronted and Little Bee-eaters, Redbilled Woodhoopoe, Blackcollared and Yellowfronted Tinker Barbets, Blackheaded Oriole, Yellowbellied Bulbul, Redfaced Cisticola, Tropical Boubou, Collared Sunbird and Yellow White-eye (in the northern parts).

Trees along the major rivers may reach 20 m in height and include *Acacia albida*, *A. nigrescens*, *Croton megalobotrys*, *Schotia brachypetala*, *Ficus sycomorus* and *Lonchocarpus capassa*. On the banks are occasional *Phoenix reclinata* palms, *Phragmites* reeds and the dominant grass is *Panicum maximum*. The woodland in the river valleys which includes *Acacia xanthophloea* and *A. karroo* quickly change to tree and bush savanna and on the slopes of hills these are stunted and sparse. These savannas are diverse and may be principally *Combretum*, *Acacia*, *Grewia*, mopane, *Terminalia*, *Rhigozum*, *Kirkia* or *Commiphora*. Acocks (1988) subdivides such areas in South Africa into different Bushveld types of which the following also occur west of the Limpopo River; *Grewia flava* veld, Dwarf *Terminalia-Rhigozum* veld, Dwarf *Combretum apiculatum* veld, *Dichrostachys-Acacia* veld, *Panicum maximum-Acacia karroo* veld. In this zone *Boscia albitrunca*, *Sclerocarya caffra* and *Adansonia digitata* are common trees in patches and *Dichrostachys cinerea* and *Ziziphus mucronata* are common in bushy form. Birds typical of this wooded savanna are Little Banded and Gabar Goshawks, Coqui, Crested and Swainson's Francolins, Redcrested Korhaan, Doublebanded Sandgrouse, Cape Turtle Dove, Laughing and Namaqua Doves, Meyer's Parrot, Grey Lourie, African, Black and Diederik Cuckoos, Scops, Pearlspotted and Spotted Eagle Owls, Fierynecked Nightjar, Redfaced Mousebird, Striped Kingfisher, Lilacbreasted and Purple Rollers, African Hoopoe and Scimitarbilled Woodhoopoes, Grey and Yellowbilled Hornbills (Redbilled in mopane areas mostly), Pied and Crested Barbets, Greater Honeyguide, Bennett's, Goldentailed and Cardinal Woodpeckers, Monotonous and Sabota Larks, Redbreasted and Greater Striped Swallows, Forktailed Drongo, Ashy and Southern Black Tits, Arrowmarked Babbler, Kurrichane and Groundscraper Thrushes, Whitethroated and Whitebrowed Robins, Longbilled Crombec, Bleating Warbler, Rattling Cisticola, Chinspot Batis, Paradise Flycatcher, Redbacked Shrike, Puffback, Threestreaked Tchagra, White and Redbilled Helmetshrikes, Brubru, Glossy and Greater Blue-eared Starlings, Whitebellied Sunbird, Redbilled Buffalo Weaver, Whitebrowed Sparrowweaver, Greyheaded Sparrow, Redheaded and Masked Weavers, Cutthroat Finch, Melba Finch, Paradise Whydah, Yelloweyed Canary and Goldenbreasted Bunting.

EASTERN *ACACIA* ZONE

The Eastern *Acacia* Zone is a transitional area between the mainly broadleafed hardveld vegetation and the open *Acacia* savanna of the Kalahari proper and it fits into White's (1983) category 35a. It has vegetation characteristics similar to those of the Northern Kalahari Savanna. It is identified for the birdwatcher because true Kalahari species commonly occur alongside tropical woodland species. In this area, even along the drainage lines, there is much less profusion of vegetative growth although tongues of rich woodland reach into the zone along some of the major rivers such as at Serowe, Paje, Kalamare, Mahalapye and Shoshong, immediately adjacent to semidesert savannas. There are also good stands of woodland on Kalahari sand between Molepolole and Lephephe and again between Serowe and Mmashoro. This woodland is mainly *Acacia* and *Terminalia* but includes *Combretum* and mopane. In the savanna *Sclerocarya*, *Peltophorum*, *Boscia*, *Dichrostachys*, *Grewia* and *Ziziphus* are common. Trees may reach 10 m in height in woodland patches but most of the vegetation is in bush form elsewhere. There are no typical species for this transitional zone.

The birdwatcher needs to be aware that he may find both Arrowmarked and Pied Babbler, Chinspot and Pririt Batis,

Whitebrowed and Kalahari Robin, Rattling and Tinkling Cisticola, Puffback and Whitecrowned Shrike, Violeteared and Blue Waxbill, Pintailed and Shafttailed Whydah, Yelloweyed and Yellow Canary adjacent to each other in the wooded savanna. The eastern edge of the Kalahari is not a clearly defined line (see page 20).

NORTHERN KALAHARI SAVANNA

The Northern Kalahari Savanna is a transition similar to that described for the Eastern *Acacia* Zone. It fits into White's (1983) category 35a and its separation from the southern Kalahari savanna as White has done for the vegetation is reflected in the change in bird species-composition. Knowledge of rainfall and a view of the vegetation will show how the two zones separate. The zone is delineated to some extent by the 400-mm rainfall isohyet and the influence of the Intertropical Convergent Zone may also be significant. Travelling north of Ghanzi in the west towards the Kuke fence (at 21°S), one is immediately struck by the sudden appearance of broadleafed woodland some 30 km south of the Kuke fence. Mopane is not reached until one arrives near Maun. Travelling west to east along the Kuke fence, the vegetation is mainly *Acacia* tree and bush savanna but significant patches of broadleafed woodland, particularly *Combretum*, appear. *Terminalia sericea*, which is almost exclusively a bush or bushy tree rarely exceeding 4 m south of Ghanzi, is more commonly a tree 5–7 m tall north of Ghanzi. The current maps of bird distribution do not show this subdivision well except for the occasional hint from old records as far south as Ghanzi. During this survey, most of the tropical savanna species reached only as far south as Tale Pan (2022D) and this is attributed to the drought period. One should not be surprised to find tropical woodland species reaching as far south as Ghanzi in high-rainfall cycles. Species which indicate this northern Kalahari zone as an occasional extension of the tropical woodland are Greenspotted Dove, Lesser Masked Weaver, Yellowthroated Sparrow, Cape Sparrow (northern limit), Whitebellied Sunbird, Redbilled Oxpecker, Plumcoloured Starling (regular to Ghanzi and also occasionally in western woodland), White Helmetshrike, Puffback, Neddicky, Rattling Cisticola, Bleating Warbler, Whitebrowed Robin and Kurrichane Thrush. The western part of this region, particularly north of the Kuke fence, has been poorly studied.

Lake Ngami is an area of major ornithological interest occurring in this zone. It should be considered primarily as a southern appendage of the Okavango Delta because its importance lies mainly in its waterbirds. In particular its shores and contours are highly suited to migrant waders in very large numbers. Common birds are Dabchick, most herons and egrets, Sacred and Glossy Ibises, African Spoonbill, Purple Gallinule, Moorhen, Redknobbed Coot, African Jacana, Painted Snipe, Blacksmith Plover, Blackwinged Stilt, and Greyheaded Gull.

Usually in large numbers at some times of the year are Hottentot Teal, Pygmy Goose, Knobbilled, Whitefaced and Fulvous Ducks, Wood, Marsh and Curlew Sandpipers, Greenshank, Little Stint, Ruff and Whitewinged Tern. Sparse but regular are Curlew and Blacktailed Godwit and rarely the Bartailed Godwit. Breeding in very large numbers are Whiskered Tern and Redwinged Pratincole and sometimes both species of pelicans and flamingos. There are past records of Bittern, Palmnut Vulture, Corncrake, Baillon's Crake, African Skimmer, Copperytailed Coucal, European Sedge Warbler and Yellow Wagtail and a recent record of European Marsh Harrier. From 1982 to September 1989 the lake was completely dry—partly as a result of drought but partly also of the closure of a bund in the western outlet of the delta. The area around the Lake to the north is characterized by large, 15-m-high, *Acacia erioloba* trees. Near the shores at the eastern end is a dense thicket of *Acacia mellifera* to which Olivetree Warblers migrate, and the western end is a floodplain where Richard's Pipit is abundant at some times of the year.

The Northern Kalahari Savanna also extends north to the Nxai Pan area as far as the southern border of the Chobe National Park and forms a northern extension for several Kalahari birds such as Secretarybird, Greater Kestrel, Pale Chanting Goshawk, Orange River Francolin, Kori Bustard, Black Korhaan, Spotted Dikkop, Doublebanded Courser, Burchell's Sandgrouse, Pied Barbet, Fawncoloured Lark, Black Crow, Cape Penduline Tit, Pied Babbler, Anteating Chat, Kalahari Robin, Barred Warbler, Blackchested Prinia, Marico and Chat Flycatchers, Pririt Batis, Crimsonbreasted Shrike and Scalyfeathered Finch. Examples of tropical woodland savanna species alongside these Kalahari species in the Nxai Pan area are Bateleur, Fierynecked Nightjar, Greater Honeyguide, Rufousnaped Lark, Black Cuckooshrike, Southern Black Tit, Arrowmarked Babbler, Blackeyed Bulbul, Whitebrowed Robin, Rattling Cisticola, Chinspot Batis, Puffback and Plumcoloured Starling.

SOUTHERN KALAHARI SAVANNA

The Southern Kalahari Savanna is an open *Acacia* and thorn savanna. It supports a ground cover of grasses which grow to 1,5 m after rain but is a dry, crisp mat of dead grass for several months of the year. Trees rarely grow taller than 5 m except in patches of woodland along drainage lines and occasionally where rocks reach near the surface. There are also large treeless areas of grassland and scrub. The latter often consists mainly of *Catophractes alexandri* or *Grewia flava* but a variety of plants such as *Rhigozum trichotomum* and the creeper *Tribulus terrestris* occur, with *Bauhinia tomentosa* along disturbed areas such as tracks and cutlines. *Terminalia sericea* and *Grewia flava* are very evident as bushes and woody shrubs and it is unusual to find the former as a tree. The principal Acacias are listed by White (1983) in Category 44.

This is the true Kalahari and falls mainly within the boundaries of the central Kalahari Game Reserve—referred to as central Kalahari in the species text. The typical bird species are Ostrich, Greater Kestrel, Pale Chanting Goshawk, Orange River Francolin, Redcrested and Black Korhaans, Doublebanded Courser, Burchell's Sandgrouse, Rufouscheeked Nightjar, Pied Barbet, Fawncoloured, Spikeheeled Lark, Greybacked Finchlark, Black Crow, Cape Penduline Tit, Pied Babbler, Redeyed Bulbul, Anteating Chat, Kalahari Robin, Yellowbellied Eremomela, Desert and Tinkling Cisticolas, Blackchested Prinia, Tit Babbler, Marico and Chat Flycatchers, Pririt Batis, Crimsonbreasted and Whitecrowned Shrikes, Marico Sunbird, Great Sparrow, Scalyfeathered Finch, Redheaded Finch, Violeteared Waxbill, Shafttailed Whydah, Yellow Canary and Larklike Bunting.

Interesting features of this zone for the birdwatcher are the fossil valleys, pans and woodland patches. The woodland patches are used by Palaearctic migrants such as the two species of Redfooted Kestrels and the European Golden Oriole and by large nesting colonies of Wattled Starling. The fossil valleys support different types of vegetation. The eastern section of Deception is a wide, open short grassland supporting species such as Rednecked Falcon, Pallid and Montagu's Harriers, Temminck's and Doublebanded Coursers, both species of finchlarks and Pinkbilled Larks, whereas in the western section there are many patches of woodland. The Okwa Valley near Xade and west to the Ghanzi road is 10 m deep in places and lined with the remnants of riverine forest with trees up to 8 m tall. Birds which are often found in these deeper wooded valleys and along drainage lines elsewhere in this zone include Orange River Francolin, Kurrichane Buttonquail, Bronzewinged Courser, Grey Lourie, Marsh Owl, Spotted Eagle Owl, Rufouscheeked Nightjar, Spikeheeled Lark, Ashy Tit, Fantailed Cisticola, Burntnecked Eremomela, Melba Finch, Violeteared Waxbill and Yellow Canary.

Most pans in this zone are shallow and usually covered with short grass. Birds often seen on them or nearby are Whitebacked and Lappetfaced Vultures, Blackshouldered Kite, Tawny Eagle, Blackbreasted Snake Eagle, Steppe Buzzard, Redbilled Francolin, African Quail, Helmeted Guineafowl, Kori Bustard, Crowned Plover, Temminck's Courser, Rufousnaped and Redcapped Larks, as well as those species common on other Kalahari pans. Mammals are frequently seen and it appears as though sick or dying mammals remain within the vicinity of pans which may explain the regular presence of vultures.

This zone continues south to the Molopo River which is discussed in the next section as a separate zone for birdwatching purposes.

The Southern Kalahari Savanna also contains grassland plains which are well represented between Jwaneng and Sekhoma but extend north to Serowe and south to the Molopo Valley. Some of these plains cover about 50 km^2 and are treeless. West of Jwaneng some are subdivided by strips of stunted *Acacia* woodland about 50 m wide. Typical species of these plains are Ostrich, Secretarybird, Greater Kestrel, Redfooted Kestrels, Blackshouldered Kite, Blackbreasted Snake Eagle, Black Korhaan, Burchell's Sandgrouse, European Swift, Fawncoloured, Clapper and Rufousnaped Larks, Black Crow, Anteating Chat, Desert Cisticola, Richard's Pipit and Scalyfeathered Finch. Marico and Chat Flycatchers, Titbabbler, Yellowbellied Eremomela, Lesser Grey and Redbacked Shrikes, Brubru and Greyheaded Sparrows are often found in the wooded strips; Blackchested Prinia and particularly Rufouseared Warbler occur in woody scrub and shrubs on the edges of the plains.

MOLOPO RIVERINE ZONE

The Molopo Riverine Zone is botanically part of the southern Kalahari but because of the influence of the river valley (albeit scarcely flowing in most years) the vegetation is denser and more profuse and the trees taller. It is also a farmland and relatively well populated. Dams and bends in the river trap water which may remain on the surface for most of the year in the eastern sector. The zone is defined as the strip along the Molopo River from west of Phitsane Molopo to Tshabong and McCarthysrus. The river valley appears to act as a funnel for some eastern woodland species to extend westwards. Examples of such species are Gymnogene, Black Cuckoo, Striped Cuckoo, Scops Owl, Redbilled Woodhoopoe, Whitebacked Mousebird and Sabota Lark. Uncommon birds in the central part are Dabchick, Redbilled Teal, Black Duck, Giant Eagle Owl, Plainbacked Pipit and Common Waxbill. Interestingly common along the valley are Brown Snake Eagle, Gabar Goshawk, Redbilled Francolin, Spotted Dikkop, Whitefaced and Spotted Eagle Owls, Rufouscheeked Nightjar, Striped Kingfisher, Swallowtailed Bee-eater, Purple Roller, Scimitarbilled Woodhoopoe, Grey and Yellowbilled Hornbills, Clapper Lark, Redbreasted Swallow, Pied Babbler, Groundscraper Thrush, Familiar Chat, Icterine and Willow Warbler, Bleating Warbler, Spotted Flycatcher, Pririt Batis, Brubru, Crimsonbreasted Shrike, Threestreaked Tchagra, Burchell's Starling, Marico Sunbird, Redbilled Buffalo Weaver, Cape and Great Sparrows, Melba Finch, Blackcheeked and Violeteared Waxbills, Shafttailed Whydah, Blackthroated and Yellow Canaries and Goldenbreasted Bunting.

WESTERN WOODLAND

The Western Woodland has an almost identical botanical composition to the Southern Kalahari Savanna and quite clearly belongs to White's (1983) Category 44. However, its strucural characteristics have different proportions from those of the Southern Kalahari. Even if these differences do not hold true in all climatic cycles, for the present it appears worthy of mention as a separate ornithological zone. The major differences are that there are fewer open areas devoid of trees, fewer areas of scrubland, more closed woodland (often in extensive stands), fewer isolated patches and strips of woodland and the open woodland mostly has mature trees with good ground cover of grass except in the overgrazed areas. It has several pans and is used for cattle farming between Ghanzi and Ncojane near the Namibian border. East of Ncojane, it is well populated around Hukuntsi, Lehututu and Tshane. There is poor recovery of grass in overgrazed areas around some villages. South of these villages it is barely populated and mainly reserved for game management in the Gemsbok National Park. Access to the southwestern areas of this zone is difficult.

The zone is roughly defined as the region north of the dunelands and west of the main road from Werda via Khakhea, Morwamosu and Kang to Ghanzi.

True Kalahari bird species occur throughout this zone. Some of the factors which suggest that it should be considered as a separate avifaunal zone include: trees large enough for Bateleur to nest, the presence of most woodland accipiters, Redbilled Francolin, African, Black and Striped Cuckoos, Scops and Giant Eagle Owls, Redbilled Woodhoopoe, Bennett's and Bearded Woodpeckers, and the presence of Neddicky, Rattling Cisticola and Blue Waxbill near Kang, and White Helmetshrike near Takatshwane. There are also records of Grey Lourie and Melba Finch in the western portion, a considerable distance from their usual range. The zone contains open parklike areas of *Acacia erioloba* in which Redbilled Woodhoopoes have been found. During the rainy season there is a sea of grasses about 1 m tall under the trees and in open spaces in which Common Quail and Kurrichane Buttonquail occur. Thorn savanna occurring in patches is utilized particularly by Pied Babbler, Crimson-breasted Shrike, Pririt Batis, Titbabbler, Threestreaked Tchagra, Brubru and Violeteared Waxbill. Scalyfeathered Finches nest almost exclusively in *Acacia mellifera* bushes in parts of this area. The zone may also be different from the central Kalahari because of the influence of man, cattle and, therefore, some surface water. Cattle and humans are excluded from the central Kalahari Game Reserve.

The numerous pans of this zone have many forms. Some are covered with short grass with only a few bare patches, some are totally devoid of vegetation and in character are like salt flats except that the base is usually calcrete. In this latter category are most of the larger pans such as Ukwi (12 km in circumference), Tshane and Kang. Some pans have moist areas near the centre where the grass is green and supports such species as Fantailed Cisticola. Others are covered in grass and woody scrub. Most pans are surrounded by woodland and the intervening ecotone—the apron—consists of clumps of grass, woody shrubs and small, 1 m bushes which is a good habitat for Spikeheeled Lark, Yellowbellied Eremomela, Rufouseared Warbler, Blackchested Prinia, Chat Flycatcher and Lesser Grey Shrike. This ecotone is a distinctive microhabitat. Bird species which commonly occur on denuded pans are Ostrich, Lanner Falcon, Greater Kestrel, Black Crow and several species of vultures. On sparsely vegetated pans the characteristic avifauna includes Doublebanded Courser, Redcapped Lark, Capped Wheatear, Anteating Chat, and it is rare to find a pan without one of these species. When wet, both denuded and vegetated pans may support large numbers of waterbirds such as Abdim's Storks, Threebanded and Blacksmith Plovers, Little Stint, Greenshank, Wood Sandpiper, Blackwinged Stilt and White-winged Tern.

SOUTHEAST GRANITIC ZONE

The Southeast Granitic Zone is not a botanical subdivision. It is an area where hills, mountains and gorges occupy a significant portion of the area and influence the diversity of birds seen there. Birds typical of this subdivision are Cape Vulture, Black Eagle, Rock Kestrel, Gymnogene, Freckled Nightjar, Alpine Swift, Shortclawed Lark, Shorttoed Rock Thrush, Familiar Chat, Mocking Chat, Barthroated Apalis, Striped Pipit, Bushveld Pipit and Orangethroated Longclaw, most of which are dependent on the rocks and hills. Others depend on the wooded valleys or the open short grassland plains in the Good Hope and Phitsane Molopo areas, which also have parklike areas of *Acacia erioloba*. These last two areas fall into Acocks's (1988) veld types of *Acacia erioloba*, Savanna and the Dry *Cymbopogon-Themeda* Veld. Rare birds of the zone include Blue Crane, Whitebellied Korhaan, Sicklewinged Chat, Levaillant's Cisticola, and scarce are Jackal Buzzard, Bokmakierie and Sociable Weaver. The underlying avifauna is similar to the Eastern *Acacia* Zone, i.e. a mix of Kalahari and hardveld species.

DUNELAND

The Duneland of the southwestern corner of Botswana is the true desert region of the country. Its vegetation is well described in White's Category 56. Species which are generally restricted to this zone are Pygmy Falcon, Jackal Buzzard, Ludwig's Bustard, Palewinged Starling, Dusky Sunbird and Sociable Weaver. Typical of the Nosssob Valley are Ostrich, Secretarybird, Lanner Falcon, Rednecked Falcon, Rock Kestrel, Booted Eagle, Bateleur, Tawny Eagle, Gymnogene, Kori Bustard, Black Korhaan, Namaqua Sandgrouse, Whitefaced Owl, Greybacked Finchlark, Black Crow, Familiar Chat, Rufouseared Warbler, Bokmakierie, Burchell's Starling, Whitebrowed Sparrowweaver, Cape Sparrow, Scalyfeathered Finch, Redheaded Finch, Yellow Canary and Larklike Bunting. The Rosyfaced Lovebird may also reach this area at the eastern limit of its range.

Definitions of terms

The following terms have been used in categorising habitats:

FOREST

A continuous stand of trees exceeding 10 m in height with overlapping crowns. In Botswana such stands are usually only found in riparian situations and most of the trees are evergreen.

WOODLAND

Stands of trees forming a canopy at about 8 m or more whose crowns may be touching but not overlapping (closed woodland) or whose canopies cover more than 50% of the ground surface (open woodland).

TREE SAVANNA

Areas where there are trees with canopies at 5 m or more in height and with clearly defined boles but with open grassland, bushes or shrubs occupying more than 50% of the surface.

BUSH SAVANNA

Areas of open grassland where bushes or bushy trees up to 5 m in height occupy more than 40% of the surface. There may be a few trees with canopies but these do not dominate the vegetation.

SCRUB SAVANNA

Areas of grassland where shrubs or stunted woody plants less than 1,5 m in height occupy 40% or more of the surface. There may be a few scattered bushes or trees but usually none.

THORN SAVANNA

Areas where the majority of trees or bushes bear thorns but a significant number are not *Acacia*.

ACACIA SAVANNA

Tree or bush savanna where *Acacia* species predominate.

GRASSLAND

Open areas of grassland devoid of trees where other vegetation occupies less than 40 % of the surface but usually much less. Grassland plains are areas where grass covers 95% or more of a flat surface.

THICKET

Areas of dense bushy or woody vegetation rarely exceeding 2 m in height often in the understorey of woodland which is difficult to penetrate. It may occur as isolated stands of a densely clustered species, e.g. *Acacia mellifera* in bush form.

Human factors

Botswana has a population of just over 1 million people, approximately 1,7 people per square kilometre of the country. Most of the people live in the east between Francistown and Ramatlabama and the remainder of the country is very sparsely populated. The principal industries are cattle ranching, mining and tourism. Most of the tourism occurs in the northern wetlands around Maun, the Okavango Delta, Savuti and Chobe. Diamonds are mined at Orapa, Letlhakane and Jwaneng, coal at Morupule and base metals at Selebi Phikwe. Most of the mining activity has commenced over the past 25 years. In 1988 a soda-ash process started at Sua Pan. Botswana became an independent country in 1965 and a great deal of beneficial human development has occurred since then. Many of the villages on the eastern route to the north have become well established towns with modern houses, schools and clinics. Gaborone, the capital, is a major city with a university and Francistown is a large town with several hotels and a modern provincial hospital. The two large dams supplying water to these towns are important ornithological areas in these regions. Within the past 10 years a tarred road has been completed from Ramatlabama to Kasane and major tarred roads now lead off this road to some towns and villages.

The benefits of these major developments on the avifauna appear at this stage clearly to outweigh the destruction of habitat which has been necessitated. Examples of bird species which utilize manmade environments and which have benefited as a result of these artificial habitats are given below.

Buildings

Rock Pigeon, Barn Owl, Whiterumped Swift, Greater Striped Swallow, Cape Wagtail, Redwinged Starling, House and Cape Sparrow.

Tall structures (warehouses, mining superstructures, cooling towers, electricity pylons)

Rock Kestrel, Greater Kestrel, Lanner Falcon, Peregrine Falcon, Rock Pigeon, Little Swift, Rock Martin, Pied Crow, Redwinged Starling.

Gardens (including outhouses and fenceposts)

Cape Turtle and Laughing Doves, Diederik Cuckoo, Speckled and Redfaced Mousebirds, Brownhooded Kingfisher, Crested Barbet, Arrowmarked Babbler, Kurrichane and Olive Thrushes, Familiar Chat, Heuglin's Robin, Neddicky, Fiscal and Paradise Flycatchers, Cape Wagtail, Plumcoloured and Glossy Starlings, Marico, Whitebellied and Black Sunbirds, Cape White-eye, Cape and Greyheaded Sparrows, Scalyfeathered Finch, Lesser Masked and Masked Weavers, Bronze Mannikin, Blackcheeked Waxbill.

Waste sites (abattoirs, rubbish tips, fishing and game camps, road kills)

Marabou Stork, Whitebacked and Lappetfaced Vultures, Yellowbilled Kite, Tawny Eagle, Bateleur, Pied Crow.

Fields (football, airfields, fallow land)

Blackheaded Heron, Cattle Egret, Secretarybird, Greater, Lesser and Redfooted Kestrels, Blackshouldered Kite, Blackbreasted Snake Eagle, Montagu's Harrier, Crowned Plover, Temminck's Courser, Yellowthroated Sandgrouse, Rufousnaped, Shortclawed and Redcapped Larks, Banded Martin, Capped Wheatear, Desert Cisticola, Richard's Pipit, Lesser Grey Shrike, Cutthroat Finch, Rock Bunting.

Dams and sewage ponds

Great Crested Grebe, Dabchick, Whitebreasted Cormorant, Hamerkop, Hadeda Ibis, South African Shelduck, Redbilled and Cape Teal, Maccoa Duck, Black Crake, Redknobbed Coot, most Palaearctic waders, Avocet and Blackwinged Stilt, Greyheaded Gull, Whitewinged Tern, Rock Pigeon (dam wall), Brownthroated Martin, Fantailed Cisticola, Cape Reed, European Sedge and African Marsh Warblers, Cape and Pied Wagtails, Red Bishop, Common and Orange-breasted Waxbills, Rock Bunting (dam wall).

Roads (including culverts, telephone wires, ditches and borrow pits)

White Stork, Lesser Kestrel, Yellowbilled and Blackshouldered Kites, Steppe Buzzard, Pale Chanting Goshawk, Spotted Eagle Owl (also often a roadkill victim), European, Greater Striped and Redbreasted Swallows, House Martin, Shorttoed Rock Thrush, Anteating Chat, Rock Bunting. Migrating and soaring birds also benefit from the thermal uplift from roads.

Boreholes, wells and cattle troughs

Cattle Egret, Cape Turtle and Laughing Doves, Pied Crow, Glossy Starling, Whitebrowed Sparrowweaver, Greyheaded Sparrow, Masked Weaver, Redbilled Firefinch, Blue Waxbill, Blackthroated Canary.

It is difficult at present to think of a bird species that has come under threat as a result of the human progress in Botswana. This may be mainly because ample alternative areas are available and birds are ousted rather than threatened. A great many species, as in the lists above, have benefited and been able to enlarge previous ranges and populations. An important factor has been the dependence of humans on water, for themseves and their cattle, and with the discovery of more water, its availability to other animals and birds and to an increase and/or change in the vegetation by gardening or farming. If the current balance survives, there is no reason to think that birds will not continue to thrive. The future danger lies in overpopulating with people or cattle and consequent irreversible destruction of the vegetation. Most of Botswana's cattle are well adapted to browsing when the sparse grass is seasonally depleted. An excess of browsing may eventually remove important protection to the ground vegetation and cause failure of the following season's growth of grass. Some areas of Botswana already show these effects, apparently more as a result of goats and sheep than cattle. However, these important factors are well known to the planners and agencies who control these matters in Botswana and are under their constant attention and surveillance.

Before ending the discourse on the human factors affecting bird distribution, it is necessary to highlight species which are particularly under threat, i.e. endangered species. Two species are under threat of extinction as a direct result of human interference. The Wattled Crane is threatened by failure to breed because of the physical presence of people at its breeding sites. This is a problem which responsible people should easily be able to correct, either through their own behaviour or through that of others whom they witness behaving irresponsibly towards these conspicuous birds. With the aid of binoculars close approach (within 100 m) is simply not necessary to observe this large species well. Preferably one should not stop but move in a direction away from the birds if they are thought to be breeding. The second endangered species is the Cape Vulture. Botswana is one of the last refuges for this species which is almost extinct in many areas of southern Africa. The main hazard to this species has been the use of poisoned bait for predators (e.g. jackals). Slaughter by farmers who consider the species a threat to stock animals has also played a role in the decline of the species. A protection programme, including education and the provision of vulture 'restaurants', is in progress. So far the results are promising but there is still much to be achieved before the threat of extinction is eradicated. Both these species appear in the Red Data Books for birds in southern Africa and in continental Africa. Other species which are considered vulnerable are Slaty Egret, White-backed Night Heron, Saddlebilled Stork, Ayres' Eagle, Black Harrier and Shortclawed Lark. There are 59 other species which are considered at risk (Penry, 1986).

In terms of global conservation the most precious diamond in Botswana today is the Okavango Delta—not the commercial or gem variety that is found in the kimberlite pipes. If it was made of crystalline carbon like the commercial diamonds, it would be indestructible and there would be less concern for its long term future. However, its dazzling sparkle is consumable water and its carbon form is the organic matter of plants and animals which are combustible and edible. In addition, concerned people should regard the Kalahari as a very fragile ecosystem and irreversible changes should be avoided for the sake of future generations who may know better how to care for it than we do today.

Chapter 4

The avifauna

This chapter gives a broad overview of the avifauna of Botswana starting with the relationship of the birds to the Afrotropical and other biogeographical regions of the world. An analysis is made of the composition of the avifauna and the influence of certain factors, habitats, food and migration on the diversity and distribution of species. All species are allocated to major subdivisions of the Afrotropics in a list of families and finally a map of species diversity within Botswana is given.

Botswana belongs to the Afrotropical region which stretches from south of the Sahara desert to the Cape of Good Hope and which is also known as Subsaharan Africa. Within this biogeographical region 16 vegetation zones are listed by White (1983) which vary in character from the moist evergreen forests of West Africa to the bare sand dunes of the Namib desert. The existence of over 2 000 species of birds in the region is a reflection of the large diversity of the vegetation and climatic influences. Although some species are very restricted in their range because of the limitation of suitable habitat, representatives of the majority of families are present throughout the region. This biogeographical concept emphasizes that Botswana and its birds do not exist in isolation. It is not the local factors alone which have a major influence on and determine where and why a particular species occurs in a particular area of continental Africa. To obtain a better understanding of the factors which govern the distribution of birds some knowledge of evolution, speciation, adaptive radiation, migration and the continental isolation of species is desirable. To elaborate on these topics is not within the scope of this discussion.

Birds of other continents and other biogeographical regions occur within the Afrotropics even as far south as Botswana, both on the basis of annual migration and as an extension of global range under existing circumstances. The largest group is from the Palaearctic (Eurasia) on annual migration. There are scientific arguments which state that such birds originated from Africa rather than *vice versa*.

The occurrence of some of Botswana's birds in other regions of the world deserves a brief mention. More than 50 species which are mainly sedentary (i.e. not migrants) in Botswana also occur in North Africa and southern or eastern Europe, e.g. Dabchick, White Pelican, Purple Heron, Little Egret, Cattle Egret, Squacco Heron, Night Heron, Glossy Ibis, Shorttoed Eagle (Blackbreasted Snake Eagle), Lanner Falcon, Rock Kestrel, Baillon's Crake, Common Moorhen, Purple Gallinule, Blackwinged Stilt, Redwinged Pratincole, Scops Owl, Hoopoe, Redcapped (Shorttoed) Lark, Richard's Pipit, and Fantailed Cisticola. Nearly 60 of the sedentary species in Botswana also occur as far away as Southeast Asia —their distribution from southern Africa may or may not be continuous through northern tropical Africa, the Arabian peninsula and the Indian subcontinent. In addition to most of the species included above, Southeast Asia has such species as Woollynecked Stork, Kurrichane Buttonquail, Blackshouldered Kite, Little Banded Goshawk, Bat Hawk, Painted Snipe, Grass Owl, and Tawnyflanked Prinia which most African birdwatchers would consider as African birds. The above examples have included birds at the species level; at the family level of taxonomy there is even greater association with Europe and Asia.

Composition

The 496 main species occurring in Botswana are represented by 285 nonpasserine and 211 passerine species from 47 and 25 families respectively. Sixty-four species (13%) are endemic to southern Africa, most of which are dry-savanna species, and 104 (21%) are species which occur in diverse habitats in other parts of the continent. The largest subdivision (42% of the avifauna) is made up of 208 species of tropical woodland savanna affinities which create the high diversity seen in the northern and eastern parts of the country. One fifth of the avifauna is composed of migrants, from the Palaearctic (12%) or intra-African (8%).

Influence of water

There are 144 species (29% of the avifauna) from 37 families which are dependent on water habitats for feeding and breeding, of which 99 are traditionally called waterbirds (grebes, pelicans, cormorants, herons, ibises, flamingos, ducks, crakes, waders, terns, etc.), and the others include such species as African Fish Eagle, African Marsh Harrier, Pel's Fishing Owl, most kingfishers, reed and marsh warblers, wagtails, longclaws, etc. These species are found mainly in the north and east of the country but several members of these families may move temporarily to Kalahari pans under wet conditions. In contrast, only about 108 species (22%) are common in the Kalahari woodland savanna, of which only 56 fall into the category of true Kalahari species, i.e. species of dry Kalahari woodland or savanna which rarely occur in other habitats.

Influence of the Kalahari

It is well known that the Kalahari is not a true desert. Nevertheless the semidesert conditions on the Kalahari sand beds in Botswana discourage some species of adjacent habitats from occurring there. A rough measure of the effect of this intangible barrier can be obtained by counting the number of species of tropical woodland-savanna affinities which occur south of 22°S and west of 25°E. During the drought period of this survey, 76 (36%) of the tropical woodland-savanna species which occur in Botswana were recorded regularly in the true Kalahari and up to 42% at least rarely. Widely represented families from this group are diurnal birds of prey, doves, owls and cuckoos—most of the latter being present for only part of the year. It is estimated that about another 30 species might occur in the central and southern Kalahari in wet climatic cycles bringing the tropical woodland-savanna component to over 50% of those available. Just over half of the Palaearctic migrants arriving in Botswana have been recorded in the central Kalahari, twice as many of these being nonpasserine as passerine (21 compared with 10). Of the intra-African migrants nearly 32% occur in the central Kalahari.

Species richness

An analysis of the number of species recorded in each square (Figures 3 & 24) shows that northern Kalahari squares (all those with Kalahari vegetation north of 22°S) have a potential to support about 150 species and 225 species at the extremes of dry and wet cycles respectively. On the same basis central and southern Kalahari squares (south of 22°S and west of 25°E) can support about 125 and 175 species respectively. The mixed broadleafed and *Acacia* woodlands of the northern Kalahari represent an extension of tropical woodland-savanna in high rainfall years and even the southern Kalahari will support some species extensions under those conditions. To these extensions of woodland species can be added the dispersal of waterbirds to Kalahari pans to give a significant increase in the species diversity of the Kalahari in contrasting climatic cycles. This species diversity is seen in the artificial habitats of some towns in the southern Kalahari. Squares of the northern wetlands with permanent water have a capacity for more than 400 species and squares with woodland, savanna, riparian forest and water habitats, such as the Boteti, Nata, and Limpopo drainage about 350 species.

Migration

Botswana accommodates very large numbers of migrants in the summer months. The breeding behaviour and vocalizations of intra-African migrants such as Yellowbilled Kite, Wahlberg's Eagle, the cuckoos, Woodland Kingfisher, Monotonous Lark, Greater and Lesser Striped Swallows and Paradise Flycatcher are major features of the birdwatching scene in this period of the year. Most of the intra-African migrants arrive in Botswana to breed, the main exception to breeding being Abdim's Stork. Nearly all the migrants from the north are present in the summer months from October to March, which is the nonbreeding period for the Palaearctic birds. Some migrants, such as the Yellowbilled Kite, arrive in August, while others, such as European Marsh Warbler, not until December, but most arrive in October and November. Many of the migrants occur in Botswana on passage to areas further south but a portion of nearly all of these visitors remains in Botswana for the austral summer. Current evidence suggests that Botswana is an important wintering region for such species as Western Redfooted Kestrels, Caspian Plover, European Golden Oriole, Olivetree Warbler, Whitethroat, Icterine Warbler, Spotted Flycatcher, Redbacked Shrike and Lesser Grey Shrike. Only three species (Fiscal Flycatcher, Fairy Flycatcher and Fiscal Shrike) are known to be regular austral winter visitors to Botswana from southern regions.

Palaearctic representatives of many families frequently outnumber the resident species. The European Swallow is the most frequently recorded swallow in the country in any year. European Swift, Willow Warbler, Redbacked Shrike and Lesser Grey Shrike are seen in larger numbers than residents of the same families and habitats. Although the resident Greater Kestrel is common throughout the country, the visiting Lesser Kestrels appear in much larger numbers as do the other visiting falcons such as Eastern and Western Redfooted Kestrels. Many of the aerial-feeding migrants such as kestrels and swifts arrive with the weather fronts that bring the first rain of the summer and subsequent appearances often coincide with termite-alate emergences. The visiting species which appear in largest numbers are insectivorous or aquatic.

Food

Species which feed primarily on insects form the largest portion of the total avifauna. The figures shown in Table 11 detail the dietary analysis of 537 species (all the main species and most of the rarities):

Table 11 Dietary categories of birds in Botswana

Feeding type	No. of spp.	% of total
Insectivores	243	45
Aquatic	80	15
Granivores	69	13
Carnivores	67	13
Piscivores	33	6
Omnivores	22	4
Frugivores	15	3
Nectar feeders	8	1

In the above groups the classification is based on what is known of the principal dietary factor. The analysis is not based on evidence from Botswana. Carnivores include species feeding principally on reptiles such as snakes and lizards. Piscivores have been shown separately from species which feed principally on other aquatic fare such as frogs, aquatic plants, molluscs, crustaceans, etc.

Biogeography

The birds of Botswana are presented in Table 12 in systematic family order and arranged into the biogeographical subdivisions in which they mainly occur on the African continent. The allocation of a species to such subdivisions does not mean that the species may not occur in other habitats or other geographical areas. The list includes only the 496 species shown in the main maps of this work. The subdivisions are as follows:

Southern endemic

Species which occur only in southern Africa, although some have extensions into Angola and Zambia. Most are dry-savanna species but some are primarily woodland or wetland

Figure 24 **Species richness: graphic**

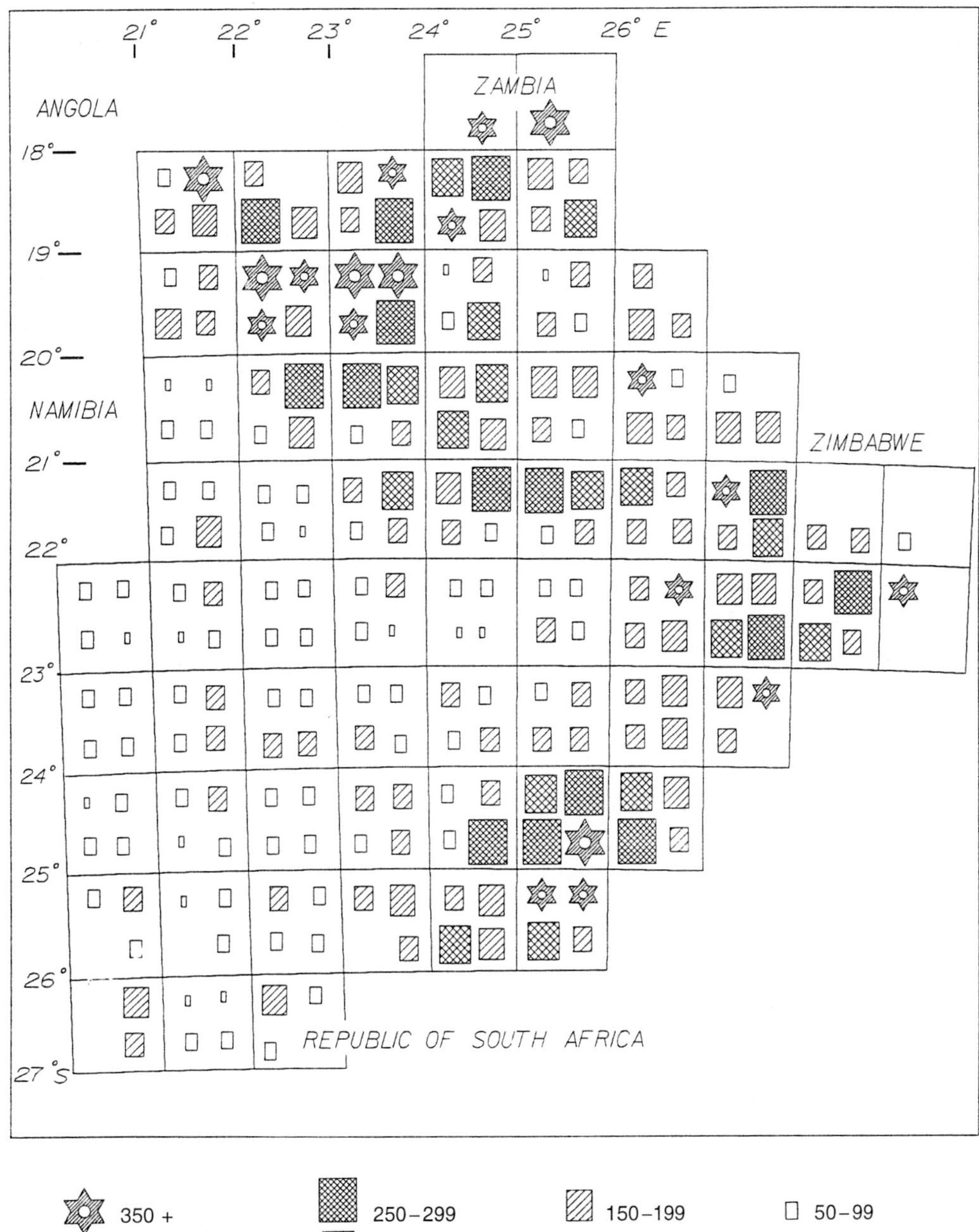

species or they occur in a diversity of habitats. Examples—Bradfield's Hornbill, Sabota Lark.

Dry savanna

Species which occur principally in dry savanna in other parts of the continent as well as southern Africa. Examples—Ostrich, Secretarybird.

Tropical woodland savanna

Species whose range is mainly in the woodland or wooded grassland of tropical Africa. It includes species of tropical wetland and riparian forest. There is no species in Botswana which occurs only in forest elsewhere on the continent except Pel's Fishing Owl and Western Banded Snake Eagle which occur mainly in riparian forest. Riparian forest is considered to be a form of tropical woodland for the purpose of this classification. Examples—Goliath Heron, Bateleur, Wattled Crane, Water Dikkop, Redeyed Dove, Scops Owl, Terrestrial Bulbul, Heuglin's Robin, African Marsh Warbler, Chinspot Batis, Greyheaded Sparrow, Blue Waxbill.

Palaearctic migrant

Species which breed in the Palaearctic region and visit Botswana in the nonbreeding season. Examples—Willow Warbler, Redbacked Shrike.

Intra-African migrant

Species whose migratory movements occur regularly on an annual cycle within continental Africa. Examples—African Cuckoo, Paradise Flycatcher.

Pan-African

Species which occur regularly in more than one zone of the African continent although not necessarily throughout the continent. Examples—Grey Heron, Crowned Plover, Barn Owl, Lilacbreasted Roller.

Table 12 **Birds of Botswana in their main biogeographical subdivisions**

Abbreviations are: S.E. = Southern Endemic; D.S. = Dry Savanna; TW/S = Tropical Woodland Savanna; P.M. = Palaearctic Migrant; A.M. = Intra-African Migrant; Pan Afr. = Pan African. Priorities in this classification are: P.M. > A.M. > Pan Afr. > TW/S > D.S. > S.E. The classification of species associated with water into an aquatic category is unhelpful in a biogeographical sense and has not been used.

Nonpasserines	S.E.	D.S.	TW/S	P.M.	A.M.	Pan Afr.
Struthionidae – Ostrich	–	1	–	–	–	–
Podicipedidae – Grebes	–	–	–	–	–	3
Pelecanidae – Pelicans	–	–	–	–	–	2
Phalacrocoracidae – Cormorants	–	–	1	–	–	2
Ardeidae – Herons, Bitterns	1	–	13	(1)	1	3
Scopidae – Hamerkop	–	–	–	–	–	1
Ciconiidae – Storks	–	–	6	1	1	–
Plataleidae – Ibises, Spoonbill	–	–	4	–	–	–
Phoenicopteridae – Flamingos	–	–	–	–	–	2
Anatidae – Ducks	3	–	8	–	–	5
Sagittariidae – Secretarybird	–	1	–	–	–	–
Accipitridae – Vultures	1	–	4	–	–	–
– Eagles, Hawks, etc	1	–	12	6	2	4
– Sparrowhawks, etc	–	1	6	–	–	1
– Harriers	–	–	1	2	–	–
Falconidae – Falcons, Kestrels	–	2	2	5	–	3
Phasianidae – Francolins	3	–	3	–	–	3
Turnicidae – Buttonquail	–	–	1	–	–	–
Gruidae – Cranes	1	–	2	–	–	–
Rallidae – Crakes, Gallinules	–	–	6	1	3	1
Heliornithidae – Finfoot	–	–	1	–	–	–
Otididae – Bustards	1	1	1	–	–	2
Jacanidae – Jacanas	–	–	2	–	–	–
Rostratulidae – Painted Snipe	–	–	1	–	–	–
Charadriidae – Plovers	–	–	2	3	–	7
Scolopacidae – Sandpipers, Snipe	–	–	1	13	–	–
Recurvirostridae – Avocet, Stilt	–	–	–	–	–	2
Burhinidae – Dikkops	–	1	1	–	–	–
Glareolidae – Coursers, Pratincole	1	1	2	1	1	1
Laridae – Gulls, Terns	–	–	–	1	–	3
Rynchopidae – Skimmer	–	–	–	–	1	–
Pteroclidae – Sandgrouse	2	1	1	–	–	–
Columbidae – Pigeons, Doves	1	–	4	–	–	3
Psittacidae – Parrots	–	–	2	–	–	–
Musophagidae – Louries	–	–	1	–	–	–
Cuculidae – Cuckoos, Coucals	–	–	2	1	9	1
Tytonidae – Barn Owl	–	–	–	–	–	1
Strigidae – Owls	–	–	3	–	–	6
Caprimulgidae – Nightjars	–	–	3	1	2	1
Apodidae – Swifts	–	–	1	1	5	–
Coliidae – Mousebirds	1	–	1	–	–	1
Trogonidae – Trogon	–	–	1	–	–	–
Alcedinidae – Kingfishers	–	–	4	–	2	3
Meropidae – Bee-eaters	–	–	2	2	1	1
Coraciidae – Rollers	–	–	1	1	1	2
Upupidae – Hoopoes	–	–	1	–	–	2
Bucerotidae – Hornbills	1	1	3	–	–	1
Lybiidae – Barbets	1	–	3	–	–	–
Indicatoridae – Honeyguides	–	–	3	–	–	–
Picidae – Woodpeckers	–	–	3	–	–	1
TOTALS	18	11	119	40	29	68

Table 12 *(Continued)*

Passerines	S.E.	D.S.	TW/S	P.M.	A.M.	Pan Afr.
Alaudidae – Larks	7	3	1	–	1	1
Hirundinidae – Swallows, Martins	–	–	3	3	7	2
Campephagidae – Cuckooshrikes	–	–	2	–	–	–
Dicruridae – Drongo	–	–	–	–	–	1
Oriolidae – Orioles	–	–	1	1	1	–
Corvidae – Crows	1	–	–	–	–	1
Paridae – Tits	1	–	1	–	–	–
Remizidae – Penduline Tits	1	–	1	–	–	–
Timaliidae – Babblers	2	–	2	–	–	–
Pycnonotidae – Bulbuls	1	–	3	–	–	–
Turdidae – Thrushes, Chats, Robins	5	–	8	1	1	3
Sylviidae – Warblers	3	2	8	8	–	1
– Cisticolas	–	1	8	–	–	–
– Prinias	2	–	1	–	–	–
Muscicapidae – Flycatchers	5	–	5	1	1	–
Motacillidae – Wagtails	–	–	1	1	–	1
– Pipits	–	–	2	1	–	4
– Longclaws	1	–	1	–	–	–
Laniidae – Shrikes	–	1	–	2	–	1
Malaconotidae – Bush Shrikes	2	–	7	–	–	1
Prionopidae – Helmet Shrikes	1	–	2	–	–	–
Sturnidae – Starlings	1	1	3	–	1	2
Buphagidae – Oxpeckers	–	–	2	–	–	–
Nectariniidae – Sunbirds	2	–	4	–	–	–
Zosteropidae – White-eyes	1	–	1	–	–	–
Ploceidae – Weavers, Sparrows	4	–	2	–	–	4
– True Weavers	–	–	6	–	–	2
– Bishops, Widows	1	–	2	–	–	2
Estrildidae – Waxbills, Finches	1	3	6	–	–	4
Viduidae – Whydahs, Widowfinches	1	–	3	–	–	3
Fringillidae – Canaries	1	–	3	–	–	1
– Buntings	2	–	–	–	–	2
TOTALS	46	11	89	18	12	36
OVERALL TOTALS	**64**	**22**	**208**	**58**	**41**	**104**

Explanation of map symbols

Symbol	Frequency	Description
■	50% or more	Very common
⊠	10,1–49,9%	Uncommon to common
□	10% or less	Sparse or rare
✡	Not analysed	Pre-1980 records only

OSTRICH — 1

Struthio camelus

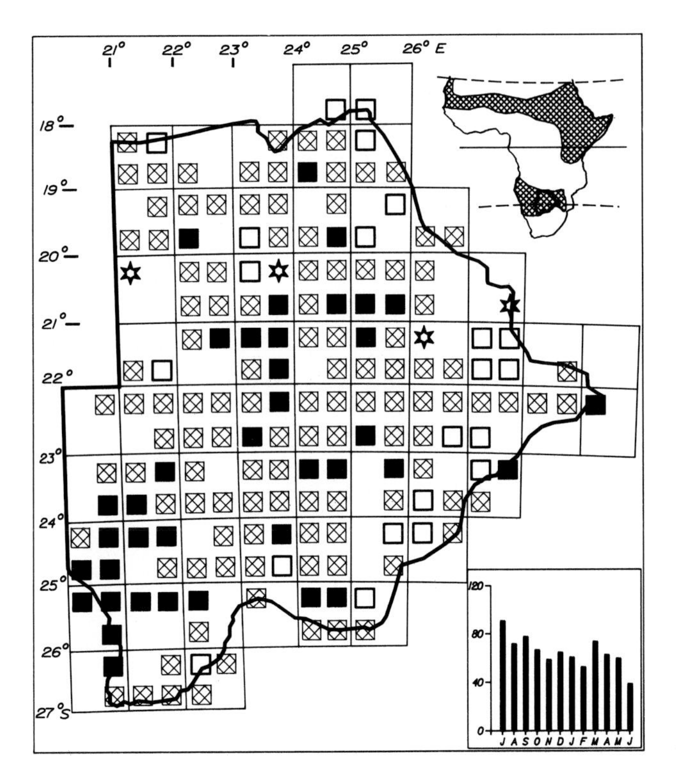

Status

A common to locally very common resident throughout the country. It is least common in the populated areas of the east and well-wooded areas of the northeast. Its apparent absence just north of Ghanzi (2121D) is probably an artefact in this survey. Its absence at times in suitable habitat may be as a result of breeding behaviour. Egglaying is known in all months from December to April and from June to August. It may breed throughout the year in years of good rainfall.

Habitat

Open tree and bush savanna, grass plains, duneland, Kalahari scrub savanna, open areas on the edges of woodland in the north and west. Frequently found on pans in the west when denuded or covered in short grass.

Analysis

177 squares (77%) Total count 815 (0,46%)

GREAT CRESTED GREBE — 6

Podiceps cristatus

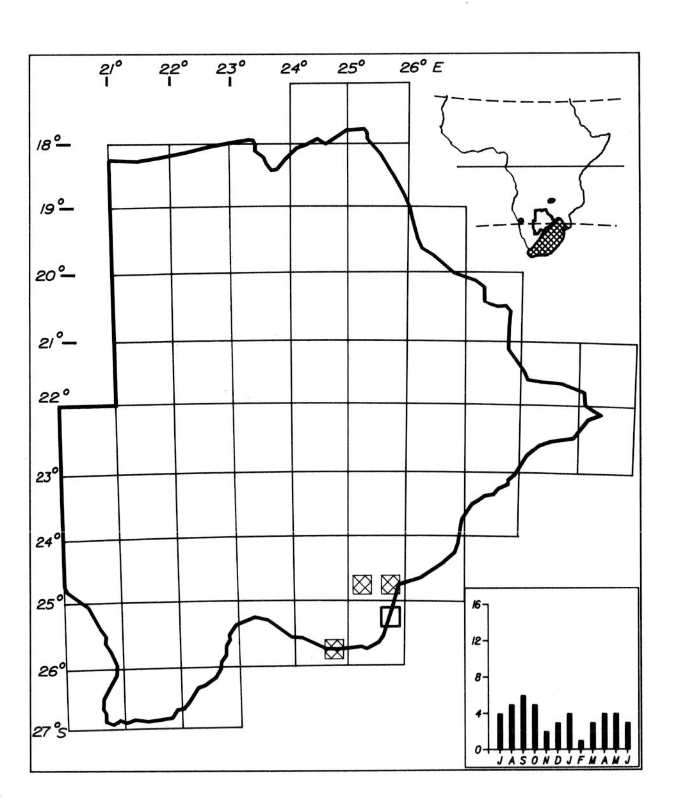

Status

Sparse to common resident of the southeast at Kanye, Gaborone and Lobatse, to as far north as Moshaneng. Records are too few for movement analysis but local movements to and from the adjacent Transvaal are likely. Breeds at most known sites. Not recorded by Smithers (1964) but recorded by Beesley (1976) in January 1974. An extension of range from adjacent Transvaal may therefore have occurred in the past 25 years. Egglaying February, May, June, July. Usually in pairs.

Habitat

Known mainly from mature manmade dams with well-vegetated edges, some emergent vegetation and large areas of open water.

Analysis

4 squares (2%) Total count 48 (0,027%)

BLACKNECKED GREBE

7

Podiceps nigricollis

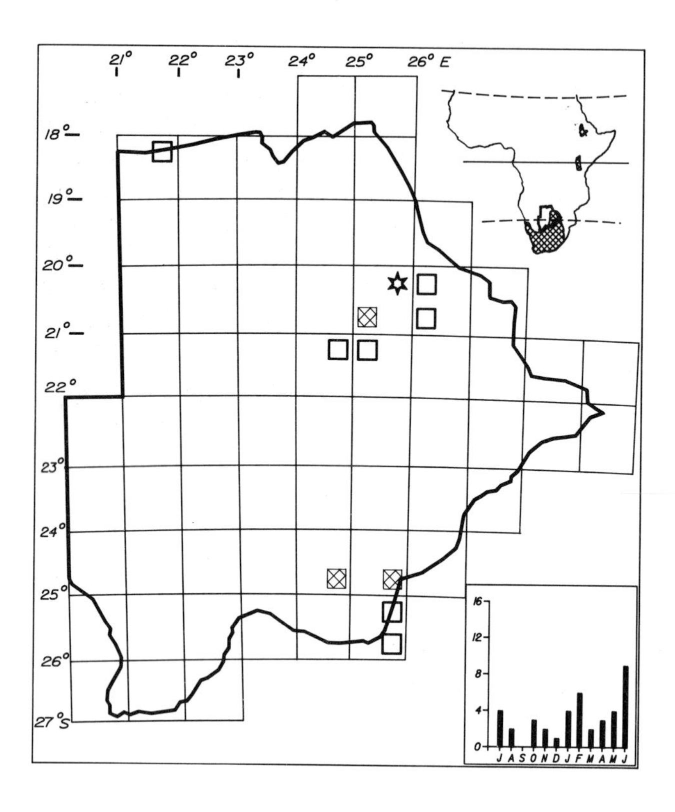

Status

Sparse to uncommon resident of the southeast. Breeds on the Makgadikgadi Pans at least in some years. Although recorded in all months of the year except September, it does not stay at one locality for more than six months. Dates and localities suggest that it is present at breeding locations from December to May and moves away to other local waters from June to November. Egglaying January and February. Usually in small groups of 3–10 birds.

Habitat

Shallow saline pans, dams and sewage ponds. Requires floating vegetation for nesting.

Analysis

11 squares (5%) Total count 42 (0,023%)

DABCHICK

8

Tachybaptus ruficollis

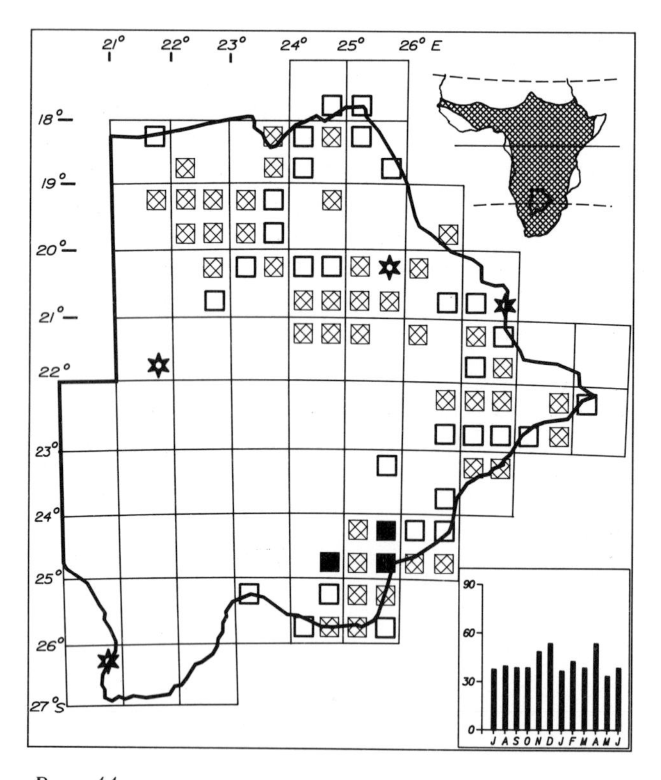

Status

Common to very common resident of the southeast. Sparse to common in the north and east. Absent from the remainder of the country where it is likely only as a vagrant in years of high rainfall. The species is present throughout the year on permanent waters but the population is constantly changing as evidenced by the changing proportion of birds in various stages of breeding, nonbreeding and juvenile plumage. Considerable local movements occur especially to and from seasonal waters. Egglaying December to May. Usually in pairs or small groups.

Habitat

Breeds on mature ponds, dams, lakes, lagoons and inundated pans with floating or emergent vegetation, also floodplains and slow-moving rivers. When not breeding utilises any body of open water including saline pans and commonly at sewage ponds.

Analysis

78 squares (34%) Total count 542 (0,3%)

WHITE PELICAN

Pelecanus onocrotalus

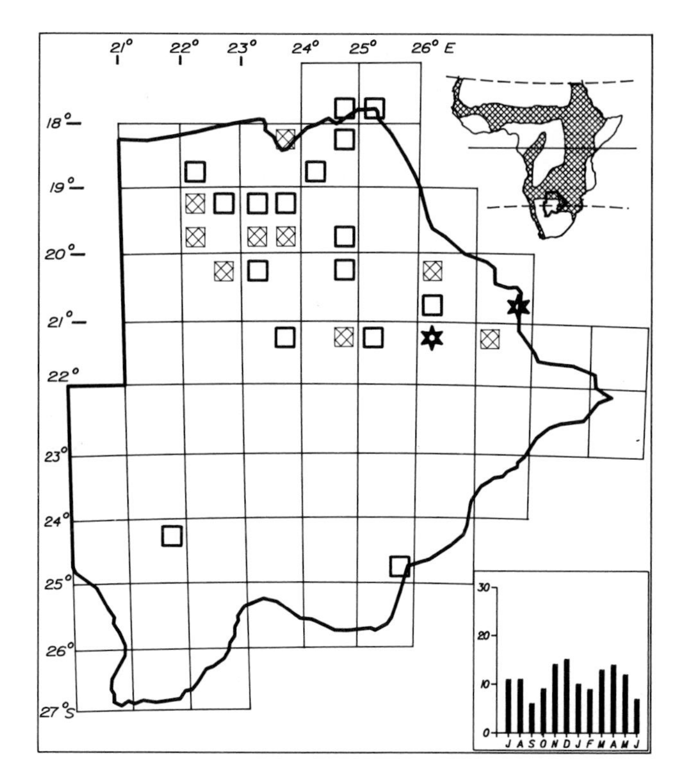

Status

Sparse to fairly common resident of the north to as far south as Mopipi and Shashe dams. A vagrant in the southeast. It is present throughout the year but was not recorded regularly at any locality during this survey when known breeding localities such as Lake Ngami and Makgadikgadi Pans were dry. Nomadic; dependent on the availability of fish. Feeds in tight flocks on the water. Egglaying March, April, May, June, July. Usually in small groups of 2–10 birds, sometimes solitary or in larger groups of up to 100 birds when breeding. There were several thousand breeding birds at Nata in July 1979; 4 000 at Lake Ngami in 1981.

Habitat

Large expanses of shallow water, freshwater lakes, saline pans, large dams.

Analysis

27 squares (12%) Total count 143 (0,08%)

PINKBACKED PELICAN

Pelecanus rufescens

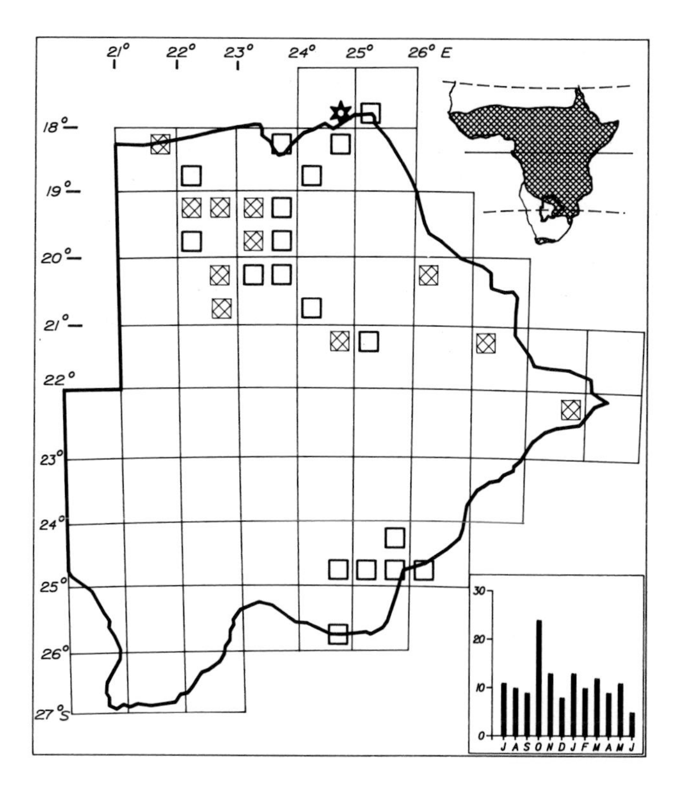

Status

A sparse to fairly common resident in the north. In the southeast it is a sparse to uncommon vagrant (more common than the White Pelican). It is present throughout the year but has been recorded most widely in October. Nomadic; not recorded at any locality for more than four consecutive months in this survey. Regular and predictable at Shakawe, Okavango Delta and Shashe Dam. Has bred on the Chobe River, Okavango Delta and Makgadikgadi Pans. Egglaying April and May. Usually in small groups of 2–5 birds, but larger congregations occur.

Habitat

Lagoons, floodplains, dams, lakes, pans, quiet wide reaches on rivers. Often on large stretches of water but it utilises smaller impoundments than the White Pelican. Perches readily in trees, sometimes on dry floodplains up to 30 km from water.

Analysis

30 squares (13%) Total count 143 (0,08%)

WHITEBREASTED CORMORANT

Phalacrocorax carbo

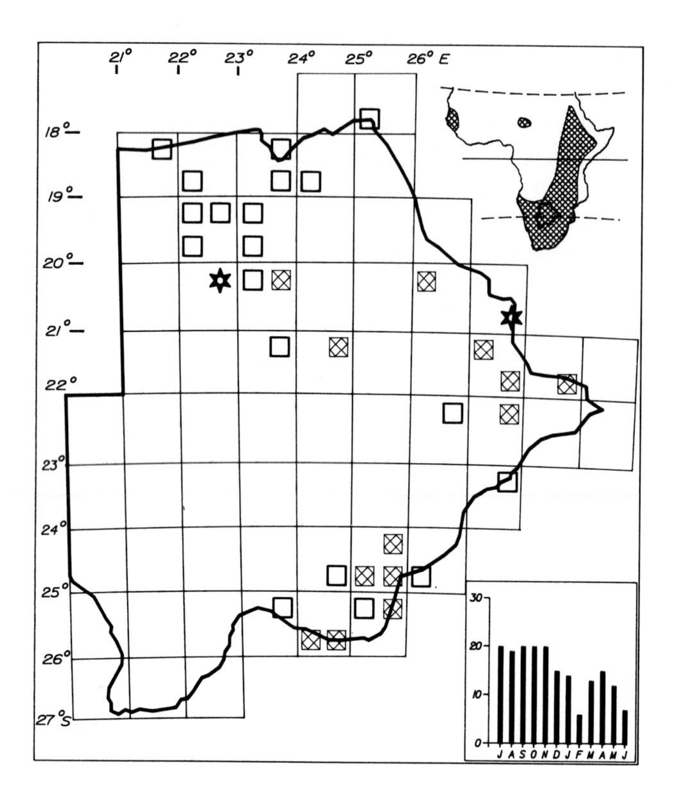

Status

Sparse to fairly common resident of the Okavango, Linyanti and Chobe river systems. Common at the Nata River delta. It is sparse to common at localities along the Shashe River, Serowe, Mopipi and in the southeast. Mainly sedentary and localised within a short radius of its breeding locality. Some movements may occur to and from neighbouring countries. Egglaying March. Usually solitary or in small groups of 2–6 birds, except when breeding colonially.

Habitat

Lakes, dams and lagoons, particularly those with dead trees standing in the water in which it builds its stick nest. Also on quiet stretches on large rivers, flooded pans and estuaries with large pools.

Analysis

34 squares (15%)   Total count 202 (0,11%)

REED CORMORANT

Phalacrocorax africanus

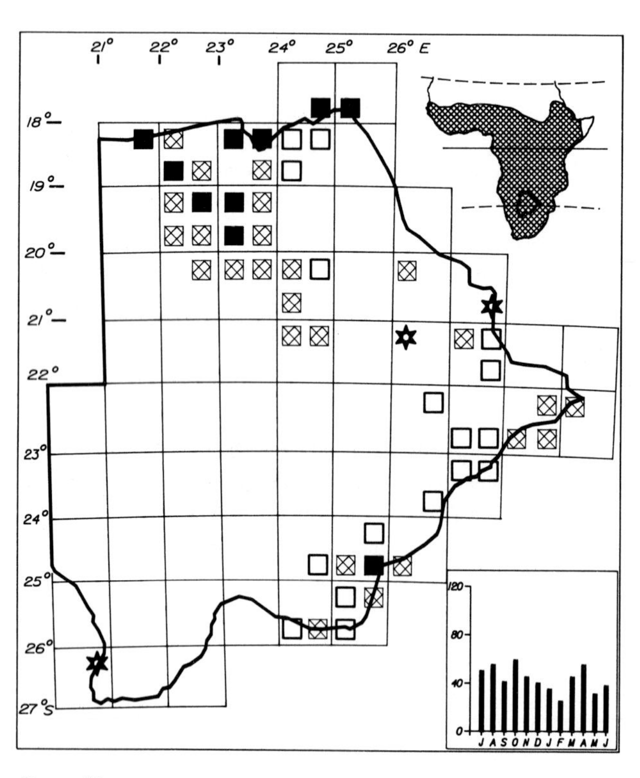

Status

Very common resident of the Okavango Delta and the Linyanti and Chobe river systems. Sparse to locally common in the east and southeast. Breeds communally with other water birds in the north. Numbers are much fewer in the east and southeast where breeding is suspected but not proven.

Habitat

Rivers, lakes, lagoons, large and small dams, flooded pans, sewage ponds. May occur on seasonal floodwater on floodplains or small catchments on farms and irrigation schemes.

Analysis

55 squares (24%)   Total count 553 (0,31%)

DARTER 60

Anhinga melanogaster

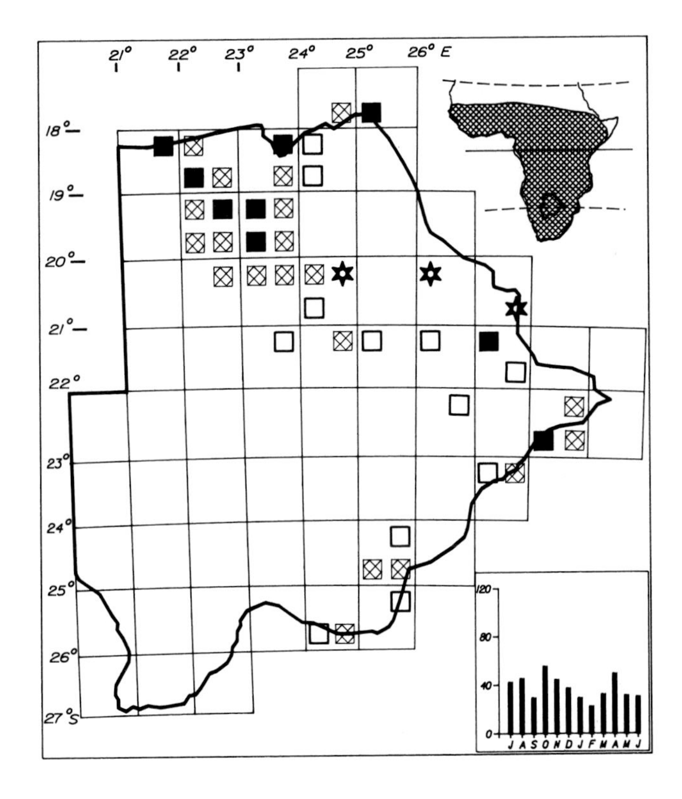

Status

Common to very common resident of the Okavango, Linyanti, Chobe, Boteti and Nhabe river systems to as far south as Mopipi and Orapa. Sparse to locally common along the Shashe, Limpopo and Molopo rivers and occasionally at dams at Kanye, Gaborone and Lobatse. Egglaying August. Usually in pairs or solitary, but several birds are usually present at one locality.

Habitat

Large rivers, lagoons, large dams and other expanses of open or vegetated waters.

Analysis

44 squares (19%) Total count 476 (0,26%)

GREY HERON 62

Ardea cinerea

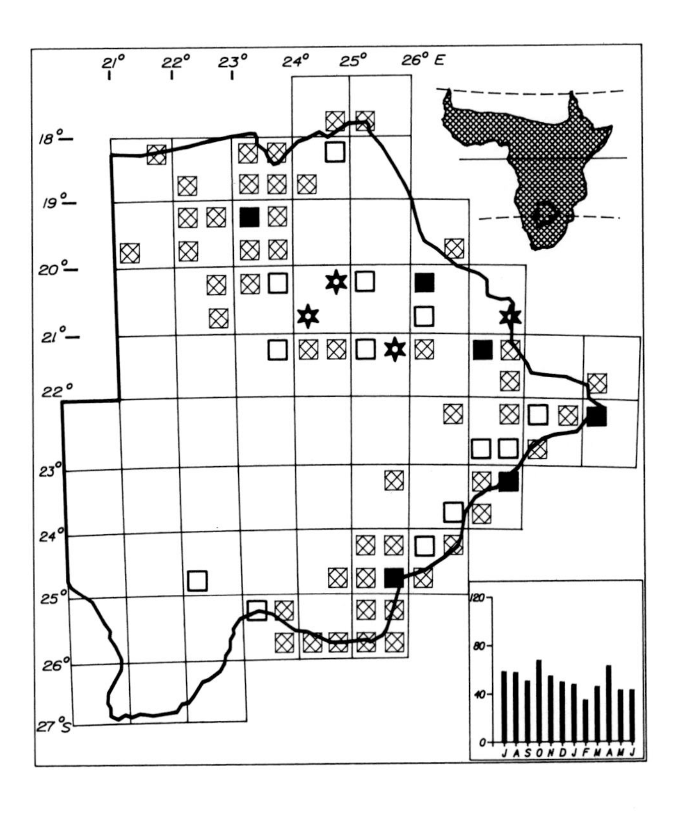

Status

A fairly common to locally very common resident throughout the northern and eastern regions. Extends westwards along the Molopo River to as far as Bray. Vagrant at Mabuasehube—it may be more common on pans in the west during wet cycles. Seasonal movements occur with changing conditions. Egglaying March, August and December. Usually solitary or in pairs. In the Okavango Delta it roosts and breeds with other herons.

Habitat

Edges of lagoons, lakes, rivers, pools and dams, in shallow water usually where fringes are vegetated. Also on wet and flooded pans, sewage ponds, floodplains, marshes and seasonally flooded depressions in savanna.

Analysis

69 squares (30%) Total count 668 (0,37%)

BLACKHEADED HERON

Ardea melanocephala

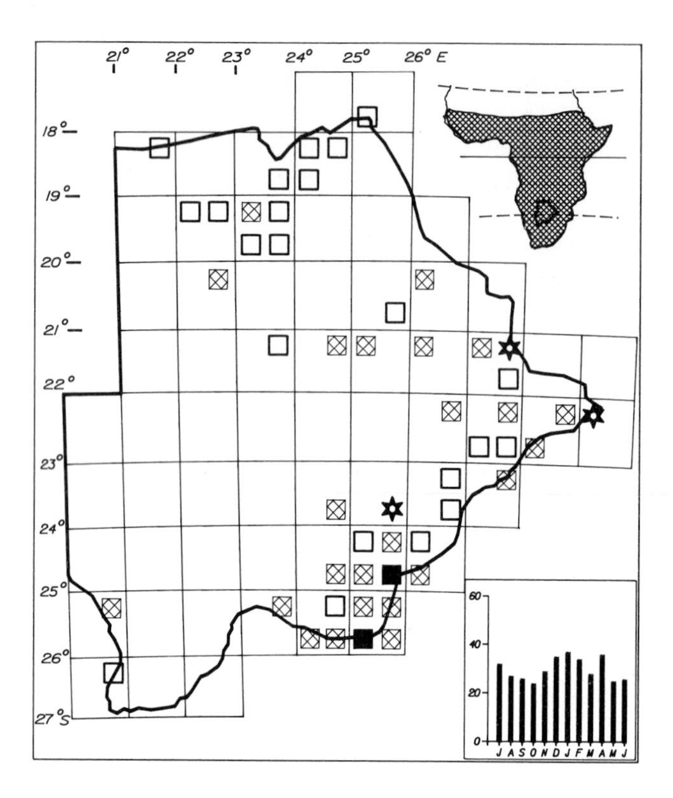

Status

Common to very common resident of the southeast. Fairly common in the east and southern areas of the Makgadikgadi Depression. Sparse in the northern wetlands. Sparse to uncommon in the Nossob Valley. Egglaying July, August, October, November and December. Solitary or pairs.

Habitat

Grassland and open savanna in river valleys, floodplains and marshes—mainly when dry. Sometimes on the apron of dry pans. Often in cultivated or fallow fields and at sewage ponds.

Analysis

51 squares (22%) Total count 385 (0,21%)

GOLIATH HERON

Ardea goliath

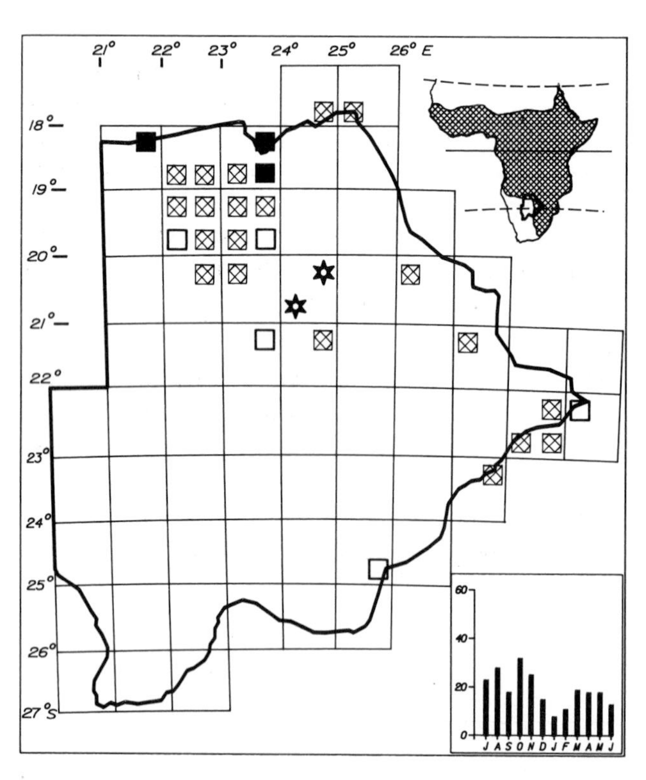

Status

Sparse to locally common resident of the Okavango, Linyanti and Chobe river systems. Sparse to fairly common at isolated water habitats such as Mopipi, Shashe Dam, Nata Delta, along the Limpopo River, and on dams in the Gaborone area. Mainly sedentary. Solitary or in pairs.

Habitat

Shallows of wide perennial rivers, large lagoons and dams. Usually in open areas of clear shallow water unlike the Purple Heron which is more often in or near vegetation.

Analysis

30 squares (13%) Total count 238 (0,13%)

PURPLE HERON

Ardea purpurea

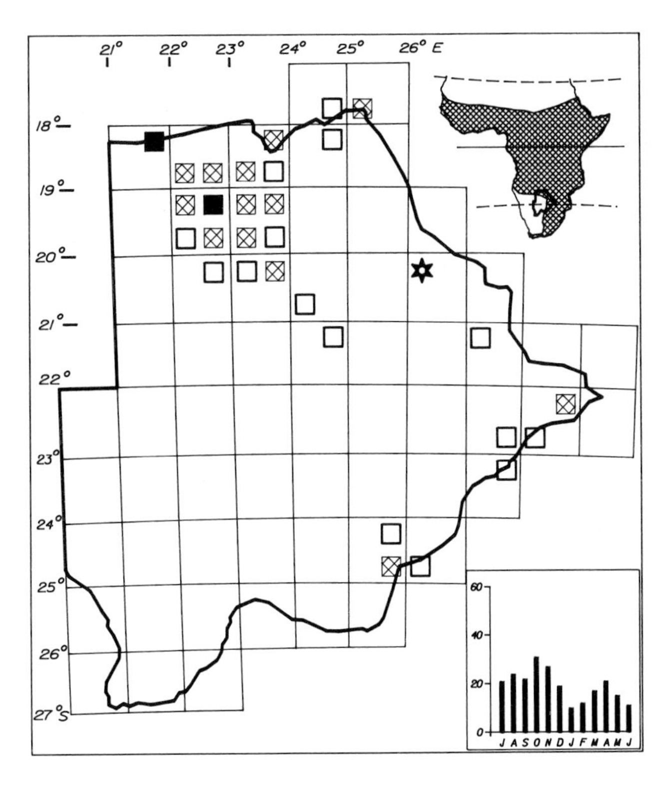

Status

Common to very common resident of the Okavango and Linyanti river systems. Sparse to fairly common on the Chobe, Shashe and Limpopo rivers and sparse in the southeast around Gaborone and Molepolole. Local movements occur with changing habitat conditions. Egglaying August. Breeds in isolation or as few pairs in mixed heronries. Usually solitary or in pairs.

Habitat

Lagoons, marshes, reed beds, waterlogged depressions and ditches with reeds on floodplains and in river valleys, dams and pans with tall aquatic vegetation.

Analysis

31 squares (13%) Total count 237 (0,13%)

GREAT WHITE EGRET

Egretta alba

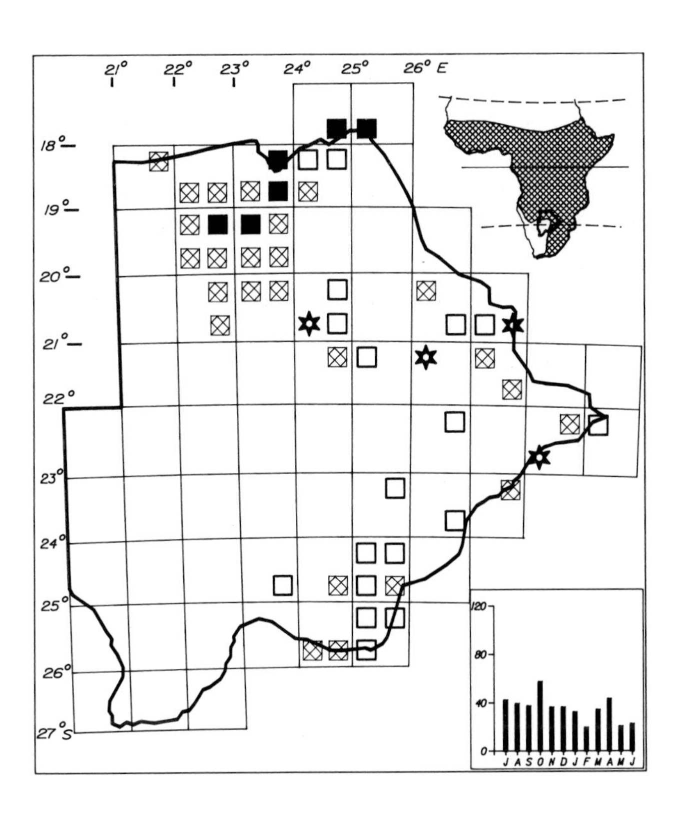

Status

Common to very common resident of the Okavango, Linyanti and Chobe river systems south to Mopipi. Sparse to fairly common in the east and southeast. Extends westwards along the Molopo River to Boshoek and may occur as a wanderer on western pans as at Khakhea. Breeding—with young in March. Usually solitary or in small loose congregations.

Habitat

Shallow waters of lagoons, dams, pans and rivers, mostly in areas with floating and fringing vegetation. Occasionally on wet floodplains and large waterlogged depressions in grassland.

Analysis

52 squares (23%) Total count 451 (0,25%)

LITTLE EGRET

Egretta garzetta

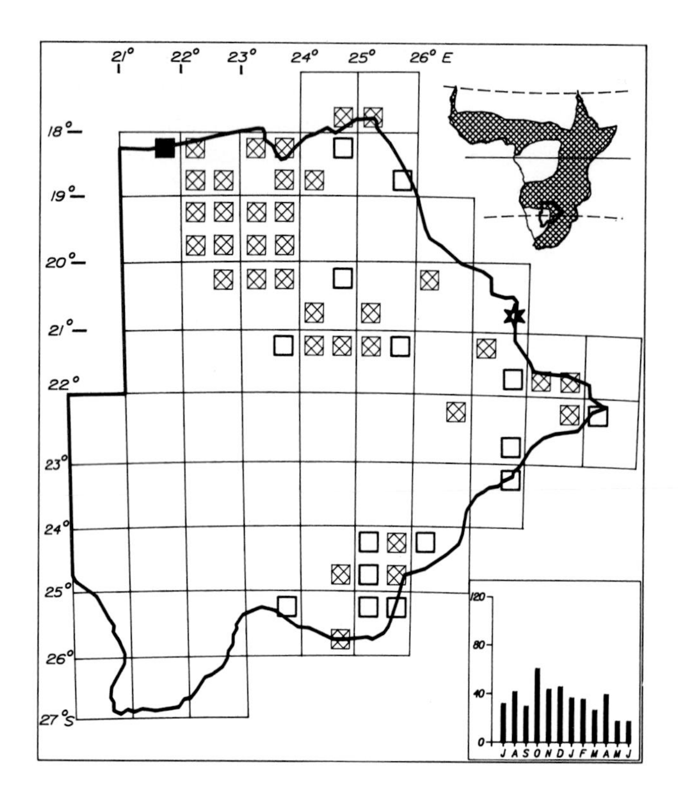

Status

Fairly common to common resident of the Okavango, Linyanti, Chobe and Boteti river systems and in the east in the Nata, Shashe and Limpopo drainage. Sparse to common on the water catchment areas of the southeast and eastern Molopo River. Local movements occur which are most clear in the southeast where birds move in during summer from September to March but may be present in smaller numbers throughout the year on larger waters such as Gaborone Dam. Egglaying January and February. Usually solitary, but more than one bird may be present at one locality.

Habitat

Edges of lagoons, lakes, dams, pans, sewage ponds, rivers, seasonally flooded depressions in river valleys.

Analysis

52 squares (23%) Total count 445 (0,25%)

YELLOWBILLED EGRET

Egretta intermedia

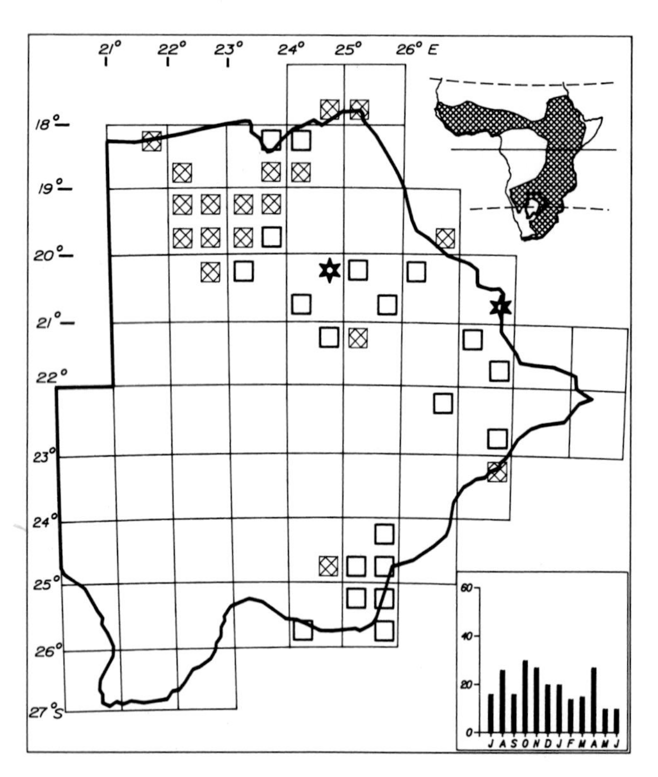

Status

Uncommon to common resident of the Okavango, Linyanti, Chobe and Nata river systems. Sparse in the southeast and uncommon along the Limpopo drainage. Occurs throughout the year in the northern wetlands. In the southeast it is mainly a summer visitor. Usually solitary or in small loose groups.

Habitat

Marshes, inundated floodplains, seasonally flooded grassland, shallows of well-vegetated dams, lagoons and pans. Occasional visitor at sewage ponds where it feeds mainly on adjacent wet or marshy ground.

Analysis

40 squares (17%) Total count 244 (0,13%)

BLACK EGRET

Egretta ardesiaca

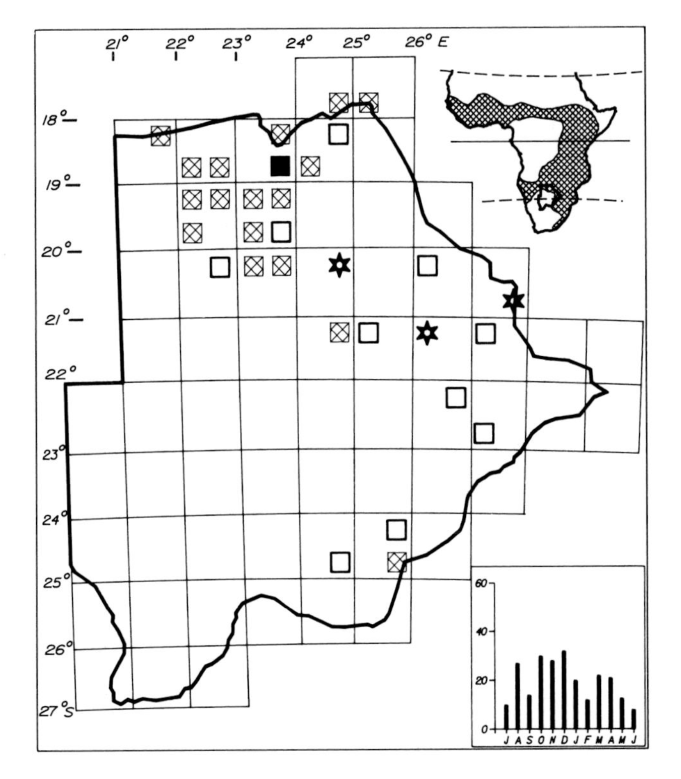

Status

Fairly common to common resident of the Okavango, Linyanti, and Chobe river systems to as far south as Mopipi. A sparse visitor elsewhere in the north and east including the Makgadikgadi Pans. A summer visitor from September to May in all areas away from the permanent waters of the north. Often solitary but at suitable sites it may be present in congregations up to 40 birds. Roosts and breeds communally in mixed heronries in the Okavango Delta. Local and seasonal movements occur.

Habitat

Lagoons and inundated floodplains where it feeds in shallows on the edge of deeper water. Less common on the edges of slow moving rivers, vegetated pans and dams.

Analysis

31 squares (13%) Total count 243 (0,13%)

SLATY EGRET

Egretta vinaceigula

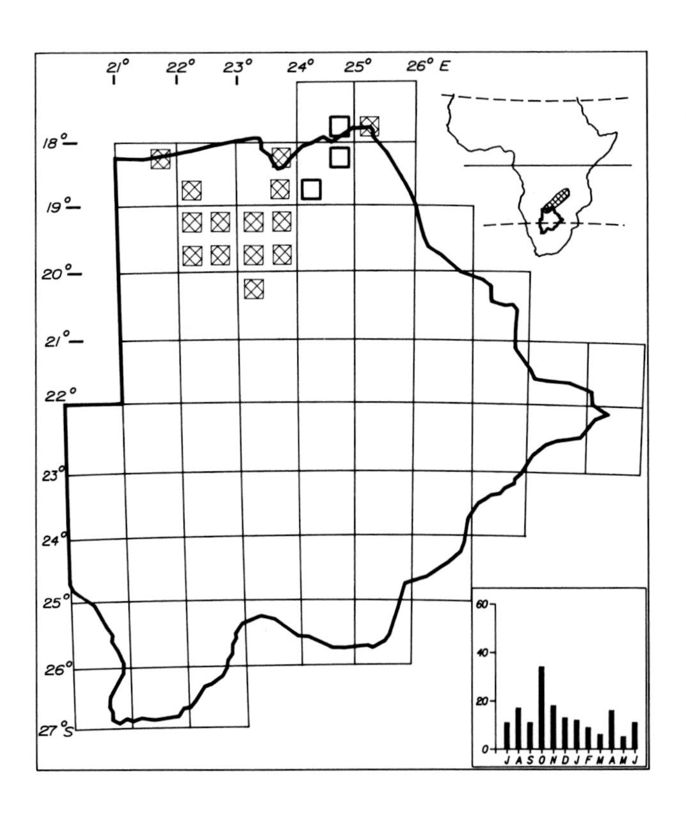

Status

Uncommon to locally common resident of the Okavango, Linyanti and Chobe river systems to which it is confined. A scarce African egret with restricted range whose habitat requires conservation for its survival. Local and international movements occur but these have been poorly studied. The few Zimbabwe records fall between August and November. There is also likely to be some movements to and from Zambia. Egglaying February and March. Breeds communally March to June with other herons, particularly Rufous-bellied Heron. It also breeds in high rainfall years in northern Namibia. Usually solitary when feeding, but small groups also occur.

Habitat

Lagoons and floodplains where it feeds in shallower water than Black or Little Egrets. Often in very wet mud left by receding water and in between clumps of aquatic vegetation.

Analysis

17 squares (7%) Total count 165 (0,09%)

CATTLE EGRET

Bubulcus ibis

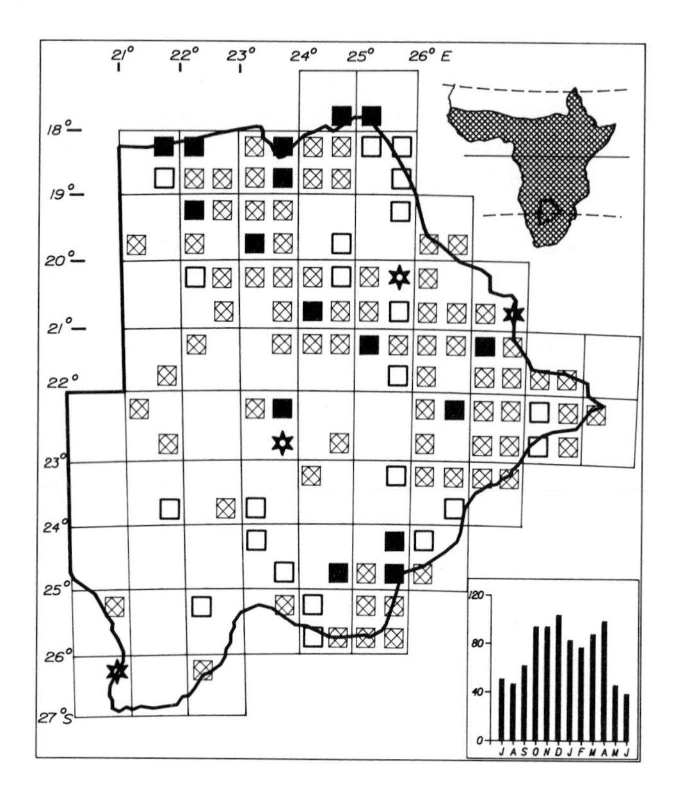

Status

Sparse to locally very common resident of the north and east. In central and western regions it is sparse to uncommon mainly in the populated areas and near cattle posts. Seasonal movements are evident and some intra-African movement is suspected but no clear pattern is known. Present throughout the year. Egglaying January. Usually in flocks of 5–50 birds when foraging; roosting and breeding sites often have several hundred birds.

Habitat

Grassland, pastures, cultivation, tree and bush savanna and a wide variety of nonaquatic habitat in populated and farming areas. Also at pans, dams, sewage ponds, shores of rivers and lagoons, and marshes. Breeds communally in trees or reed beds near water.

Analysis

114 squares (50%) Total count 949 (0,53%)

SQUACCO HERON

Ardeola ralloides

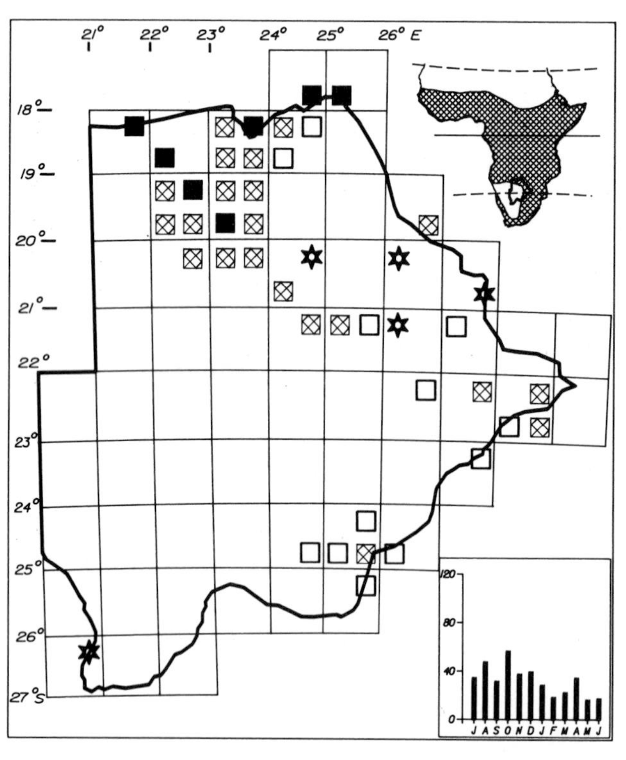

Status

Common to very common resident of the Okavango, Linyanti and Chobe river systems. Sparse to uncommon in the east and southeast. Most records away from permanent water fall between September and April. Two subspecies are thought to occur—resident *A. r. paludivaga*, and Palaearctic *A. r. ralloides* as a nonbreeding summer migrant. Usually solitary or in pairs, may also occur in loose flocks of up to 50 birds in suitable areas of wetland.

Habitat

Reedbeds and other tall aquatic vegetation on the margins of lagoons, floodplains, rivers, dams, seasonal pans and in river valleys. Occasionally perches on riparian bushes and trees.

Analysis

44 squares (19%) Total count 405 (0,22%)

GREENBACKED HERON

Butorides striatus

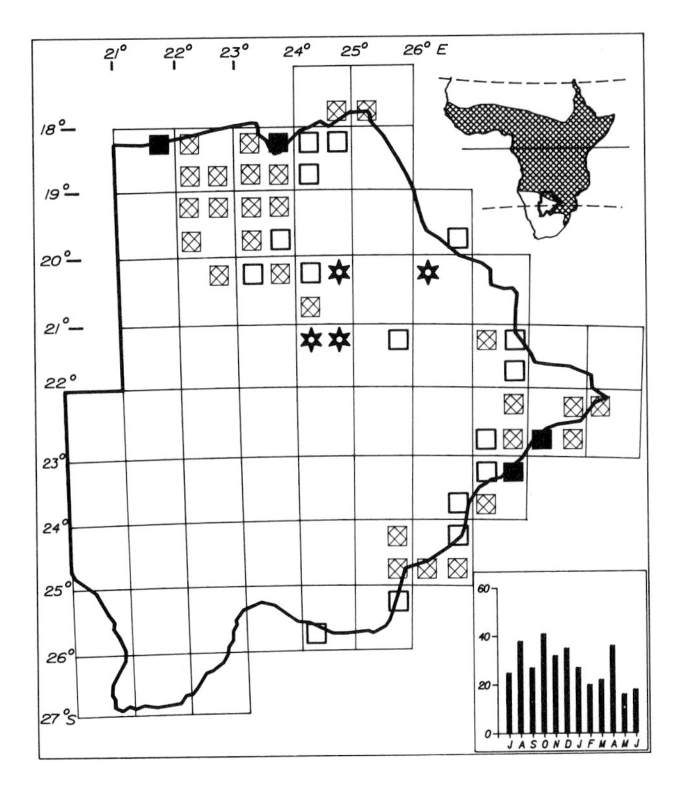

Status

Uncommon to locally very common resident of the Okavango, Linyanti, Chobe, Shashe, Limpopo and Ngotwane river systems. Extends seasonally and in higher rainfall years to other tributaries and water catchments. Egglaying all months March to June and September. Solitary or in pairs which take up territory along one stretch of river when breeding.

Habitat

Rivers and streams particularly where there is fringing forest, woodland or tall vegetation along the banks. Also lagoons, dams (Shashe and Gaborone) and other impoundments with good cover on the margins.

Analysis

52 squares (23%) Total count 352 (0,19%)

RUFOUSBELLIED HERON

Butorides rufiventris

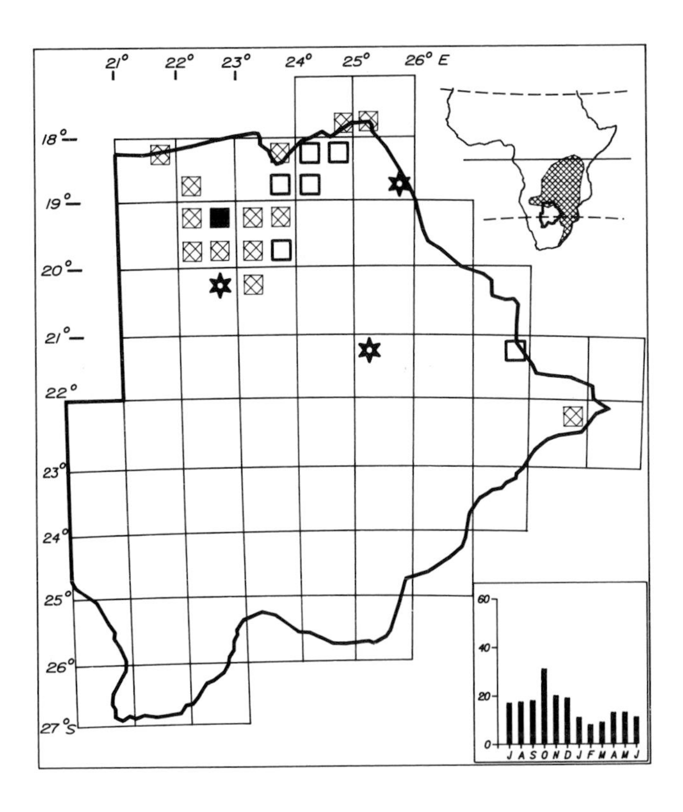

Status

Uncommon to locally very common resident of the Okavango, Linyanti and Chobe river systems to which it is mainly confined. There may be seasonal movements; the two records in the east occurred in January and December. In the past it was recorded from Mpandamatenga, Orapa and Lake Ngami and may occur in such localities again in high rainfall years. Egglaying February, March and October. Solitary or in pairs. Roosts and breeds communally with other herons.

Habitat

Marshes, swamps, lagoons and floodplains with reedbeds and tall aquatic vegetation.

Analysis

23 squares (10%) Total count 206 (0,12%)

BLACKCROWNED NIGHT HERON

Nycticorax nycticorax

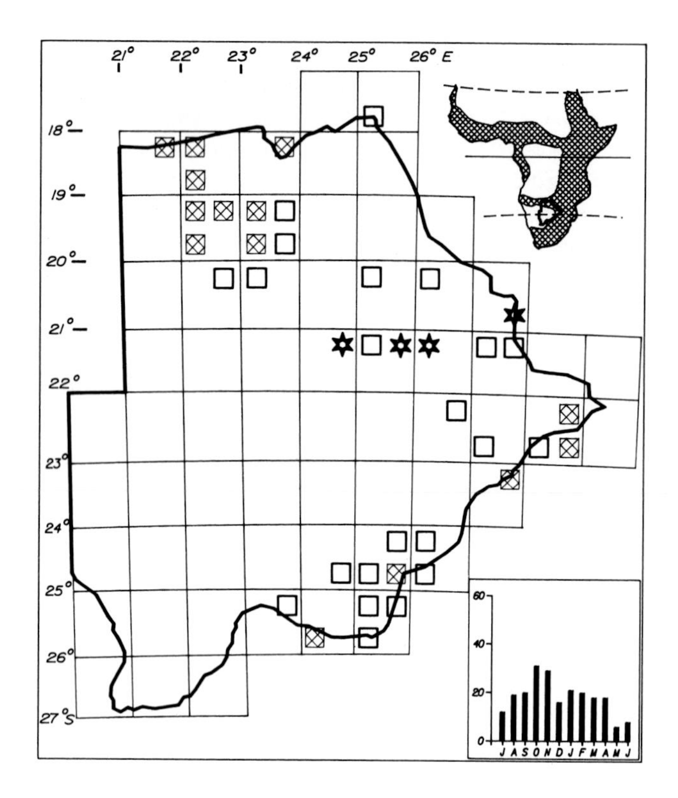

Status

Sparse to fairly common resident of the Okavango, Linyanti, Chobe, Nata, Shashe, Limpopo, Ngotwane, and Molopo rivers. Rare at adjacent inland waters during this survey. Sparse or absent on seasonally dry stretches of river. Mainly sedentary but local movements occur. Crepuscular and nocturnal and easily overlooked. Active nest January. Usually in small groups or family parties.

Habitat

Bushes and trees overhanging rivers, lagoons, swamps, marshes, dams and rarely pans where it lies up during the day. Also in reedbeds. Readily flies over savanna, forest, gardens and woodland adjacent to its aquatic habitat.

Analysis

40 squares (17%) Total count 231 (0,13%)

WHITEBACKED NIGHT HERON

Gorsachius leuconotus

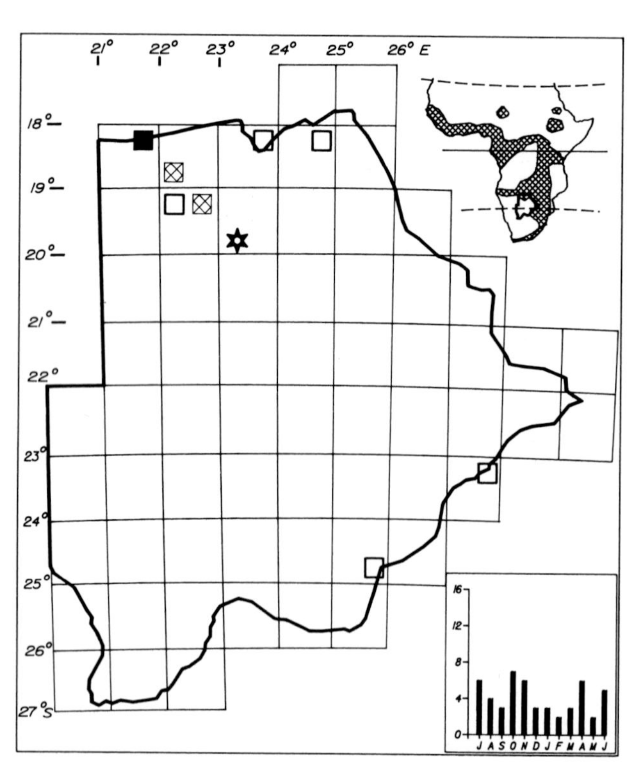

Status

Uncommon resident of the Okavango River. Rare on the Linyanti, Chobe, Limpopo and Ngotwane rivers from each of which it has been recorded from one locality only. Known from several sites near Shakawe and Mohembo whence most records are derived. Poorly known and the distribution warrants more intensive study. Egglaying August. Solitary or in pairs.

Habitat

Confined to rivers with overhanging bushes and trees. Usually on quiet stretches of slow moving water with denser riparian vegetation than the Blackcrowned Night Heron.

Analysis

9 squares (4%) Total count 53 (0,03%)

LITTLE BITTERN 78

Ixobrychus minutus

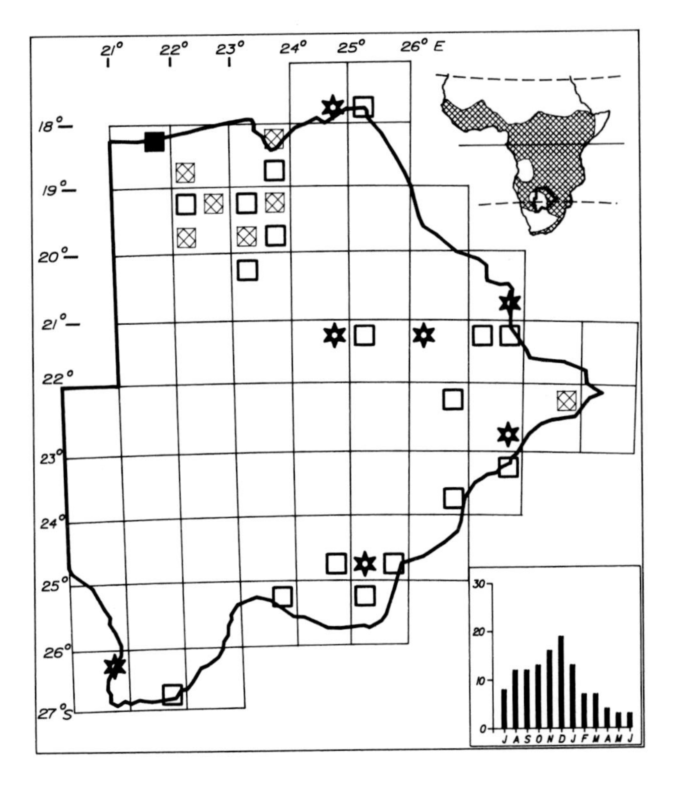

Status

Sparse to fairly common resident race *I. m. payesii* is augmented by Palaearctic subspecies *I. m. minutus* probably from December to March as in adjacent Zimbabwe and South Africa. (There was no coordinated attempt to distinguish the subspecies in this survey.) Most records fall between July and January. Occurs in most of the river systems of the north, east and southwest and on inland waters when suitably wet conditions prevail. Further study of breeding, movements and subspecies is indicated to clarify its status in Botswana. Solitary or in pairs.

Habitat

Rivers, swamps, marshes, dams and lagoons with fringing reedbeds and tall aquatic vegetation. Also in similar vegetation on seasonally inundated floodplains and pans.

Analysis

30 squares (13%) Total count 128 (0,07%)

DWARF BITTERN 79

Ixobrychus sturmii

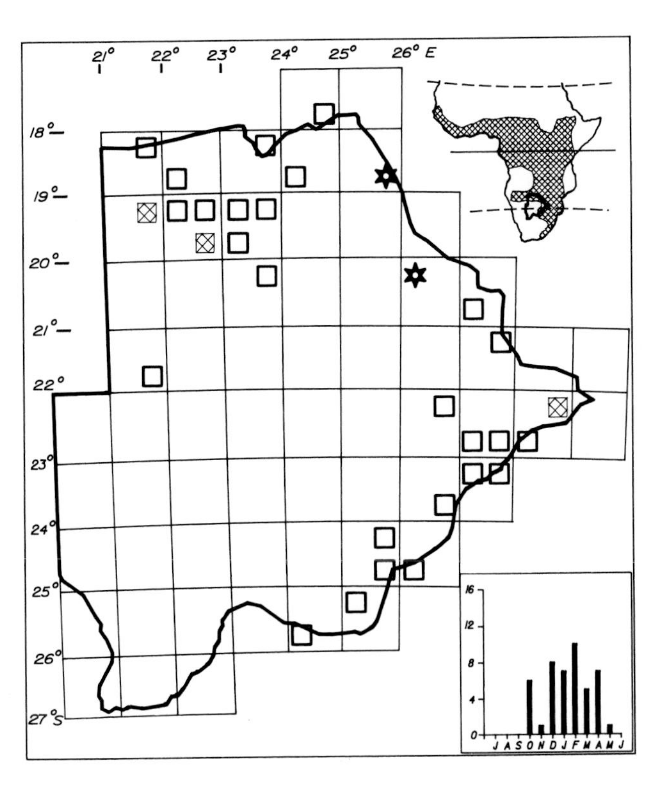

Alternative name

Rail Heron

Status

Sparse to uncommon breeding intra-African migrant to the northern and eastern regions between October and April. Unlikely to occur in central and western areas because of the lack of suitable breeding habitat in arid conditions. Commoner in high rainfall years. May extend further west into woodland in periods wetter than those covered by this survey. Active nest December. Usually in pairs or solitary.

Habitat

Seasonally flooded depressions in savanna and woodland into which it moves opportunistically to breed. Breeds in trees, bushes and large logs in or over water. Also occurs at ditches and pools on the side of roads and on floodplains.

Analysis

30 squares (13%) Total count 52 (0,03%)

BITTERN

80

Botaurus stellaris

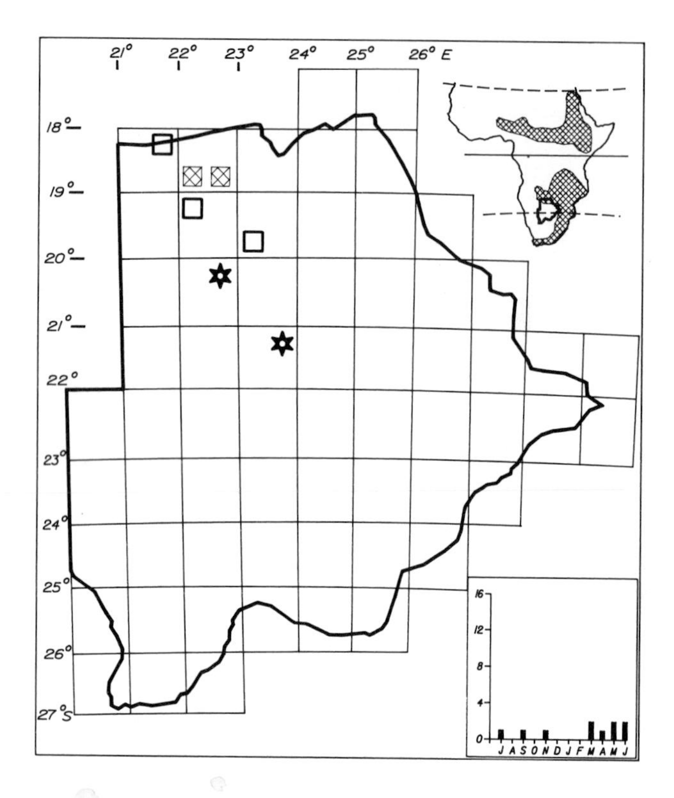

Status

Sparse to uncommon resident of the Okavango Delta. Elusive and easily overlooked. Its true status is uncertain. There are past records from Lake Ngami and Mopipi Dam and unconfirmed records from Shashe Dam and Gaborone. It may be more widespread in high rainfall years and some local movements may occur. Usually solitary.

Habitat

Reedbeds and tall aquatic vegetation in swamps and marshes.

Analysis

7 squares (3%) Total count 12 (0,006%)

HAMMERKOP

81

Scopus umbretta

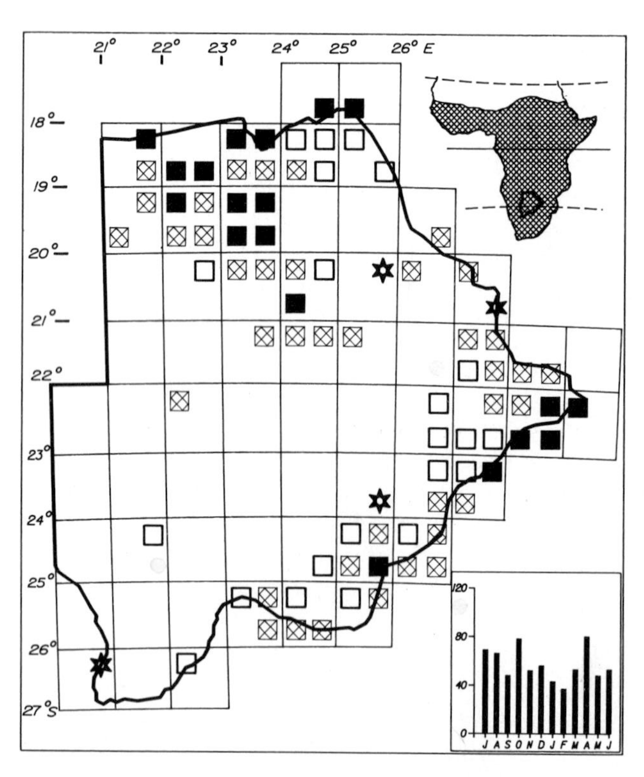

Status

Very common resident of the Okavango, Linyanti, Chobe and Limpopo rivers. Sparse to common in the drainage of these rivers and along the Boteti, Nata, Shashe, Ngotwane, Marico and Molopo river systems. Rare on pans in the west where it may occur more commonly in high rainfall years. Active nests November and January. Usually solitary or in pairs, sometimes in small groups.

Habitat

Rivers, lakes, lagoons, dams, ponds, marshes and flood-plains. Occasional at flooded grassland, seasonally wet pans and dry river beds.

Analysis

84 squares (37%) Total count 712 (0,4%)

WHITE STORK 83

Ciconia ciconia

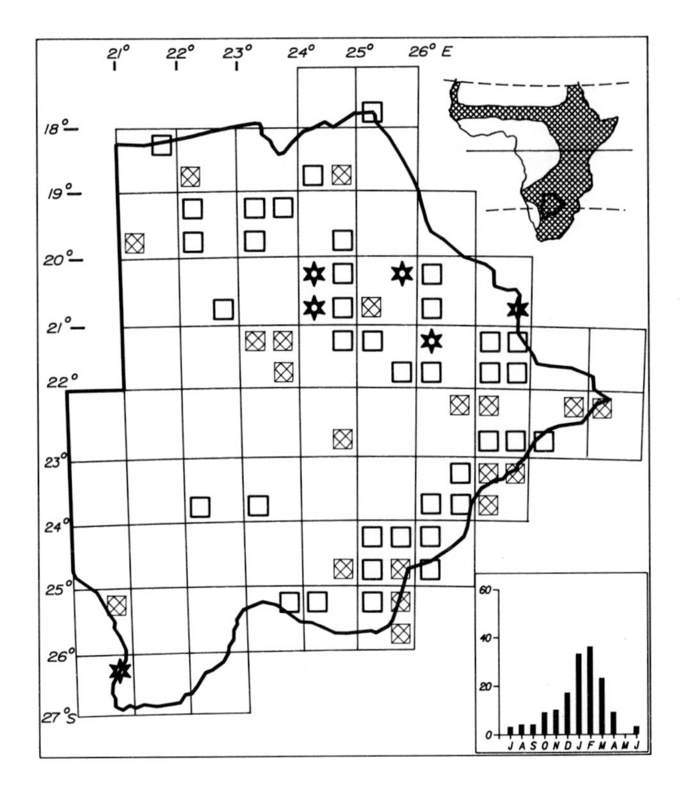

Status
Sparse to fairly common Palaearctic migrant mainly to northern and eastern areas but extending further west in some years. Present from October to April. Peak numbers occur in January and February. Winter records of flocks up to 80 birds are uncommon but have been seen at several localities. These may represent post breeding dispersal of South African breeding birds. Usually in flocks of 20–80 birds, solitary birds and large flocks up to 400 also occur.

Habitat
Open grassland, floodplains (wet or dry), open tree savanna, cultivation, flooded depressions including borrow pits. Often near water, also occurs in semidesert scrub on migration.

Analysis
64 squares (28%) Total count 171 (0,09%)

BLACK STORK 84

Ciconia nigra

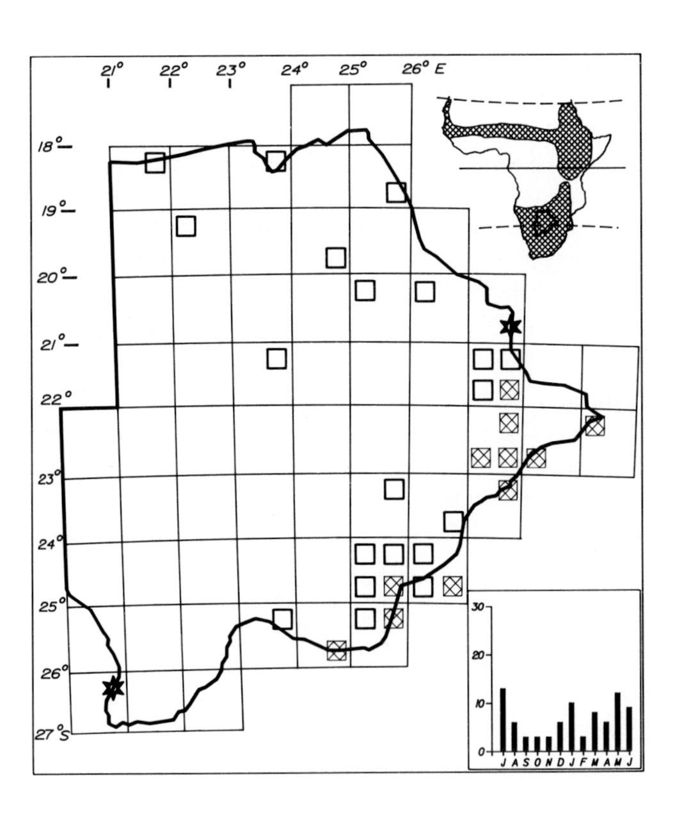

Status
Sparse to fairly common resident of the hills and valleys of the east and southeast from Francistown to Ootse and Lobatse. Rare elsewhere. The northern records are most likely to be wanderers from the eastern breeding grounds but possibly also from the Victoria Falls gorges. Palaearctic birds have not been recorded further south than Tanzania. Egg-laying in all months from May to August. Usually solitary, in pairs at breeding sites.

Habitat
Breeds on cliffs and gorges in hills and mountains. Forages at nearby rivers, marshes, dams and pools in dry river beds.

Analysis
33 squares (14%) Total count 99 (0,06%)

ABDIM'S STORK

Ciconia abdimii

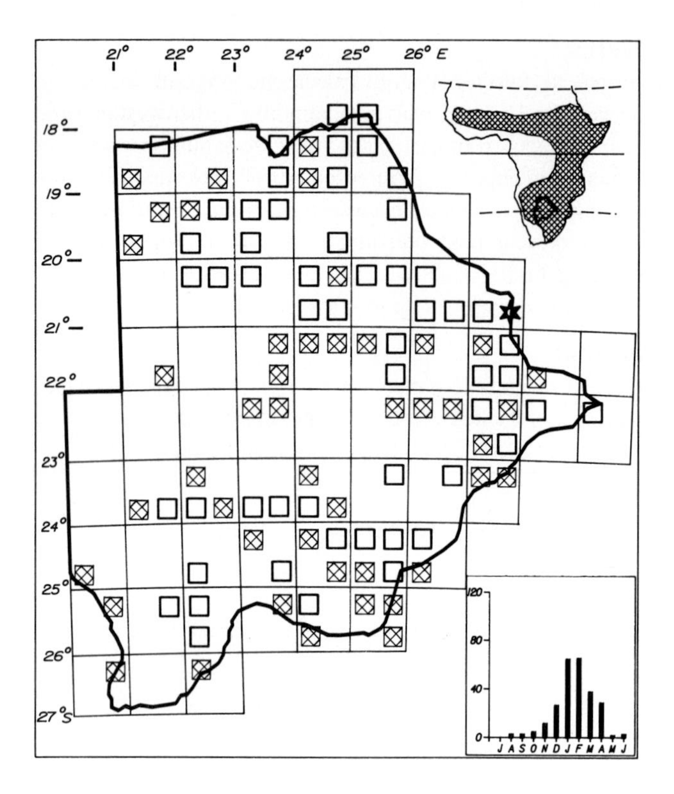

Alternative name
Whitebellied Stork

Status
Sparse to fairly common nonbreeding intra-African migrant to all regions of the country mainly from November to April. Some birds arrive in late August and some depart as late as early June in some years. Peak numbers occur in January and February. During drought years records were confined mainly to the east and north. Sightings in central and western areas increased markedly from 1987 onwards. Usually in flocks of 20–200 birds, but flocks of 500–1000 birds are not unusual.

Habitat
Short grassland on floodplains, plains, edges of pans, open tree and bush savanna, cultivation, farmland, suburban areas. Also in semidesert on migration. Most commonly near water.

Analysis
101 squares (44%) Total count 276 (0,15%)

WOOLLYNECKED STORK

Ciconia episcopus

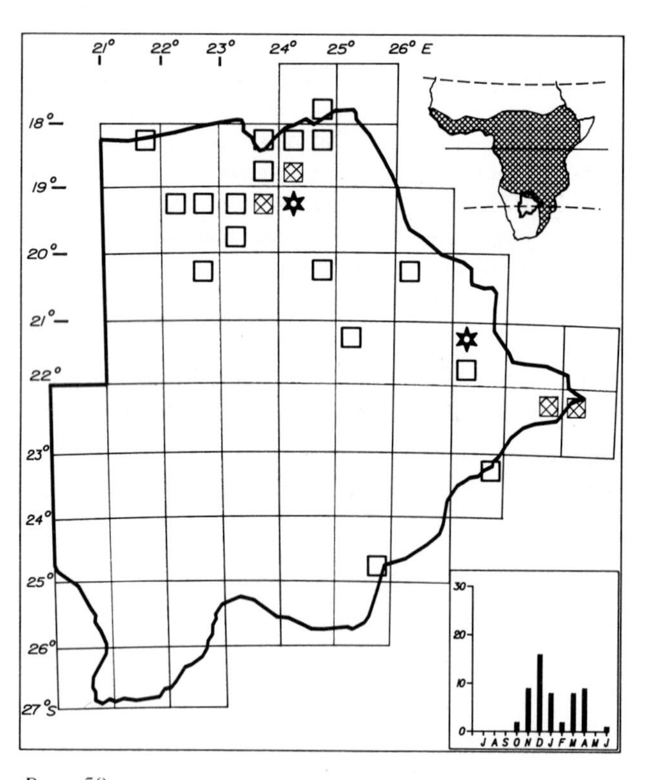

Status
Sparse and uncommon in the north and east. Its status is uncertain. There may be a small resident population but it is likely that most records are of birds wandering from breeding localities in Zimbabwe and Zambia. Most records fall between November and April. The breeding season in Zimbabwe is August to October. The June record hints at breeding in the Tuli region where there is suitable breeding habitat. Usually solitary or in pairs, but flocks of 10–100 birds have been reported.

Habitat
Grassland near water, floodplains, marshes, edges of lakes, dams and lagoons. Usually in well-wooded areas, on edges of forest or in tall riparian trees.

Analysis
23 squares (10%) Total count 60 (0,03%)

OPENBILLED STORK

87

Anastomus lamelligerus

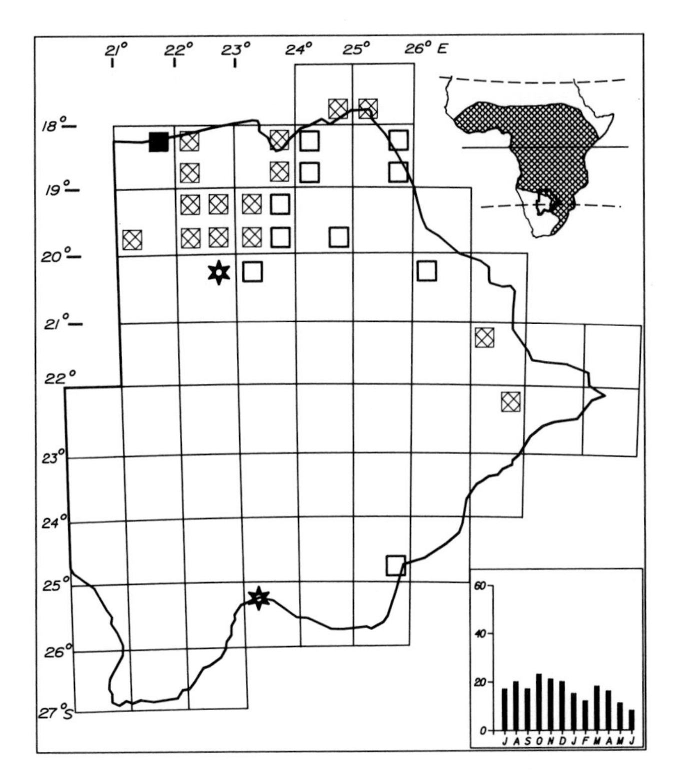

Status

Sparse to locally common resident and intra-African migrant of the Okavango, Linyanti and Chobe river systems. Very common at Shakawe. Regular at Shashe Dam, sparse at Nata, and recorded once from near Sefophe and once at Gaborone. Present in all months. Considerable movements occur but their extent and nature are not understood. Has bred on the Makgadikgadi Pans in the past but there are no recent breeding records. Usually in flocks of 20–200 birds.

Habitat

Lagoons, floodplains, marshes, dams and slow-moving rivers; often where there are large trees, particularly dead trees in water. Feeds in clear shallow water on molluscs.

Analysis

28 squares (12%) Total count 207 (0,12%)

SADDLEBILLED STORK

88

Ephippiorhynchus senegalensis

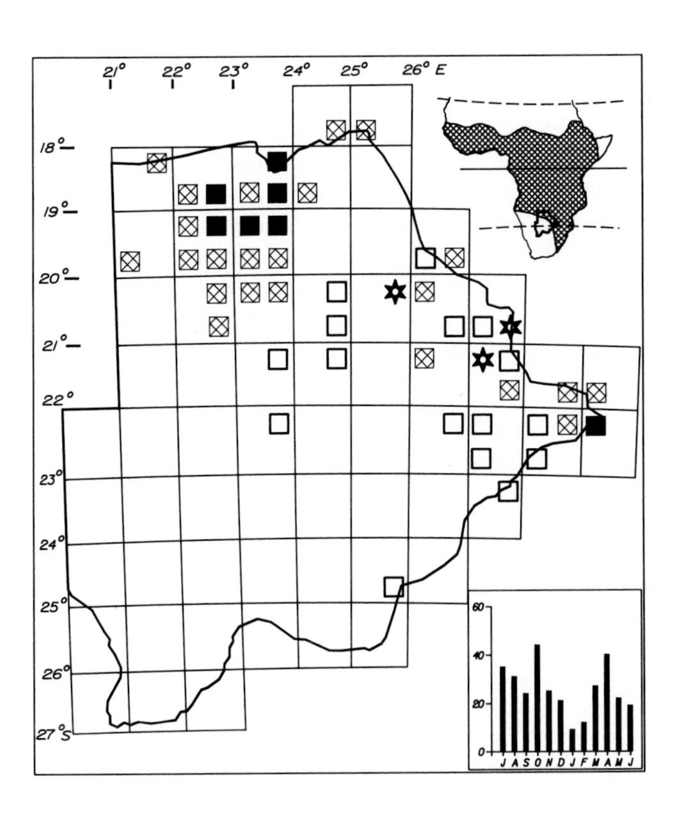

Status

Common to very common resident of the Okavango, Linyanti and Chobe river systems. Sparse to locally common in the east—very common in the area of the Shashe/Limpopo confluence. It wanders widely to nearby areas in the north and east and occurs irregularly as far south as Gaborone. Egglaying May, June and January. Solitary or in pairs.

Habitat

Marshes, lagoons, wet floodplains, vegetated edges of quiet rivers, dams, pans and at pools in dry river beds. In the east it is mainly confined to river valleys.

Analysis

49 squares (21%) Total count 322 (0,18%)

MARABOU STORK 89

Leptoptilos crumeniferus

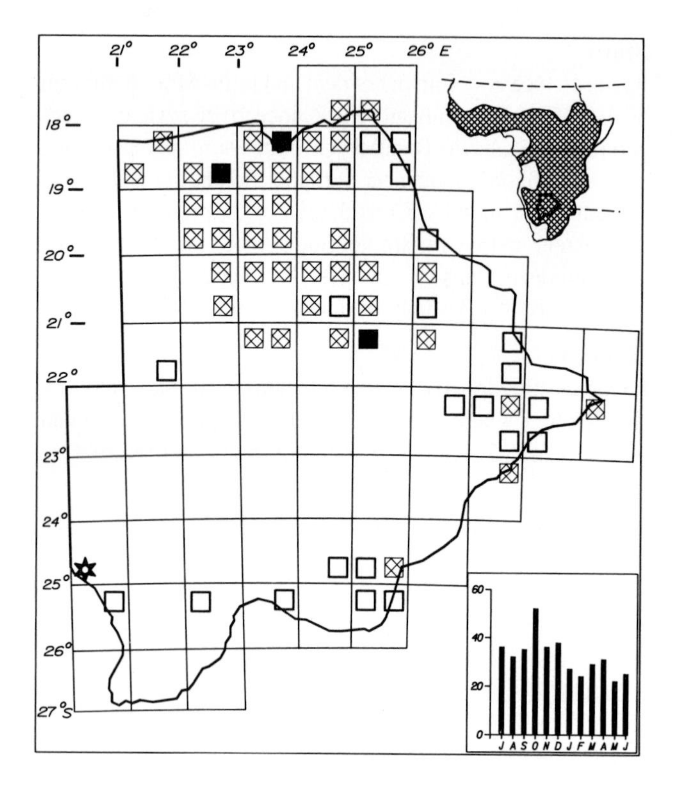

Status

Common to very common north of 22°S. Sparse to fairly common in the east and southeast. Very sparse in the west in game reserves (Mabuasehube and Nossob Camp) and in cattle farming areas (Bray and Ghanzi). Breeds in the Okavango Delta in some years. Egglaying August. Usually in groups, occasionally solitary.

Habitat

Open tree and bush savanna mainly in the high rainfall areas. Commonest where game animals are plentiful. Also associated with human habitation for offal as at fishing camps, rubbish dumps, temporary and permanent abattoirs, intensive cattle ranching in the calving season. Apparently absent from many central and western areas even where game is plentiful.

Analysis

64 squares (27%) Total count 405 (0,22%)

YELLOWBILLED STORK 90

Mycteria ibis

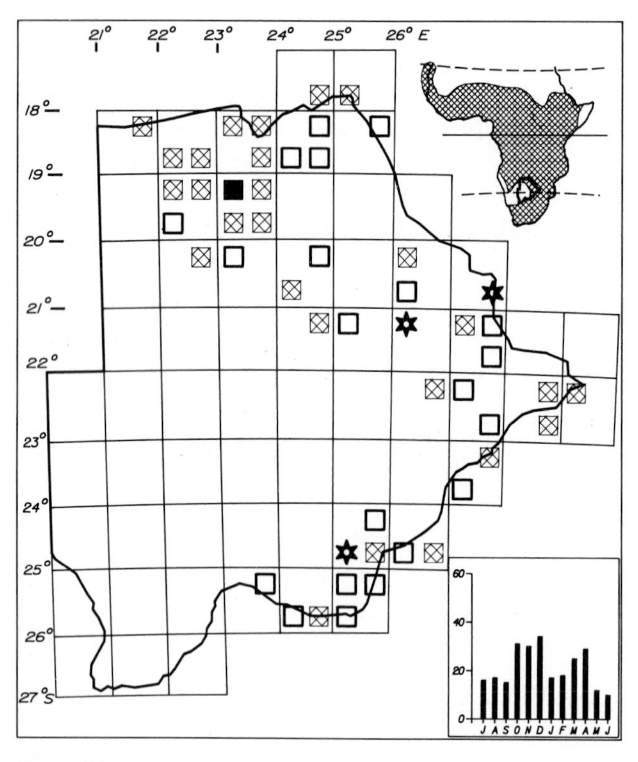

Status

Fairly common to locally common in the Okavango, Linyanti and Chobe regions. Elsewhere in the north, east and southeast it is sparse to uncommon. It is mainly a nonbreeding summer migrant arriving in July and August with peaks in October to December and again in March and April. Present at some localities throughout the year and part of the population may be resident as it breeds in northern areas in some years. Egglaying August. Usually in small groups, occasionally solitary.

Habitat

Shallow waters of rivers, dams, pans, lagoons, flooded grassland, swamps and marshes.

Analysis

50 squares (22%) Total count 272 (0,15%)

SACRED IBIS

91

Threskiornis aethiopicus

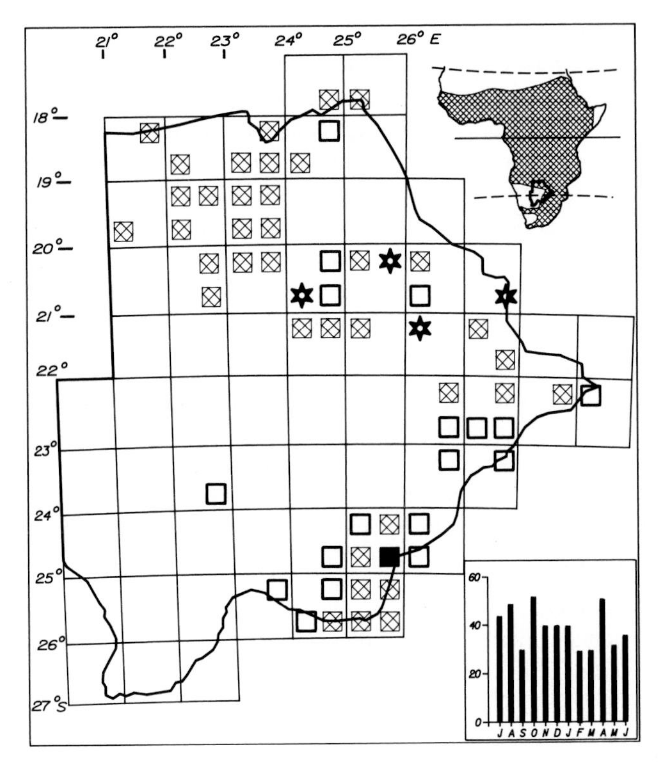

Status

Fairly common to common resident of the north, east and southeast. In the northwest it extends as far south as Lake Ngami and Tale Pan, and along the Boteti River to Rakops, Mopipi and Orapa. In the east it is less predictable and generally sparse south of Selebi Phikwe. In the southeast it has been recorded in every month at Gaborone for years on end and is sparse to fairly common in nearby areas west to Jwaneng and Bray. Once at Kang. Usually in small or large flocks, occasionally solitary.

Habitat

Marshes, lagoons, rivers, dams, lakes, sewage ponds. Also flooded grassland, irrigation, cultivation, farms, cattle posts, villages, rubbish dumps, airfields and football fields. Has adapted to a wide variety of manmade habitats.

Analysis

59 squares (26%) Total count 485 (0,27%)

GLOSSY IBIS

93

Plegadis falcinellus

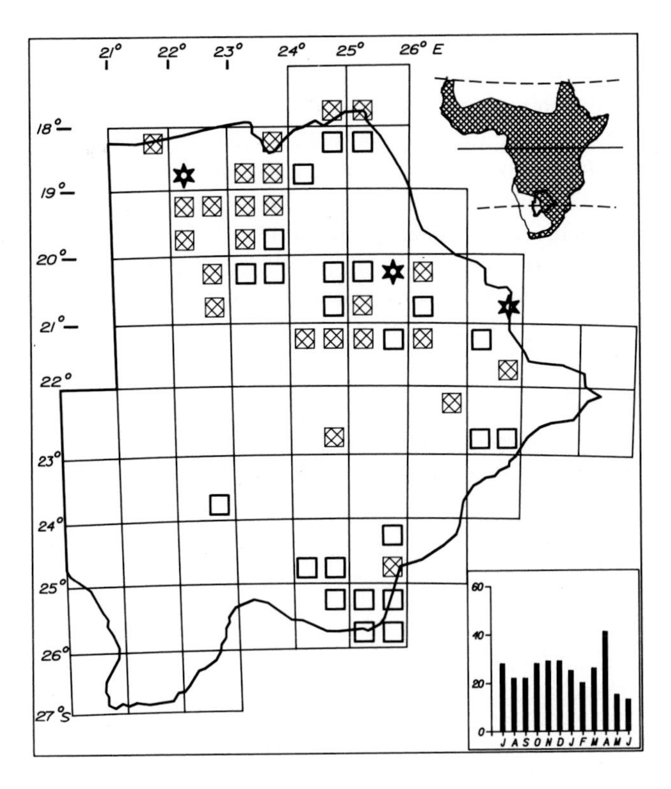

Status

Fairly common to common resident of the northern wetlands to as far south as Tale Pan, Rakops, Mopipi and Orapa. South of 22°S it is sparse and of irregular occurrence. Local and international movements occur but are complex with no regular pattern from year to year. Numbers are fewest in May and June. The population of this species has been increasing in southern Africa since the 1950s. Usually in small groups of 4–15 birds, sometimes solitary or in large flocks of several hundreds in the northern wetlands.

Habitat

Marshes, wet grassland, edges of lakes, lagoons, flooded pans and sewage ponds. Feeds by probing in soft mud.

Analysis

50 squares (22%) Total count 313 (0,17%)

HADEDA IBIS

Bostrychia hagedash

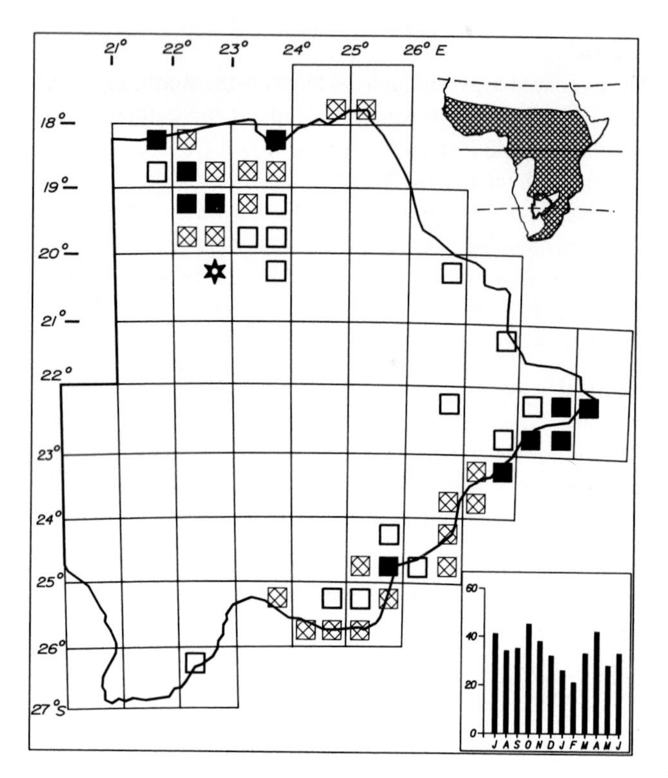

Status

Common to very common resident of the Okavango Delta and along the Linyanti and Chobe rivers. Very common along parts of the Limpopo River and at Gaborone and fairly common elswhere in the Limpopo, Ngotwane and Molopo drainage to as far west as Bray. Recorded once at each of Tshabong, Serowe and Francistown. Mainly sedentary but with some local movements. Occurs at the western limit of its range in southern Africa. Egglaying October and November. Usually in groups of 4–20 birds.

Habitat

Tall riparian trees, riverine forest and adjacent grassland and marshes. Confined to the major river systems. In the southeast it has adapted well to manmade habitats such as farm dams, irrigation, cultivation, exotic plantations of *Eucalyptus*, electricity pylons, golf courses and some gardens.

Analysis

46 squares (20%) Total count 418 (0,23%)

AFRICAN SPOONBILL

Platalea alba

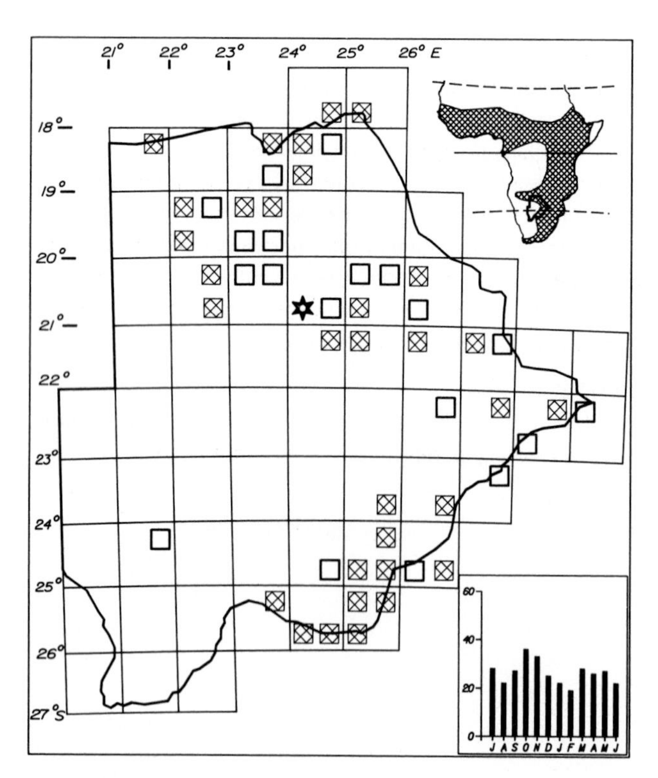

Status

Sparse to fairly common resident of the north and east. Regular only at few localities and nowhere very predictable. Commonest at Shakawe, Moremi, Kasane, Mopipi, Orapa, Nata River mouth, Shashe Dam and Gaborone Dam. Highly nomadic; moving with changing water levels to suitably wet conditions. In the southeast birds probably occur as part of the seasonal movements of populations in the western Transvaal and Orange Free State where they are considerably more common. Egglaying July, August and April. Solitary or in small groups.

Habitat

Shallow waters of lakes, lagoons, dams and pans. Also marshes, wet floodplains and sewage ponds.

Analysis

52 squares (23%) Total count 336 (0,19%)

GREATER FLAMINGO 96

Phoenicopterus ruber

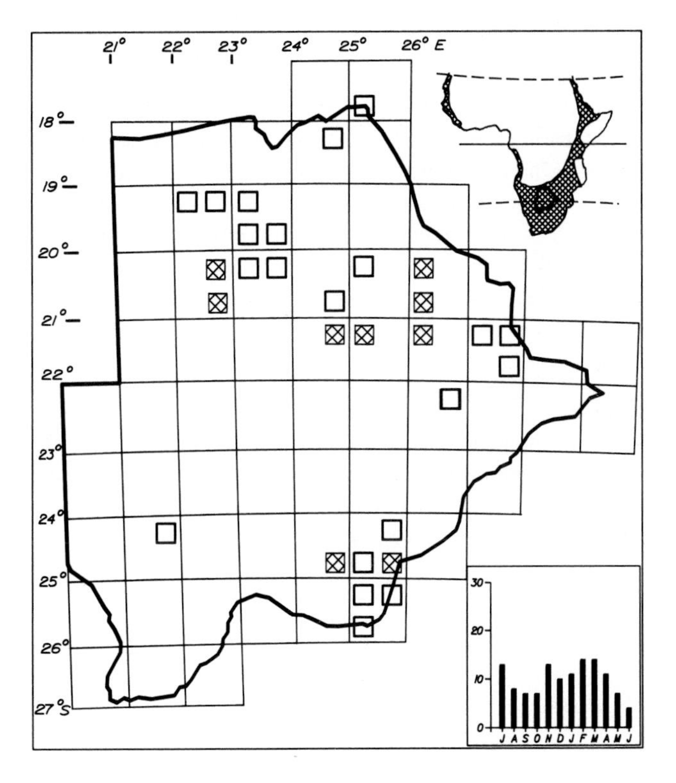

Status

In some years an abundant breeding resident of the Makgadikgadi Pans from where the largest African breeding colony (15 000 birds) has been recorded. In most recent years sparse to fairly common in that area. Elsewhere it is sparse or uncommon and unpredictable. Considerable movements occur to neighbouring countries. Mass migration to the Orange River mouth and to the Rift Valley has been postulated but movement to and from the Etosha Pan, Orange Free State and southern Transvaal is probably more regular and common. Present in all months. Usually in small flocks of 5–50 birds, but can be solitary.

Habitat

Large bodies of shallow water such as pans, lakes and dams. Occasionally at sewage ponds. Feeds in deeper water than the next species.

Analysis

30 squares (13%) Total count 129 (0,07%)

LESSER FLAMINGO 97

Phoeniconaias minor

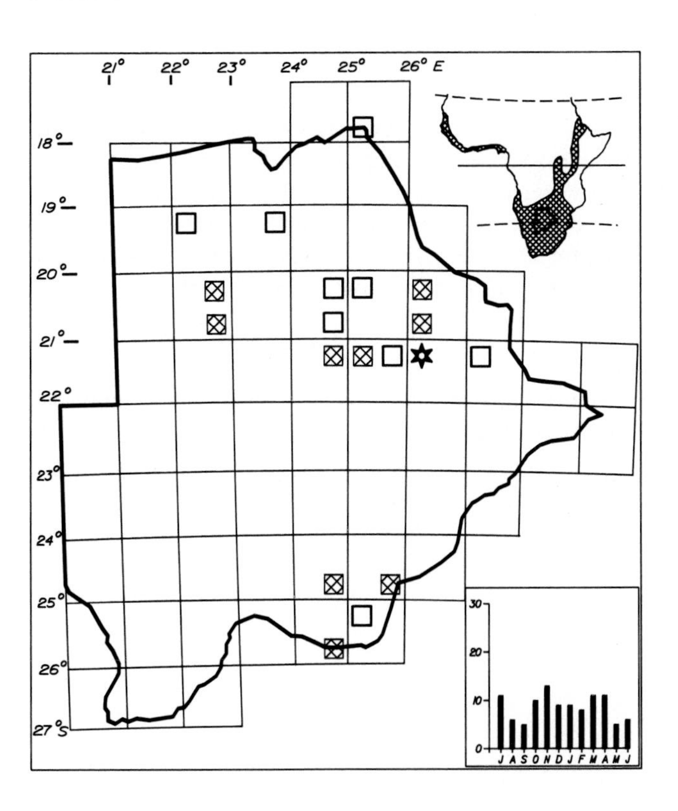

Status

Sparse to uncommon resident mainly in the Makgadikgadi Pans but also at few other suitable localities such as Lake Ngami, Tale Pan, Linyanti, Mopipi Dam, Orapa, Gaborone and Jwaneng. Like the Greater Flamingo it may breed in the Makgadikgadi in some years and be present in thousands but during this survey most suitable sites have been dry due to drought. Both local and intra-African movements occur with changing habitat conditions. Recorded in all months. Occurs in flocks of 100–500 birds at large expanses of water but in small flocks of 5–50 birds at smaller sites.

Habitat

Shallow and alkaline or brackish water in pans, lakes, estuaries, large lagoons and sewage ponds. Feeds on algae in shallow water near the shore. Does not compete for food with the Greater Flamingo and they are often seen together.

Analysis

19 squares (8%) Total count 109 (0,06%)

WHITEFACED DUCK

Dendrocygna viduata

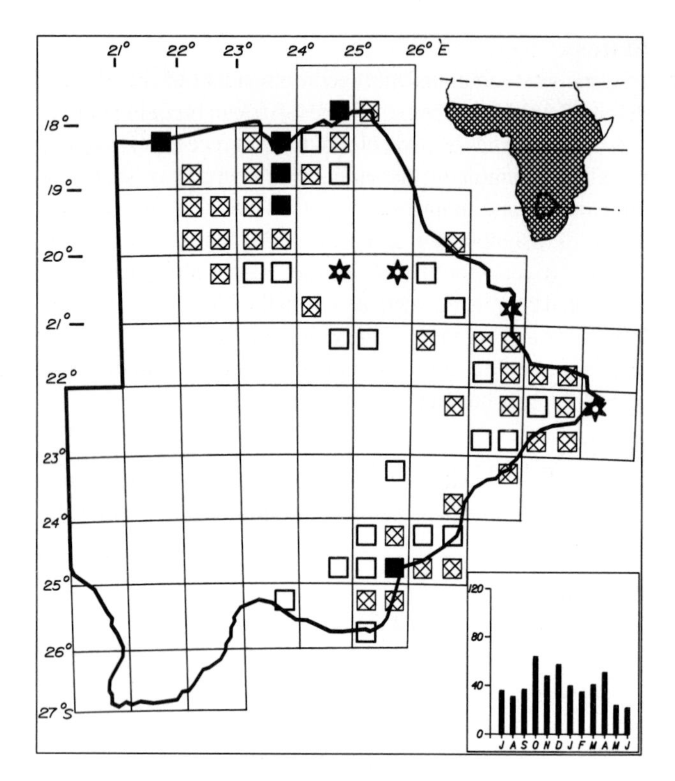

Status

Common to very common resident of the Okavango, Linyanti and Chobe river systems. Sparse to fairly common in the east south of Francistown. Very common in the Gaborone area. Extends along the Molopo River to as far west as Bray. It occurs throughout the year. Local and international movements occur with influxes in October to December and decreased numbers in May and June. Egg-laying all months from November to February. Usually in small groups of 5–20 birds, but gathers in hundreds at suitable feeding sites.

Habitat

Shallow water and vegetated or muddy edges of lagoons, dams, pans and lakes, marshes, seasonally flooded grassland and sewage ponds. Nests in trees in woodland or in grassland and aquatic vegetation near water.

Analysis

64 squares (28%) Total count 516 (0,29%)

FULVOUS DUCK

Dendrocygna bicolor

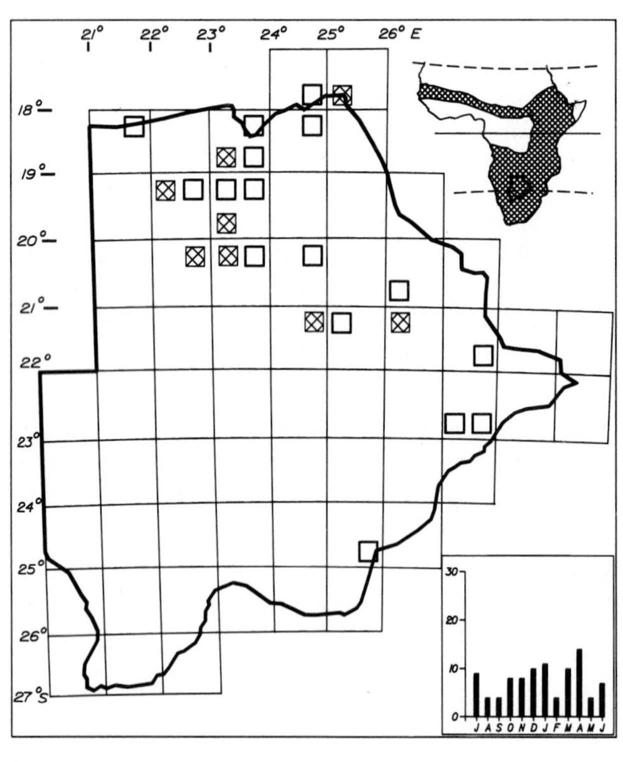

Status

A sparse to fairly common resident in the Okavango, Linyanti and Chobe river systems which occurs patchily to as far south as Mopipi and Orapa and the eastern Makgadikgadi. It is a vagrant elsewhere. Present throughout the year with local movements to suitable habitats as water levels change. Egglaying October. Usually in small groups of 5–20 birds, rarely gathering in such large numbers as the Whitefaced Duck.

Habitat

Vegetated lagoons, pans, dams, ponds and floodplains. Rarely at sewage ponds.

Analysis

23 squares (10%) Total count 99 (0,05%)

WHITEBACKED DUCK 101

Thalassornis leuconotus

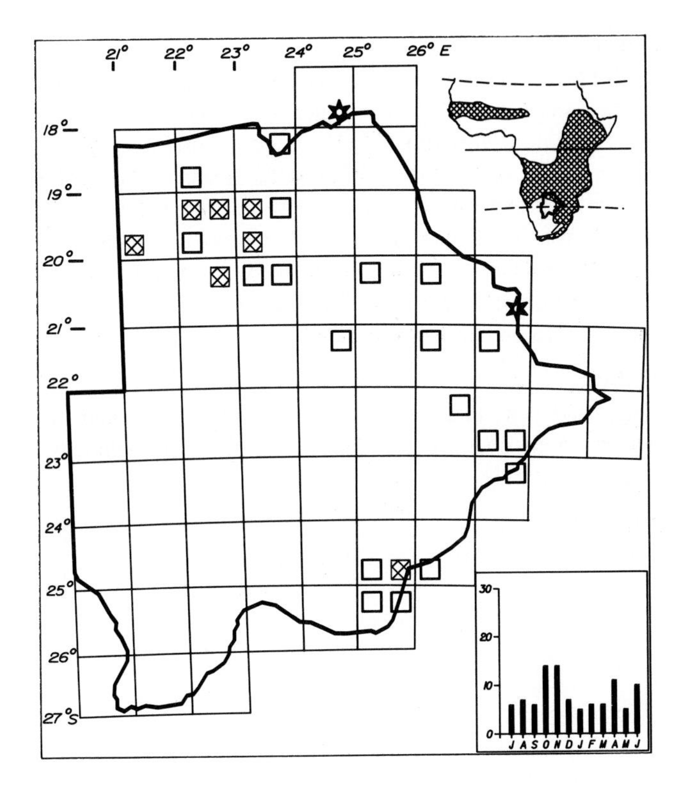

Status

Sparse to locally common resident of the Okavango Delta where it is present throughout the year. Elsewhere it is a scarce and temporary visitor after periods of prolonged rain or at the end of the rainy season. However, in years of good rainfall it may remain throughout the year in some areas away from the Okavango. Mainly sedentary but seasonal movements occur. Egglaying March. Usually in pairs.

Habitat

Lagoons, ponds, lakes and dams with floating and emergent aquatic vegetation.

Analysis

28 squares (12%)   Total count 112 (0,06%)

EGYPTIAN GOOSE 102

Alopochen aegyptiacus

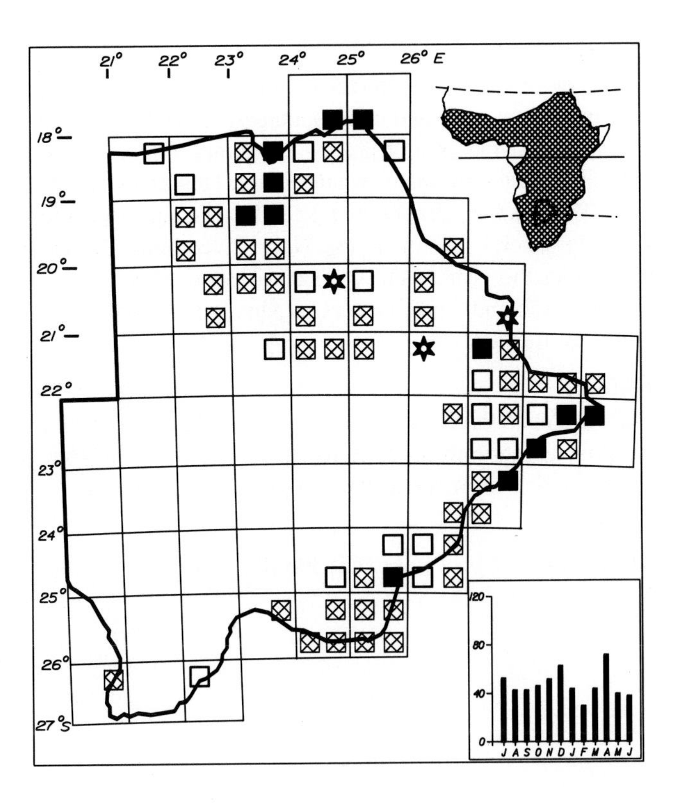

Status

Common to very common resident of the Okavango, Linyanti, Chobe, Shashe and Limpopo river systems. Sparse to fairly common in adjacent areas of the north and east. Common in the eastern parts of the Molopo River but unpredictable west of Bray where it may be commoner in high rainfall years. Present throughout the year with little fluctuation in numbers overall. Egglaying June, July and all months from September to February. Usually in pairs, but may gather in flocks of 20 to several hundred birds when not breeding.

Habitat

Marshes and floodplains. Also edges of lagoons, lakes, dams and rivers. Pans when flooded or with small pools, inundated grassland, cultivation and irrigated fields; sewage ponds. Grazes on grasslands and in fields. Roosts and nests in trees.

Analysis

76 squares (33%)   Total count 612 (0,34%)

SOUTH AFRICAN SHELDUCK

103

Tadorna cana

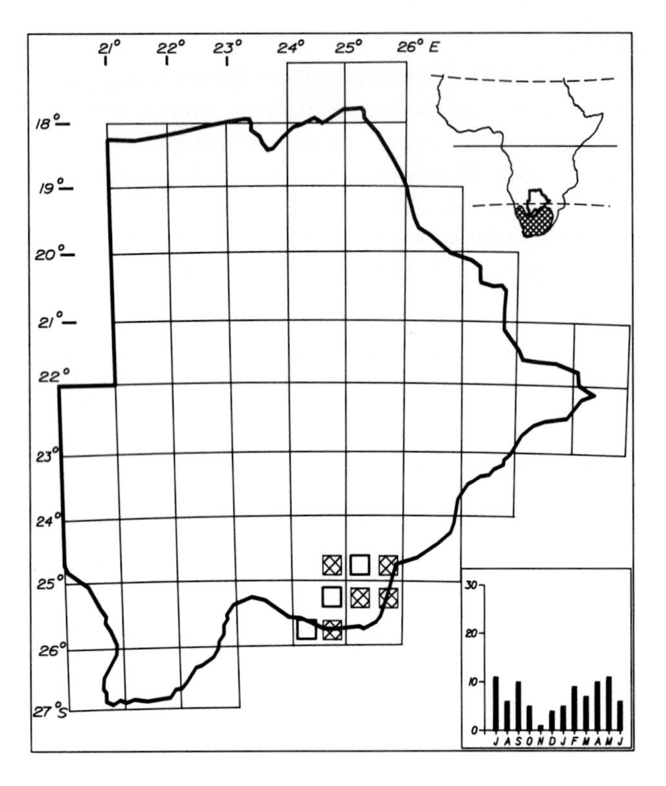

Status

Sparse to fairly common resident restricted to the southeast to as far north as Moshaneng and Gaborone. Occurs in Botswana at the northern limit of its southern African range. Recorded in all months in Gaborone. Otherwise mainly present from March or April to October with much fewer records from November to January. Fewer are present during the summer moult. It is present in Botswana mainly during the breeding season (May to September). Egglaying June and July. Usually in pairs.

Habitat

Dams and pans usually with adjacent open grassland or marsh—at Moshaneng the dam is mostly surrounded by woodland. Rarely at sewage ponds.

Analysis

8 squares (3%) Total count 91 (0,05%)

YELLOWBILLED DUCK

104

Anas undulata

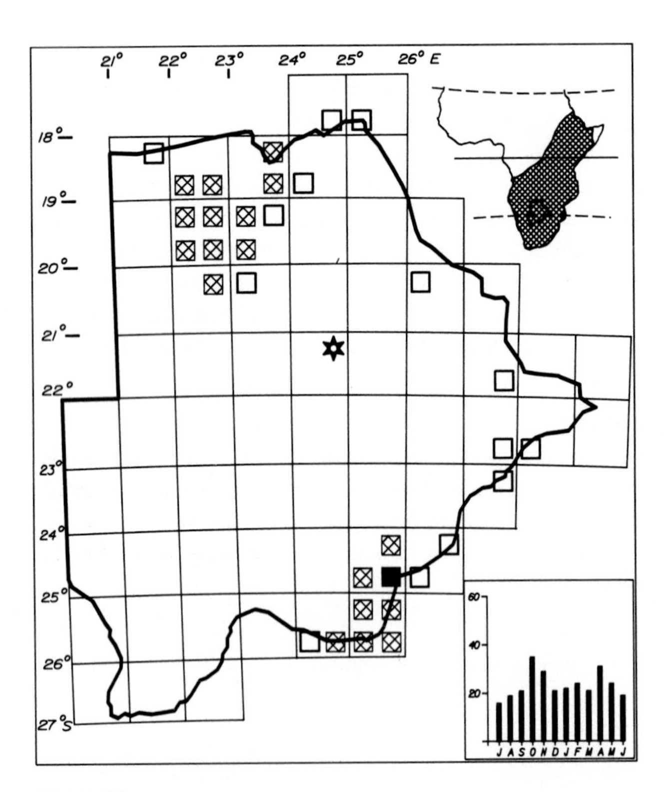

Status

Fairly common to common resident of the Okavango Delta, the Linyanti drainage and in the southeast. Elsewhere sparse and unpredictable as a casual or temporary visitor during local migratory movements when suitable habitat conditions exist. The alkaline pans of the Makgadikgadi do not appear to suit this species except at the Nata River inlet. Present throughout the year in the Okavango and southeast. Egg-laying March and April. Occurs in pairs when breeding, in groups or flocks at other times.

Habitat

Dams, lagoons, seasonally flooded pools and pans, marshes, floodplains and sewage ponds; usually with muddy substrate in deep and shallow water and an open sloping shoreline.

Analysis

34 squares (15%) Total count 306 (0,17%)

AFRICAN BLACK DUCK

Anas sparsa

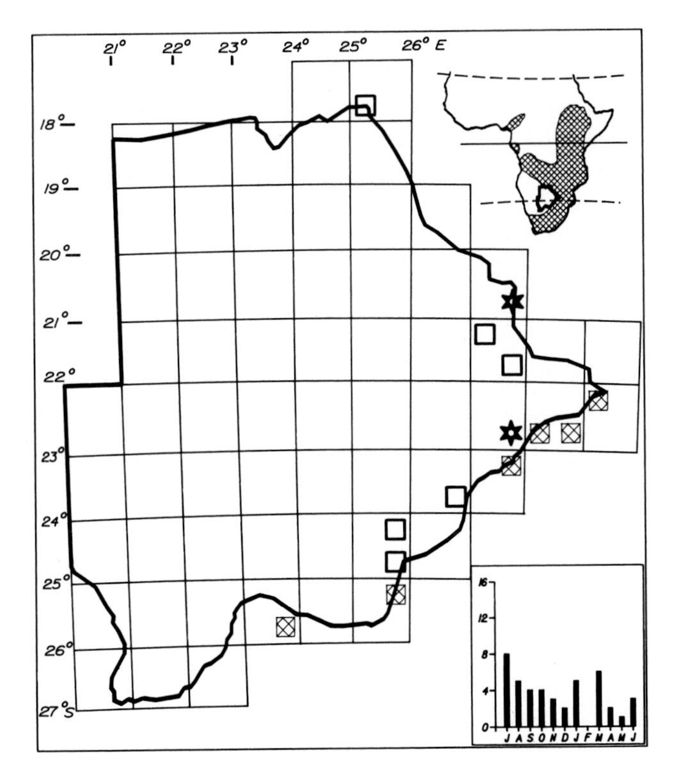

Status
A locally common resident on stretches of the Limpopo and Ngotwane rivers to as far south as Lobatse. It is sparse along the Shashe and Molopo drainage but may occur more frequently in years of good rainfall on these and other rivers. Recorded only once at Kasane. Egglaying December on the Motloutse River. Occurs in pairs.

Habitat
Quiet backwaters, remnant pools and seasonal catchments along well-wooded rivers and streams. Usually in the main course of the river, occasionally on adjacent waters such as dams and ponds.

Analysis
14 squares (6%) Total count 49 (0,02%)

CAPE TEAL

Anas capensis

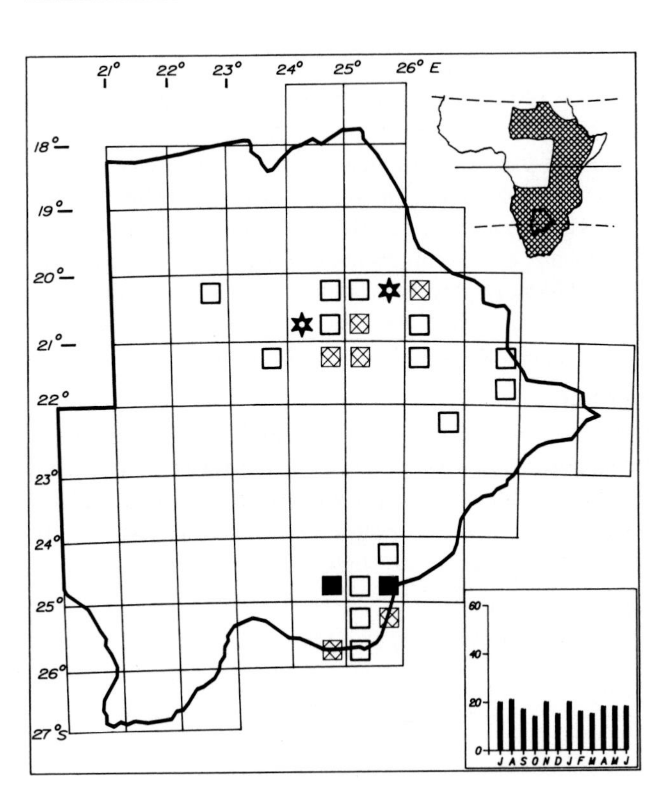

Status
Sparse to locally very common resident of the southeast. Common on the Makgadikgadi Pans in years of good rainfall but virtually absent there during drought. Not recorded from the major river systems nor the Okavango Delta but may occur on suitable waters of these areas as part of local movements in some years. Breeds throughout the year if suitable conditions prevail. Egglaying May to July and all months from November to March. Usually in pairs or small groups.

Habitat
Brackish or saline waters of dams, pans, sewage ponds and pools; preferring permanent or mature and stable conditions. Less frequently on fresh water dams and lagoons. Breeds in adjacent grassland.

Analysis
24 squares (10%) Total count 222 (0,12%)

HOTTENTOT TEAL

107

Anas hottentota

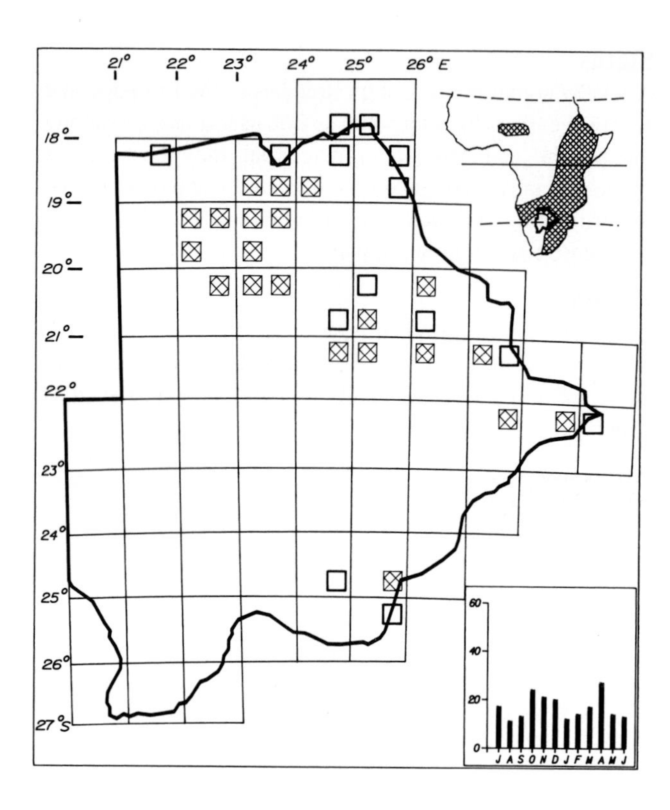

Status

Fairly common resident of the Okavango Delta south to Lake Ngami and in the northeastern wetlands of Linyanti, Savuti, Mopipi, Makgadikgadi, Orapa and Shashe. Sparse on northern rivers. Uncommon on some permanent waters of the southeast. Local and international movements occur but no regular pattern is currently evident. Egglaying May. Occurs in pairs or small groups.

Habitat

Lagoons, dams, lakes, marshes, swamps and flooded grassland; usually on waters with emergent aquatic vegetation in which it breeds. Occasionally at sewage ponds.

Analysis

35 squares (15%) Total count 211 (0,12%)

REDBILLED TEAL

108

Anas erythrorhyncha

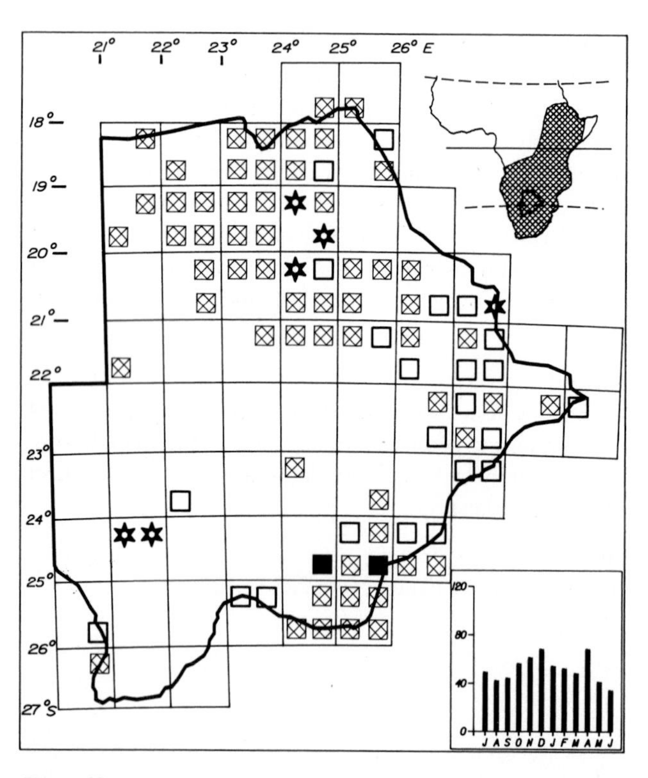

Status

Common resident of the north and east. Largely absent in central and western regions during this survey but it may be regular on pans in these areas in wetter years. Uncommon along the Molopo and Nossob rivers but probably regular there when these rivers have flowed. Local and long distance movements occur—to Zambia, Zimbabwe, Angola and South Africa. Egglaying all months from January to April. Occurs in pairs when breeding, otherwise usually in small flocks, sometimes large flocks of thousands.

Habitat

Lagoons, pans, dams, marshes, swamps, flooded grassland, floodplains and sewage ponds; favours those with submerged aquatic vegetation and adjacent dense grass, reeds or sedges for breeding. Breeding occurred at Jwaneng sewage ponds within one year of construction.

Analysis

90 squares (39%) Total count 661 (0,37%)

CAPE SHOVELLER 112

Anas smithii

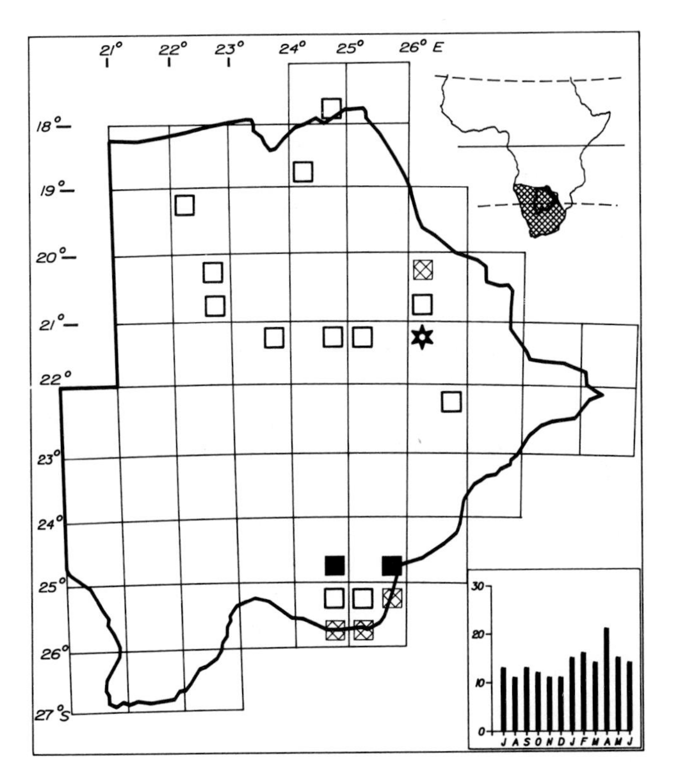

Status

Fairly common resident of the southeast where it is locally very common at Jwaneng and Gaborone. Further north it is sparse and uncommon mostly as a temporary visitor. In higher rainfall years it may occur in larger numbers and more regularly at sites such as Lake Ngami, Mopipi and Nata River mouth and breed there. Present throughout the year in the southeast where it breeds. Local and international movements occur to South Africa, Zimbabwe and as a vagrant to Zambia. Egglaying July, August, October and February. Occurs in pairs or small groups.

Habitat

Pans, dams, estuaries and sewage ponds with shallow brackish or saline water. Less often on temporarily flooded grassland. Breeds in dense vegetation near water.

Analysis

19 squares (8%) Total count 177 (0,1%)

SOUTHERN POCHARD 113

Netta erythrophthalma

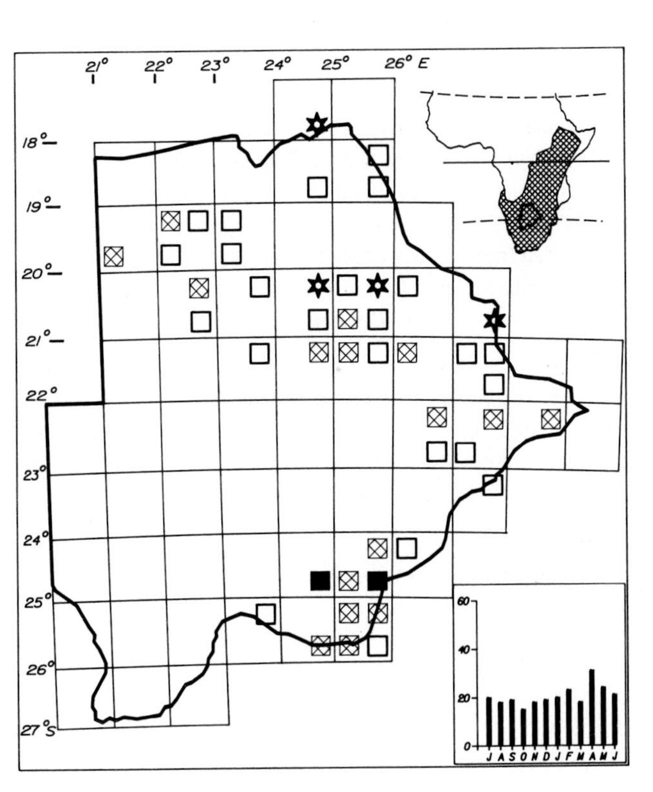

Alternative name

Redeyed Pochard

Status

Common resident of the southeast—very common at Jwaneng and Gaborone. Sparse to fairly common in the east and north. It is present throughout the year to as far north as Serowe. Further north most records fall between March and October suggesting off season dispersal from breeding grounds in the Transvaal (main breeding December to April). It is known to migrate long distances e.g. to Mozambique and possibly Malawi. May breed in the north in years of good rainfall (Clark 1980). Egglaying February (two records) at Lobatse. Solitary, in pairs or small groups of 4–30 birds.

Habitat

Dams, lagoons, pans and sewage ponds with deep water. Less often on vegetated waters such as swamps and flooded grassland.

Analysis

46 squares (20%) Total count 271 (0,15%)

PYGMY GOOSE

Nettapus auritus

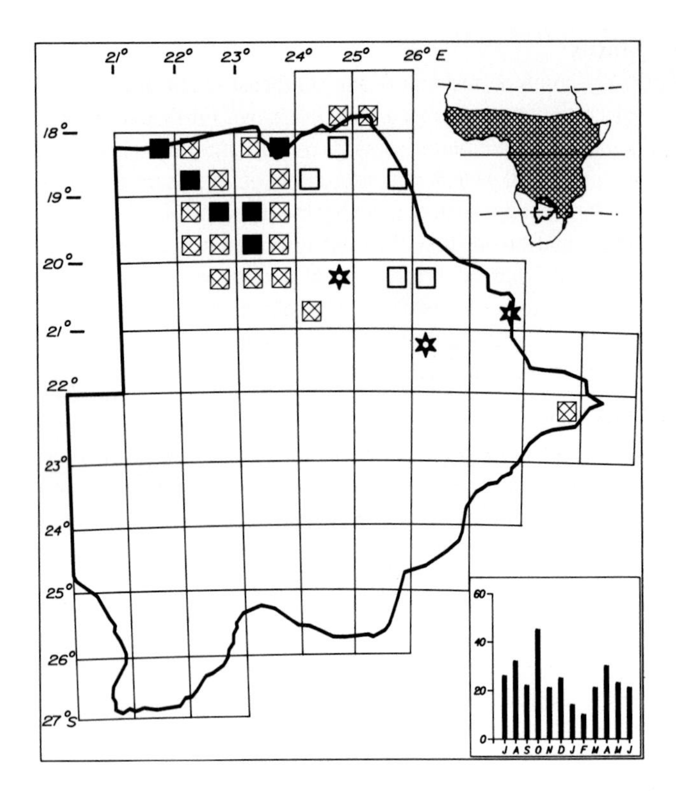

Status

Common to very common resident of the Okavango, Linyanti and Chobe river systems. Very uncommon in the east and northeast during this survey but past records suggest it is more widespread there in wetter years. Mainly sedentary but local movements occur. Egglaying December and January. Nearly always in pairs.

Habitat

Swamps, lagoons and pans with floating aquatic vegetation. Rarely strays from patches of vegetated waters amongst which this brightly coloured duck is well camouflaged. Also on quiet stretches of seasonal rivers such as the Nhabe and Boteti and well-vegetated lakes and dams. Nests mainly in trees and old nests of Hamerkop.

Analysis

30 squares (13%) Total count 300 (0,17%)

KNOBBILLED DUCK

Sarkidiornis melanotos

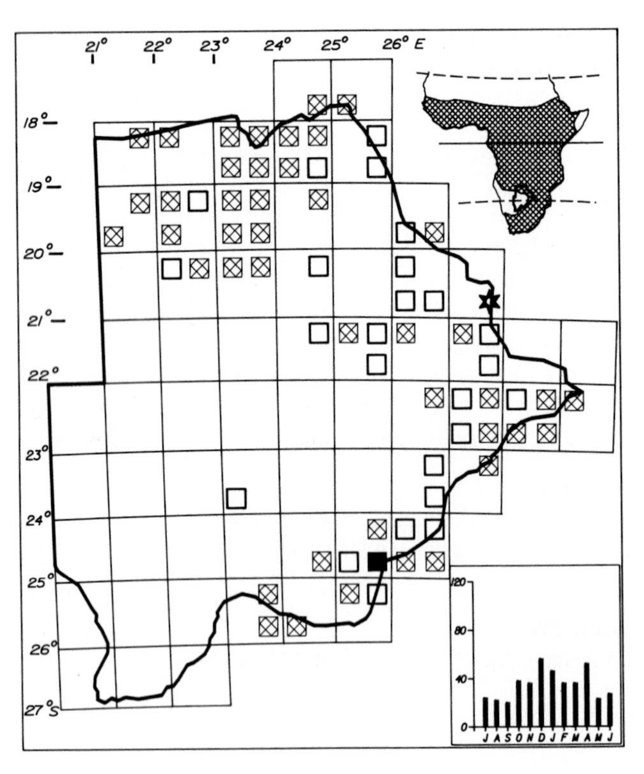

Status

Fairly common to common resident of the north and east on all the major river systems and extending west along the Molopo River to as far as Bray. Likely to occur on central and western pans in wetter years. It is present in all months throughout its range but numbers are highest from October to April when they are augmented by birds from other countries either on passage or temporarily resident. Occurs in small or large flocks, solitary and in pairs.

Habitat

Marshes, seasonally flooded grassland and savanna, floodplains, dams, lakes, lagoons and ponds; usually but not always in waters with emergent grass or aquatic vegetation. Feeds in shallows or nearby grasslands.

Analysis

10 squares (30%) Total count 430 (0,24%)

SPURWINGED GOOSE

116

Plectropterus gambensis

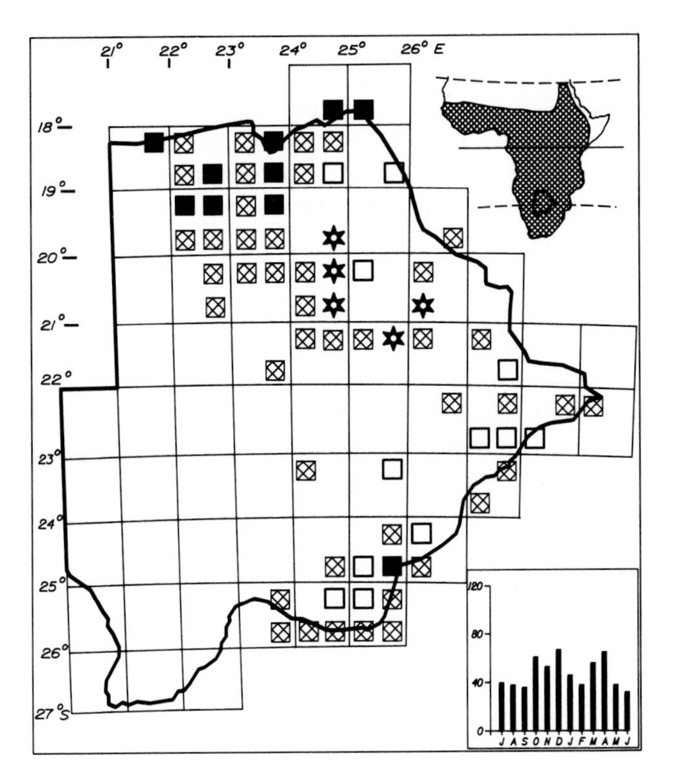

Status

Very common on the floodplains of the north and common in the adjacent wetlands south to Lake Ngami, Deception Pan, and Mopipi. Patchily distributed in the east and common to very common in the southeast. Extends westwards along the Molopo River to as far as Bray. Egglaying January and March. Occurs in large flocks of several hundred birds and as smaller groups, pairs or solitary.

Habitat

Floodplains, marshes, flooded grassland and edges of dams, pans, lakes and occasionally sewage ponds. Feeds on tubers and plants in water and grass in cultivation, fallow fields or nearby grassland.

Analysis

70 squares (30%) Total count 603 (0,34%)

MACCOA DUCK

117

Oxyura maccoa

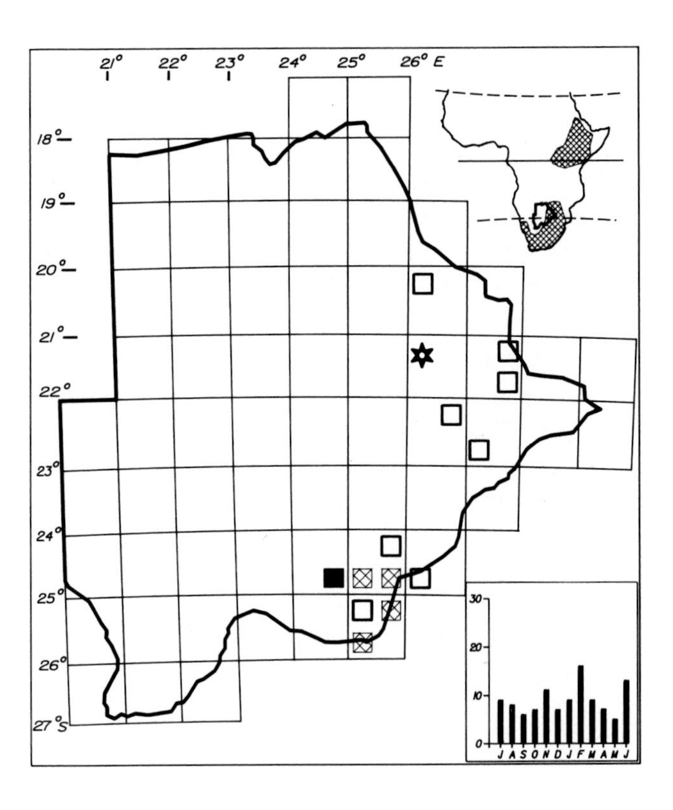

Status

Fairly common resident of the southeast where it breeds regularly. Very common at Jwaneng. Further north in the east it is sparse and unpredictable but it also breeds in this area. Present throughout the year although local and international movements to Zimbabwe and South Africa occur. It is likely to be more common on the Makgadikgadi Pans in high rainfall years. Egglaying December, March and April. Occurs in pairs, solitary or in small groups sometimes of one sex.

Habitat

Dams large and small, flooded pans, and sewage ponds; favours brackish or saline waters.

Analysis

14 squares (6%) Total count 115 (0,06%)

SECRETARY BIRD

Sagittarius serpentarius

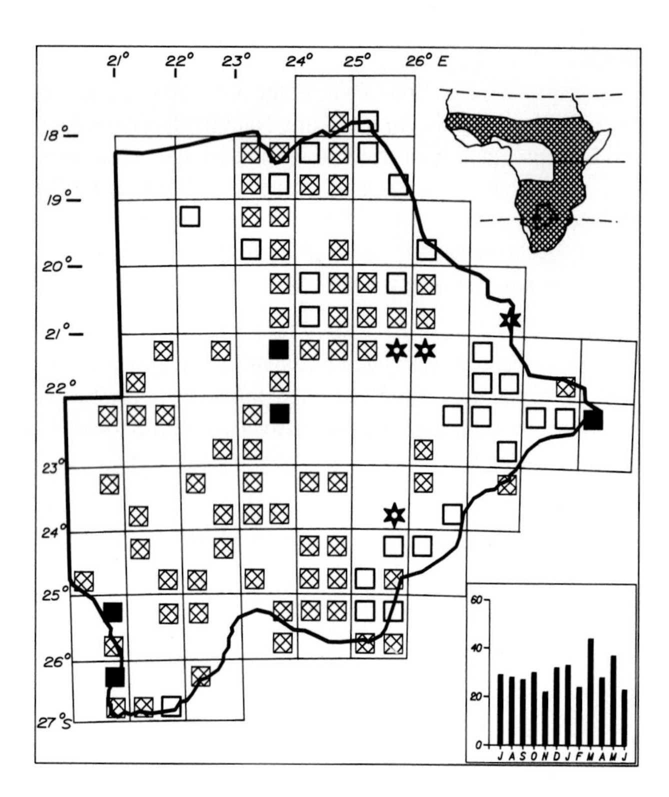

Status

Fairly common to common resident throughout except in the east where it is generally sparse and in the northwest where it is largely absent. Locally very common in Deception Valley and along the Nossob Valley. Mainly sedentary. Egglaying October to January, April and July. Usually in pairs but often in threes or solitary.

Habitat

Grassland plains, open savanna with widely scattered trees or bushes, dry floodplains, fossil valleys, edges of pans. Usually but not always in long grass. Nests on flat-topped trees particularly *Acacia* utilising stunted trees (3m high) if no taller ones are available as in parts of the Makgadikgadi.

Analysis

104 squares (45%)   Total count 422 (0,24%)

HOODED VULTURE

Necrosyrtes monachus

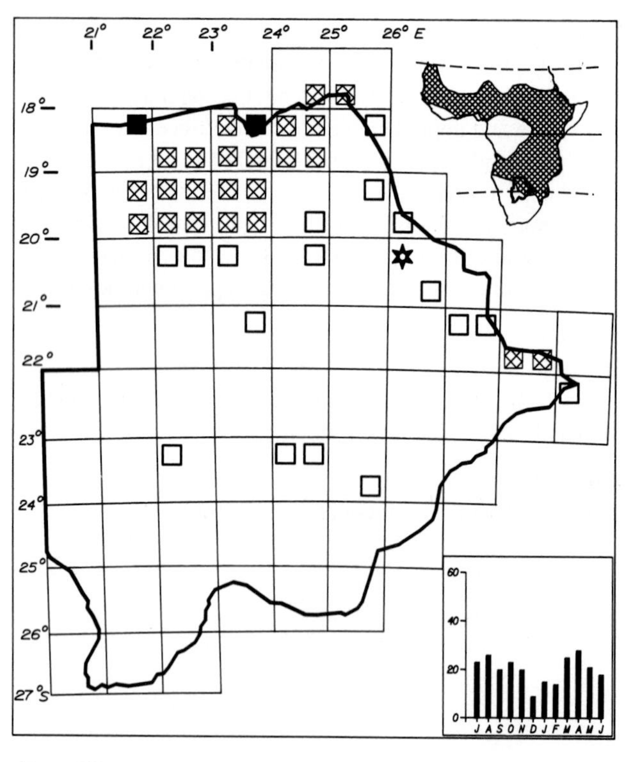

Status

Common to very common resident in the Okavango Delta and adjacent wooded terrain. Elsewhere it is sparse or uncommon and such records appear to represent wanderers prior to breeding (March to June) to as far south as the southcentral Kalahari woodlands. Further study is required to assess the status of these wanderers. Active nests April and June. Usually solitary, rarely in groups at a carcass.

Habitat

Well-developed woodland and adjacent tree savanna in high rainfall regions. Usually not far from water. May wander in the nonbreeding season to drier woodland.

Analysis

43 squares (19%)   Total count 254 (0,14%)

CAPE VULTURE

Gyps coprotheres

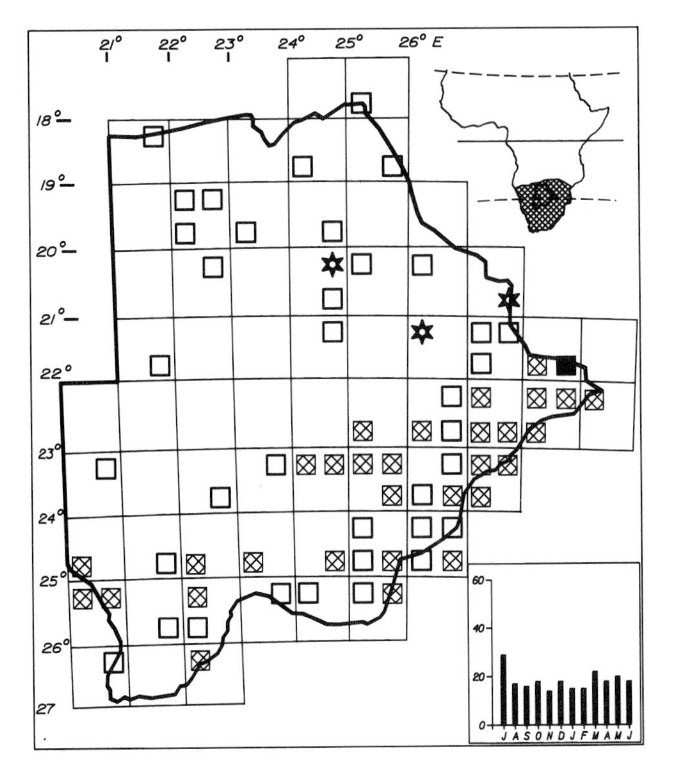

Status

Uncommon to fairly common resident of the east and southeast where viable breeding colonies still exist. It is a species in decline, regarded as seriously endangered and in need of protection. Nonbreeding adults wander to the west and into Zimbabwe, Transvaal and Cape Province. Few, mainly juveniles, migrate into northern Botswana and rarely southern Zambia. Sparse north of 21°S. Egglaying April to July. Usually in small groups with or without other vultures, in colonies at breeding sites.

Habitat

Mountainous and hilly country for breeding. Breeds on cliffs and steep faces. Forages over a wide range of savanna, wooded hills, farmland, and semidesert.

Analysis

71 squares (31%) Total count 240 (0,13%)

WHITEBACKED VULTURE

Gyps africanus

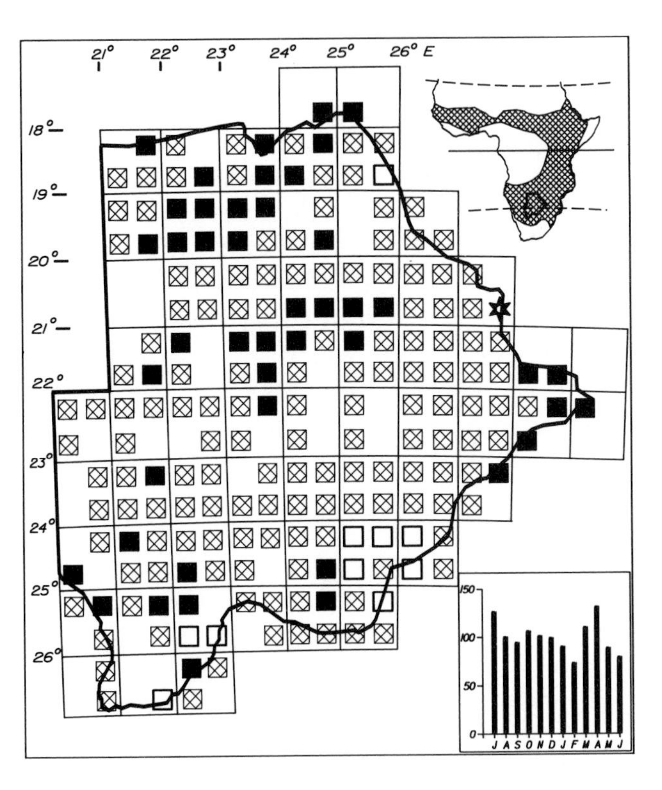

Status

Common to very common resident throughout except in densely populated areas. Commonest in well-wooded regions. A proficient and successful scavenger. Usually the commonest vulture at a carcass. Nests and roosts in trees. Egglaying May to July. Solitary or in small groups when soaring, in groups of up to 50 birds at a carcass.

Habitat

Tree and bush savanna, woodland, farmland. Forages over any habitat including towns where it sometimes congregates near abattoirs.

Analysis

197 squares (86%) Total count 1222 (0,68%)

LAPPETFACED VULTURE

Torgos tracheliotus

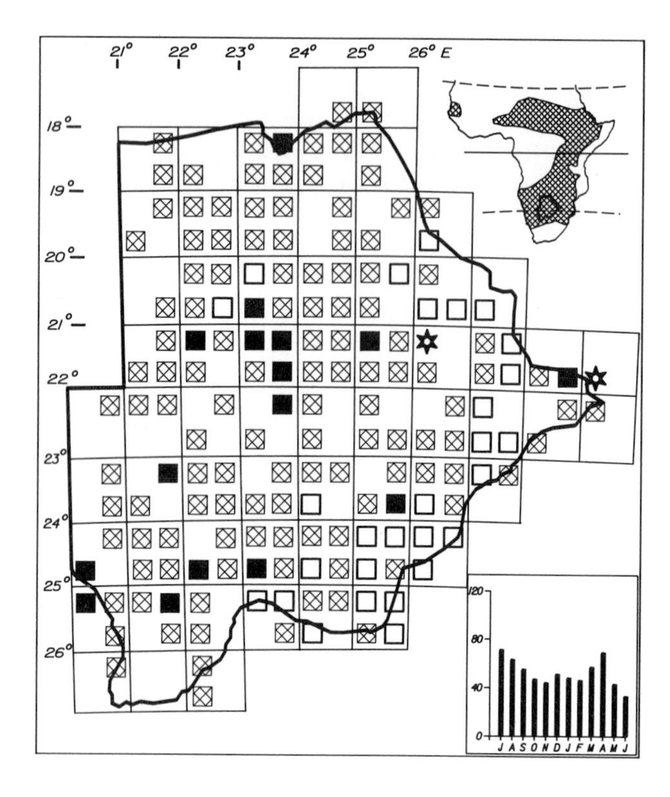

Status

Common to very common resident throughout most of the country—less common in the east. Dominates other vultures at carcasses by virtue of its large size. Egglaying all months May to August. Usually solitary or in very small numbers in flight and is less gregarious than the Whitebacked Vulture.

Habitat

Tree and bush savanna, woodland, farmland. In moist and arid conditions like the Whitebacked Vulture.

Analysis

166 squares (72%) Total count 701 (0,39%)

WHITEHEADED VULTURE

Trigonoceps occipitalis

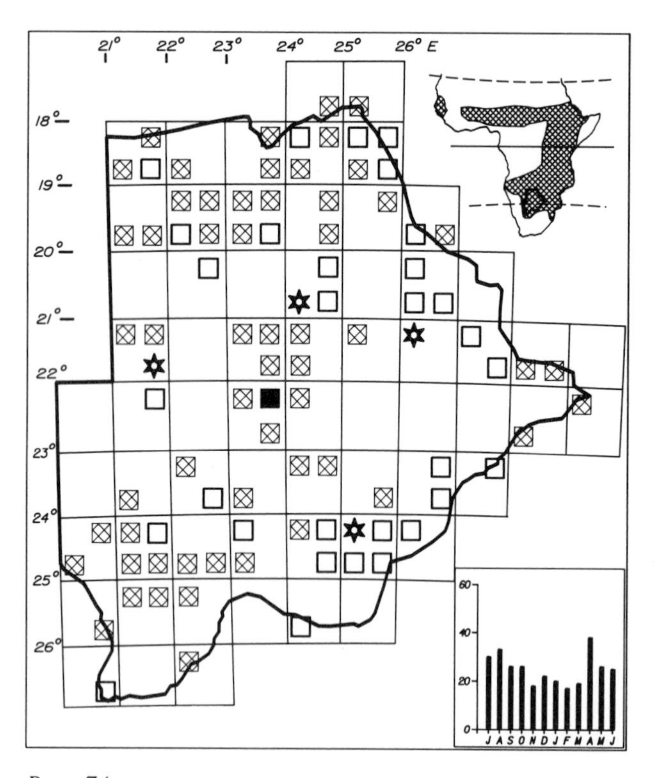

Status

Sparse to fairly common resident throughout but is unpredictable. Apparently holds large breeding territories. Current distribution suggests that it is commonest in the least-inhabited areas such as the central Kalahari Game Reserve and Kalahari Gemsbok National Park. Egglaying August. Mostly solitary and often in the company of other vultures.

Habitat

Any woodland, tree and bush savanna. Infrequently near habitation and cultivation.

Analysis

93 squares (40%) Total count 333 (0,19%)

BLACK KITE

Milvus migrans migrans

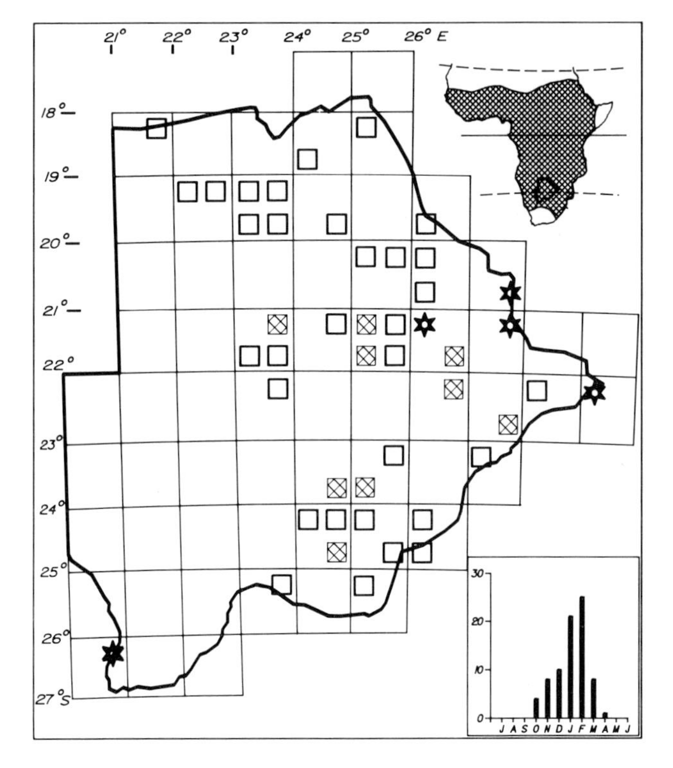

Status

Sparse to uncommon Palaearctic migrant to northern and eastern areas. Recorded once pre-1980 in the Nossob Valley. Recorded between October (4th) and March with one April (2nd) record. Highest numbers occur between December and February and most sightings occur in February coinciding with peak northward passage of birds from South Africa. Easily confused with Yellowbilled Kite particularly when immature. Usually in groups or flocks of 5–40 birds, occasionally with the Yellowbilled Kite.

Habitat

Woodland, tree and bush savanna, dead trees on floodplains or in semidesert scrub or bush savanna. Much less common than the Yellowbilled Kite near human habitation. May occur over any habitat in the north and east.

Analysis

46 squares (20%) Total count 90 (0,05%)

YELLOWBILLED KITE

Milvus migrans parasitus

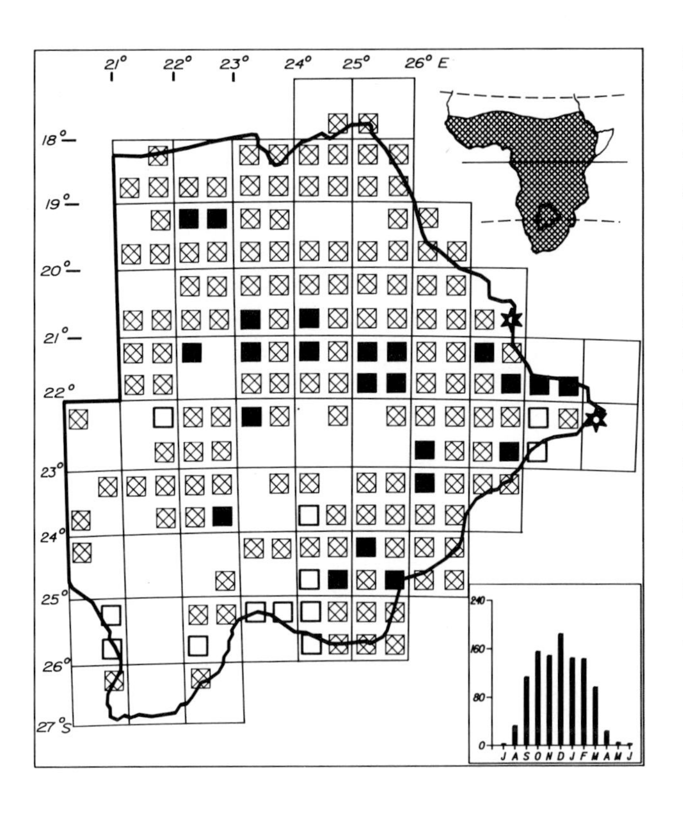

Status

Common to locally very common breeding intra-African migrant in all regions except the southwest where it is mainly very sparse. Generally less common in the west and central areas even near habitation. The first arrivals appear in early August (1st) reaching peak numbers from October to February. Departs March and April with a few birds remaining until early May. Egglaying September to November, mainly October. Usually in pairs.

Habitat

Any type of woodland; more common in broadleafed woodland. Tree and bush savanna, towns, villages, farms and cattle posts. Not as strongly associated with human habitation in Botswana as it is in the northern parts of its African range. Frequently forages for carrion on roads and perches on telegraph poles. Nests inside the canopy of medium to large trees.

Analysis

168 squares (73%) Total count 1092 (0,61%)

BLACKSHOULDERED KITE

Elanus caeruleus

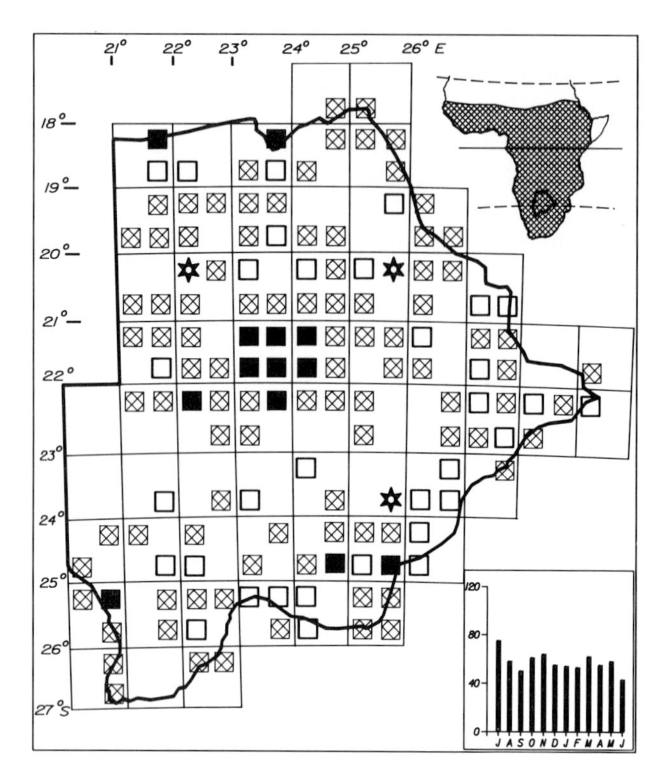

Status

Fairly common to common resident in all regions. Widespread but in some areas thinly distributed or temporarily absent. May be locally very common, as at Deception Valley and parts of the Okwa Valley. Distribution depends on the availability of rodents which constitute 90% of its diet. Numbers may fluctuate from year to year. Occasionally very localised—several pairs have been found breeding in one locality and none of this species seen in any nearby area. Egglaying March and May. Usually solitary or in pairs, occasionally gathers in flocks.

Habitat

Open tree and bush savanna, edges of woodland, semidesert bush and scrub savanna, fossil valleys, edges of pans, floodplains and plains. Quarries its prey by hovering and hunts on the wing or by perching on exposed trees, posts and telegraph poles.

Analysis

143 squares (62%) Total count 714 (0,39%)

CUCKOO HAWK

Aviceda cuculoides

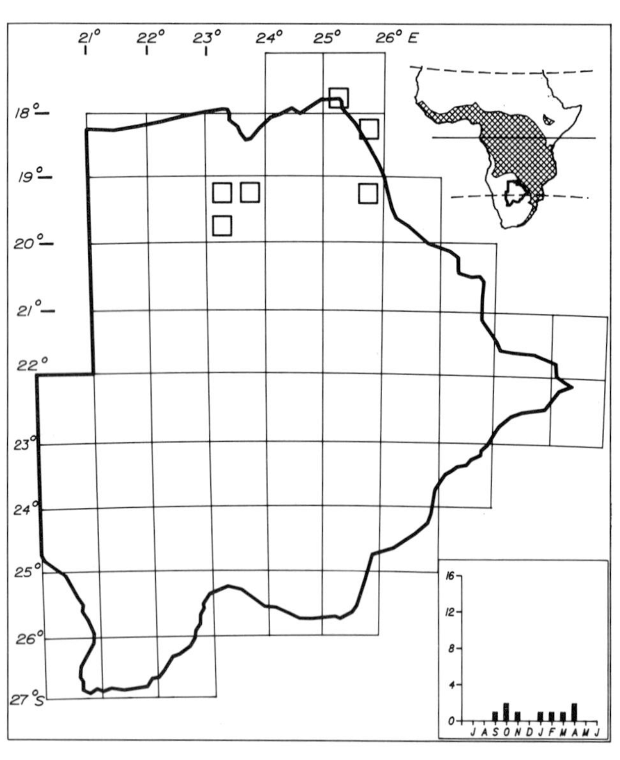

Status

Very sparse resident in the north and northeast whose true status is not known. Recorded in most months between September and April which is the breeding season in Zimbabwe. It may be an overlooked resident or a scarce breeding visitor in some years. A record from 2126A in February 1974 has not been substantiated. Occurs in Botswana at the western limit of its range. Solitary.

Habitat

Mature broadleafed woodland in the high rainfall areas.

Analysis

6 squares (3%) Total count 8 (0,004%)

BAT HAWK

129

Macheiramphus alcinus

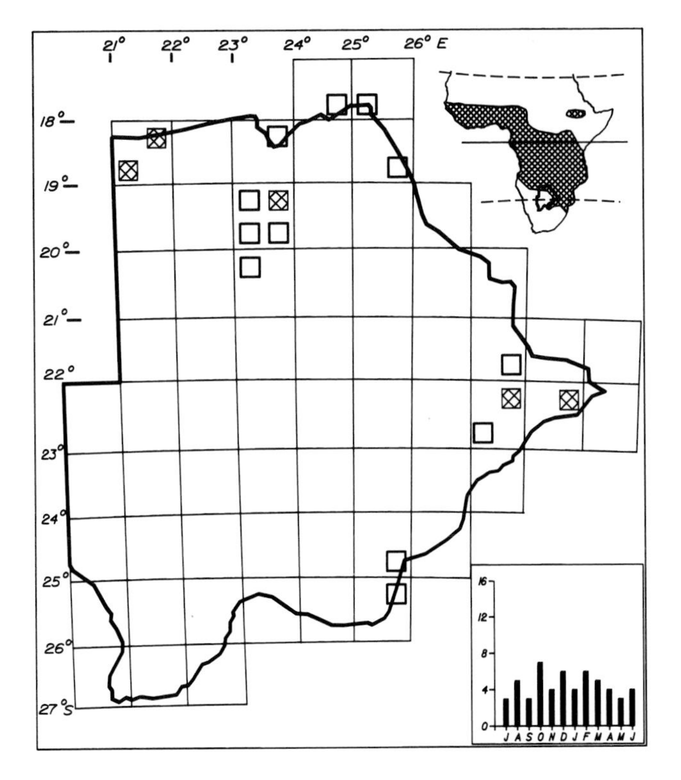

Status

Sparse to uncommon resident in widely scattered localities in the north and east. Most regularly reported from Shakawe and Maun. It is easily overlooked unless its presence is known or suspected. Usually seen at dusk. Poorly known and merits special study in Botswana. Usually solitary.

Habitat

Large trees in riparian forest or dense woodland. Hunts at dusk over open savanna or water. Small bats are usually resident within its territory.

Analysis

16 squares (7%) Total count 56 (0,03%)

BLACK EAGLE

131

Aquila verreauxii

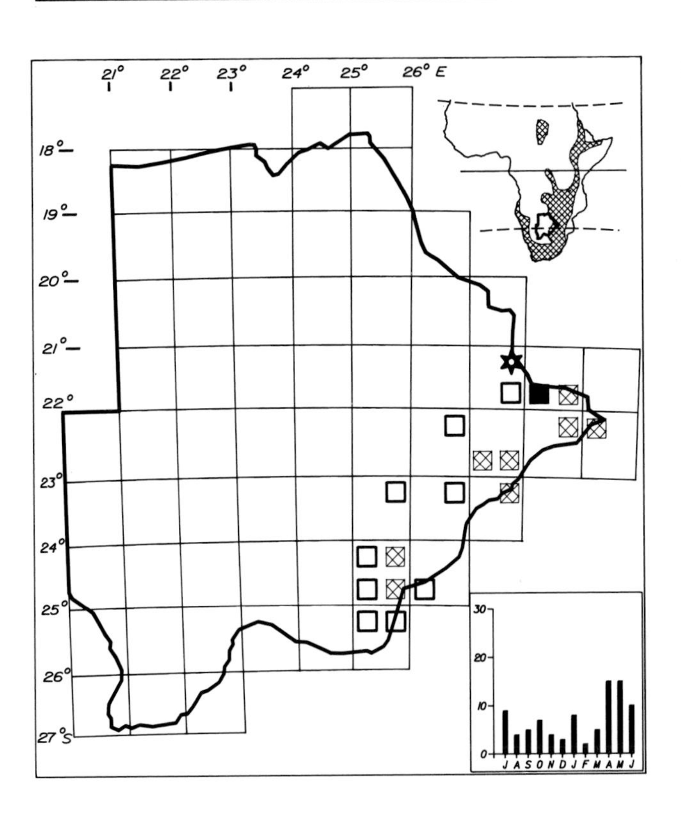

Status

Sparse to locally common resident in hilly areas of the east and southeast. In the southeast it is sparse to fairly common in the granite hills around Kanye, Lobatse, Gaborone and Molepolole. In the east it is common in hills around Bobonong and Selebi Phikwe and in the Tswapong Hills. Recorded once at Kasane and once at Gomare although it has not been seen at the nearby Tsodilo Hills. Usually solitary or in pairs.

Habitat

Hills and mountains with cliffs, gorges or large boulders. Old nests have been found even on small cliff faces. Forages over rocky areas for its main prey, the rock hyrax, and over savanna and farmland for other small mammals and game birds.

Analysis

21 squares (9%) Total count 98 (0,05%)

TAWNY EAGLE

132

Aquila rapax

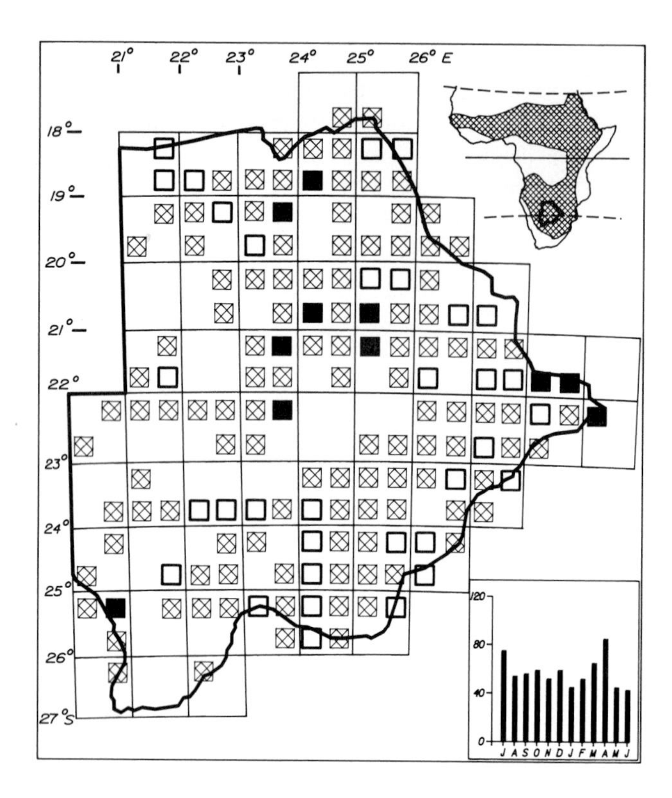

Status

Fairly common to common resident in all regions of the country. Very common in some areas such as Savuti, eastern Moremi, western Makgadikgadi, Deception Valley, northern Tuli Block and Nossob Valley all of which areas are rich in game animals. Egglaying April, June and July. Solitary or in pairs.

Habitat

Broadleafed and *Acacia* woodland and tree savanna, semidesert bush savanna, farmland. Nests in tall trees usually in the crown. Perches often on the top of trees.

Analysis

155 squares (67%) Total count 735 (0,41%)

STEPPE EAGLE

133

Aquila nipalensis

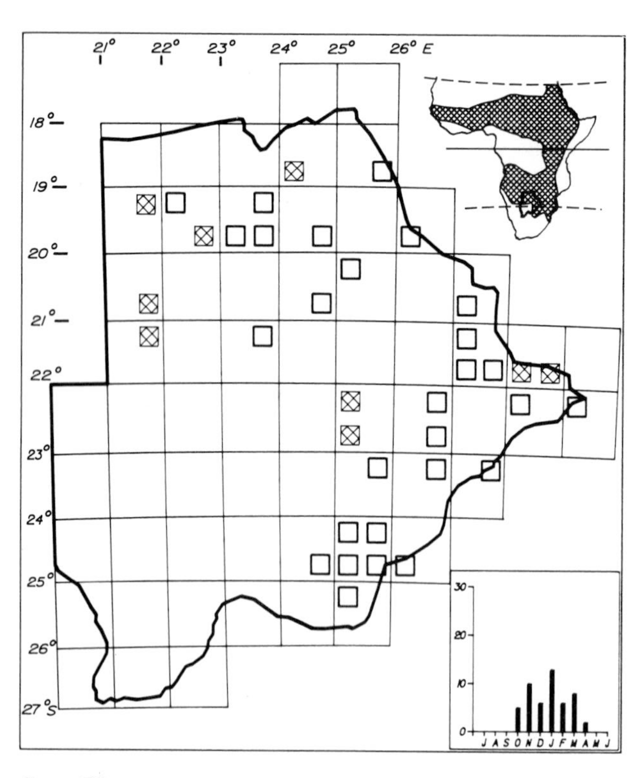

Status

Sparse to uncommon Palaearctic migrant to the northern and eastern regions of the country. Recorded between October (6th) and April (2nd). There are insufficient records to draw conclusions on movement patterns. Irwin (1981) suggests that Zimbabwean birds move west into Botswana from February onwards. A flock of 50 birds at Mashatu (2229A) on 27 January 1986 appears to lend support to the above. Usually in small numbers of 1–5 birds, but larger flocks are occasionally seen.

Habitat

Edges of woodland, open tree savanna, semidesert bush savanna. Its habitat is very similar to that of the Tawny Eagle to which it is closely related and with whose young it is sometimes confused. The two species occasionally occur together particularly at carcasses. It is found on the ground more frequently than the Tawny Eagle.

Analysis

37 squares (16%) Total count 53 (0,029%)

LESSER SPOTTED EAGLE

134

Aquila pomarina

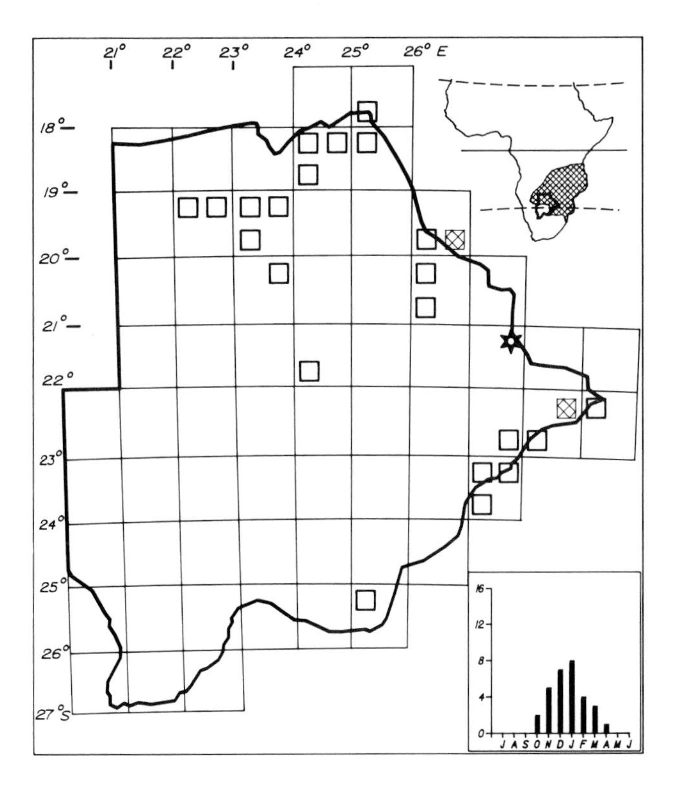

Status

Very sparse to uncommon Palaearctic migrant to wooded areas of the north and east. Some birds may wander south of the Tropic of Capricorn in some years as they do rarely into the adjacent Western Transvaal. Its range in Botswana represents the southern limit of its migration at these longitudes. Easily overlooked and may be more common than reported in the northeastern woodlands. Usually solitary, but small loose groups can occur.

Habitat

Mature broadleafed woodland and tree savanna. Occasionally on the edge of riparian forest. Its current distribution suggests that it occurs mainly in the major river valleys.

Analysis

25 squares (11%) Total count 34 (0,018%)

WAHLBERG'S EAGLE

135

Aquila wahlbergi

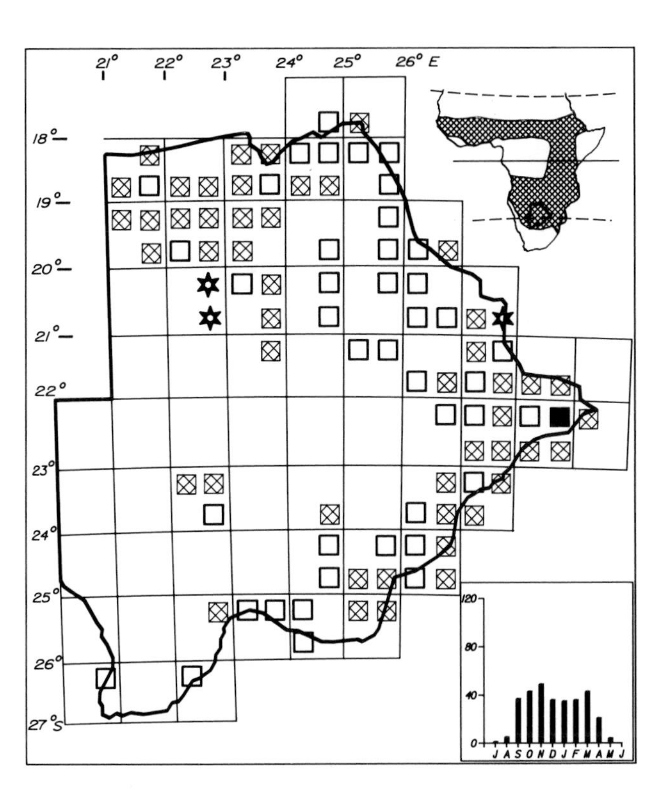

Status

Sparse to locally common breeding intra-African migrant in the north and east. Extends westward uncommonly into Kalahari woodland near the Kutse Game Reserve, along the Molopo River and into the western woodlands around Kang. It may be found more regularly in these latter regions in high rainfall years. Present mainly from September to March with early arrivals in August (23rd) and late departures in May (11th). Egglaying October. Solitary or in pairs.

Habitat

Any woodland but mainly broadleafed. Usually in areas with tall mature trees as occurs along river valleys, drainage lines, near pans and in some Kalahari woodland patches. Also occurs in tree savanna and around human habitation, such as farms, towns and villages, where there are large trees.

Analysis

95 squares (41%) Total count 341 (0,19%)

BOOTED EAGLE

Hieraaetus pennatus

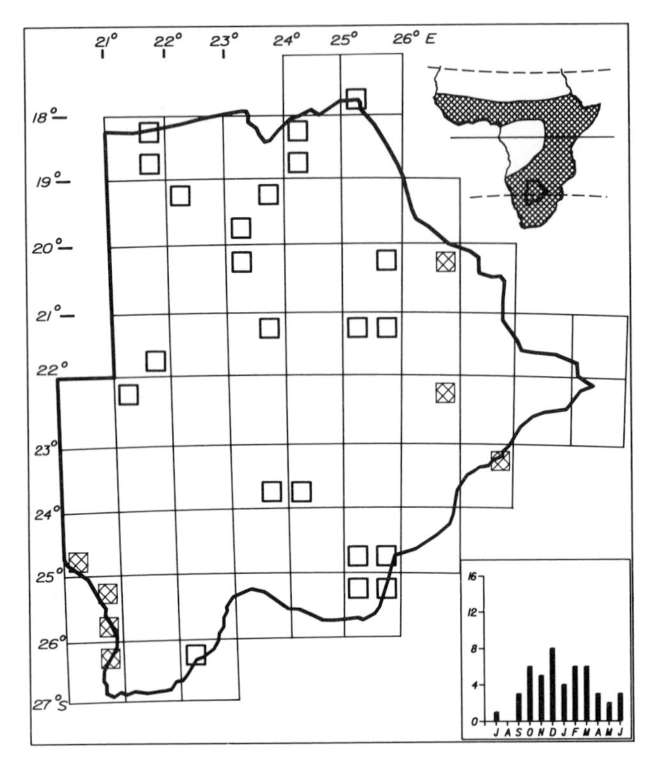

Status

Sparse Palaearctic migrant in the north and east. In the southwest, particularly in the Nossob Valley, birds of the South African breeding population are present throughout the winter breeding season and are thought to breed there. Isolated winter records have also been reported from Nxamaseri, Serowe, Kutse Game Reserve and Kanye. Based on dates Palaearctic birds, from October to March, occur as far south as Tshabong and Lobatse. It appears as if both populations may occur anywhere in the country, very sparsely and at different times of the year. Solitary and in pairs in the southwest, solitary or in loose groups elsewhere.

Habitat

Palaearctic birds occur over a wide variety of habitat, mainly woodland and tree savanna in high rainfall areas. Southern breeding birds occur in arid tree and bush savanna including bare sand dunes but are mainly concentrated in winter in the dry Nossob River valley.

Analysis

29 squares (13%) Total count 54 (0,03%)

AFRICAN HAWK EAGLE

Hieraaetus spilogaster

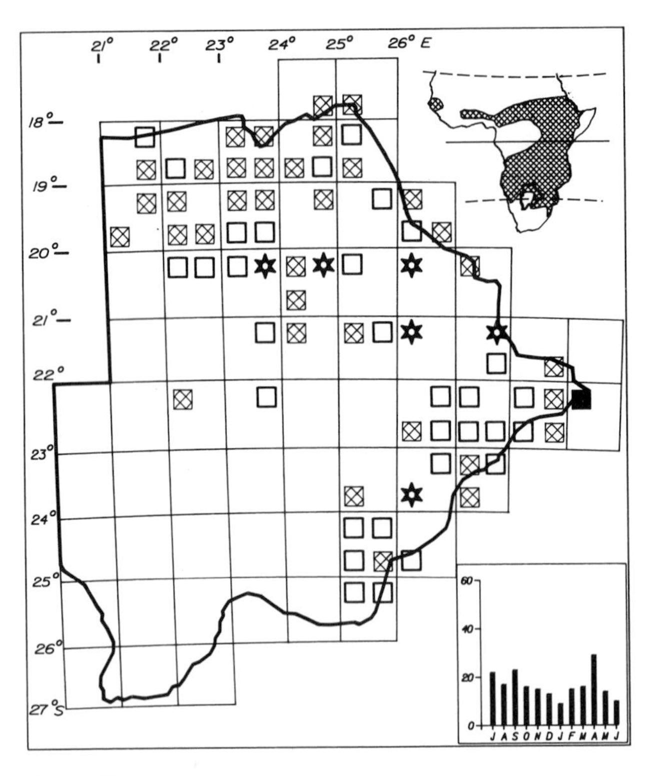

Status

Fairly common to common resident north of 20°S. Elsewhere in the north, east and southeast it is sparse to fairly common. Occurs in Botswana at the southern limit of its continental range at these longitudes. Solitary or in pairs.

Habitat

Woodland and tree savanna mainly in high rainfall areas. Sparse in semidesert *Acacia* savanna. Often found in hilly areas, near isolated hills and along river valleys.

Analysis

73 squares (32%) Total count 225 (0,125%)

AYRES' EAGLE

Hieraaetus ayresii

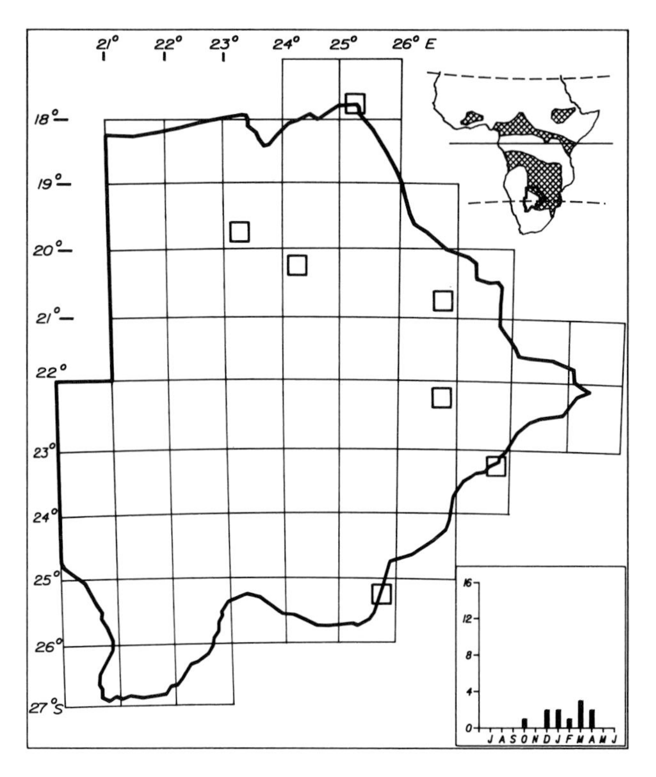

Status

Very scarce. Status uncertain. Its occurrence in Botswana may only be the western edge of a migration of the species near the southern limit of its range in Africa. All records fall between December and March. There is no evidence of residence or of breeding although immature birds have been recorded. It is a species which is poorly known and whose movements are little understood in many areas of Africa. Usually solitary.

Habitat

Mature dense woodland, wooded rocky hills, edges of riparian forest.

Analysis

7 squares (3%) Total count 12 (0,006%)

LONGCRESTED EAGLE

Lophaetus occipitalis

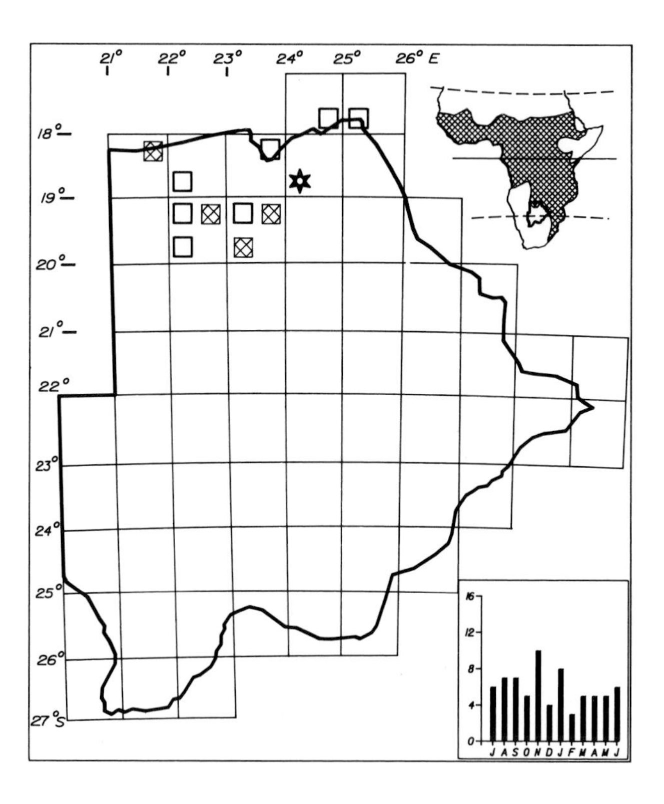

Status

Sparse to locally fairly common resident of the Okavango, Linyanti and Chobe river systems. Most frequently recorded at Shakawe and Maun. The total population of this species in Botswana is unlikely to be more than 200 pairs. Not well known and warrants special study. It may be vulnerable as a species of restricted range and habitat in Botswana. Occurs at the southern limit of its continental range at these longitudes. Solitary or in pairs.

Habitat

Edges of woodland particularly those bordering floodplains and marshes, tree savanna in river valleys. Often perches prominently on dead trees or posts in open areas in search of its rodent prey.

Analysis

12 squares (5%) Total count 76 (0,04%)

MARTIAL EAGLE

Polemaetus bellicosus

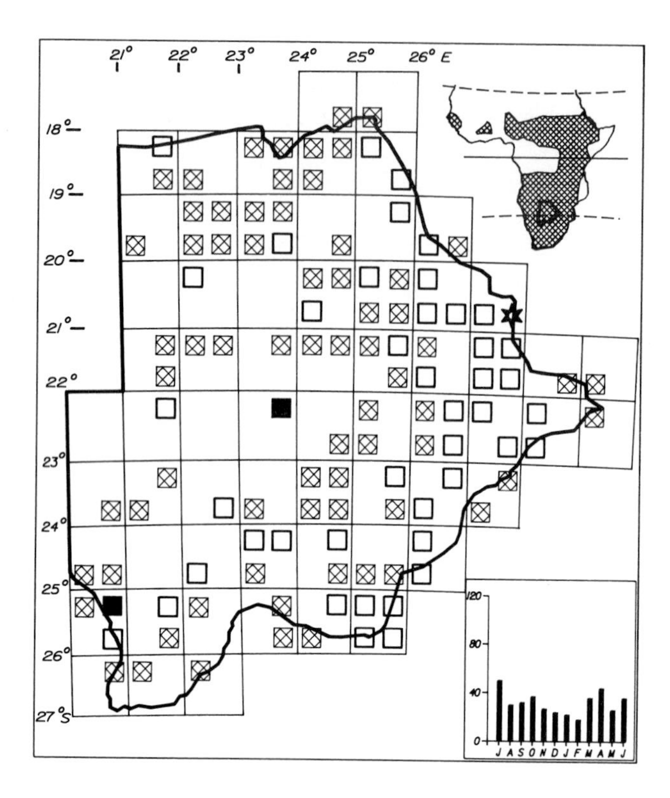

Status

Sparse to fairly common resident throughout the country. Recorded as common in widely scattered localities—Kasane, Savuti Marsh, Maun, Deception Valley, eastern Okwa Valley, Jwaneng and Nossob Camp—generally unpredictable. Egglaying in all months from May to August. Solitary or in pairs.

Habitat

Open woodland, tree and bush savanna, grassland, and quite frequently at Kalahari pans. Hunts a wide variety of prey in open country—guineafowl, francolins and small mammals.

Analysis

115 squares (50%) Total count 411 (0,23%)

BROWN SNAKE EAGLE

Circaetus cinereus

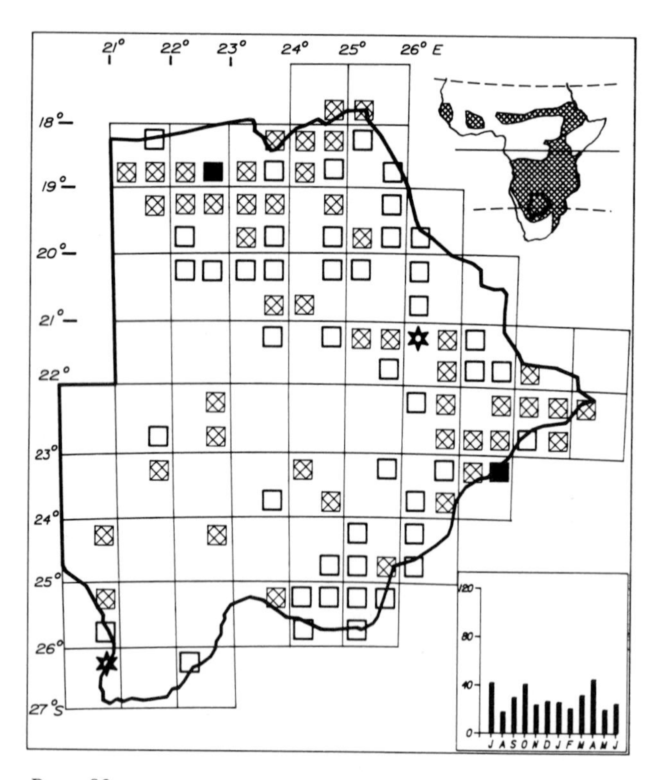

Status

Sparse to common resident mainly in the north and east. In central and western areas it is thinly distributed but may be more common in wooded areas than current records indicate. Most frequently recorded in the Okavango, Linyanti and Chobe regions of the north and in the Tuli Block in the east. Egglaying November. Usually solitary.

Habitat

Any woodland including pure *Acacia* and mopane but commonest in broadleafed and mixed *Acacia* woodlands. Also in tree and bush savanna in high rainfall regions. In semidesert areas occurs near mature woodland particularly along river valleys and near pans. Less common in open savanna away from woodland than the Blackbreasted Snake Eagle.

Analysis

95 squares (41%) Total count 375 (0,21%)

BLACKBREASTED SNAKE EAGLE 143

Circaetus pectoralis

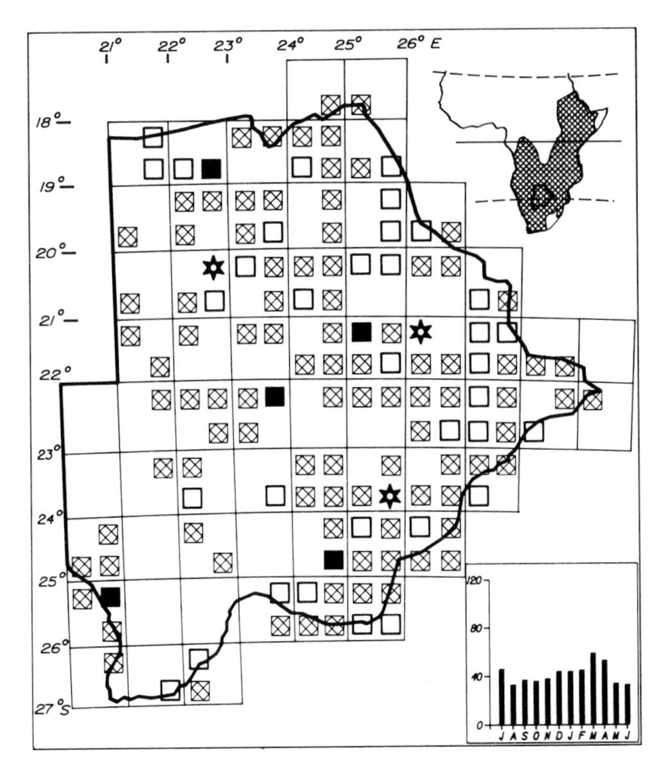

Status

Fairly common to common resident in all regions. It appears to be more thinly distributed in central and western regions but this may be due to inadequate coverage in this survey. It is common to very common in open areas such as along major fossil valleys (Deception and Okwa), the western parts of the Makgadikgadi Pans, the floodplains of the major river systems of the north, grassland plains at Serowe and Jwaneng and in the Nossob Valley. Usually solitary.

Habitat

Open tree and bush savanna, semidesert bush and scrub savanna, grassland plains, floodplains, grassy pans, farmland. Quarters its prey by hovering or from a prominent perch.

Analysis

136 squares (59%) Total count 550 (0,31%)

WESTERN BANDED SNAKE EAGLE 145

Circaetus cinerascens

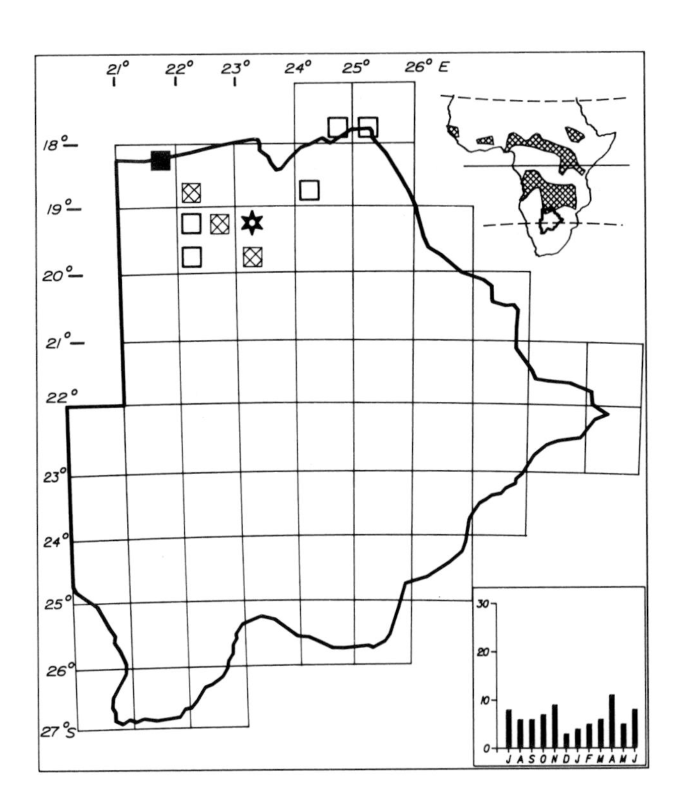

Status

Sparse to locally common resident of the Okavango Delta, Savuti and Chobe regions. Commonest at Shakawe and Chief's Island areas of the Okavango Delta. Occurs in Botswana at the extreme southern limit of its continental range. It is suspected of breeding in this area but, as elsewhere in its range, breeding evidence is difficult to obtain. Usually solitary.

Habitat

Riverine forest, dense tall woodland on the edges of swamps and marshes.

Analysis

10 squares (4%) Total count 80 (0,04%)

BATELEUR

146

Terathopius ecaudatus

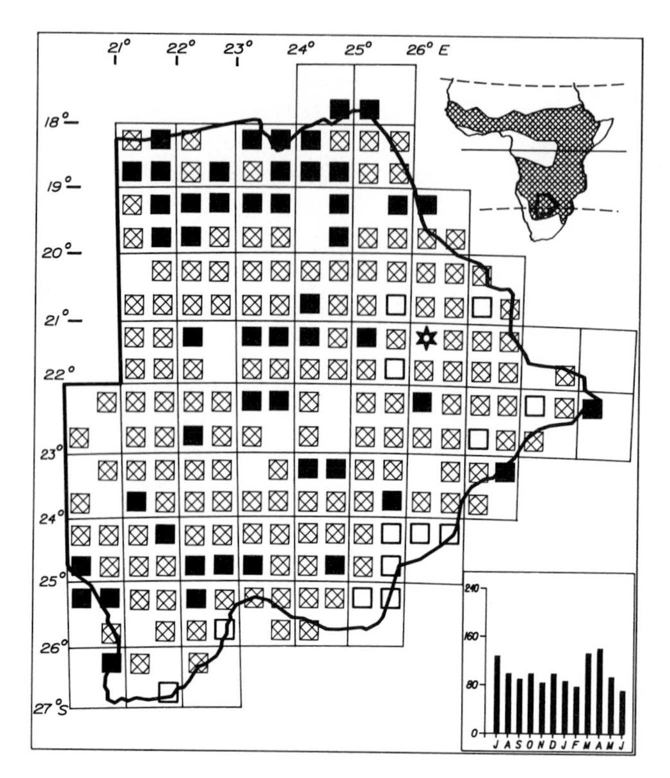

Status

Common to very common resident throughout the country. It is commonest in the game areas of the north. It is more common in Botswana than in any neighbouring country where pesticides are thought to be the main cause of its decline in the past decade. Its predilection for carcasses is thought to be a contributory factor in its being vulnerable to pesticide poisons, such as cattle dip and poison bait for vermin and predators. There may be early evidence of a decline in eastern Botswana where it is sparse in some areas. Local movements occur. Some immature birds congregate temporarily in areas such as the Nossob Valley and Savuti after leaving their natal territory. Egglaying February, May and December. Solitary or in pairs.

Habitat

Any woodland in high and low rainfall regions, tree and bush savanna. Forages over open areas including grass plains and tarred roads for live prey and carrion.

Analysis

203 squares (88%) Total count 1233 (0,68%)

AFRICAN FISH EAGLE

148

Haliaeetus vocifer

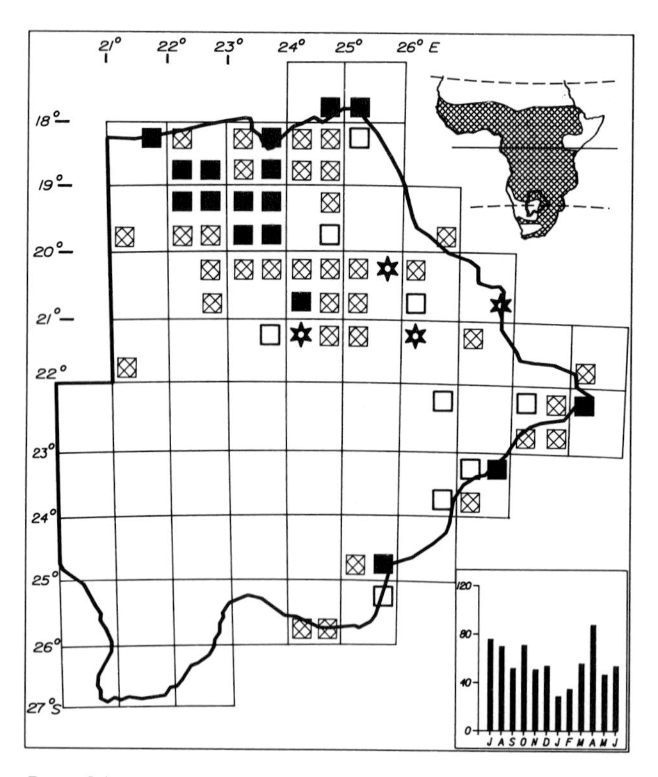

Status

Very common resident of the Okavango, Linyanti and Chobe rivers and near the confluence of the Shashe and Limpopo rivers. Elsewhere it may be locally common on some waters e.g. along the Boteti River to Mopipi Dam, at Gaborone Dam and along some stretches of the Limpopo River. Local movements occur to temporarily suitable habitat e.g. Orapa, Francistown, Lobatse and along the Molopo River. Recorded once at a roadside pool west of Ghanzi. It is not common in the Makgadikgadi region except near the Boteti and Nata rivers. Egglaying all months May to July. Solitary or in pairs.

Habitat

Rivers, lakes, lagoons and dams with fringing large trees or riparian forest. Resident only in areas with permanent water. Nomadic elsewhere.

Analysis

64 squares (28%) Total count 702 (0,39%)

STEPPE BUZZARD

Buteo buteo

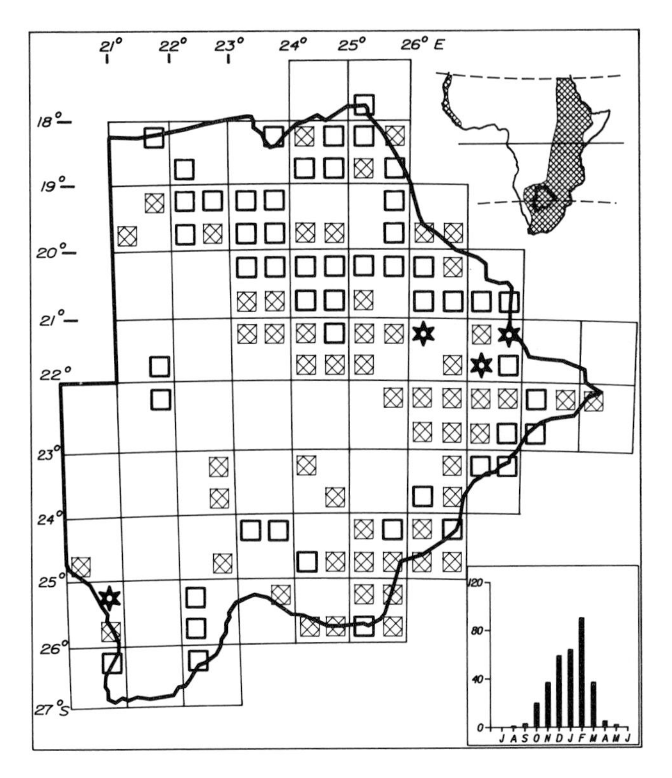

Status

Sparse to fairly common Palaearctic migrant mainly to the north and east but occurring in all regions. Its main arrival is in October (earliest date 23 August) reaching peak numbers between December and February. Departs in March and April with very few birds remaining until May. February numbers are augmented by northward migration of birds wintering further south. Usually solitary, also occurs in small groups on northward migration.

Habitat

Tree and bush savanna in high and low rainfall regions. Also on the edges of woodland and on grass plains with a few scattered trees or bushes. Perches on trees, particularly dead trees or on dead branches, and frequently on telegraph poles along main roads and on fence posts.

Analysis

110 squares (48%) Total count 337 (0,18%)

JACKAL BUZZARD

Buteo rufofuscus

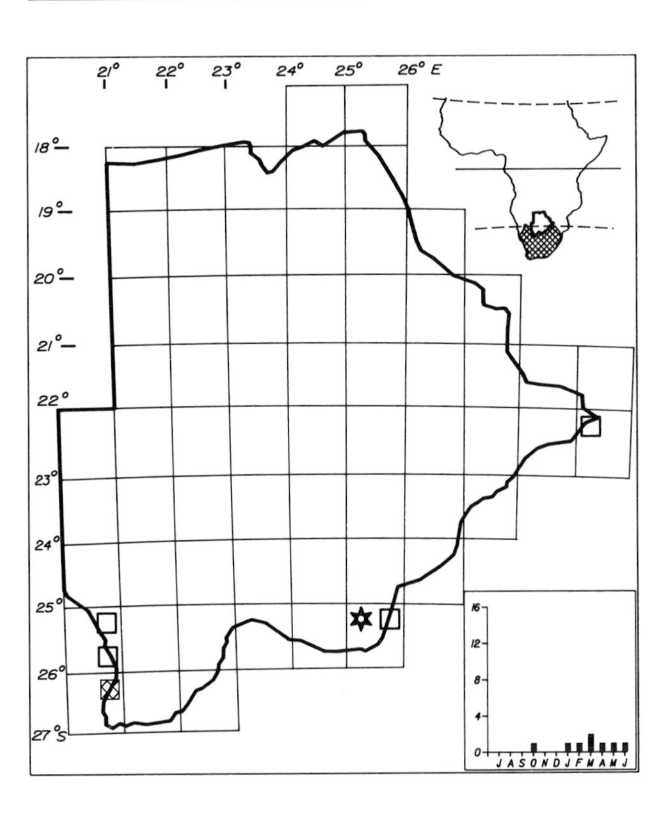

Status

Sparse vagrant in the southeast and southwest corners of the country. There are five records from Lobatse and four from the Nossob Valley. A specimen was collected from Kanye prior to this survey. Both regions are at the periphery of its known range in South Africa and have suitable mountainous or escarpment areas; it may occur more regularly than recorded here. There is one record from Pont's Drift in the east. Most records fall between January and June. Poorly known and merits special study.

Habitat

Hills and mountains in the southeast, the escarpment of the Nossob Valley and duneland in the southwest, grassland slopes and rock outcrops in open semiarid savanna.

Analysis

6 squares (3%) Total count 11 (0,006%)

LIZARD BUZZARD

154

Kaupifalco monogrammicus

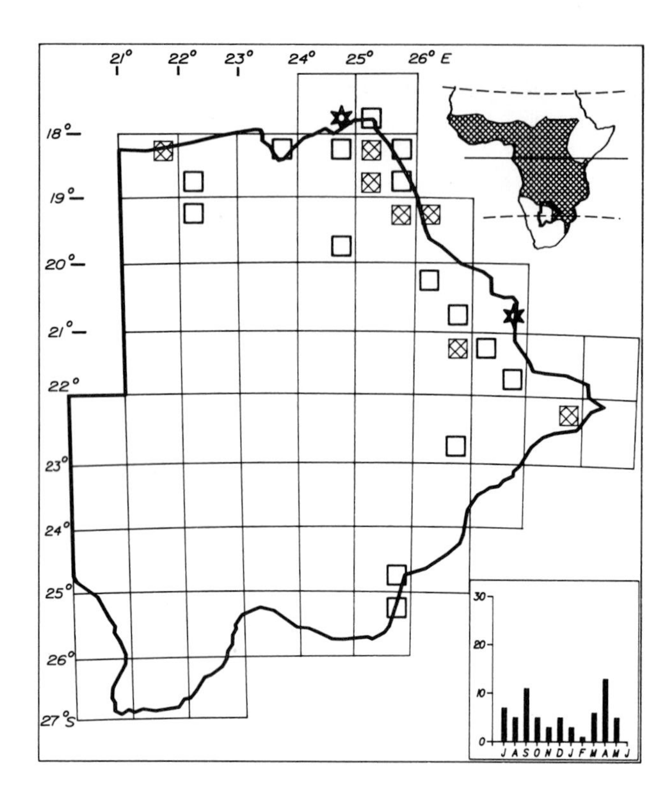

Status

Sparse to locally fairly common resident in the north and east. Rare south of 23°S where it has been recorded only at Gaborone and Lobatse. Most records are from Shakawe. It is fairly common in the northeastern woodlands and is likely to be more common than recorded in the far north. Not well known in Botswana. Solitary or in pairs.

Habitat

Broadleafed woodland and tree savanna. Mainly in well-wooded areas particularly miombo and *Baikiaea* but extends into mixed woodland and riverine bush.

Analysis

24 squares (10%) Total count 66 (0,04%)

OVAMBO SPARROWHAWK

156

Accipiter ovampensis

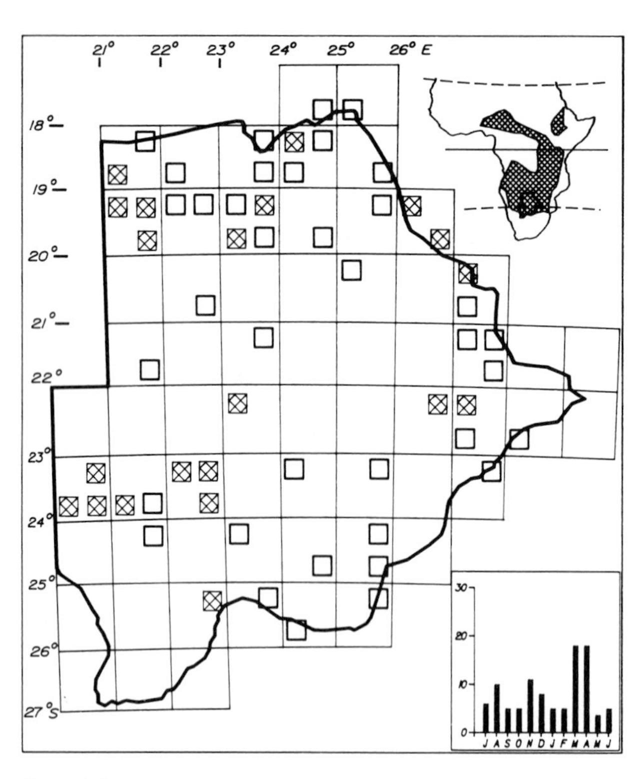

Status

Sparse to fairly common resident in all regions except the southwest where it is absent. Nowhere common but it is more common in the northwestern and western woodlands than in the east. Likely to be under recorded because of fear of confusion with other accipiters. Its true status warrants further study. Egglaying November. Solitary or in pairs.

Habitat

Acacia and mixed woodland on Kalahari sand. Also in tree savanna and in woodland along river valleys.

Analysis

58 squares (25%) Total count 109 (0,06%)

LITTLE SPARROWHAWK

Accipiter minullus

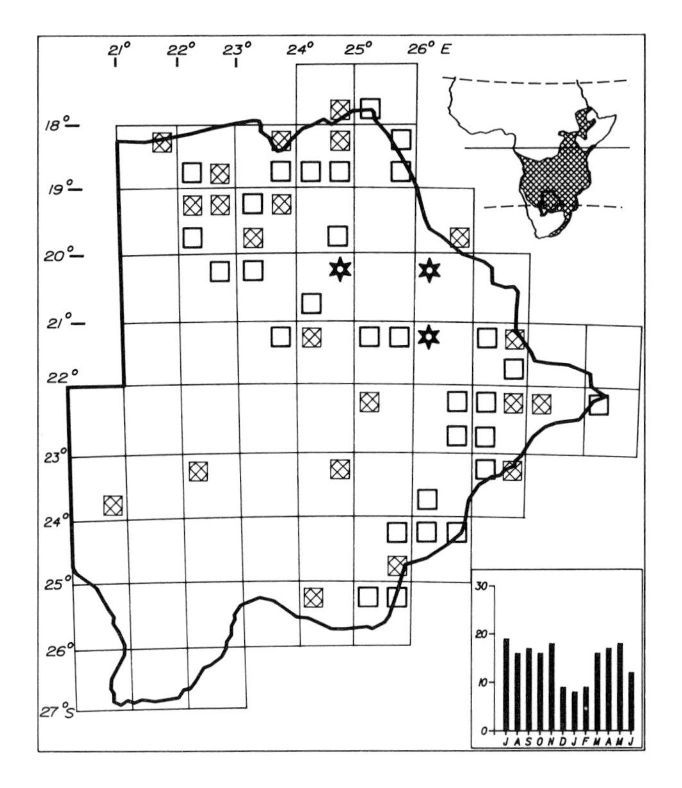

Status
Sparse to fairly common resident mainly in the north and east. Occurs patchily in central and midwestern areas but is absent from the southwest. Egglaying September and November. Solitary or in pairs.

Habitat
Broadleafed and *Acacia* woodland, edges of riparian forest, riverine bush, *Acacia* on alluvial soils, tree savanna and woodland in river valleys. Usually where there is dense bush and thicket in the understorey.

Analysis
54 squares (23%) Total count 190 (0,11%)

BLACK SPARROWHAWK

Accipiter melanoleucus

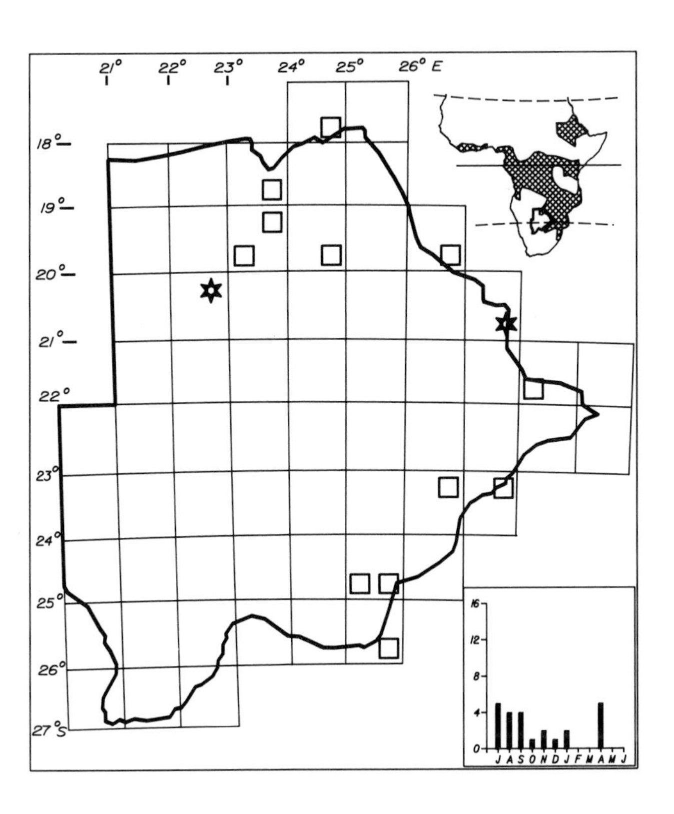

Status
Sparse resident in the north recorded once at each of the localities shown—Kasane, Savuti, eastern Moremi, Maun, Nxai Pan and Tsebanana. Also sparse in the east but recorded on three occasions at Martin's Drift. In the southeast a recent extension of range from the Transvaal into *Eucalyptus* plantations has occurred, and it now appears to be regular and breeding there. Occurs in Botswana at the western limit of its southern African range. Egglaying September. Solitary or in pairs.

Habitat
Riparian forest, dense woodland, plantations with tall trees—particularly *Eucalyptus*.

Analysis
14 squares (6%) Total count 25 (0,014%)

LITTLE BANDED GOSHAWK

Accipiter badius

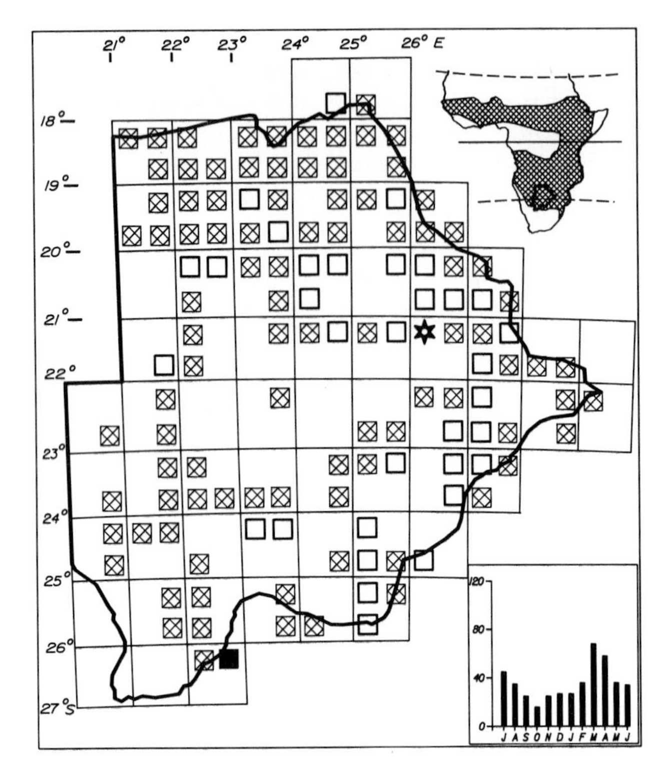

Status

Fairly common to common resident throughout the country except in the southwest corner. It occurs in Botswana at the southern limit of its continental range. It is the commonest of the small true accipiters—less common than the Gabar Goshawk. Egglaying November. Solitary or in pairs.

Habitat

Any woodland, tree and bush savanna and riverine bush, in high and low rainfall regions.

Analysis

128 squares (56%) Total count 452 (0,25%)

AFRICAN GOSHAWK

Accipiter tachiro

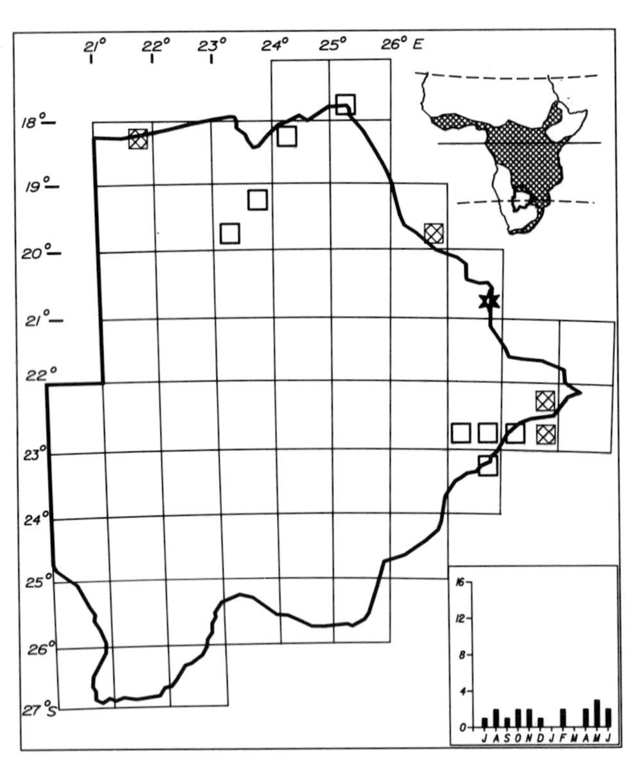

Status

Sparse to uncommon resident in the north and east. Its occurrence in Botswana has been confirmed during this survey—predicted by Smithers (1964) to occur in the east. Most frequently recorded at Shakawe and only recorded once at each of the other localities except Maun. It is present at the southern and western limits of its continental range. Reported throughout the year. Usually solitary.

Habitat

Riparian forest, mature broadleafed woodland in river valleys.

Analysis

13 squares (6%) Total count 23 (0,013%)

GABAR GOSHAWK

Micronisus gabar

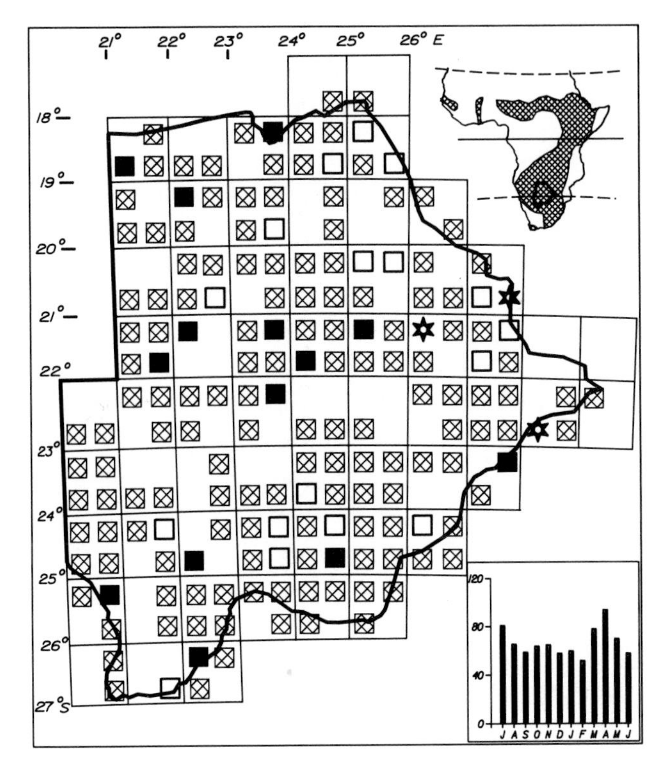

Status

Fairly common to very common resident throughout the country. It is the commonest of the small hawks. Melanistic birds are frequently reported and mixed breeding pairs of the colour morphs have been seen on several occasions. Egg-laying September, October and November. Solitary or in pairs.

Habitat

Any woodland, tree and bush savanna, semidesert scrub savanna with few scattered trees, riverine bush. Also in towns, villages, parks, gardens and cultivation.

Analysis

174 squares (76%) Total count 860 (0,47%)

PALE CHANTING GOSHAWK

Melierax canorus

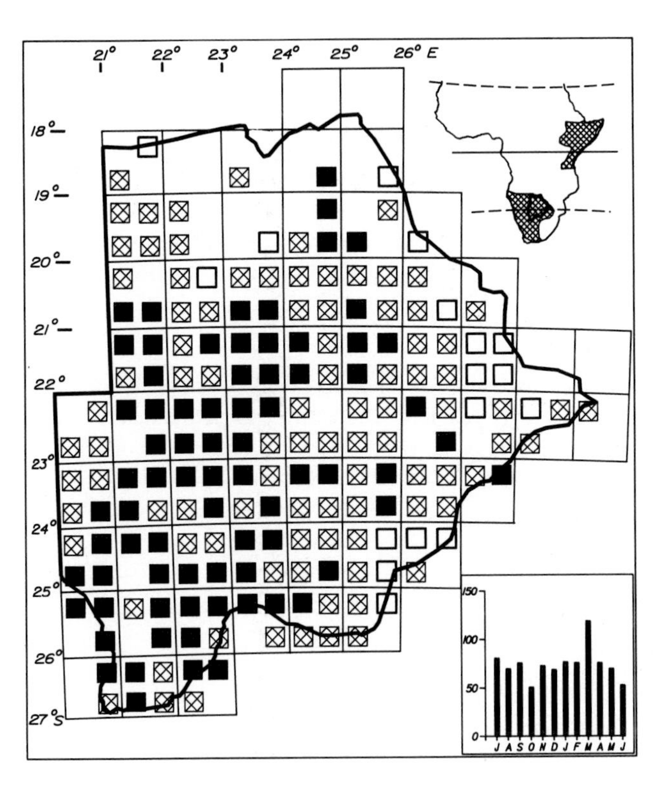

Status

Very common resident south of 20°S and west of 26°E. It is also very common in the Nxai Pan area of the north but mainly absent elsewhere in that region except in the northwestern woodlands where it is fairly common. East of 26°E it is sparse to common. It is a conspicuous and typical species of the Kalahari savannas. It occurs in Botswana at the northern limit of its southern African range at these longitudes. All egglaying records are for September. Solitary or in pairs.

Habitat

Semidesert tree and bush savanna, scrub savanna, open grassland with isolated trees, riverine bush, edges of woodland, farmland, towns and villages. Perches often on dead trees and on telegraph poles but also on live trees particularly in the crown of a tall *Acacia*. Occasionally quarters low over the ground like a harrier and sometimes rests on the ground or walks about in search of prey.

Analysis

185 squares (80%) Total count 979 (0,55%)

DARK CHANTING GOSHAWK

163

Melierax metabates

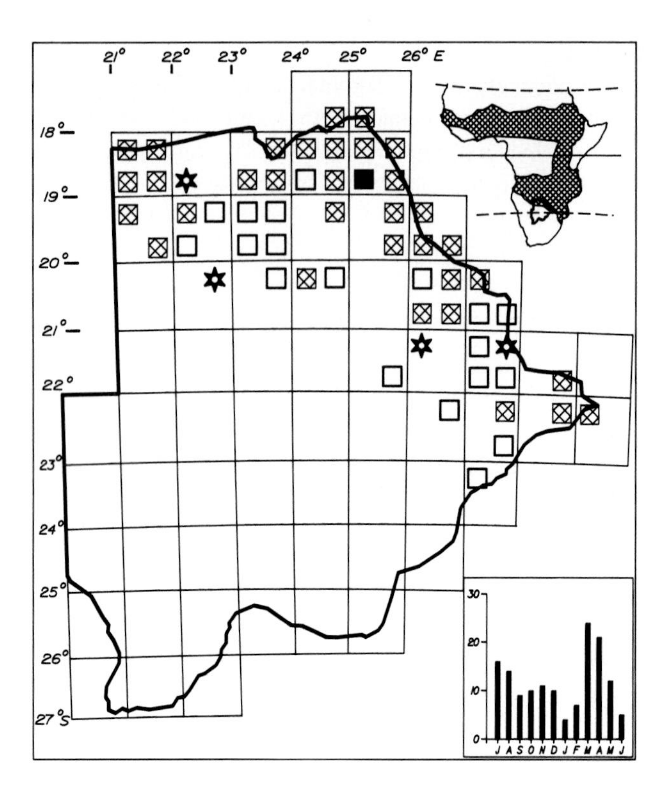

Status

Common resident of the northern and northeastern woodlands to as far south as Makalamabedi and Mosetse. Sparse in the central areas of the Okavango Delta. Also sparse in the areas around Francistown, Selebi Phikwe and Serowe but becomes common again in the northern Tuli Block. Reaches as far south as Machaneng. Its range overlaps that of the Pale Chanting Goshawk mainly in the northwest and around Nata and Francistown but they occur in different habitats. Usually solitary, occasionally in pairs.

Habitat

Broadleafed woodland, tree savanna with tall mature trees.

Analysis

57 squares (25%) Total count 150 (0,08%)

AFRICAN MARSH HARRIER

165

Circus ranivorus

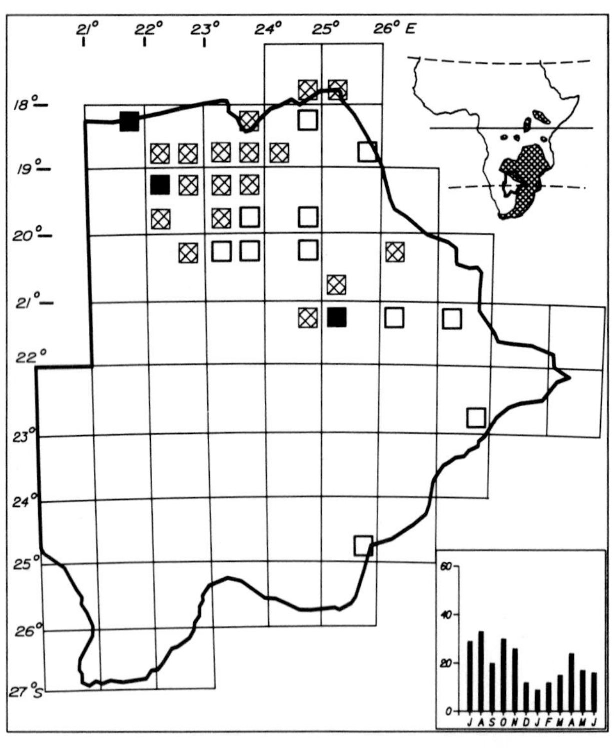

Status

Common to very common resident of the Okavango, Linyanti, Chobe and Boteti river systems south to Mopipi and Orapa. Elsewhere it has been reported very sparsely at isolated marshes in the northwest Makgadikgadi, Nata River mouth, Sua Pan, Shashe Dam, Lerala and Gaborone. It may be more widespread in temporary habitats in high rainfall years. Occurs at the western limit of its southern African range. Egglaying April. Usually solitary or in pairs.

Habitat

Marshes, swamps, floodplains, edges of dams and lagoons. Probably only resident where there is permanent water or sustained reed beds. Hunts over dry or flooded marshland, reedbeds and grassland.

Analysis

31 squares (13%) Total count 251 (0,14%)

MONTAGU'S HARRIER 166

Circus pygargus

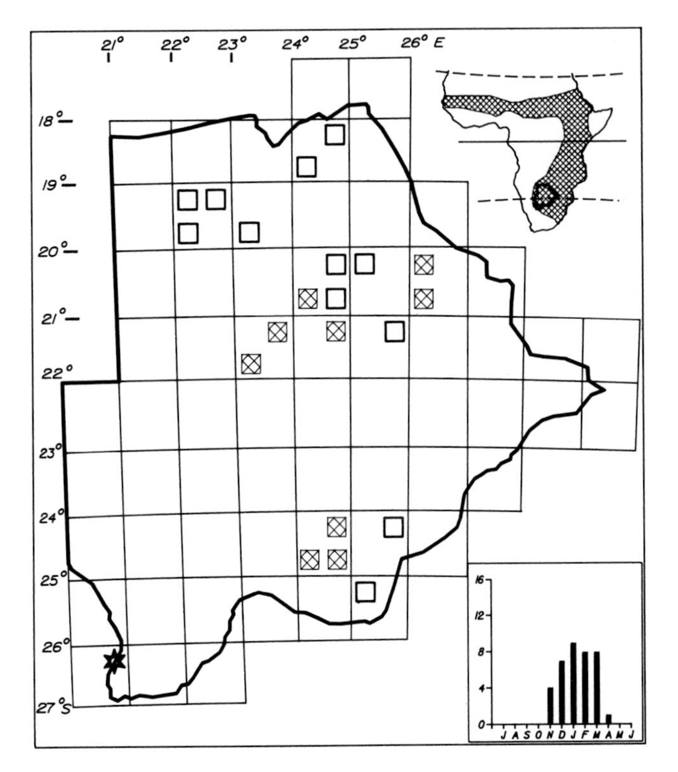

Status

Sparse to uncommon Palaearctic migrant recorded in the north and southeast. Present from November to April but mainly from December to March. At Jwaneng it was not present before December in 1980 or 1981 and this may be the normal pattern for Botswana at least in years when the rains are delayed. It is known to have stayed at one locality for three months. Usually solitary, but several birds of either sex may be present in the same vicinity.

Habitat

Open grassland plains, floodplains, semidesert scrub, fossil valleys, airfields, pans. The map shows that the species has only been recorded in the described habitats—open grassland and pans in the Makgadikgadi region, plains and airfields around Jwaneng and in the southeast, floodplains in the northern wetlands, fossil valleys as at Deception Valley, semidesert scrub south of Makalamabedi and at Lathlike.

Analysis

22 squares (10%) Total count 41 (0,02%)

PALLID HARRIER 167

Circus macrourus

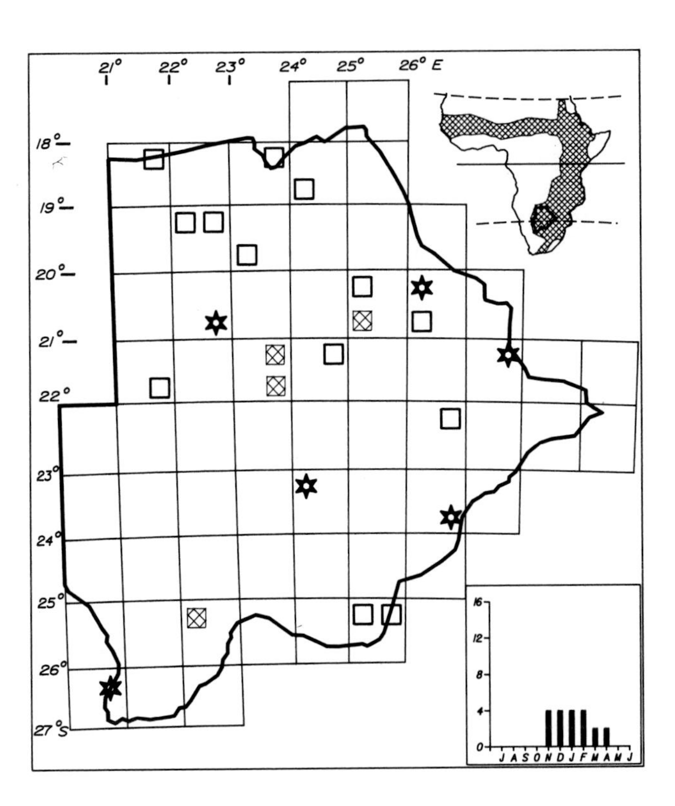

Status

Sparse to uncommon Palaearctic migrant occurring in widely scattered localities throughout the country. Less well known than the Montagu's Harrier, it has been recorded only once at each locality except Shakawe, Deception Valley and Mabuasehube. The distribution suggests that it will tolerate more arid conditions than Montagu's Harrier but they have been recorded together in similar habitat on the Makgadikgadi grassland and at Deception Valley. Usually solitary, but several birds may occur together in the same area.

Habitat

Open grassland, low scrub in fossil valleys, open areas on the edge of woodland, open sandy or stony ground such as on the edges of pans. Appears to utilise bare ground as well as shorter and less dense grass cover than Montagu's Harrier.

Analysis

23 squares (10%) Total count 29 (0,015%)

GYMNOGENE

169

Polyboroides typus

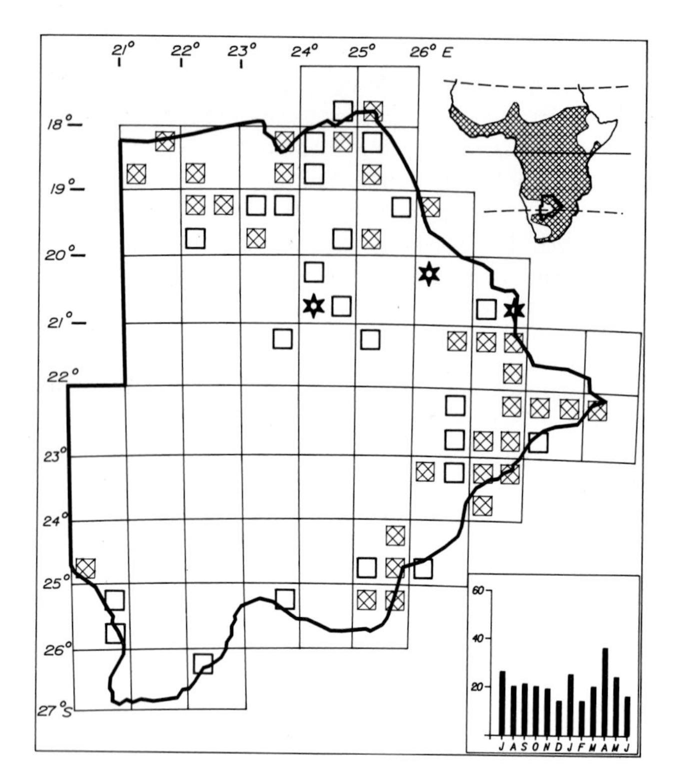

Status

Sparse to fairly common resident of the north and east. It is rare along the Molopo River but reappears in the Nossob Valley. Locally common at Shakawe, Maun and in the east and southeast. In the southeast it extends as far west as Kanye but no further. Absent in the central and western Kalahari except in the southern river valleys. Egglaying September, October and November. Solitary or in pairs.

Habitat

Dense woodland in hilly regions and in river valleys (including fossil valleys). Edges of riparian forest and riverine *Acacia*. It is often found in areas where there are nesting colonies of Ploceid weavers whose nests are raided as part of its diet.

Analysis

59 squares (26%) Total count 270 (0,15%)

OSPREY

170

Pandion haliaetus

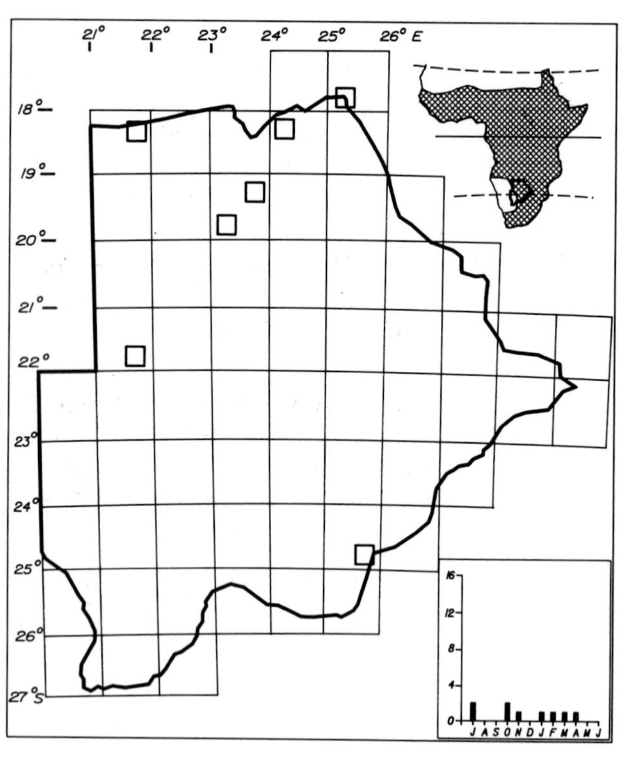

Status

A sparse Palaearctic migrant to the Okavango, Chobe and Zambezi rivers in the north where it may be more common than reported. Recorded between October (21st) and April but two records are for July—Ghanzi and Shakawe. Recorded twice in Gaborone compatible with its irregular occurrence in southern Transvaal. Immatures are known to overwinter in the southern African region in some years. Poorly known in Botswana. Solitary.

Habitat

Rivers and dams.

Analysis

7 squares (3%) Total count 9 (0,005%)

PEREGRINE FALCON

171

Falco peregrinus

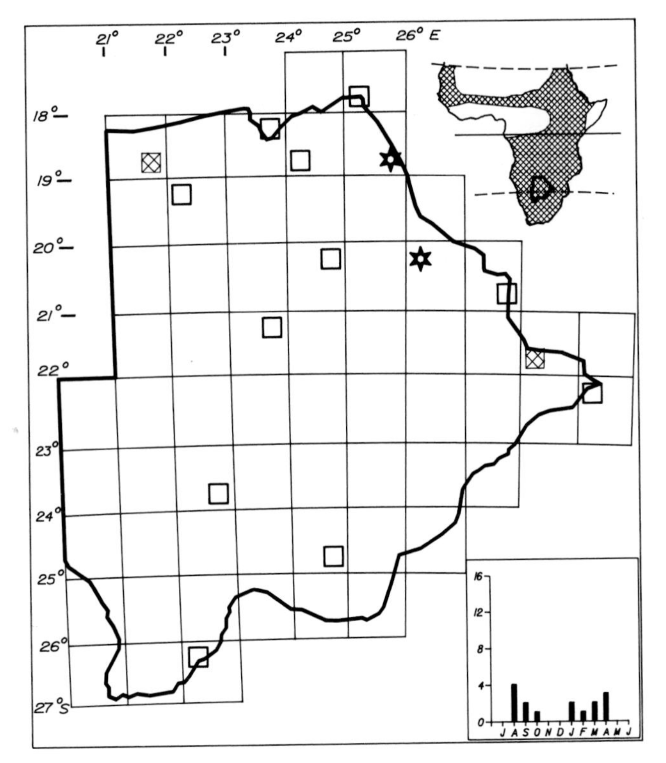

Status

Very sparse resident whose status is poorly known. The breeding population is unlikely to be much more than 50 pairs based on distribution and the limited breeding habitat available. Wanderers, some of which may be of Palaearctic origin, appear unpredictably at widespread localities. Two subspecies—resident *F. p. minor* and Palaearctic *F. p. calidus* —have been identified. No migration pattern is clear. Some of the information on this species in Botswana is derived from falconers.

Habitat

Recorded from suitable breeding habitat near cliffs, rock outcrops and steep rock faces in hilly regions. Forages and wanders over woodland, savanna and urban areas. Reported to have been temporarily attracted to tall constructions such as mine buildings and cooling towers.

Analysis

15 squares (7%) Total count 18 (0,01%)

LANNER FALCON

172

Falco biarmicus

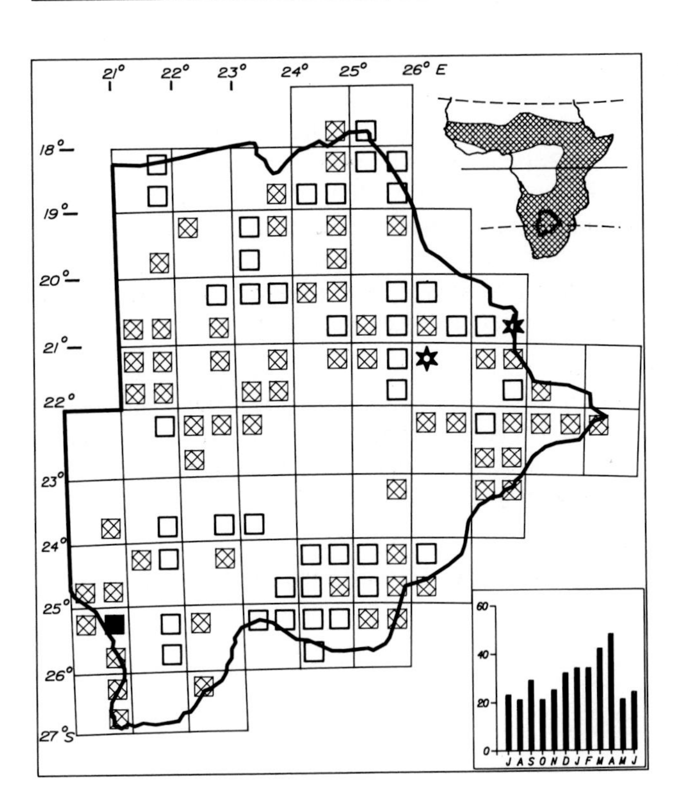

Status

Sparse to common resident throughout. Locally very common in Deception and Nossob Valleys. It is the common large falcon of the country. Movements occur, though these are as yet poorly defined, as evidenced by small groups (3–10) of immature birds seen flying together in several northern areas between February and April. Egglaying June to August, mainly June and July. Usually solitary or in pairs.

Habitat

Tree and bush savanna in lush and semidesert conditions, edges of woodland, edges of pans and plains. Stoops its prey over open areas.

Analysis

105 squares (46%) Total count 368 (0,21%)

HOBBY FALCON

Falco subbuteo

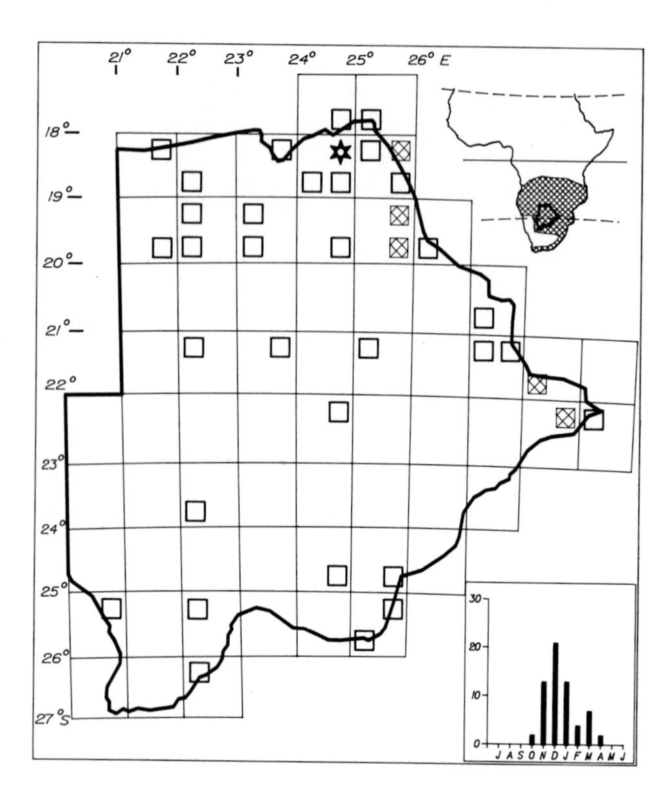

Status

Sparse Palaearctic migrant except in the northeast where it is fairly common in some years. Rare in semidesert and south of 23°S. Arrives in late October (29th) and is most numerous between November to January. Departs in March and early April (4th). Solitary or in small loose groups of 3–10 birds.

Habitat

Edges of woodland, tree savanna, plains with scattered trees, telegraph poles on roadsides. Perches on trees and poles for hunting. Often hawks at termite alate emergences or over open ground for other insects such as locusts.

Analysis

38 squares (16%) Total count 65 (0,04%)

AFRICAN HOBBY FALCON

Falco cuvierii

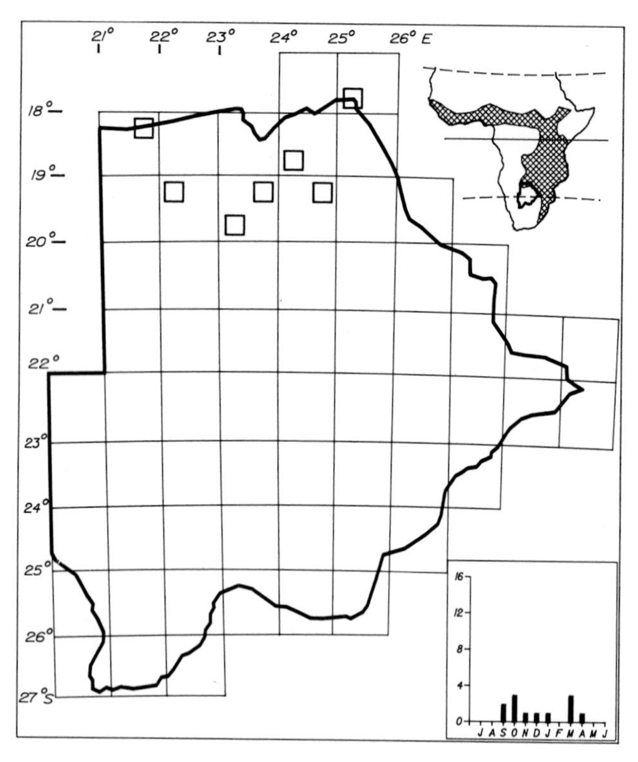

Status

Very sparse and poorly known. It is not even certain that it is a breeding resident. Confined to the north where it occurs at the southwestern limit of its continental range. It has not been recorded between May and July which is the early breeding season in adjacent countries. Normally shy and elusive. It uses the nests of other birds in which to breed. It may be that it only occurs in some years. Recorded more than once only at Shakawe and Kasane. Usually solitary.

Habitat

Broadleafed woodland and tree savanna. Hunts small birds and insects in clearings and open areas such as floodplains from tree cover on the edge of woodland or tree savanna. Usually solitary.

Analysis

7 squares (3%) Total count 15 (0,008%)

REDNECKED FALCON

Falco chicquera

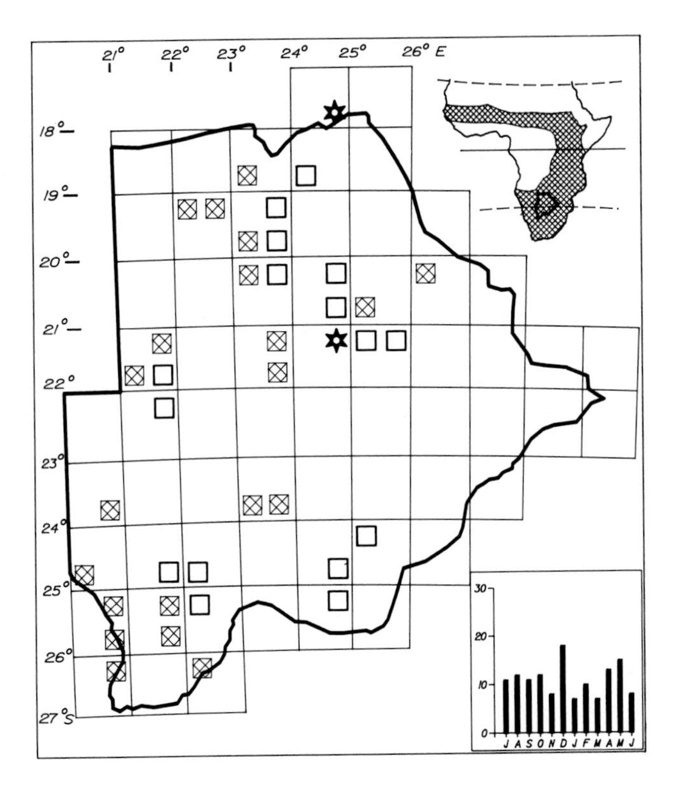

Status

Sparse to fairly common resident. Patchily distributed but apparently with population concentrations in the Okavango Delta, Makgadikgadi Pans, Ghanzi farms, southern pans and the Nossob Valley. There is no evidence of movement. Egglaying August, September and December. Usually solitary or in pairs.

Habitat

Tree savanna with tall trees, especially palms, often near water. Hunts over open areas such as pans, floodplains, farmland, grassland plains and fossil valleys. Uses palm trees with lopped or intact crowns as well as other tall trees for nesting.

Analysis

39 squares (17%)   Total count 141 (0,08%)

WESTERN REDFOOTED KESTREL

Falco vespertinus

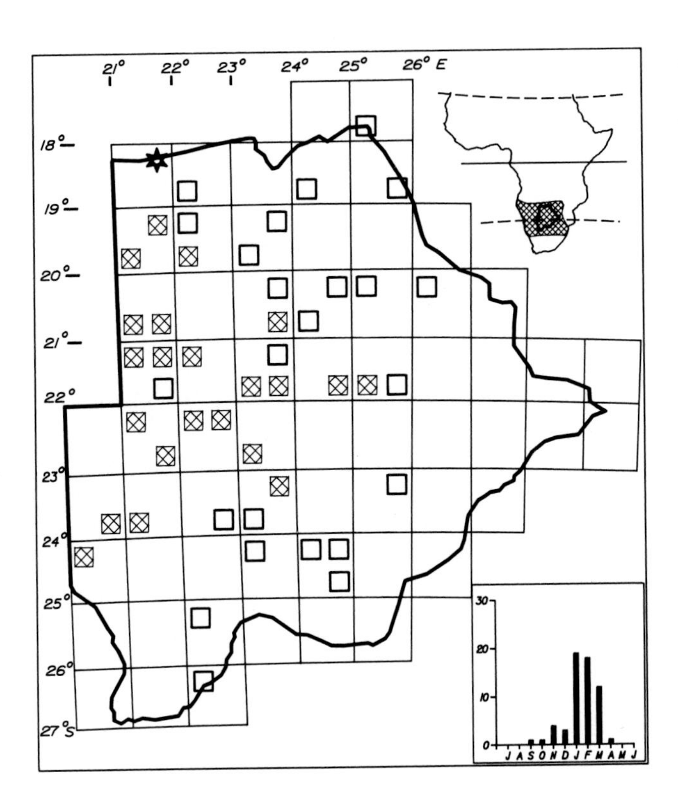

Status

Sparse to fairly common Palaearctic migrant present mainly from November to March. Some arrive in late September and some depart in early April (2nd). The majority probably remain in Botswana moving around locally with changes in the weather and the availability of food but passage birds occur to and from South Africa. Usually in small flocks of 10–30 birds, but gatherings of several hundreds occur which may be spread over a few square kilometres when resting up during the day. It is both more common and more numerous than the Eastern Redfooted Kestrel.

Habitat

Open tree and bush savanna, plains, pans and Kalahari scrub savanna. Moves about with rain and thunderstorms which appear to determine insect migrations and alate emergences.

Analysis

47 squares (20%)   Total count 61 (0,03%)

EASTERN REDFOOTED KESTREL

180

Falco amurensis

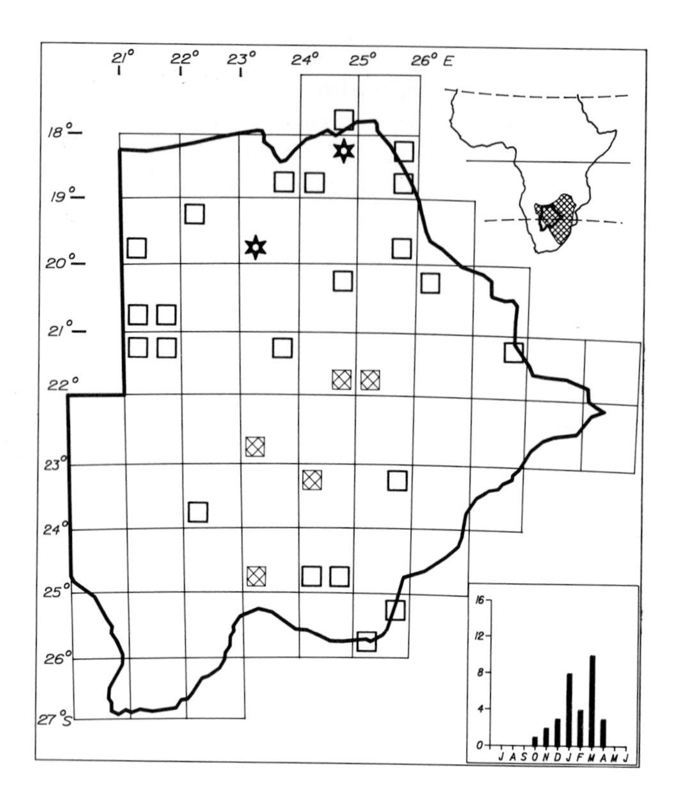

Status

Sparse to uncommon Palaearctic migrant present mainly from December to March—earliest date 29 October, latest date 4 April. Unlike the Western Redgooted Kestrel it may occur in Botswana mainly as part of a westward sweep from its main wintering grounds further east as most records fall between January and March, the start of northward return. Solitary or in small groups. Very often but not always in the company of the Western Redfooted Kestrel.

Habitat

Open tree and bush savanna, plains, Kalahari scrub savanna, edges of woodland.

Analysis

29 squares (13%) Total count 35 (0,02%)

ROCK KESTREL

181

Falco tinnunculus

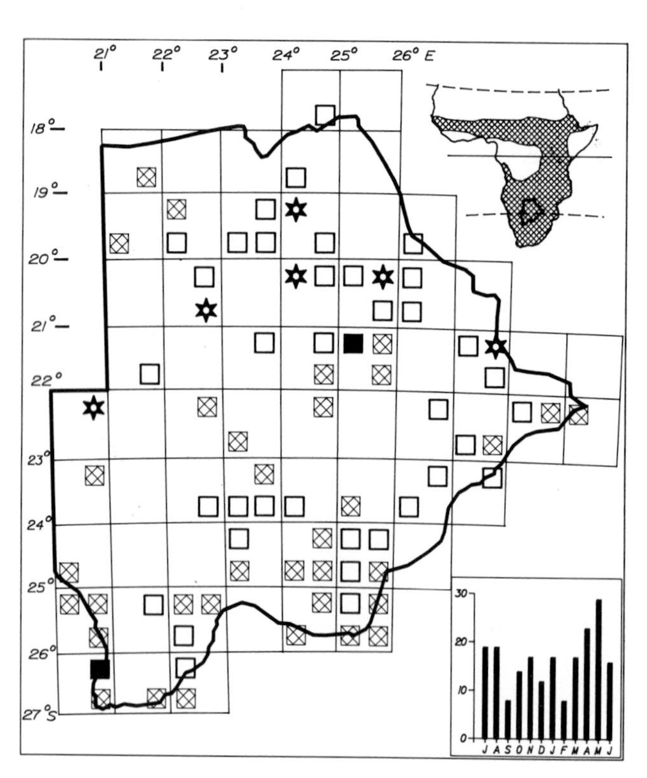

Status

Generally sparse but locally common resident. Very common at Orapa and Twee Rivieren. Elsewhere most often recorded in hilly regions such as Tsodilo, Aha and the escarpments of the east, southeast, the Nossob and western Molopo Valleys. Local movements occur but it is mainly sedentary in hilly regions. Egglaying November (one record). Usually solitary or in pairs.

Habitat

Mountainous and hilly terrain where it breeds on cliffs and rock faces and forages over surrounding woodland and savanna. In the southwest it occurs in arid semidesert on the scarps of the western Molopo and Nossob River valleys. Also in tree savanna. In populated areas it perches on telegraph poles and wires, uses tall buildings and mine rock dumps for soaring on upcurrents and may in future breed on these structures.

Analysis

79 squares (34%) Total count 224 (0,13%)

GREATER KESTREL

182

Falco rupicoloides

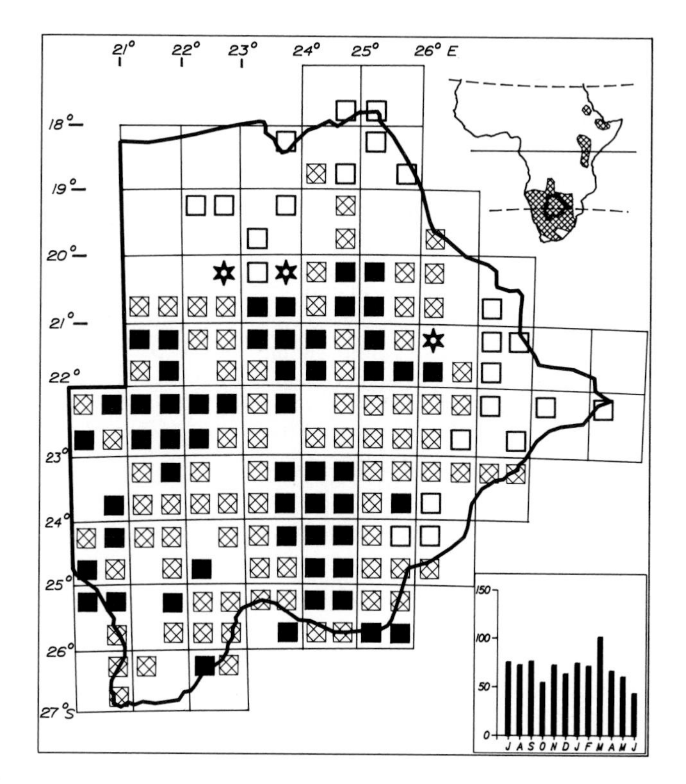

Status

Very common resident south of 20°S and west of 27°E. Sparse and uncommon in the bush and tree savannas of the east. Locally common in the Nxai Pan and Savuti areas but otherwise sparse north of 20°S. Breeds mainly in old nests of the Black Crow and the two species are often seen in close proximity. Egglaying July and all months from September to December. Usually in pairs, less often solitary.

Habitat

Grassland plains, open tree and bush savanna, Kalahari scrub savanna, edges of large pans, lightly-vegetated dunes. Generally in open areas in semidesert.

Analysis

162 squares (70%) Total count 864 (0,48%)

LESSER KESTREL

183

Falco naumanni

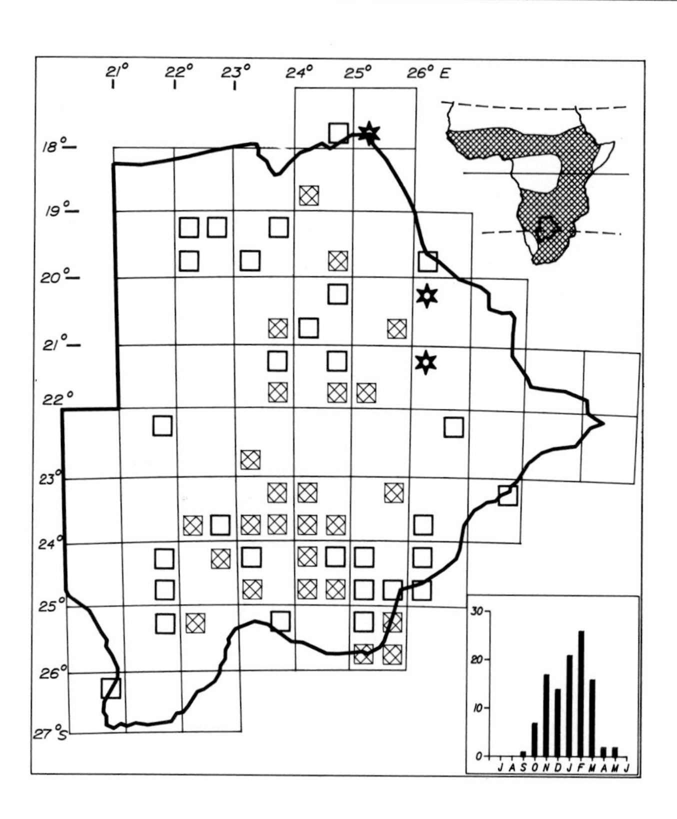

Status

Sparse to fairly common Palaearctic migrant in the south and southeast. Elsewhere it is sparse to uncommon. Arrives usually in the last week of October with the advent of the first rainstorms and is present until March (latest date 4 May). Its occurrence in Botswana may be mainly on passage—there are no reports of large roosts except in nearby Transvaal (nearest at Mafikeng). Usually seen in localised flocks of 20–200 birds, occasionally solitary or in small groups.

Habitat

Grassland plains, open tree and bush savanna in semidesert and in more lush regions, Kalahari scrub savanna, open areas of cultivation, telegraph poles and wires along main roads. Roosts in tall trees e.g. *Eucalyptus*, dead trees in the central Kalahari or on the edges of woodland.

Analysis

57 squares (25%) Total count 114 (0,06%)

DICKINSON'S KESTREL

Falco dickinsoni

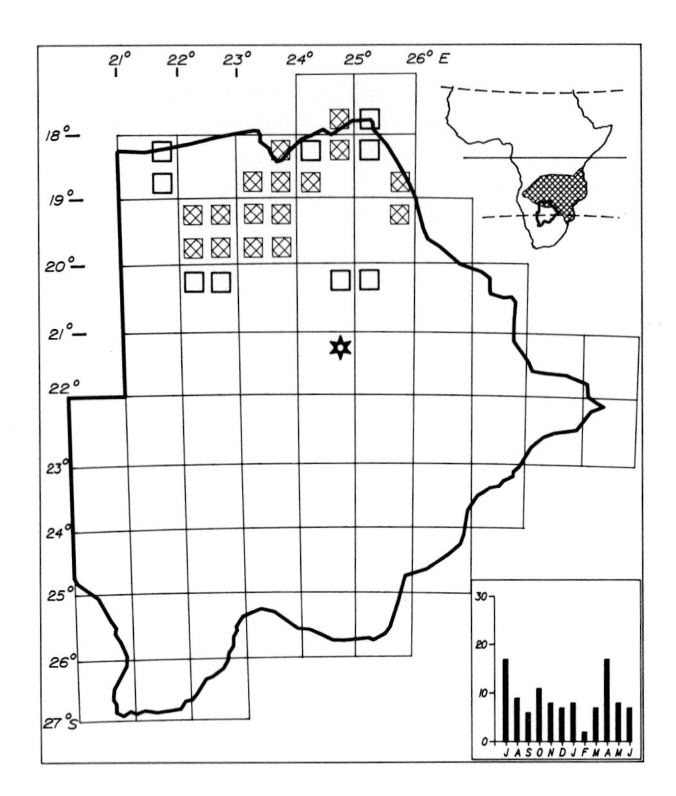

Status

Fairly common to locally common resident north of 20°S. Sparse in the northern sector of the Makgadikgadi Pans and in the northwest. Its range may have contracted a little during the drought period of this survey. Egglaying August (one record). Solitary or in pairs.

Habitat

Broadleafed woodland with mature trees, tree savanna with tall trees; usually near water sources such as pans, floodplains and dambos. Also in the vicinity of tall palms (e.g. *Hyphaene*) at the top of which it may nest.

Analysis

26 squares (11%) Total count 112 (0,06%)

PYGMY FALCON

Polihierax semitorquatus

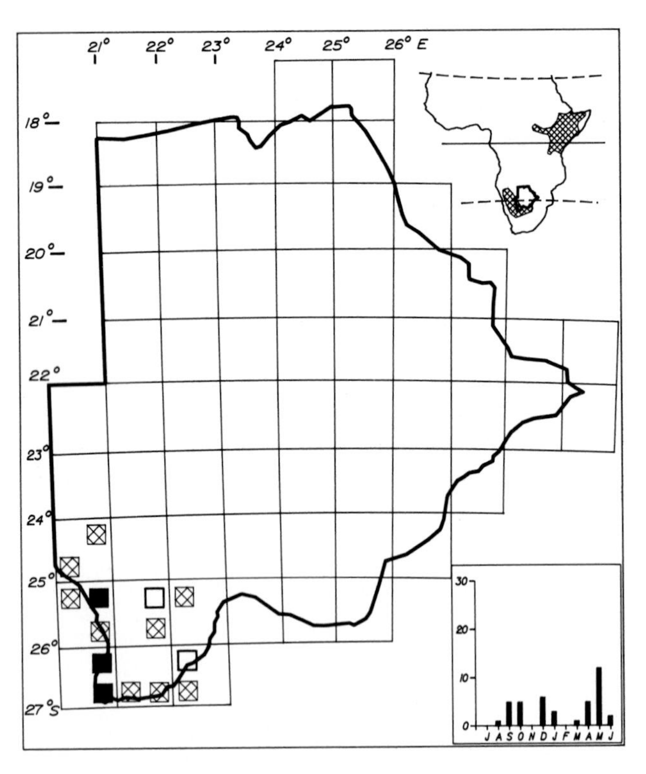

Status

Common to locally very common resident of the southwest. Mainly confined to the Gemsbok National Park and Mabuasehube Game Reserve and to the Molopo and Nossob Valleys. Rare at Tshabong. Very common in the Nossob Valley south of Nossob Camp. Occurs at the northeastern limit of its southern African range. May occur in the southeast near Good Hope where the Sociable Weaver is found. Its distribution is closely related to that of the Sociable Weaver in whose nests it breeds. Solitary or in pairs.

Habitat

Arid *Acacia* savanna, particularly *Acacia erioloba* in seasonal river valleys and in parklike savanna. Also in thorn tree savanna on well-vegetated dunes in areas with Sociable Weaver nests.

Analysis

14 squares (6%) Total count 41 (0,02%)

COQUI FRANCOLIN

Francolinus coqui

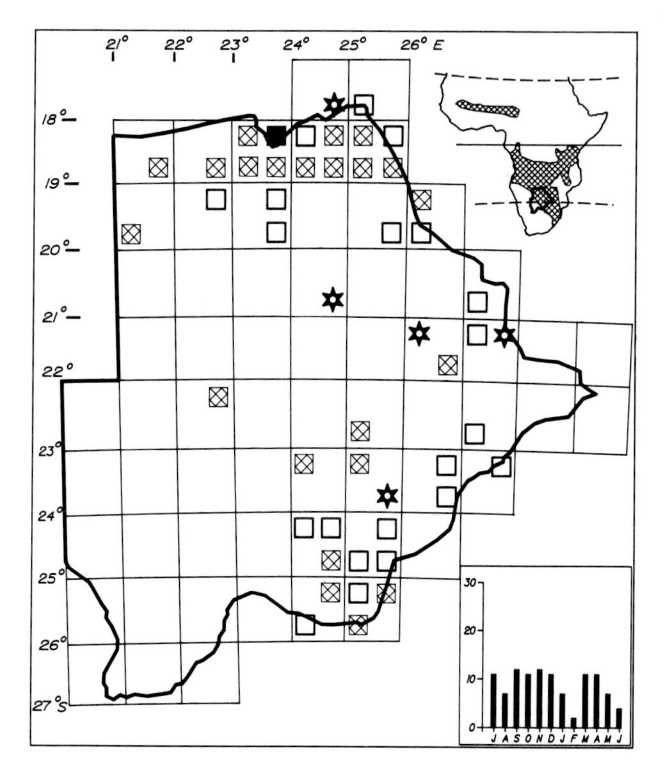

Status

Sparse to fairly common resident, recorded as very common at Linyanti. Only reported regularly in the north, elsewhere sporadically. Sparse and uncommon in the southeast extending northwards into the Kutse and central Kalahari Game Reserves. Egglaying December (one record). Occurs in pairs or small coveys.

Habitat

Wooded grassland and bush savanna. In the north it also occurs in rich woodland where there is underlying grass cover. In semidesert it has been found in extensive areas of stunted *Terminalia* savanna.

Analysis

49 squares (21%)   Total count 115 (0,06%)

CRESTED FRANCOLIN

Francolinus sephaena

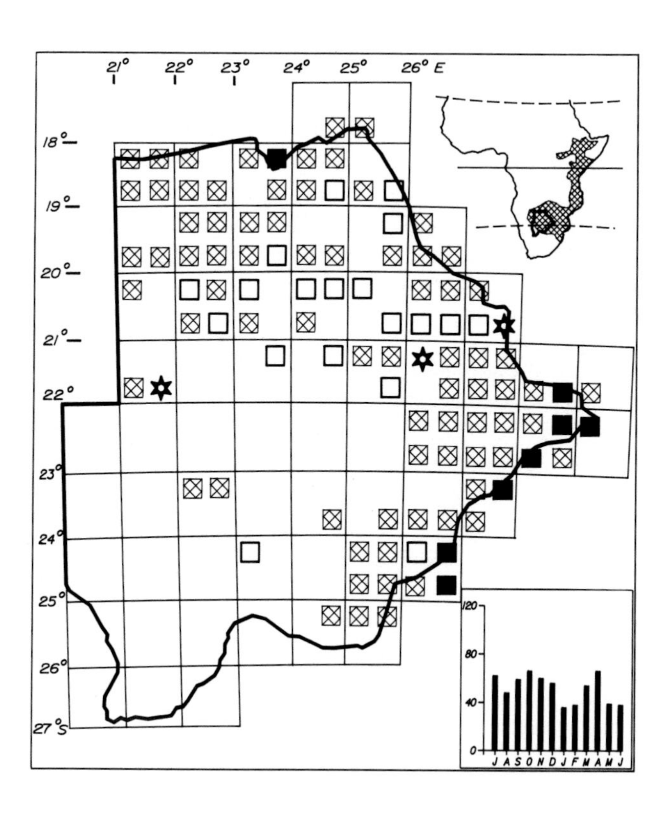

Status

Fairly common to very common resident of the north and east except in the Makgadikgadi Depression and predominantly mopane areas where it is sparse. Recorded once recently in each of the four western squares shown which are north (2 squares) and south of Kang and west of Ghanzi and where its status requires further study. Very common along the Limpopo Valley. Egglaying December. In small groups or in pairs.

Habitat

Thickets and dense vegetation in *Acacia* and broadleafed woodland, tree savanna and riverine bush.

Analysis

105 squares (46%)   Total count 621 (0,35%)

ORANGE RIVER FRANCOLIN 193

Francolinus levaillantoides

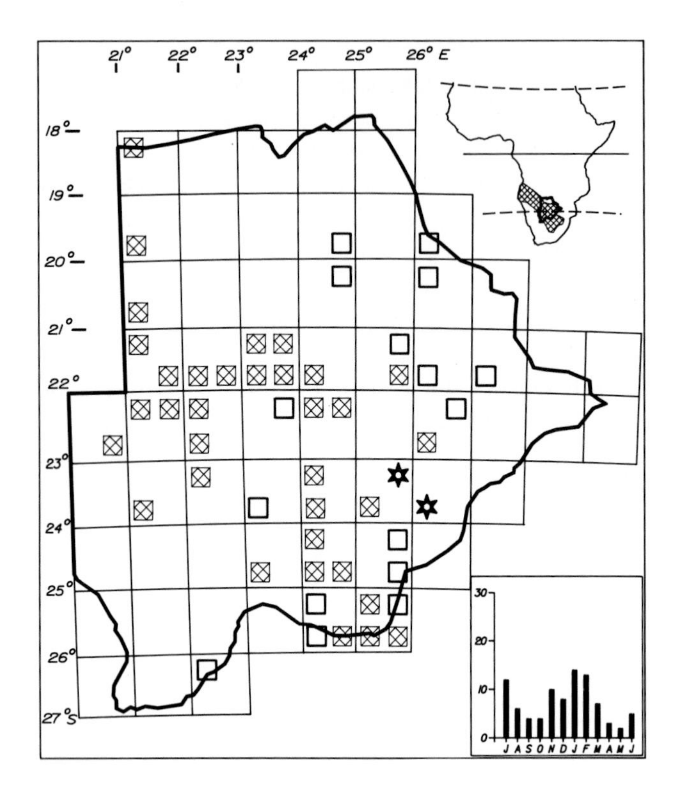

Status

Fairly common resident from the northwest to the southeast. Sparse around the periphery of the Makgadikgadi to as far north as Nxai Pan and just north of Nata. Its range may abut that of the similar Shelley's Francolin which occurs in northwest Zimbabwe but their habitats are different. Egglaying October and January. Occurs in pairs or small groups.

Habitat

Grassland and semidesert bush savanna especially along drainage lines and in fossil valleys such as Naledi, Deception and Okwa. Also in and around pans, on grass plains, *Acacia* savanna and open stunted woodland with good grass cover.

Analysis

52 squares (23%) Total count 105 (0,06%)

REDBILLED FRANCOLIN 194

Francolinus adspersus

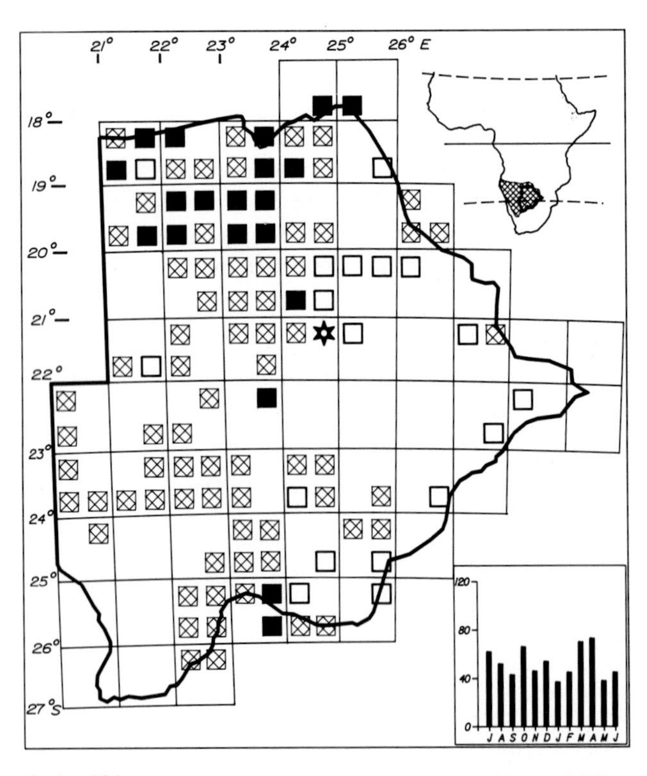

Status

Very common resident of the north and common in western and central areas. Very sparse and irregular in the east and absent in most of the southwest except along the Molopo River where it can be very common. It is the typical francolin of Kalahari sands. Egglaying in all months from March to June. Usually in loose groups of 5–10 birds, also solitary or in pairs.

Habitat

Bushes and thickets in semidesert, *Acacia* savanna, mixed woodlands and on floodplains. Sometimes far from water but also occurs very commonly in some major river valleys. Often in areas with exposed sand or only light grass cover.

Analysis

109 squares (47%) Total count 675 (0,38%)

NATAL FRANCOLIN 196

Francolinus natalensis

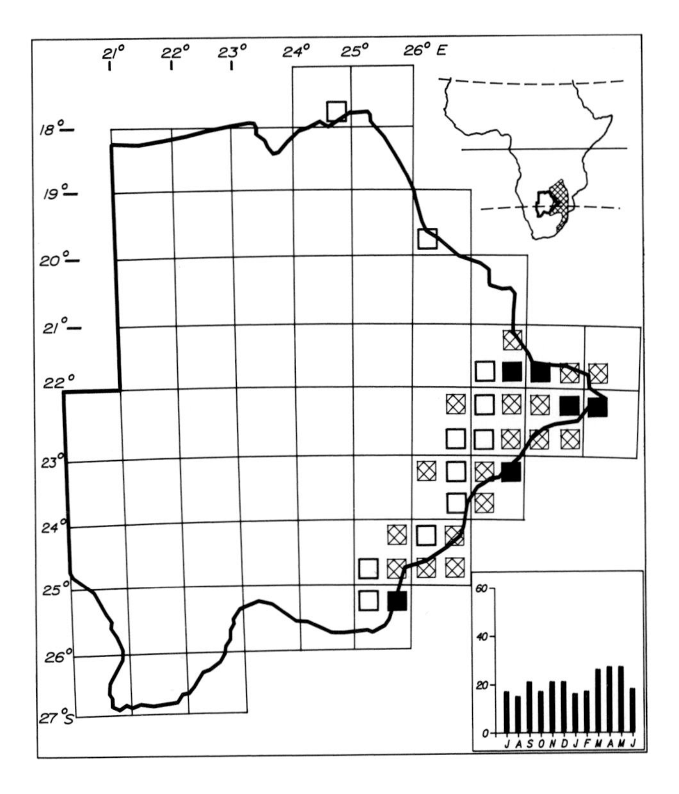

Status

Uncommon to very common resident of the east mainly south of Francistown. There is one record from just north of Nata and one west of Kasane. Apparently does not occur along the Molopo River but is very common along parts of the Shashe and Limpopo rivers. Extends west through the Tswapong Hills to the Mokgware and Shoshong hills. Egglaying August and September. Occurs in pairs or small groups.

Habitat

Bushes and dense cover along river valleys, in woodland, in riparian forest and on hills.

Analysis

34 squares (15%)   Total count 241 (0,13%)

SWAINSON'S FRANCOLIN 199

Francolinus swainsonii

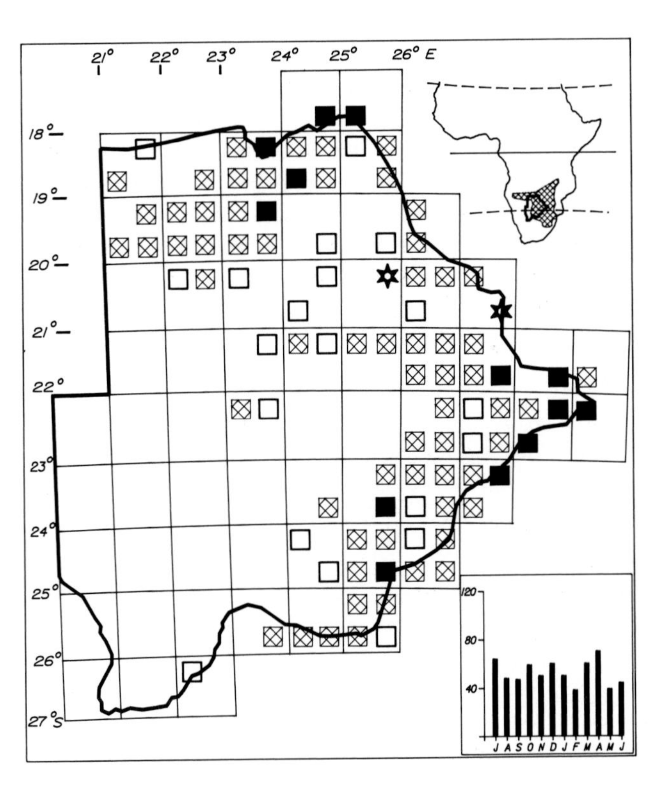

Status

Fairly common to very common resident of the north and east. Extends west along the Molopo River almost to Bray and there is one record from Tshabong. In the northern central Kalahari it occurs in the Okwa and Deception Valleys. Egglaying April, May and June. Solitary, in pairs, or in small coveys.

Habitat

Tree and bush savanna with good grass cover. In rank grass, thicket and secondary growth on the edges of woodland, cultivation, fallow lands and roadsides. Also in riverine bush and on floodplains.

Analysis

98 squares (43%)   Total count 678 (0,38%)

COMMON QUAIL

200

Coturnix coturnix

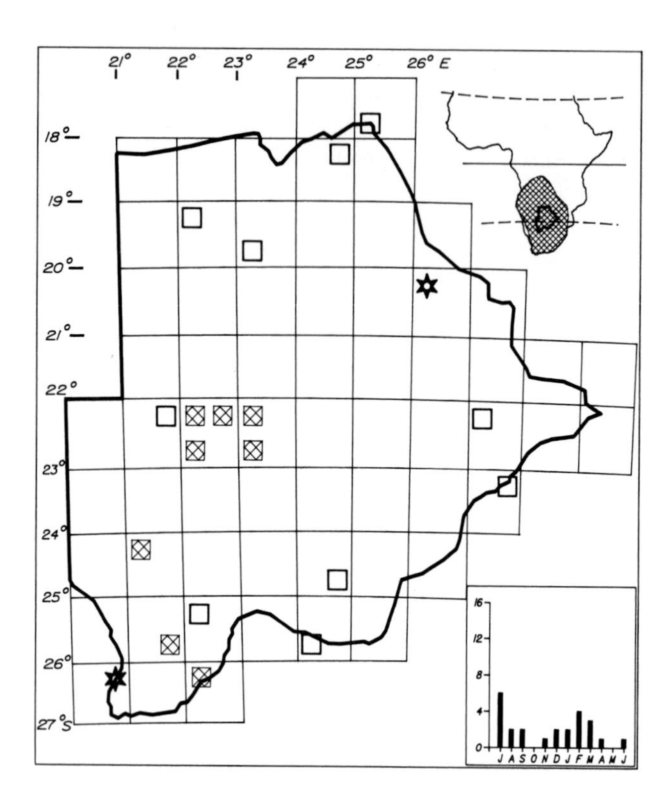

Status

Sparse to uncommon in widely scattered localities. Poorly known and warrants special study. Its occurrence in Botswana, though suspected, has only recently been proven. It may breed, at least in some years, and also occur on passage from its breeding grounds in South Africa. Several pairs were counted in March in 2421A. It is likely that only the subspecies *C. c. africana* occurs. Migratory and irruptive.

Habitat

Open or lightly-wooded savanna with fairly long grass, well-grassed pans and fossil valleys. There is an abundance of suitable breeding habitat in the acres of grassland in the central and western Kalahari in years of good rainfall.

Analysis

20 squares (9%) Total count 27 (0,015%)

HARLEQUIN QUAIL

201

Coturnix delegorguei

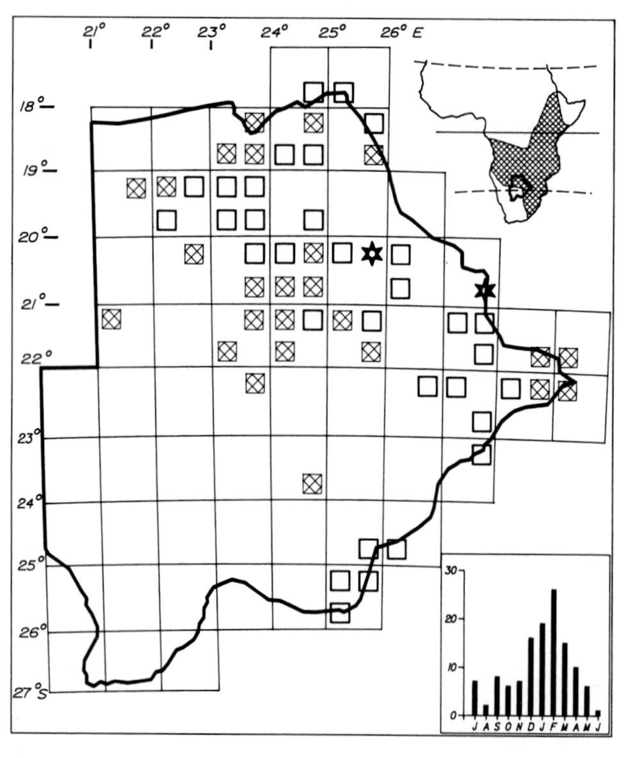

Status

Sparse to fairly common breeding intra-African migrant. Present mainly from September to May but a significant number are present throughout the year particularly in the Makgadikgadi region. In years of good rainfall it moves further into the central Kalahari and larger numbers may remain to breed. Less common in drought years. One was collected from the roof of the Anglican Cathedral, Gaborone on 17 February 1988. Egglaying January. Usually in small localised groups.

Habitat

Rank grass in moist situations including patches of black cotton soil in the central Kalahari after rain. Also on the edges of marshes and floodplains, on well-grassed pans and on fallow or cultivated lands in river valleys.

Analysis

59 squares (26%) Total count 134 (0,07%)

HELMETED GUINEAFOWL

Numida meleagris

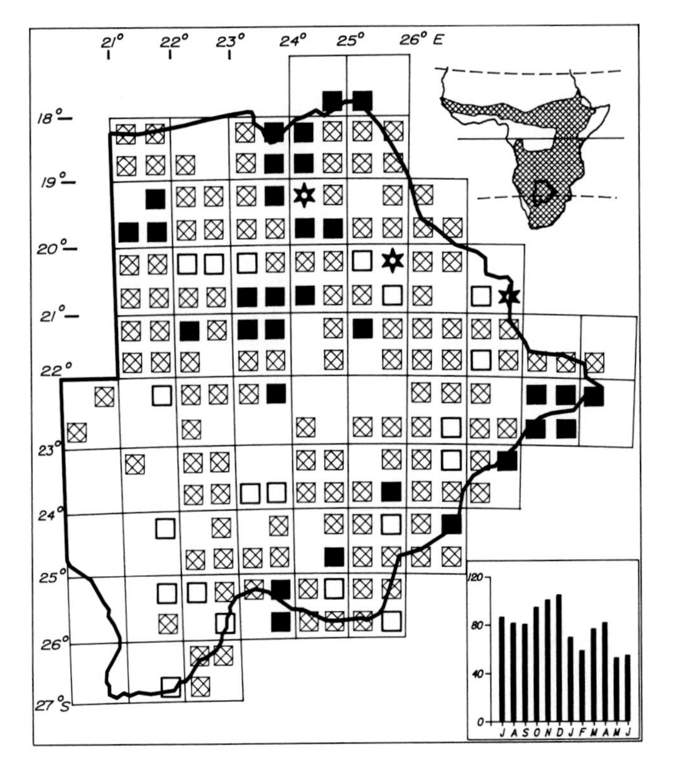

Status

Fairly common to very common resident throughout except in the southwest and parts of the central Kalahari. Local movements occur and often it is absent from suitable habitat. It is still common in inhabited areas in spite of hunting pressure. Egglaying December to February. Usually in flocks of 10–100 birds or in pairs when breeding.

Habitat

Tree and bush savanna, open grassland, cultivation, farmland, edges of woodland, riverine bush, fossil valleys, floodplains and less commonly in semidesert bush and scrub savanna.

Analysis

174 squares (76%) Total count 979 (0,55%)

KURRICHANE BUTTONQUAIL

Turnix sylvatica

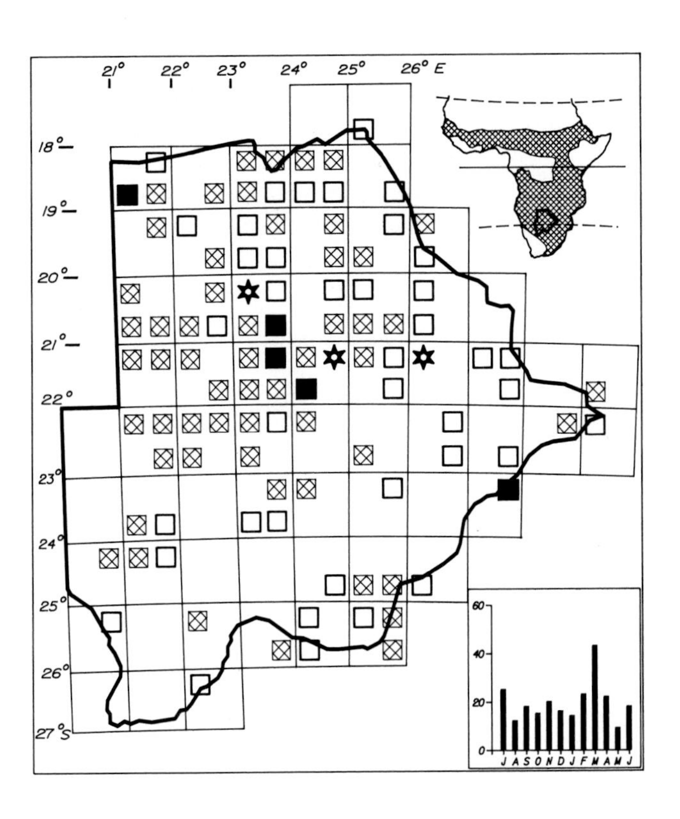

Status

Sparse to common resident throughout. It is locally very common when suitable grassland conditions prevail and when breeding. No pattern of movement is shown on analysis but local movements occur and it may be more common in some years than others. Egglaying December to February. Usually in pairs or small groups.

Habitat

Grassland, open tree and bush savanna, edges of cultivation, fallow lands; predominantly in dry conditions. Utilises tall clumpy grass as well as short and regenerating vegetation. Like the Common Quail it also is found in expanses of Kalahari grassland but nearer to woodland.

Analysis

103 squares (45%) Total count 245 (0,13%)

WATTLED CRANE

207

Bugeranus carunculatus

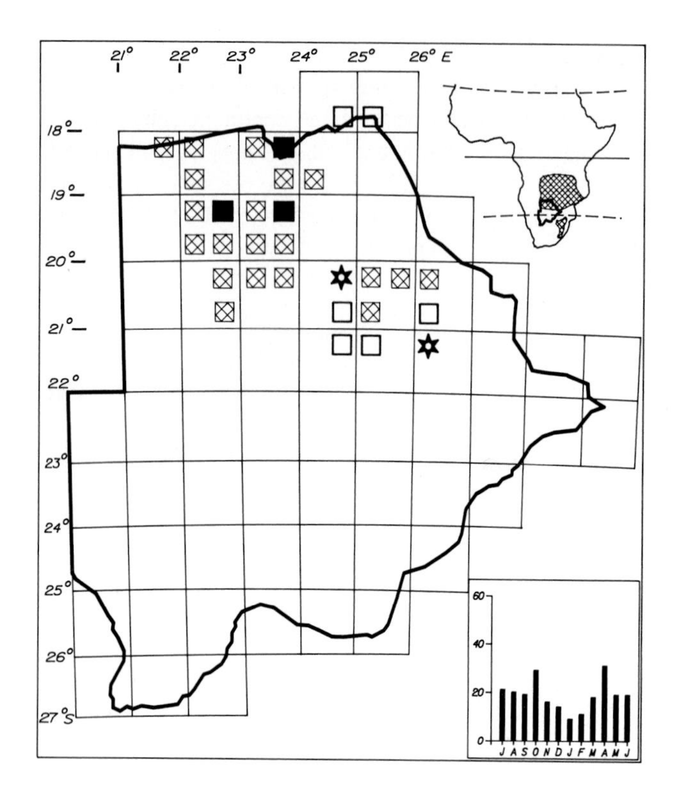

Status

Sparse to locally very common resident of the Okavango and Makgadikgadi regions. It is an endangered species which is very sensitive to physical disturbance at its breeding site. It is easily recognisable from a distance and close approach is not necessary nor desirable. Some movements may occur between this population and those in Zambia and Zimbabwe. Usually in pairs, but flocks of up to 200 birds occur.

Habitat

Marshes, floodplains, moist pans, open grassland in high rainfall regions.

Analysis

31 squares (13%) Total count 236 (0,13%)

BLUE CRANE

208

Anthropoides paradiseus

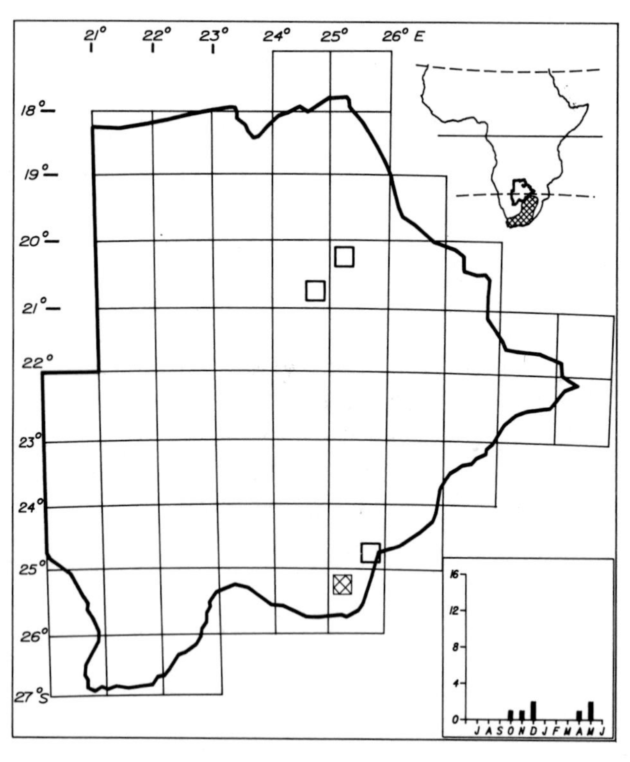

Status

An infrequent and unpredictable visitor from breeding grounds in South Africa. Recorded rarely at Makgadikgadi Pans and Gaborone and uncommonly at Good Hope. The visits to the southeast may occur more regularly than reported as the area is at the northern limit of its South African range. Usually solitary.

Habitat

Open short grassland plains and pans.

Analysis

4 squares (2%) Total count 8 (0,004%)

GREY CROWNED CRANE 209

Balearica regulorum

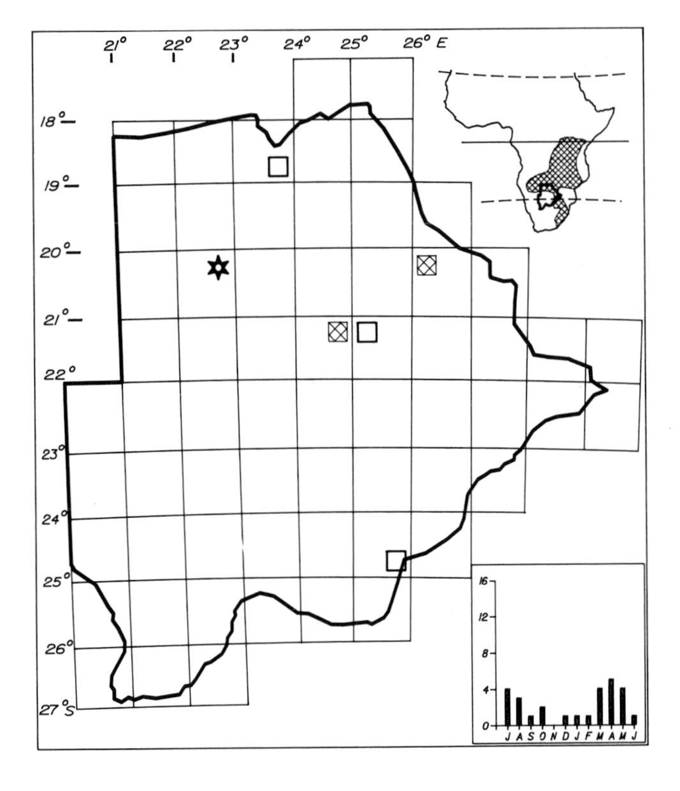

Status

Sparse and uncommon resident whose status is poorly known. The paucity of reports in this survey gives cause for concern but may be attributable to the drought. The northern wetlands of Botswana have the potential to carry a sizeable population of this species. It has been recorded more than once only at Nata and Mopipi. Present throughout the year. Usually in small groups of 3–20 birds.

Habitat

Marshes, floodplains, short grass on edges of lakes and dams.

Analysis

6 squares (3%) Total count 30 (0,016%)

AFRICAN WATER RAIL 210

Rallus caerulescens

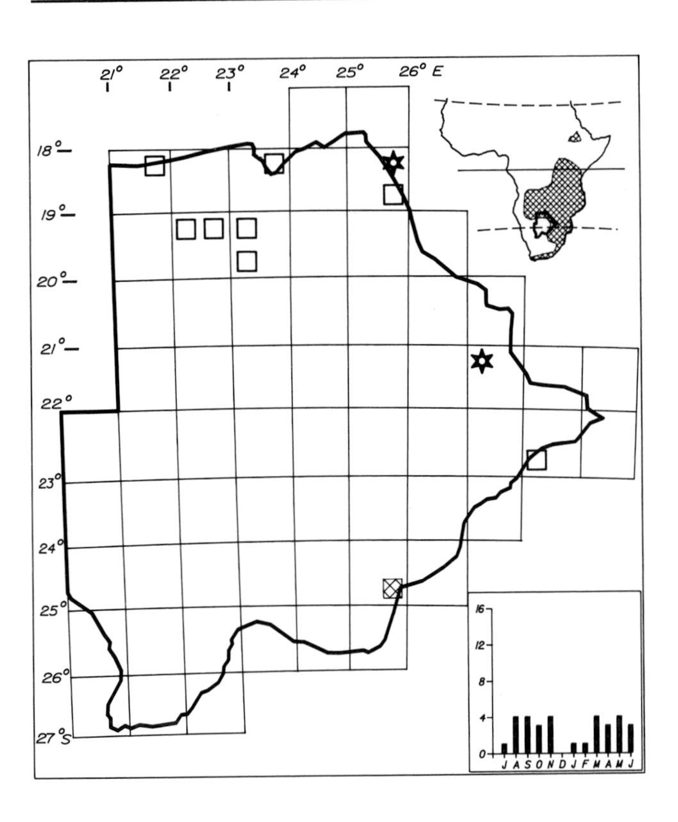

Status

Sparse and uncommon resident of the Okavango, Linyanti and Mpandamatenga areas. Elsewhere it is rare or overlooked—recorded only once along the Limpopo Valley. Locally fairly common at Gaborone. It is present throughout the year and is mainly sedentary. It is likely to be more common than current records indicate. Solitary birds usually heard and seen.

Habitat

Swamps, marshes, reedbeds even when partly dried out seasonally. Moves locally during dry months to nearby wet patches, if available, such as dam and sewage pond outlets with adequate reed cover.

Analysis

11 squares (5%) Total count 37 (0,02%)

AFRICAN CRAKE

Crex egregia

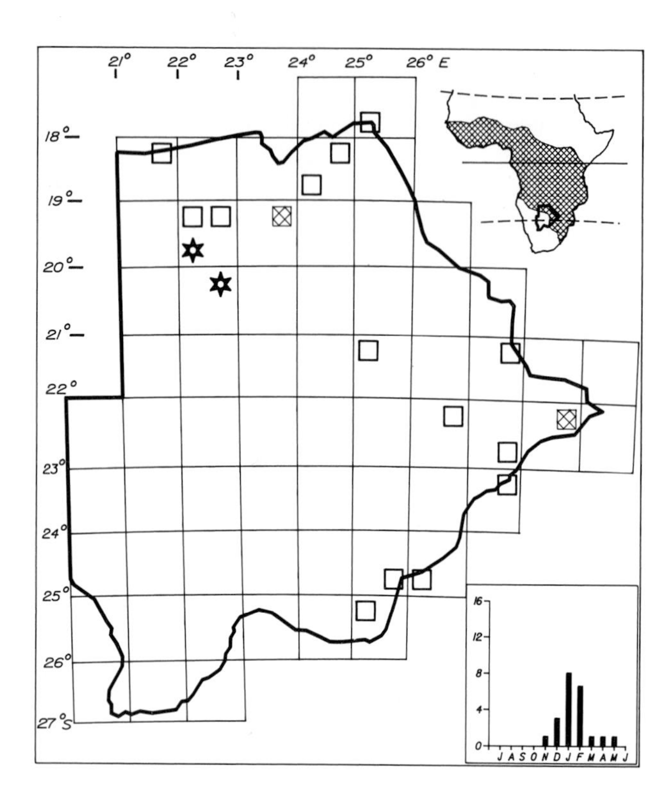

Status

Sparse to uncommon breeding intra-African migrant present from late November to May but mainly between December and February. One breeding record—for January. Occurs in widely scattered localities in the north and east and is nowhere common. Poorly known. Usually solitary.

Habitat

Marshes, seasonally wet pans and grassland, edges of floodplains; in rank grass, reeds or dense vegetation on the margins of water.

Analysis

18 squares (8%) Total count 23 (0,01%)

BLACK CRAKE

Amaurornis flavirostris

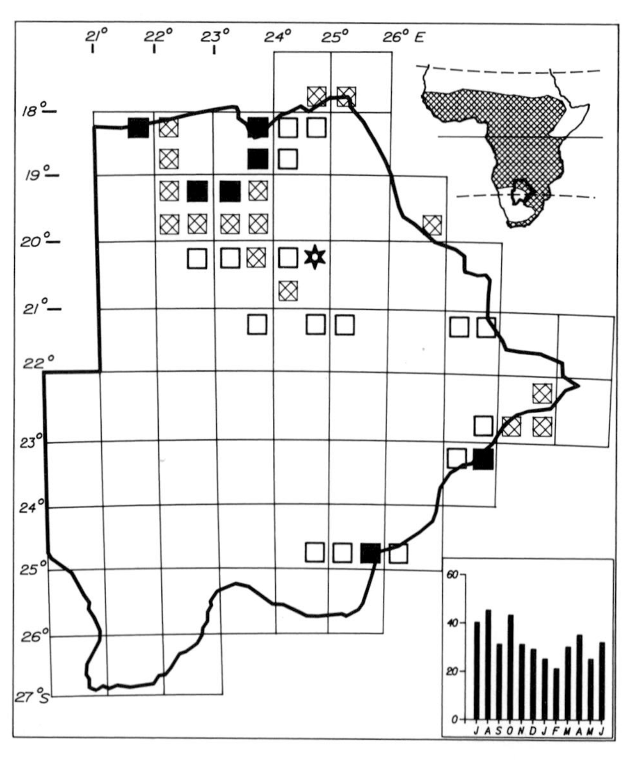

Status

Fairly common to locally very common resident of the Okavango, Linyanti and Chobe river systems. Sparse on the periphery of these regions during drought. Highly localised in the east and southeast where it is generally uncommon but it can be locally very common. Egglaying August, September, October. Solitary or in pairs, but often several birds are present at any one locality.

Habitat

Marshes, swamps, lagoons, or in aquatic vegetation on the margins of lakes, dams and rivers.

Analysis

40 squares (17%) Total count 409 (0,23%)

SPOTTED CRAKE

214

Porzana porzana

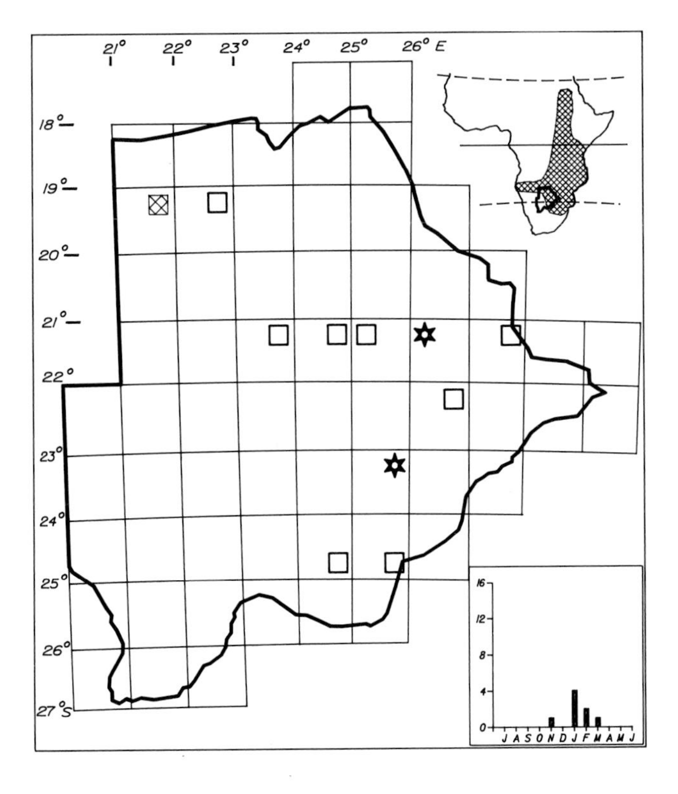

Status

A sparse Palaearctic migrant recorded between November and March with most records in January. Greater numbers are present in some years (such as 1987/1988) than others. Probably will occur mainly in the north and east. The specimen killed by flying into a building at Jwaneng in November 1987, however, suggests that it may migrate over the central Kalahari. The species remained temporarily at Orapa during January and February 1988. Usually solitary.

Habitat

Marshes, flooded grassland and seasonal pans.

Analysis

11 squares (5%) Total count 12 (0,007%)

BAILLON'S CRAKE

215

Porzana pusilla

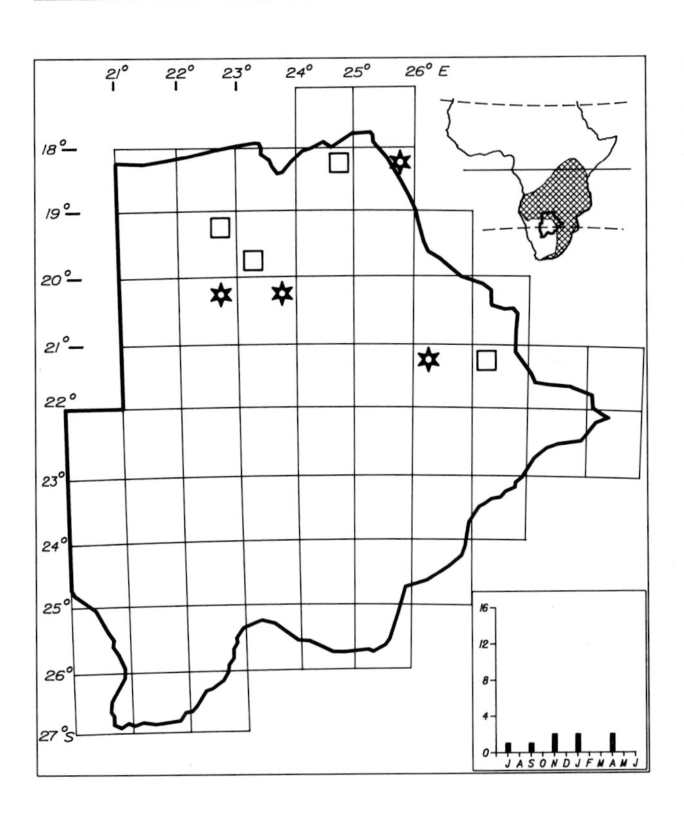

Status

A very sparse and very poorly known resident—in breeding condition in January (Smithers 1964). It is easily overlooked because of its skulking habits. A specimen was collected in January 1974 at Nthane (Ginn). Of the eleven records three are undated. Usually solitary.

Habitat

Marshes, inundated grassland, reedbeds on margins of lakes, dams and swamps.

Analysis

8 squares (3%) Total count 11 (0,006%)

REDCHESTED FLUFFTAIL

217

Sarothrura rufa

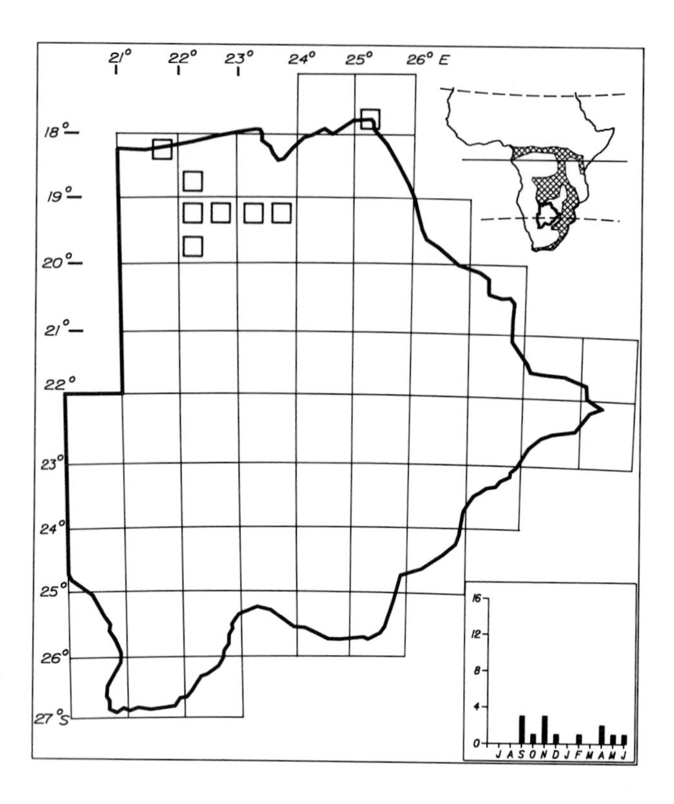

Status

Sparsely recorded resident of the Okavango Delta. Recorded once at Kasane. Poorly known. It is likely to be more common than current records indicate. The Botswana population appears to be at the southern limit of and contiguous with the Zambian population based on current knowledge. Usually heard as a solitary bird.

Habitat

Marshes, floodplains, seasonally inundated grassland, swamps. It is found mainly in seasonally growing vegetation and vegetation of medium (100–500mm) height.

Analysis

8 squares (3%) Total count 14 (0,007%)

PURPLE GALLINULE

223

Porphyrio porphyrio

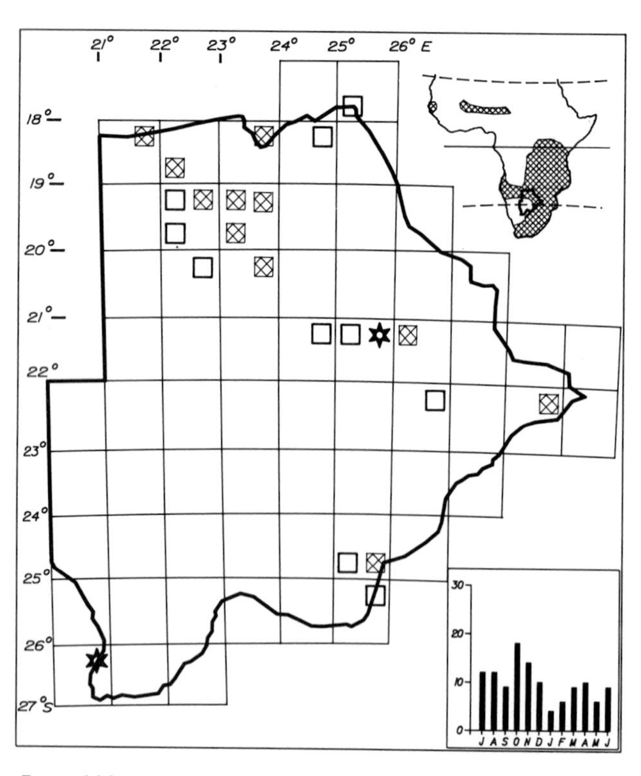

Status

Sparse to locally common resident in suitable habitat in the north, east and southeast. It is known to breed in January and July and may do so in all months of the year in some years. Local movements may occur. Solitary, but often several birds are present at one locality.

Habitat

Marshes and swamps with rushes, reeds and other tall aquatic vegetation. Also in similar cover on the margins of lakes and dams.

Analysis

23 squares (10%) Total count 132 (0,07%)

LESSER GALLINULE

224

Porphyrula alleni

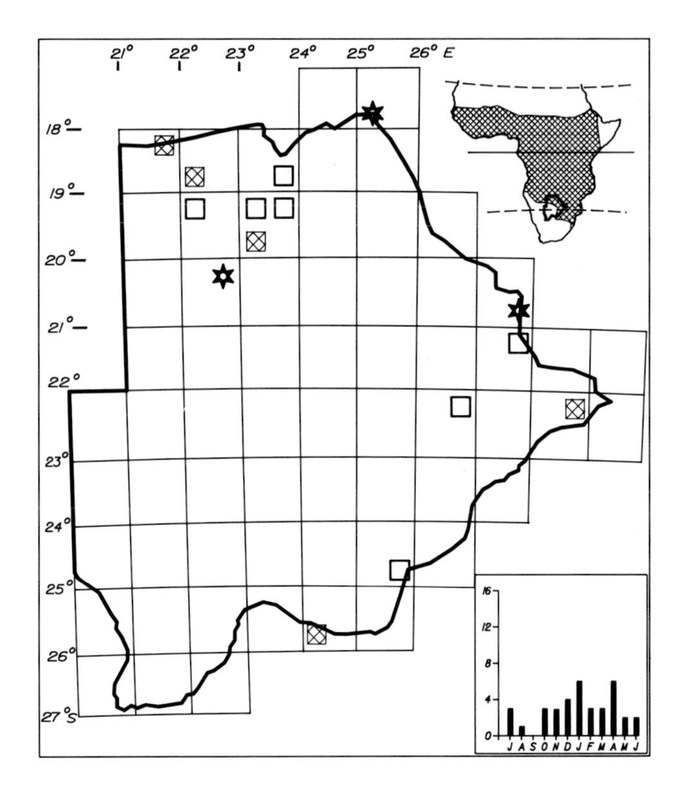

Status

Sparse to uncommon breeding intra-African migrant to the north and east of the country mainly from October to April. A few birds remain until May and June in some years and some may remain throughout the year. It occurs in Botswana at the southwestern limit of its continental movements. Egglaying November and December. Usually in pairs or solitary.

Habitat

Seasonal pans, flooded depressions in savanna, marshes, floodplains.

Analysis

15 squares (6%) Total count 41 (0,02%)

COMMON MOORHEN

226

Gallinula chloropus

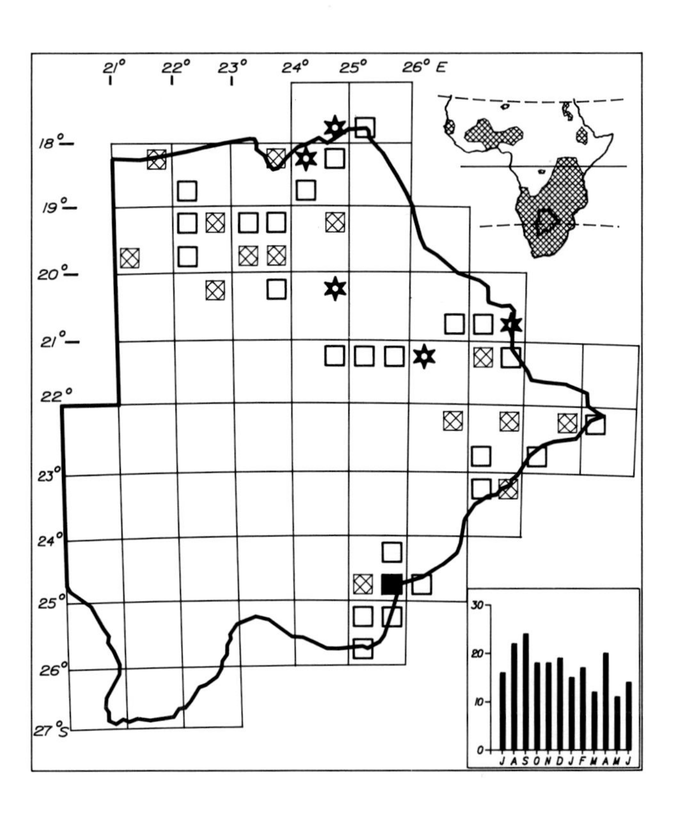

Status

Sparse to locally common resident of the north, east and southeast in suitable habitat. Recorded as very common at Gaborone. Present in all months at sites with permanent water. Local seasonal movements occur. Egglaying in June and July and possibly throughout the year in some years. Usually solitary or in pairs, often with other birds of the same species.

Habitat

Marshes, swamps, lagoons, vegetated margins of lakes, dams, ponds and rivers.

Analysis

44 squares (19%) Total count 222 (0,12%)

LESSER MOORHEN

227

Gallinula angulata

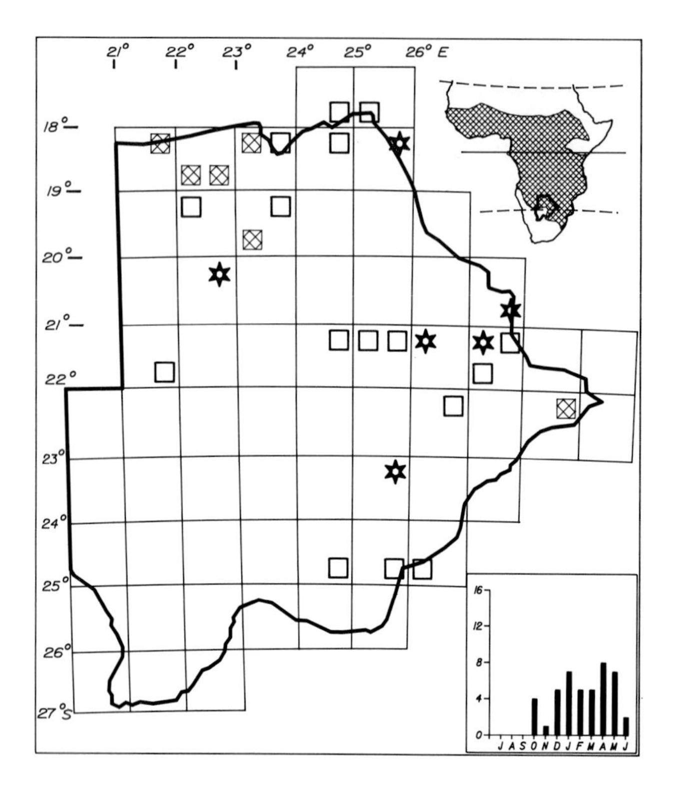

Status

Sparse to uncommon breeding intra-African migrant to the north, east and southeast. Recorded once at Ghanzi in December. Present from late October to June but mainly December to April. Like the Lesser Gallinule it occurs in Botswana at the southwestern limit of its continental distribution. One breeding record—February. Usually solitary or in pairs.

Habitat

Seasonal pans and flooded depressions in savanna, margins of well-vegetated ponds and marshes with floating and emergent aquatic plants. More likely to occur in the drier regions than the Lesser Gallinule.

Analysis

28 squares (12%) Total count 58 (0,03%)

REDKNOBBED COOT

228

Fulica cristata

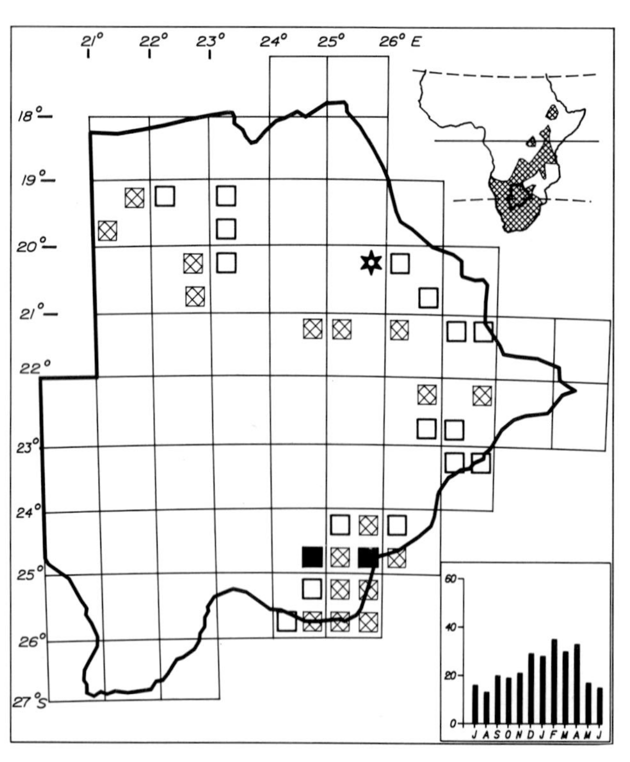

Status

Common to locally very common resident in the southeast. Sparse to fairly common in the east and north. Considerable local movements occur. No regular pattern is evident except that records and numbers are higher in the summer months. In pairs when breeding, otherwise in small or large congregations depending on the size of the habitat, e.g. 5–50 birds at sewage ponds, 100–500 birds on large dams.

Habitat

Dams, lakes, lagoons, ponds and flooded pans. Tolerates saline conditions but it is seen mainly on fresh water with floating vegetation.

Analysis

36 squares (16%) Total count 296 (0,17%)

AFRICAN FINFOOT 229

Podica senegalensis

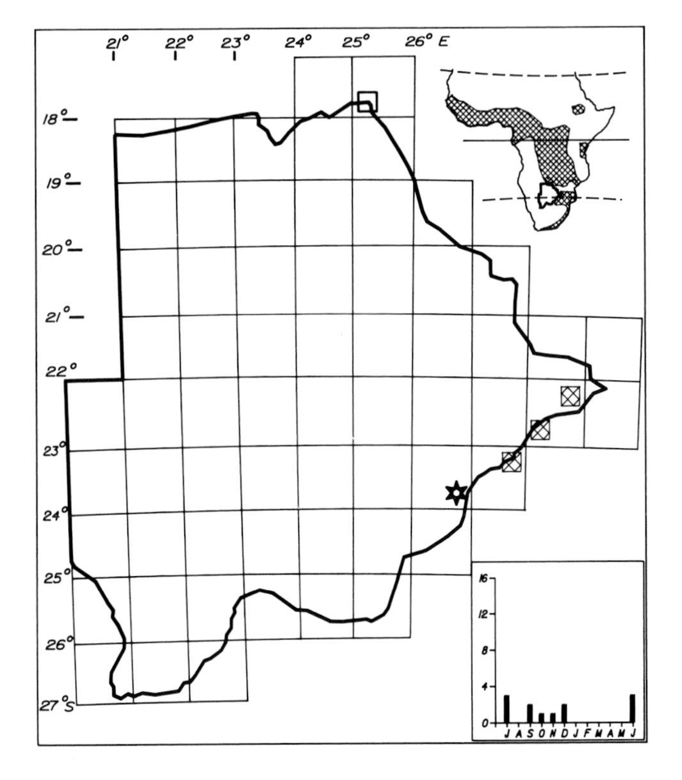

Status

Uncommon resident at the western limit of its southern African range along the Limpopo River. There is one unconfirmed record from the Okavango region. Recently reported fairly regularly from the Kasane area. Its status on the northern rivers needs investigation as it is known to occur on the Zambezi River. Easily overlooked because of its secretive habits. Solitary or in pairs.

Habitat

Large perennial rivers and their tributaries on little-disturbed stretches with well-wooded and vegetated cover on the banks. The water may be smooth and deep or turbulent and shallow over rocks.

Analysis

5 squares (2%) Total count 17 (0,009%)

KORI BUSTARD 230

Ardeotis kori

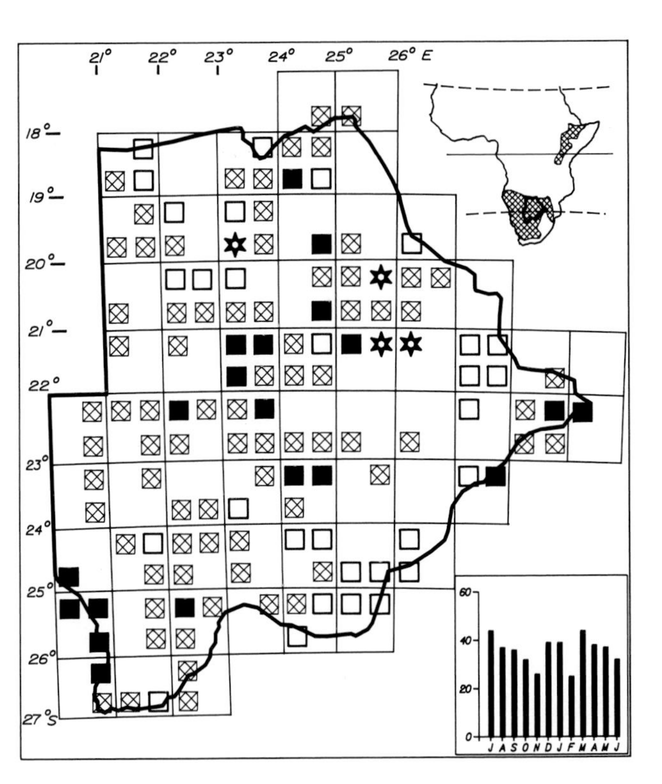

Status

Fairly common to very common resident in all regions. Sparse in most of the southeast and thinly distributed in the east except in the Tuli Block. Very common in the Nossob Valley, in some areas of the central Kalahari, at Nxai Pan and Savuti. In many areas it is unpredictable in occurrence. One egglaying record in December 1928; one large chick in November 1988. Usually solitary or in pairs.

Habitat

Grassland and open bush and tree savanna in semiarid areas. Also in duneland, fossil valleys and in wide river valleys in arid regions. Avoids woodland and areas with lots of trees. Surprisingly absent from apparently suitable grass plains in the southeast.

Analysis

130 squares (57%) Total count 487 (0,27%)

STANLEY'S BUSTARD

231

Neotis denhami

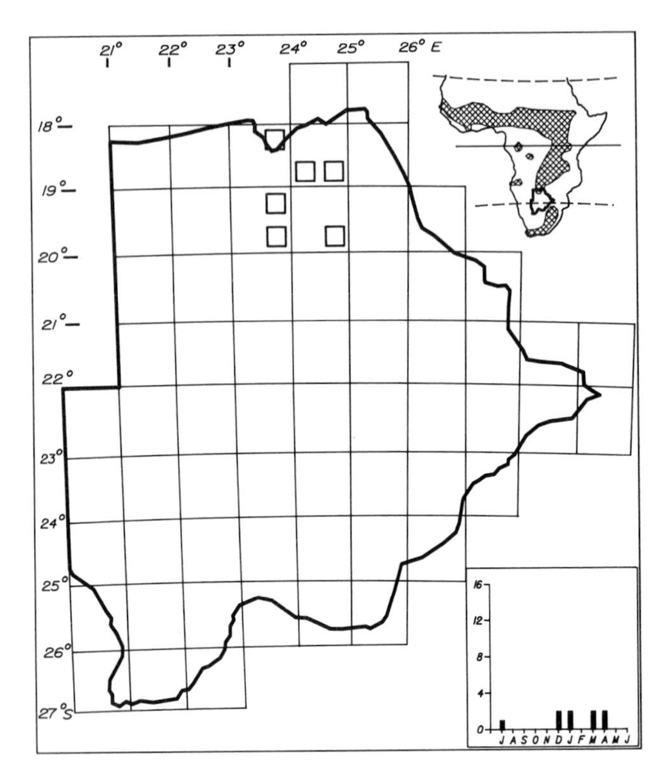

Alternative name

Denham's Bustard

Status

Very scarce. Status uncertain. Six records fall between December and April and one on 30 July. These may be intra-African migrants of the subspecies *N. d. jacksoni* at the southern limit of their range during the breeding season (February to July in Zambia) or postbreeding dispersal of the subspecies *N. d. stanleyi* from the Transvaal where it breeds from October to December. All sightings have been in the north between Linyanti and Nxai Pan. Solitary.

Habitat

Grassland and open savanna with short grass such as on floodplains, edges of marshland and airfields.

Analysis

6 squares (3%) Total count 9 (0,004%)

REDCRESTED KORHAAN

237

Eupodotis ruficrista

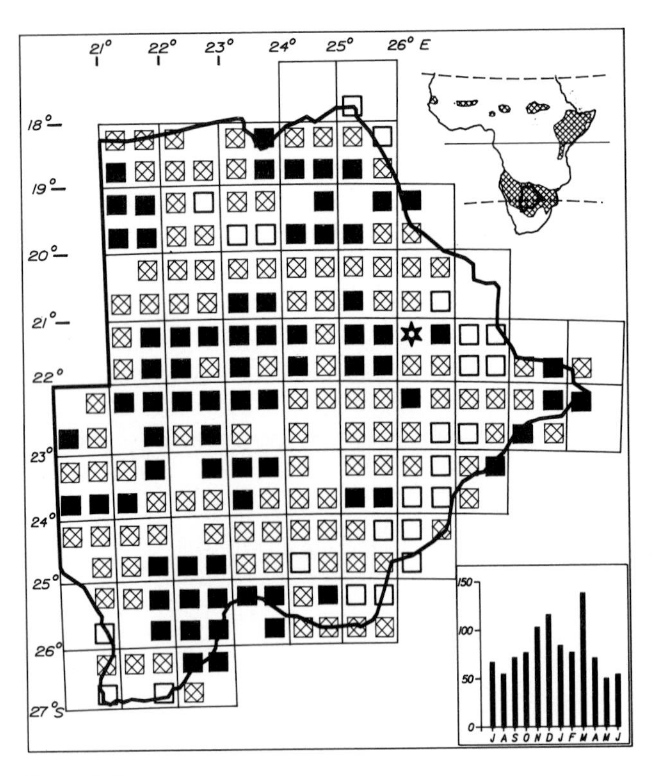

Status

Common to very common resident throughout the country. Recorded as very common in 32% of squares. It produces one of the most characteristic and frequently heard bird calls of the Kalahari desert—a prolonged descending piping call. Egglaying in October. Usually solitary; the male more commonly visible. Heard more often than seen.

Habitat

Tree and bush savanna on Kalahari sand with moderate grass cover. Most commonly in *Acacia* and *Terminalia* savanna, also in scrub savanna, mopane and edges of broadleafed woodlands. Usually not far from trees and absent from the open grassland niche occupied by the Black Korhaan.

Analysis

208 squares (90%) Total count 1032 (0,57%)

BLACKBELLIED KORHAAN

238

Eupodotis melanogaster

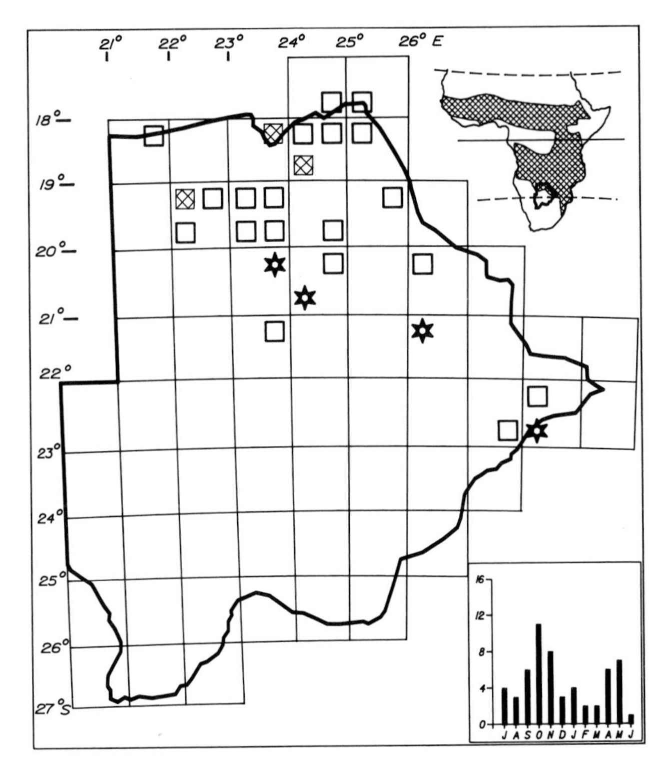

Status

Sparse to fairly common resident of the north to as far south as Deception Valley. Most frequently recorded at Gomare, Savuti and Linyanti. Reappears in the Tuli Block in the east —its status between this area and Nata needs more study. Egglaying in January, February and March. Usually solitary or in pairs.

Habitat

Long grass in open woodland, broadleafed bush and tree savanna, ecotone of plains and floodplains, edges of marshland. Occurs only in the high rainfall areas.

Analysis

24 squares (10%) Total count 64 (0,04%)

BLACK KORHAAN

239

Eupodotis afra

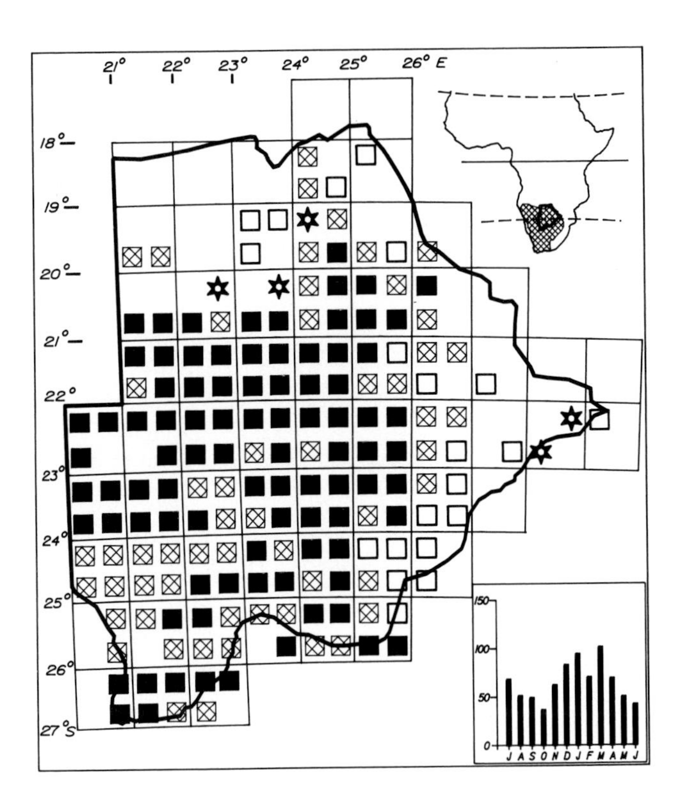

Status

Generally a very common resident within its main range south of 20°S and west of 26°E. Extends north through the Nxai Pan and Savuti regions. Sparse or uncommon east of 26°E and in peripheral areas of the north where it is mainly absent. Egglaying in October. Usually solitary; sightings are predominantly of the conspicuous males.

Habitat

Grassland plains, open savanna, semidesert scrub, grass-covered dunes, fossil valleys, usually where the grass is 0,5 m to 1 m high. Males often use termite mounds for perching and as calling posts.

Analysis

173 squares (75%) Total count 871 (0,49%)

AFRICAN JACANA

Actophilornis africanus

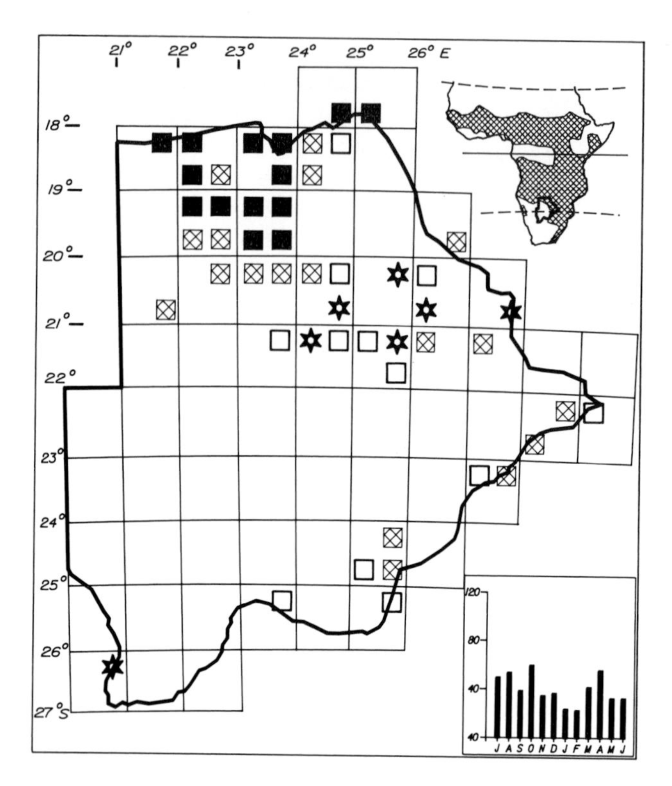

Status

Common to very common resident of the Okavango, Linyanti and Chobe river systems. Extends west along the Nhabe River to Lake Ngami and south along the Boteti River to Mopipi and Orapa. Sparse on the Makgadikgadi Pans. Sparse to locally fairly common along the Limpopo River and in the southeast. Recorded once on the Molopo River at Bray and there is a past record from the Nossob Valley. Local movements occur which are most noticeable in areas away from permanent water. Egglaying January, February and March. Usually in pairs or solitary, but several birds and sometimes flocks may be present at one site.

Habitat

Swamps, lagoons, vegetated wet pans, margins of lakes and dams; usually where there are water lilies or other floating aquatic vegetation.

Analysis

50 squares (22%) Total count 527 (0,29%)

LESSER JACANA

Microparra capensis

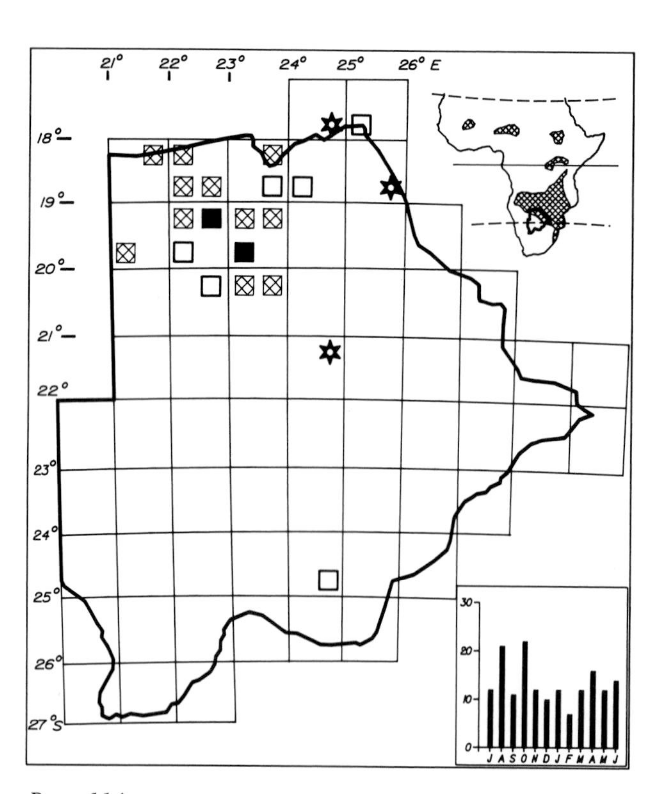

Status

Fairly common to locally very common resident of the Okavango Delta and the Linyanti River. Sparse at Savuti, Kasane, Nokaneng and Lake Ngami during this survey but it may become more common in these areas in high rainfall years. Occurs in the upper reaches of the Boteti River as far as Makalamabedi. Local movements occur and some long distance wanderings take place erratically as evidenced by the records from Sefophe (not shown) and Jwaneng. Egg-laying March. Solitary or in pairs, occasionally in small groups of up to 15 birds.

Habitat

Swamps, marshes, vegetated pans, lagoons and lakes where there is floating aquatic vegetation.

Analysis

23 squares (10%) Total count 171 (0,09%)

PAINTED SNIPE 242

Rostratula benghalensis

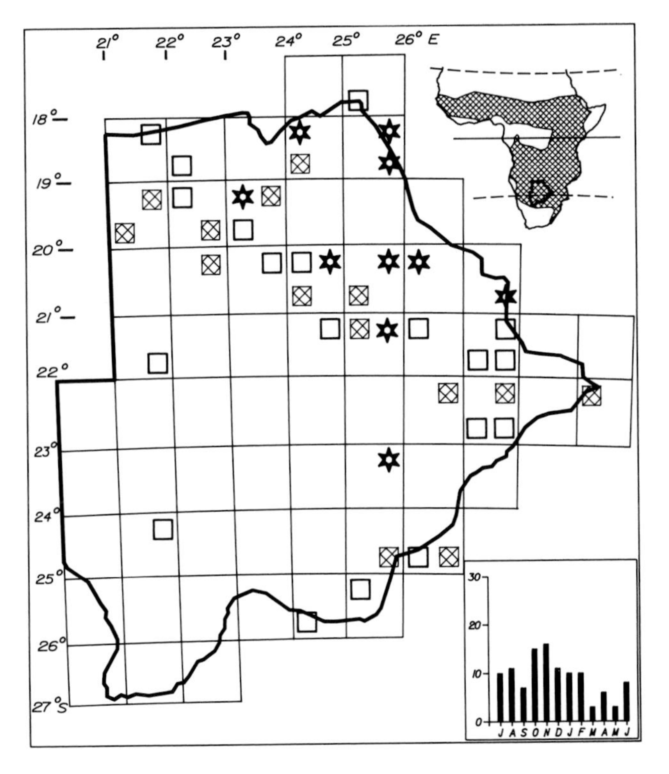

Status

Sparse to uncommon resident mainly in the north, east and southeast. Present in all months of the year but it has unpredictable local movements and part of the population may be intra-African migrants. Some east/west movement to and from Zimbabwe is suspected (Irwin 1981). One bird was found on a 10 m long muddy depression at Hukuntsi in March (a long way from its known distribution). Egglaying September, October and February. Solitary or in pairs, occasionally in groups of up to 8 birds.

Habitat

Marshes, floodplains, edges of lakes, dams and ponds, seasonal pans, reedbeds.

Analysis

43 squares (19%) Total count 129 (0,07%)

RINGED PLOVER 245

Charadrius hiaticula

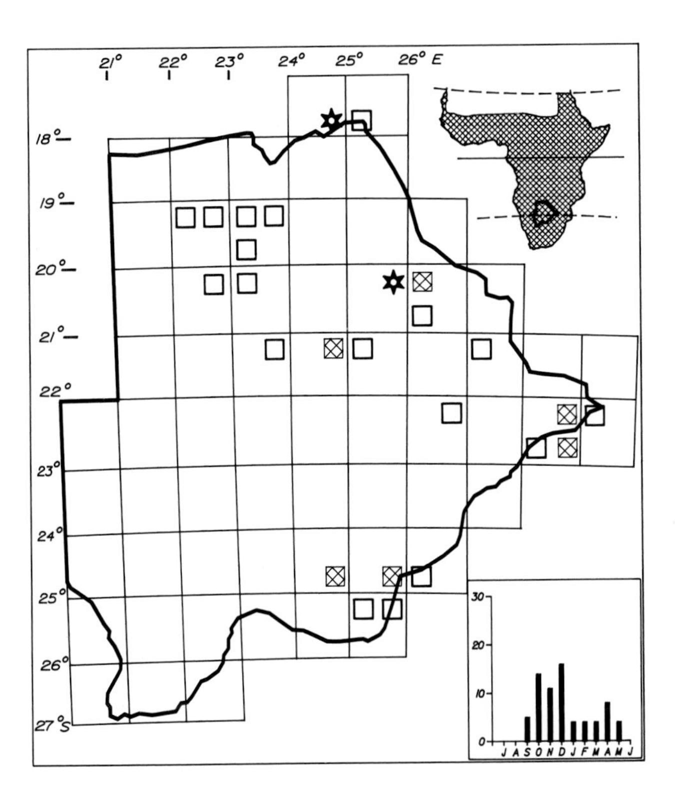

Status

Sparse to fairly common Palaearctic migrant recorded at scattered localities in the north, east and southeast. Present from late September to April with a few birds remaining until May. It is commonest between October and December. There is one record of an overwintering bird in July 1988. Birds usually remain at one locality for a short period e.g. one to ten days before moving elsewhere. Usually solitary or in small groups of 3–8 birds at one site.

Habitat

Shores of pans, dams, lakes, sewage ponds, pools and flooded depressions. Usually where there is exposed sand or mud and little or no vegetation.

Analysis

26 squares (11%) Total count 77 (0,04%)

WHITEFRONTED PLOVER

246

Charadrius marginatus

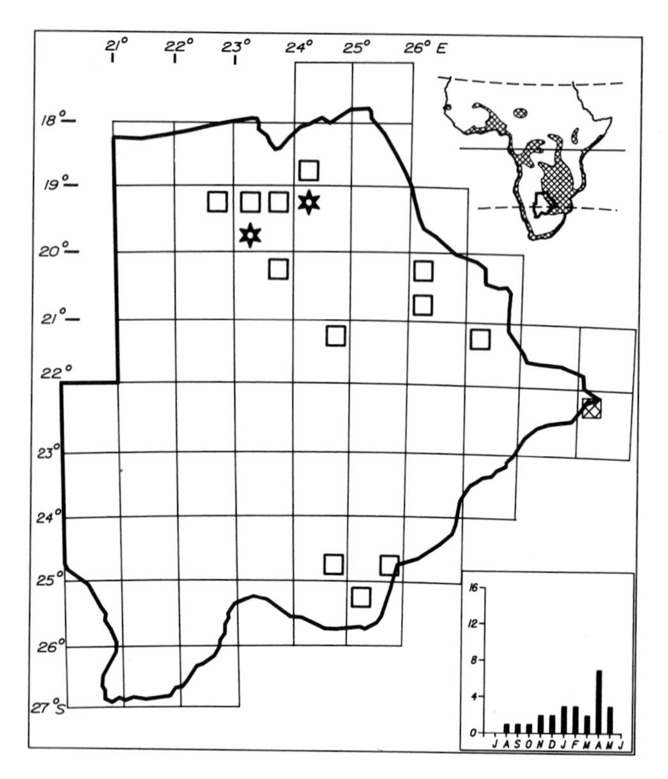

Status

Sparse resident in the north mainly in the Okavango Delta and on the Makgadikgadi Pans. Sparse and irregular in the southeast. It is not well known and its movements are poorly understood. Recorded in all months except June and July. The few available records show that it is commonest between November and May with a peak in April. In Zimbabwe it is present mainly between May and December and between March and May in the Transvaal. These inland movements are probably related but no clear pattern has yet emerged. Usually solitary or in small groups.

Habitat

Sandy shores of pans, lagoons, estuaries and dams. Occasionally at sewage ponds.

Analysis

15 squares (6%) Total count 30 (0,016%)

CHESTNUTBANDED PLOVER

247

Charadrius pallidus

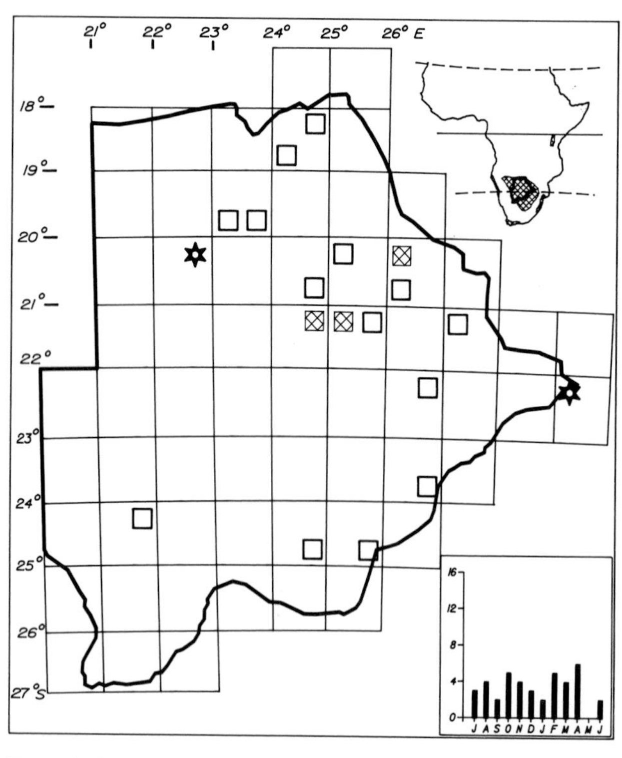

Status

Sparse to locally fairly common resident whose population in Botswana is based mainly at its breeding grounds in the Makgadikgadi region. Recorded only once at all other localities shown except Gaborone. In the southeast birds may occur as part of a local movement from pans in the western Transvaal. Recorded once in the west at Tshane Pan. Egglaying June to August and January to March—at Nata, Sua Pan, Mopipi and Orapa. Solitary or in pairs.

Habitat

Sandy shores of shallow inland waters, saline pans, brackish dams and estuaries.

Analysis

19 squares (8%) Total count 45 (0,02%)

KITTLITZ'S PLOVER

Charadrius pecuaris

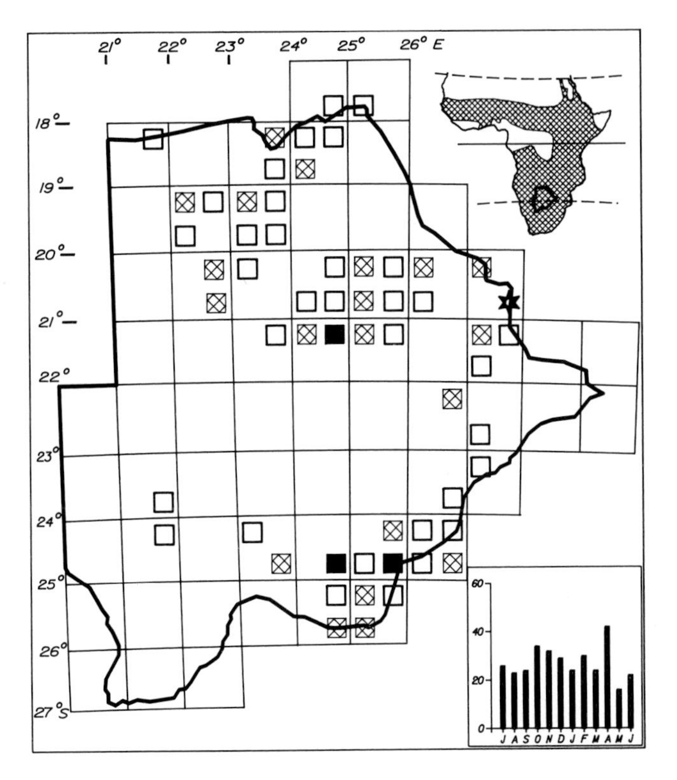

Status

Sparse to common resident of the north particularly in the Makgadikgadi region. Very common at Mopipi. Irregularly recorded in the east and sparse to locally very common in the southeast. It appears on western pans in some years. Although present in all months, considerable local movements occur with changing habitat conditions and these are likely to involve Zimbabwe and the Transvaal. There is no clear pattern but peak numbers occur in April. Egglaying all months July to November and January. Usually in pairs or small groups.

Habitat

Shores of pans, dams and lakes where there is bare sand or firm gravel. Also on short grass on earth banks, around sewage ponds, on airfields and on dry floodplains.

Analysis

57 squares (25%) Total count 341 (0,19%)

THREEBANDED PLOVER

Charadrius tricollaris

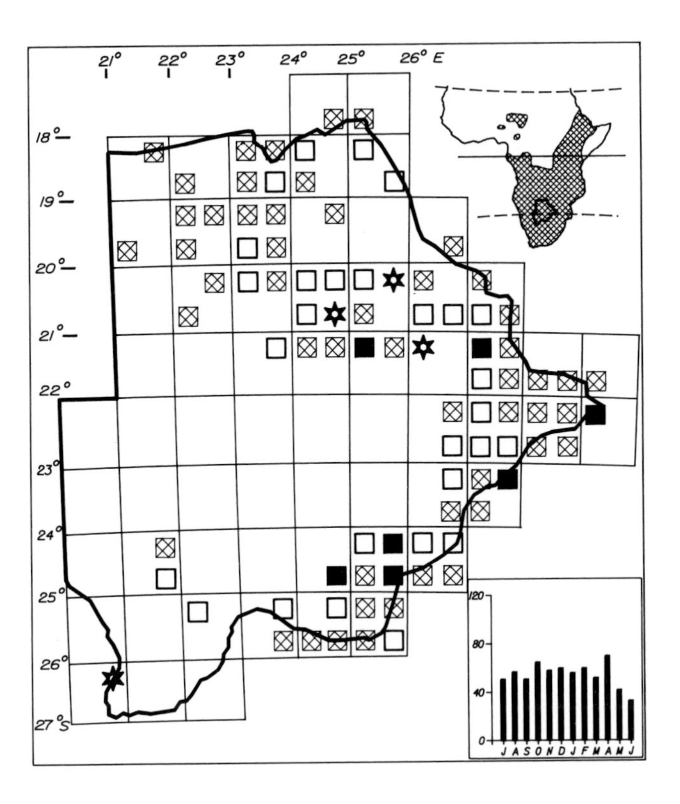

Status

Fairly common to locally very common resident of the north, east and southeast. It extends regularly west along the Molopo River to as far as Bray and occasionally turns up on pans in the west. It is the commonest and most predictable resident small plover. Local movements occur with changing habitat conditions but it is resident at some localities throughout the year. Egglaying October, February, March and April. Usually in pairs or small groups.

Habitat

Shores of pans, dams, sewage ponds, lagoons, floodplains, lakes and rivers. May be found on bare sand, gravel or rocky ground. Occasionally at flooded depressions in savanna, roadside pools and ditches.

Analysis

90 squares (39%) Total count 698 (0,38%)

CASPIAN PLOVER

252

Charadrius asiaticus

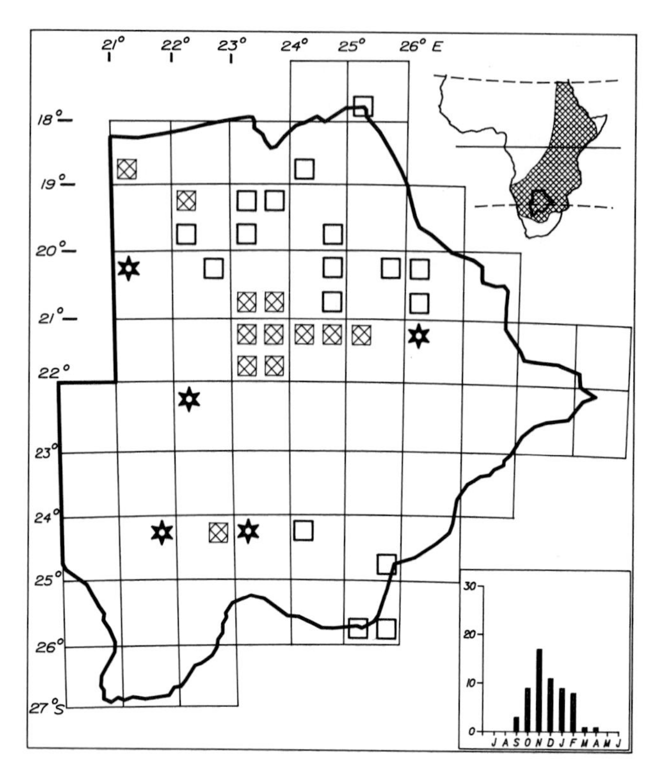

Status

Sparse to locally fairly common Palaearctic migrant mainly to areas north of 22°S. It is present between September and February (earliest date 20 September) with very few birds remaining until April. It arrives in nonbreeding dress and moults into breeding dress during its stay in Botswana. Most birds have moulted by the end of January. Its northward migration starts earlier than most other Palaearctic species. Most records are from the Makgadikgadi Pans west to Ngamiland. Usually in small groups of 10–30 birds, but there were over 100 birds from 1922C in September 1987, solitary on occasions.

Habitat

Flat open short grass, bare or recently burnt ground, in tree and bush savanna. Also in semidesert scrub savanna, open plains and on large sparsely covered Kalahari pans. Not associated with water.

Analysis

34 squares (15%) Total count 70 (0,04%)

GREY PLOVER

254

Pluvialis squatarola

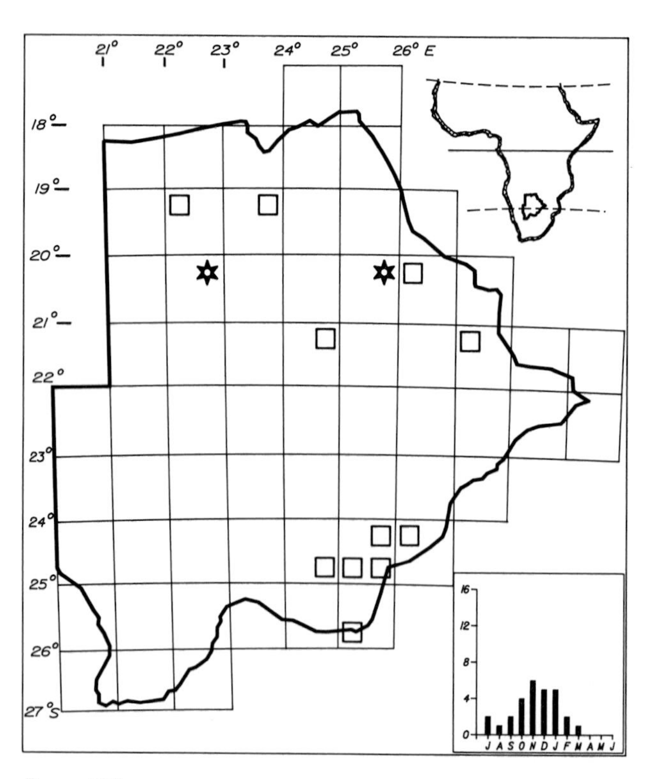

Status

Sparse Palaearctic migrant to scattered localities in the north, east and southeast. It is mainly a coastal migrant in southern Africa but there are numerous irregular inland records. Present mainly between October and January but some birds arrive in July and some remain until March. Recorded in every year except 1986. Usually solitary, but small groups of up to 6 birds may occur.

Habitat

Shores of dams, pans, sewage ponds and inlets; usually but not always where there is a stretch of sand or mud.

Analysis

13 squares (6%) Total count 34 (0,018%)

CROWNED PLOVER

Vanellus coronatus

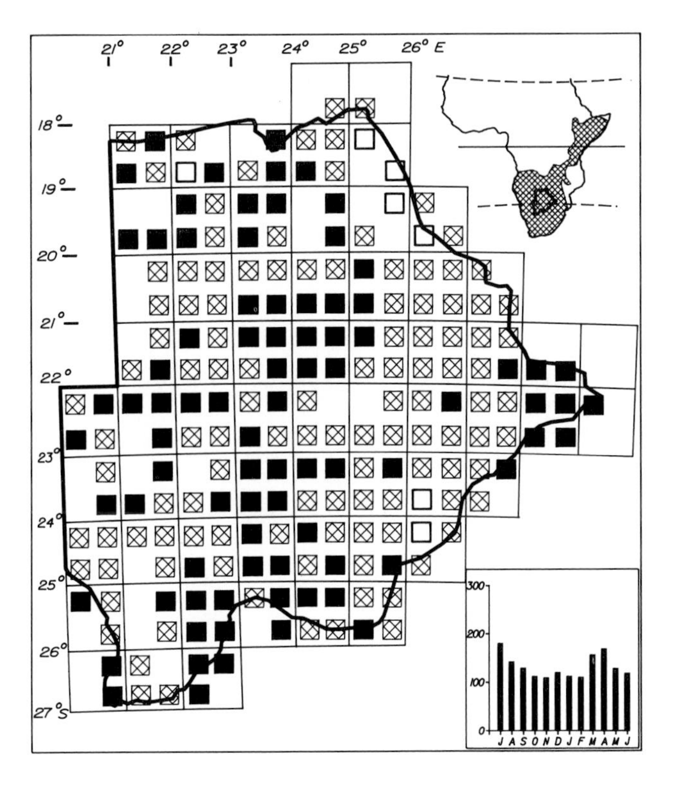

Status

Common to very common resident throughout the country. It is the commonest and most widespread plover in the territory. Recorded as very common in 36% of the squares. Present in all months of the year. Local movements occur in relation to breeding. Egglaying January to April and August to October. In pairs when breeding, usually in flocks of 10–50 birds at other times.

Habitat

Open short grassland, burnt areas and bare ground. In a wide variety of habitats from tree and bush savanna to semidesert and dry river valleys in arid regions. Also on plains, floodplains, airfields, farmland and in towns and villages. Often on pans whether grassed or bare calcrete.

Analysis

204 squares (89%) Total count 1655 (0,92%)

BLACKSMITH PLOVER

Vanellus armatus

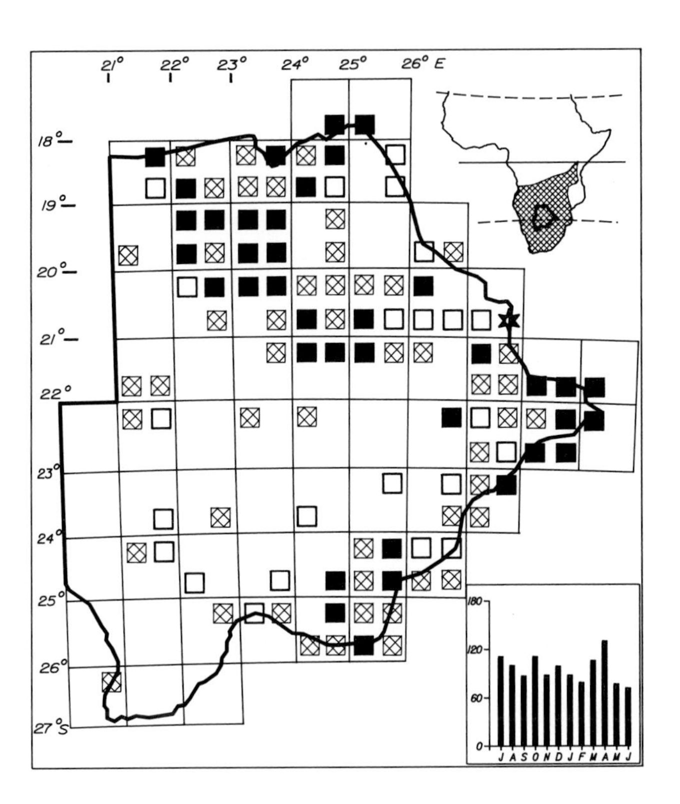

Status

Common to very common resident in the Okavango, Linyanti, Chobe, Boteti and Nata river systems. Common to very common along the Shashe and Limpopo river systems and locally in the southeast. Sparse to fairly common in suitable habitat in the west including the Nossob Valley. Egglaying March, April and all months from June to November. In pairs when breeding, in flocks of 10–30 birds at other times, but may congregate in hundreds at waterholes.

Habitat

Floodplains, marshes, edges of dams, lakes, lagoons, rivers, wet pans, sewage ponds. Usually near water and thus occasionally found in towns, villages and cattle posts near boreholes, wells, irrigation and spillways.

Analysis

111 squares (48%) Total count 1214 (0,67%)

WHITECROWNED PLOVER

Vanellus albiceps

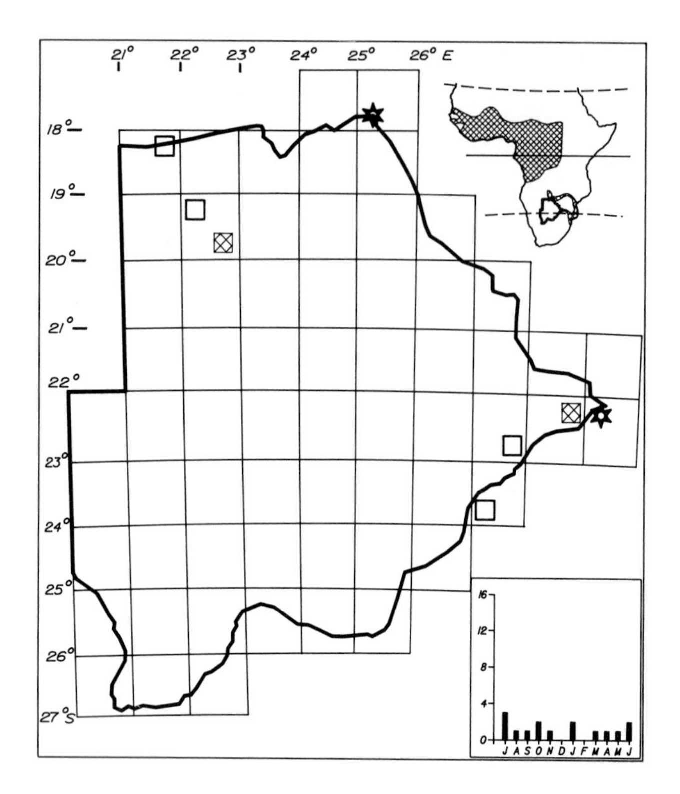

Status

Very sparse and localised resident recorded mainly on the Zambezi River at Kazungula (nine records). Recorded twice at Shakawe and Gomare and once at each of the other localities shown. It is only known from the larger tropical rivers and is unlikely to occur elswhere. The southernmost locality on the Limpopo River is near Buffel's Drift just south of the Tropic of Capricorn. Poorly known and warrants special study. Solitary or in pairs.

Habitat

Sandbanks, mud and sand shores of large rivers.

Analysis

8 squares (3%) Total count 19 (0,01%)

WATTLED PLOVER

Vanellus senegallus

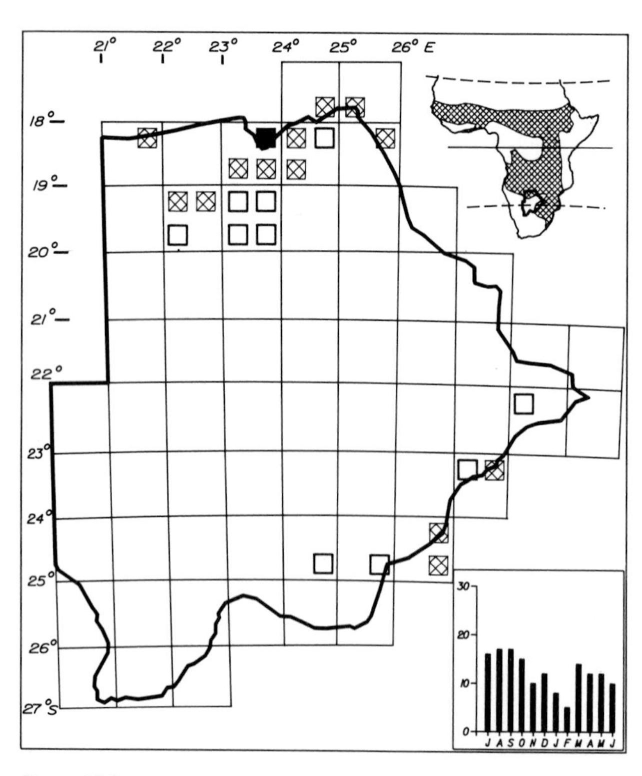

Status

Fairly common to common resident of the Okavango, Linyanti and Chobe river systems. Recorded as very common at Linyanti. Reappears in the east along the Limpopo and Ngotwane rivers where it is sparse to uncommon. Recorded twice at Jwaneng. Occurs in Botswana at the western limit of its southern African range. Present in all months of the year but local movements occur and some international migration is likely to Zimbabwe and the Transvaal. Usually in pairs, solitary or small groups of 4–10 birds.

Habitat

Floodplains, marshes, open grassland in river valleys. Infrequently at sewage ponds.

Analysis

24 squares (10%) Total count 155 (0,08%)

LONGTOED PLOVER

261

Vanellus crassirostris

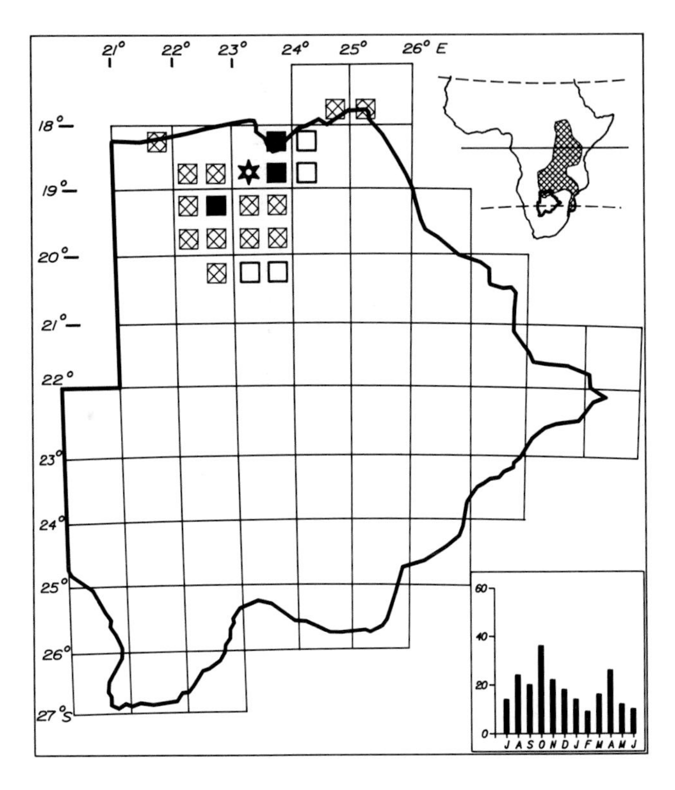

Status

Fairly common to locally very common resident of the Okavango, Linyanti and Chobe river systems to as far south as Lake Ngami. Sparse in the upper reaches of the Boteti River to nearly as far as Makalamabedi. It is confined to these regions of the north and unlikely to occur elsewhere. Mainly sedentary at permanent waters. Egglaying August. Solitary, in pairs or small groups.

Habitat

Permanent water with floating vegetation on rivers, lagoons, swamps and marshes.

Analysis

21 squares (9%) Total count 230 (0,13%)

RUDDY TURNSTONE

262

Arenaria interpres

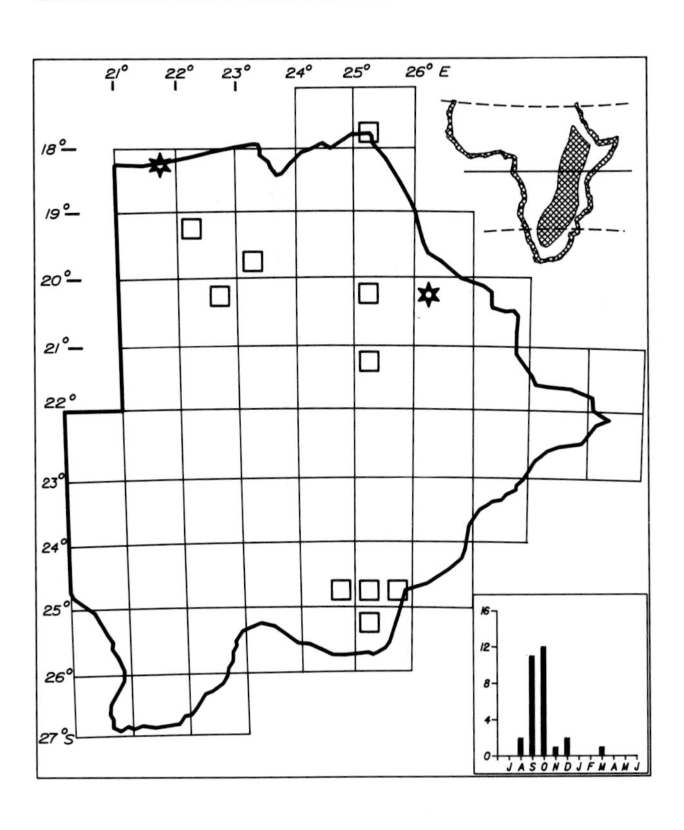

Status

Very sparse but regular Palaearctic passage migrant recorded in the north and southeast. It is principally a coastal migrant in Africa but there are regular inland records. Recorded between August (15th) and December with a peak in September. Only recorded once during northward passage— 16 April at Lake Ngami. The records are too few to make a definitive statement on its movements. Usually solitary, infrequently up to 4 birds.

Habitat

Shores of lakes, dams, pans and sewage ponds.

Analysis

12 squares (5%) Total count 32 (0,02%)

COMMON SANDPIPER

264

Actitis hypoleucos

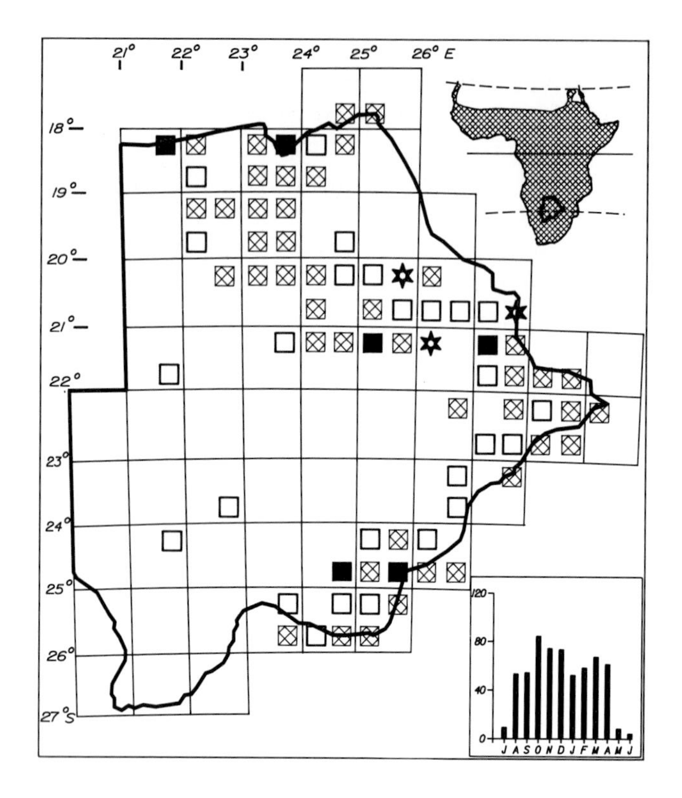

Status

Common to locally very common Palaearctic migrant to the north and east. Rare in the west during this survey but may occur regularly on western pans in higher rainfall years. Fairly common along the Molopo River to as far west as Bray. Southward passage starts in August and peaks in October. Submaximal numbers remain fairly constant between November and February before peaking on northward passage in March and April. Very few birds remain throughout the year. Usually solitary, but several birds may be present at any one site.

Habitat

Shores of dams, lakes, pans, rivers, ponds, sewage ponds and lagoons. Also at wet ditches, marshes, irrigation and occasionally puddles on untarred roads.

Analysis

78 squares (34%) Total count 621 (0,34%)

GREEN SANDPIPER

265

Tringa ochropus

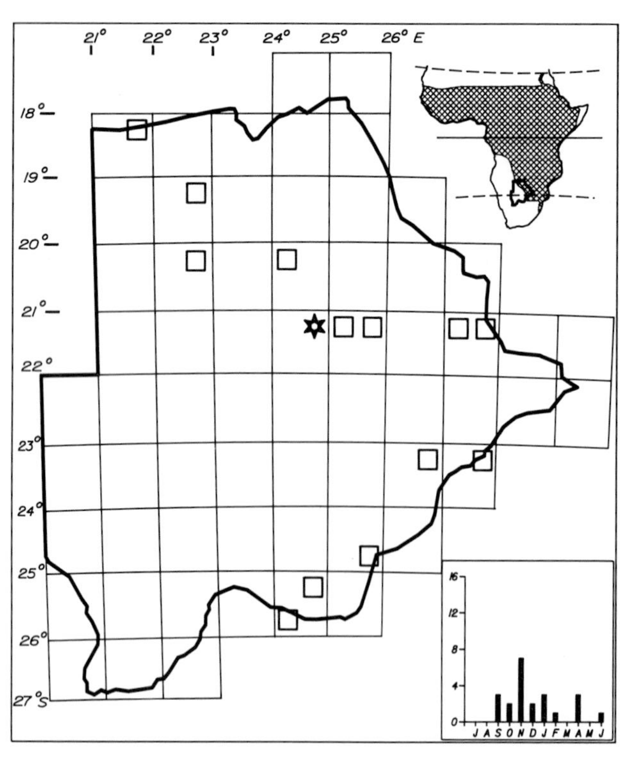

Status

Very sparse Palaearctic migrant to the north and east. Unlikely to occur in other regions of the country. Occurs in Botswana at the southwestern edge of its migratory range in continental Africa. Present between September and April. It appears to be a regular migrant but may not occur in some years. Poorly known. Usually solitary.

Habitat

Quiet shaded waters on the margins of rivers, dams and ponds. Occasionally at flooded ditches. Usually close to woodland or other vegetation giving shelter.

Analysis

14 squares (6%) Total count 23 (0,013%)

WOOD SANDPIPER

266

Tringa glareola

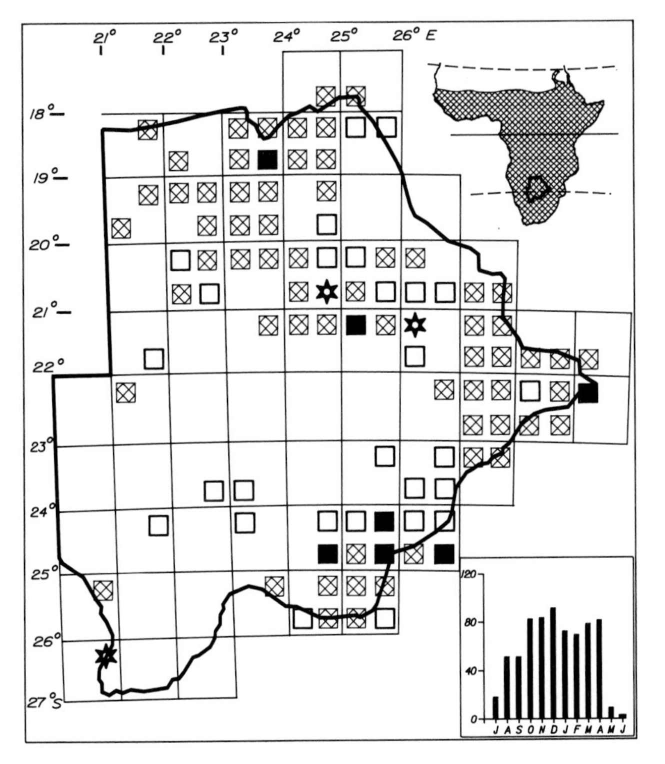

Status

Fairly common to very common Palaearctic migrant mainly to the north, east and southeast. Occurs throughout the country in suitable habitat. The commonest and most predictable of the Palaearctic waders. Southward migration starts in late July and peaks from October to December. Main northward passage occurs in March and April. About 5% of records are of birds which remain between May and July. Usually in loose gatherings of 5–30 birds or solitary, occasionally in flocks of up to 100 birds.

Habitat

Shores of dams, lakes, lagoons, ponds, inundated pans, rivers, sewage ponds. Also when suitably wet conditions prevail it is found on floodplains, marshes, seasonally flooded depressions in savanna, ditches, roadside culverts and spillways.

Analysis

102 squares (44%) Total count 721 (0,40%)

REDSHANK

268

Tringa totanus

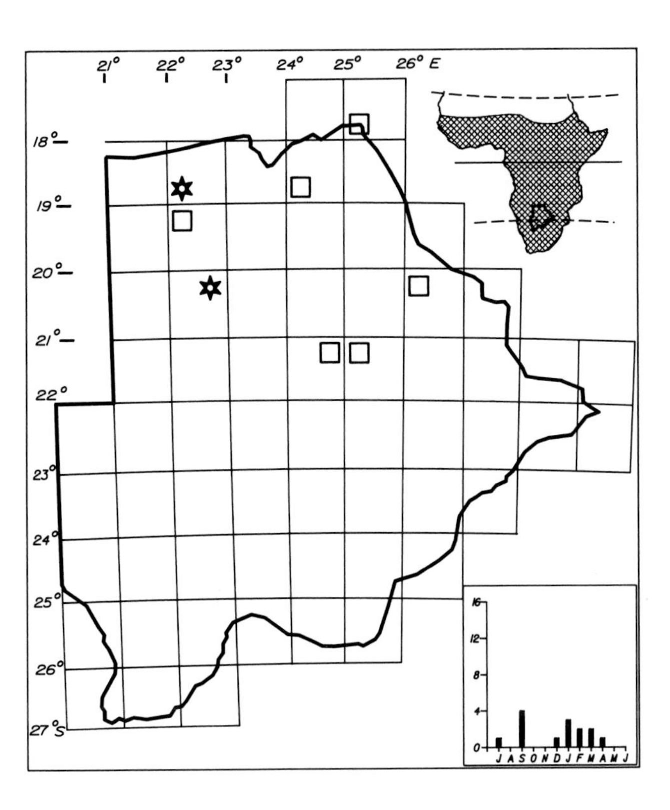

Status

Very sparse Palaearctic migrant to the northern wetlands. Recorded as far south as Mopipi, Orapa and Sua Pan. Southward passage appears to be short in September and birds are seen mainly on northward passage between January and April. There are two records of overwintering birds—at Lake Ngami on 16 June 1969 and at Nata on 8 July 1978. Smithers (1964) reports birds resident at Kanye from October to January (not December to April as reported by Wilson, 1984). Earliest date 6 September, latest date 28 April. There have been six records over the past 10 years. Poorly known. Usually solitary.

Habitat

Shores of lakes, lagoons, pans and estuaries.

Analysis

8 squares (3%) Total count 15 (0,008%)

MARSH SANDPIPER

Tringa stagnatilis

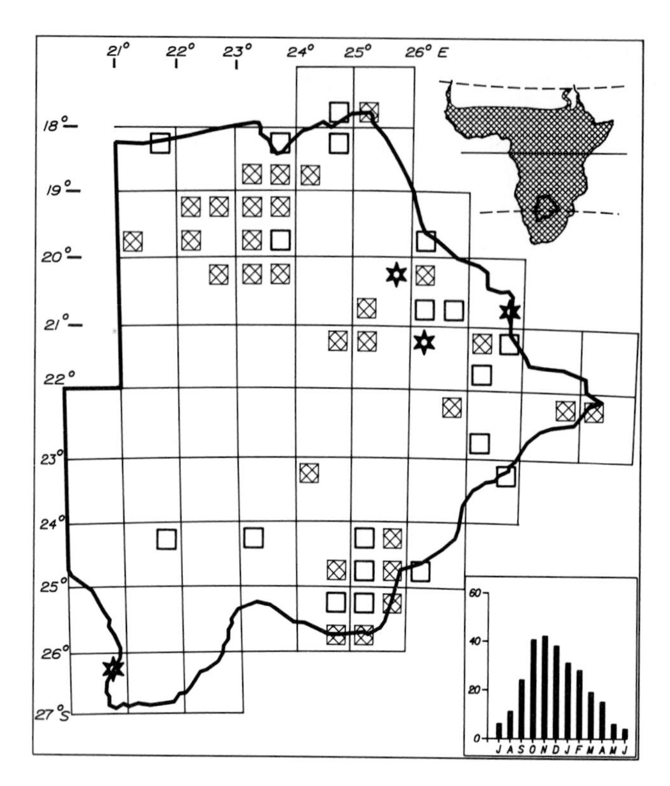

Status

Sparse to common Palaearctic migrant mainly to the north, east and southeast. When suitable habitat is available it is likely to occur throughout the country. The first arrivals appear on southward migration in late August and peak between October and December. Northward migration is not marked and records decline from January to April. A few birds remain throughout the year. Usually solitary, but several birds may be present at any one locality, occasionally in flocks of up to 50 birds.

Habitat

Edges of lakes, lagoons, dams, inundated pans, sewage ponds, rivers. Also at marshes and on wet floodplains.

Analysis

51 squares (22%) Total count 285 (0,16%)

GREENSHANK

Tringa nebularia

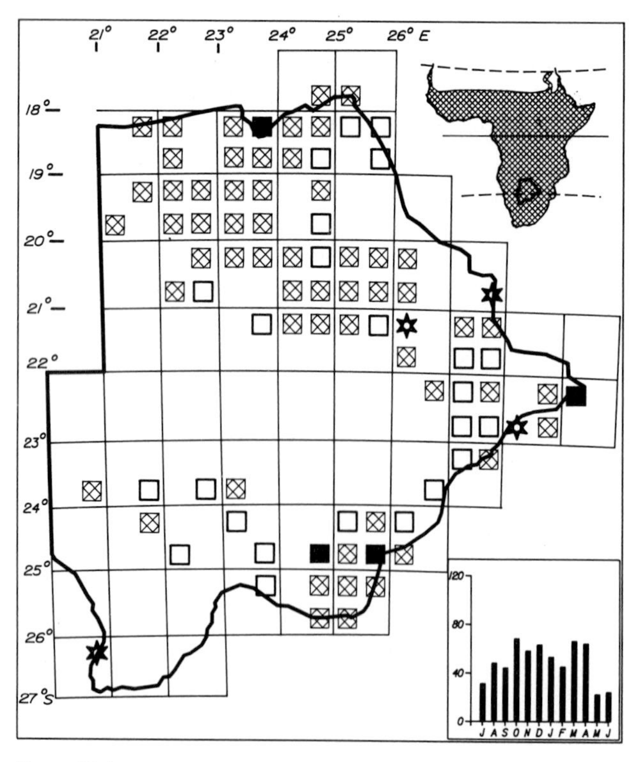

Status

Common to locally very common Palaearctic migrant mainly to the north, east and southeast. It extends into the west more commonly than any other migrant wader. Sparse to fairly common on western pans. Southward passage starts in late July and peaks in October to December. Northward passage peaks in March and April. There are more records of this species remaining throughout the year than any other Palaearctic migrant. Often solitary, but several birds may be present at any one locality, occasionally in flocks of 10–100 birds.

Habitat

Edges of dams, pans, lakes, lagoons, rivers and sewage ponds. Wades into deeper water than the Marsh Sandpiper. Also on marshes, flooded grassland and floodplains.

Analysis

88 squares (38%) Total count 620 (0,34%)

CURLEW SANDPIPER

272

Calidris ferruginea

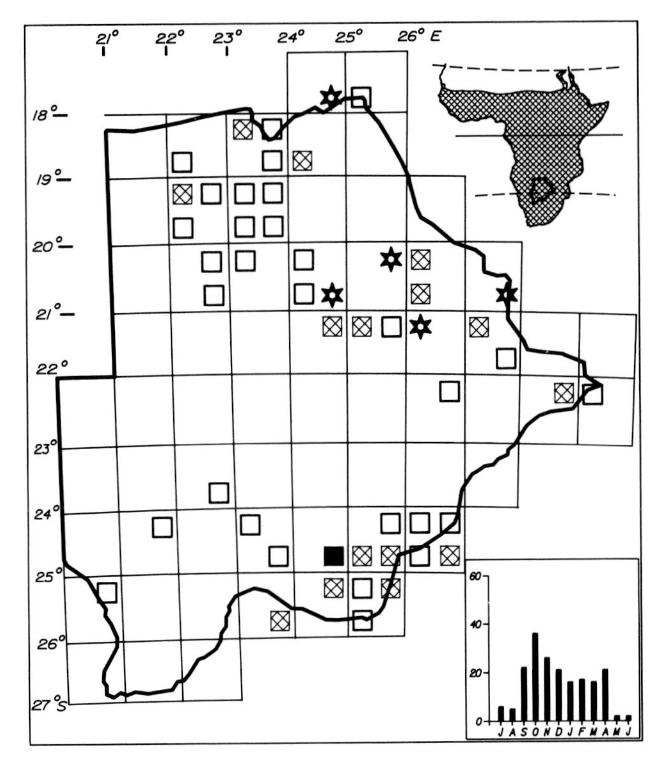

Status

Sparse to fairly common Palaearctic migrant to the north and east. Fairly common to locally very common in the southeast. Extends sparsely onto western pans. Southward migration starts in mid-August and peaks in October. Peak of northward passage is in April. A few birds remain throughout the year. It appears to be mainly a passage migrant as it is more common in areas south of Botswana, e.g. the Orange Free State and the southwestern Transvaal, than in Botswana. Birds in partial breeding dress occur in August and September and again in April. Usually in small flocks of 5–50 birds, sometimes up to 500 birds.

Habitat

Open shores of pans, dams, lakes, estuaries and sewage ponds. Occasionally on floodplains where there is sand or shingle patches—generally not on grass or vegetated areas.

Analysis

51 squares (22%) Total count 209 (0,12%)

LITTLE STINT

274

Calidris minuta

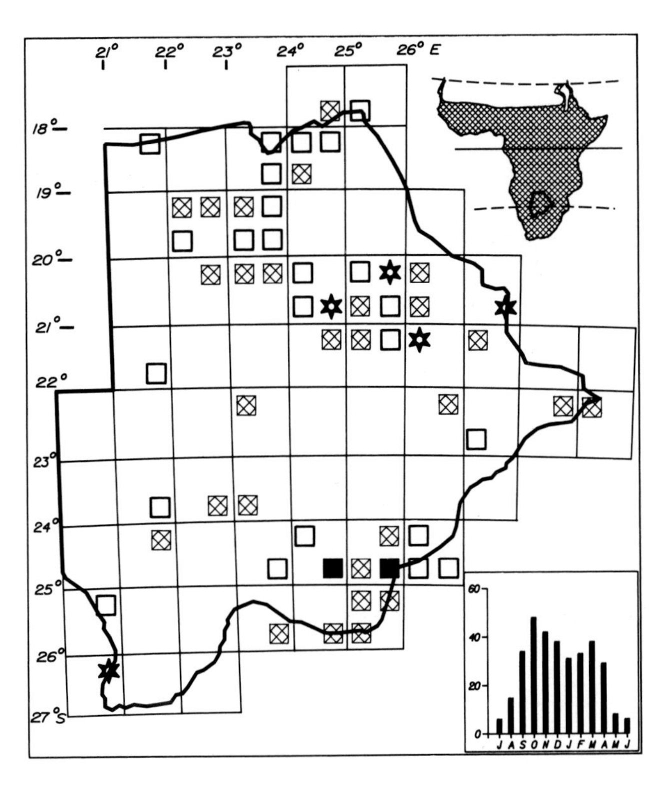

Status

Sparse to common Palaearctic migrant which occurs in suitable habitat throughout the country. Recorded as very common at Jwaneng and Gaborone. It first arrives on southward passage in mid-August and peaks during October and November. Numbers remain fairly constant until departure in March and April without clear evidence of a northward peak. Some birds remain throughout the year in some years. Usually in flocks of 6–100 birds, occasionally up to 500 or 1000 birds.

Habitat

Shores of dams, pans, lakes, sewage ponds and lagoons. Usually on mud or slime but also on sand or fine shingle.

Analysis

59 squares (26%) Total count 358 (0,19%)

SANDERLING

Calidris alba

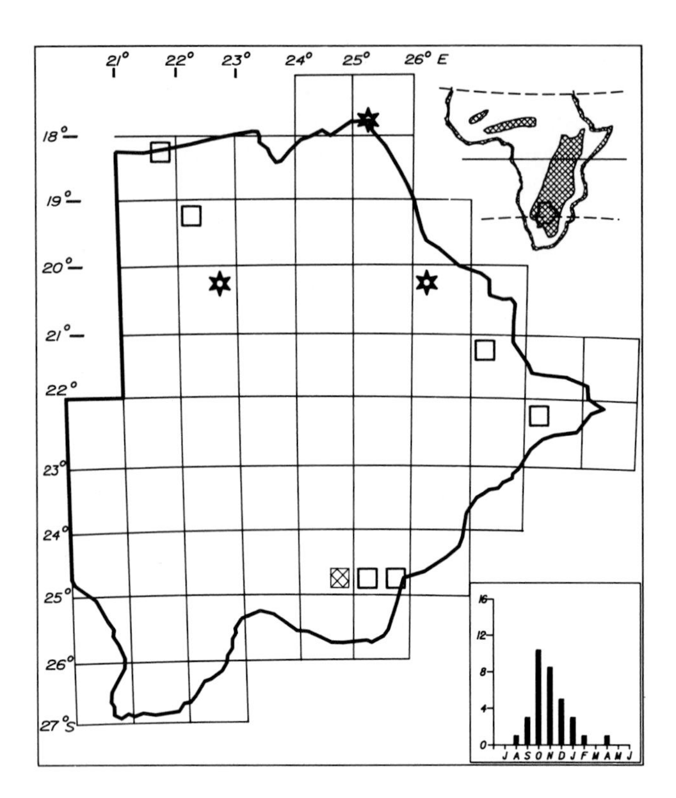

Status

Very sparse to uncommon Palaearctic migrant recorded at widespread localities in the north and east. Most of the records (79%) are from Jwaneng and Gaborone in the southeast, which suggests that it is overlooked at other suitable localities. Occurs mainly on southward passage starting in late August (30th) and reaching a peak in October and November. Records of sightings dwindle to February and the latest record was on 2 April. Usually solitary.

Habitat

Shores of dams, pans, sewage ponds and mudflats on margins of rivers.

Analysis

10 squares (4%)   Total count 36 (0,018%)

RUFF

Philomachus pugnax

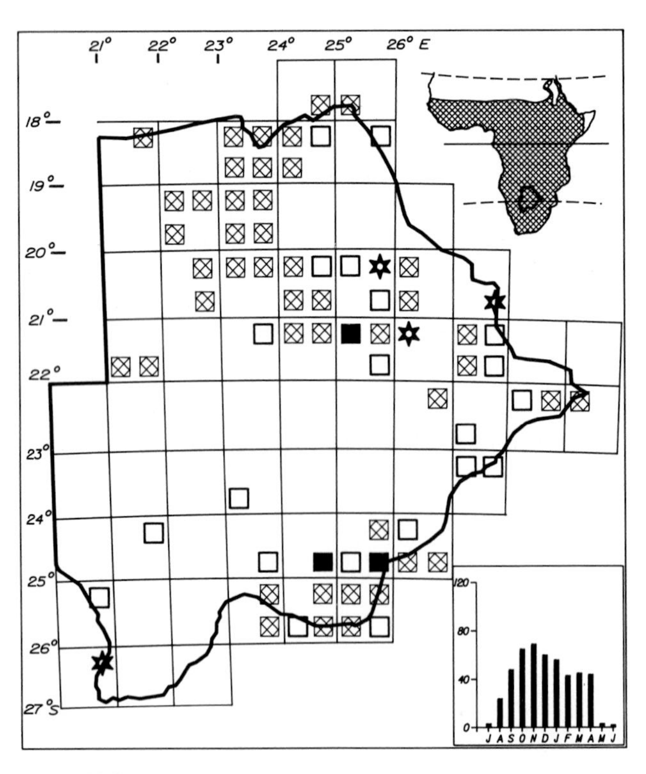

Status

Fairly common to common Palaearctic migrant to the north, east and southeast. Extends regularly west to as far as Bray and sparsely onto some western pans. Recorded as very common at Orapa, Jwaneng and Gaborone. Present mainly between August and April. Southward migration peaks from October to December and northward passage is mainly in March and April without a marked peak. About 1% of records are for May to July. Males in partial breeding dress are seen in August and September and again in April and May. Usually in flocks of 20–200 birds, but flocks of several thousands occur on large areas of suitable habitat.

Habitat

Shores of lakes, lagoons, rivers, dams, pans and sewage ponds. Also on floodplains, flooded grassland, marshes, ditches, wet drainage lines, irrigation and cultivation.

Analysis

73 squares (32%)   Total count 488 (0,27%)

ETHIOPIAN SNIPE

Gallinago nigripennis

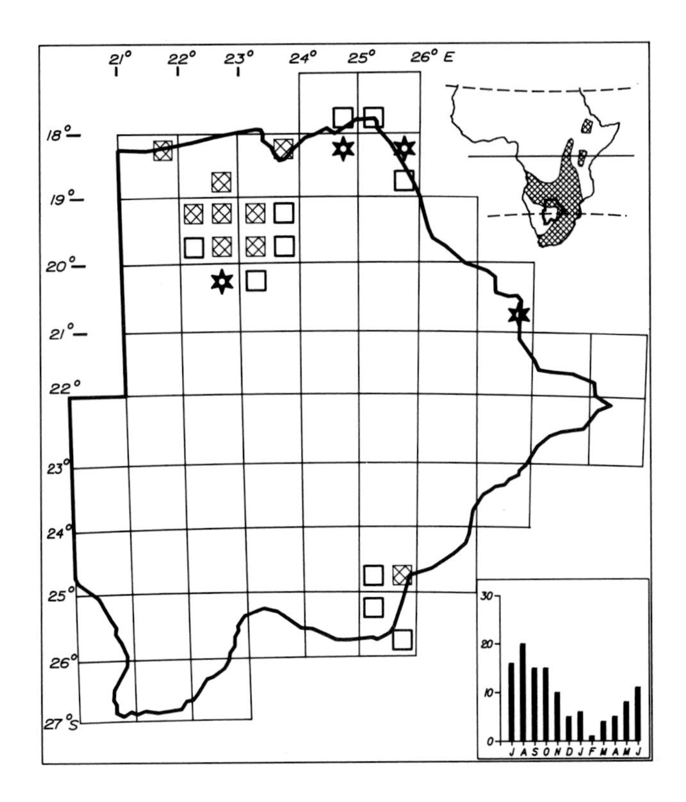

Status

Sparse to locally common resident of the Okavango, Linyanti and Chobe river systems. Sparse to fairly common in the southeast. Not recorded in the east during this survey but can be expected there in suitable habitat. Likely to be more widespread in high rainfall years. Local and international movements occur seasonally and with changes in habitat conditions. Most records fall between May and November. Egglaying October, 'drumming' in July and August. Solitary, in pairs or small groups.

Habitat

Marshes, reedbeds, flooded grassland and vegetated shores of lakes, dams and sewage ponds.

Analysis

23 squares (10%) Total count 125 (0,07%)

BLACKTAILED GODWIT

Limosa limosa

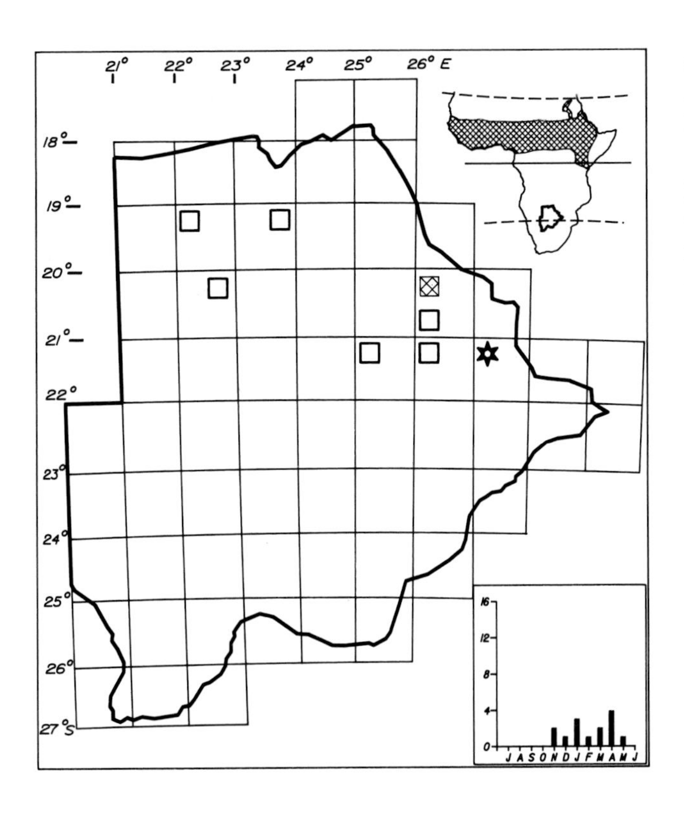

Status

A rare Palaearctic migrant to isolated localities in the north. Recorded as far south as Orapa. It normally 'winters' in the Sahel and occurs further south on the continent only in some years. Known to appear in larger numbers in some years than others and may not occur in Botswana every year. However it is likely to be more common than the Bartailed Godwit which is mainly a coastal migrant. Recorded between November and May. Eight birds recorded at Nata Delta in April 1974. Usually solitary, but often in company with other shorebirds such as ducks or waders.

Habitat

Shores of dams, lakes and pans. Also at marshes and estuaries. Sometimes in muddy inlets or creeks.

Analysis

8 squares (3%) Total count 14 (0,008%)

CURLEW

Numenius arquata

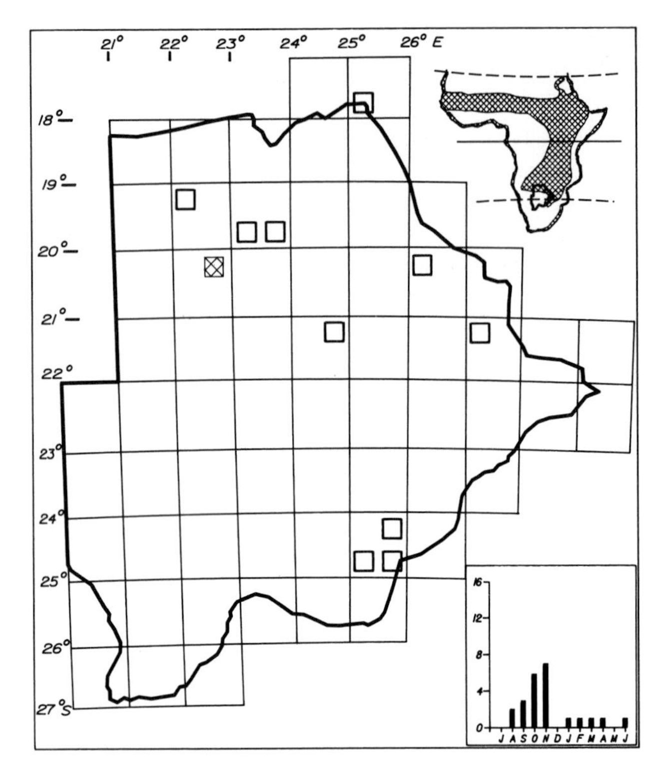

Status

Very sparse Palaearctic migrant to isolated areas in the north, east and southeast. It is mainly a coastal migrant but inland records are known from all countries surrounding Botswana. Most of the very few records fall between August and November but there are records for most months of the year. It has been recorded more frequently than the Whimbrel but the status of neither species is well understood. Usually solitary or in small groups of 2–4 birds.

Habitat

Muddy shores of lakes, pans, dams and sewage ponds. Also on floodplains and estuaries.

Analysis

11 squares (5%) Total count 24 (0,013%)

AVOCET

Recurvirostra avosetta

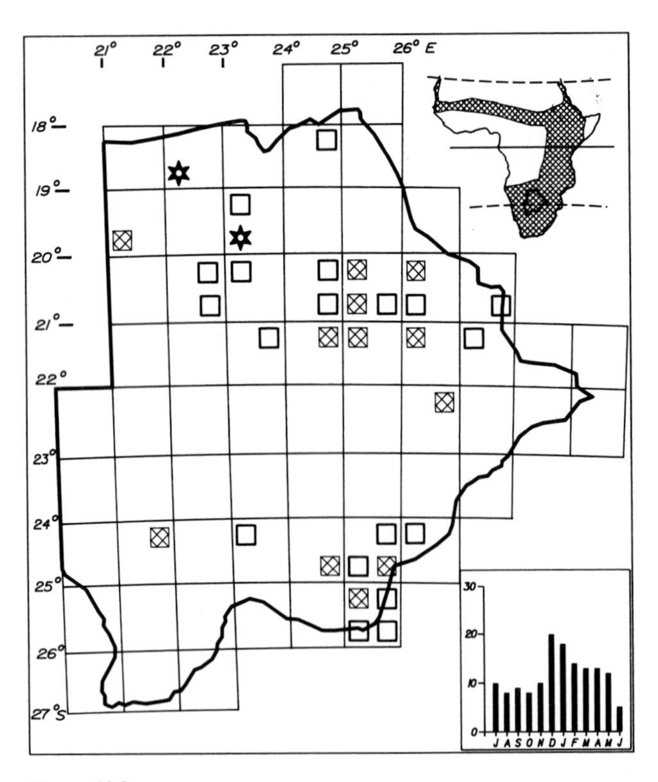

Status

Sparse to locally fairly common breeding nomad mainly in the Makgadikgadi region and in the southeast. Present irregularly in all months—mainly from December to May in the southeast and often in winter as well in the Makgadikgadi. Occurs where habitat conditions are suitable and is likely to be more widespread in years of high rainfall. Common and regular at Gaborone and Jwaneng. Egglaying May. Usually in small flocks of 4–20 birds, sometimes solitary.

Habitat

Shallow water on the edges of pans, dams, lagoons and sewage ponds.

Analysis

33 squares (14%) Total count 154 (0,08%)

BLACKWINGED STILT

Himantopus himantopus

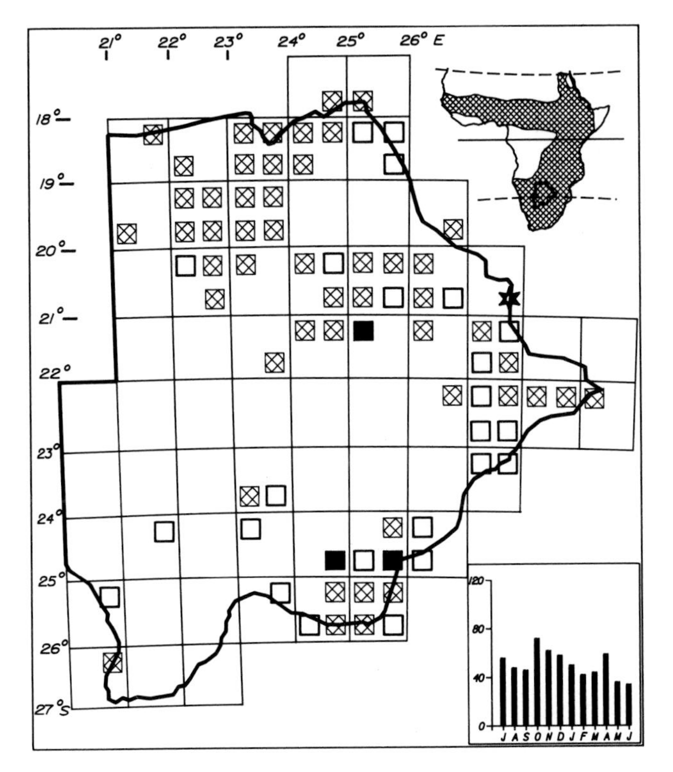

Status

Fairly common to common resident north of 22°S. Sparse to locally common in the east and southeast. Recorded as very common at Orapa, Jwaneng and Gaborone. Local movements occur with changing habitat conditions. Occurs sporadically on pans in the west and in the Nossob Valley. Likely to be more widespread in high rainfall years. Egglaying all months April to July and in October. Solitary, in pairs or flocks.

Habitat

Edges of pans, dams, lagoons, lakes and sewage ponds. Sometimes on marshes, inundated grassland and floodplains.

Analysis

78 squares (34%) Total count 628 (0,35%)

SPOTTED DIKKOP

Burhinus capensis

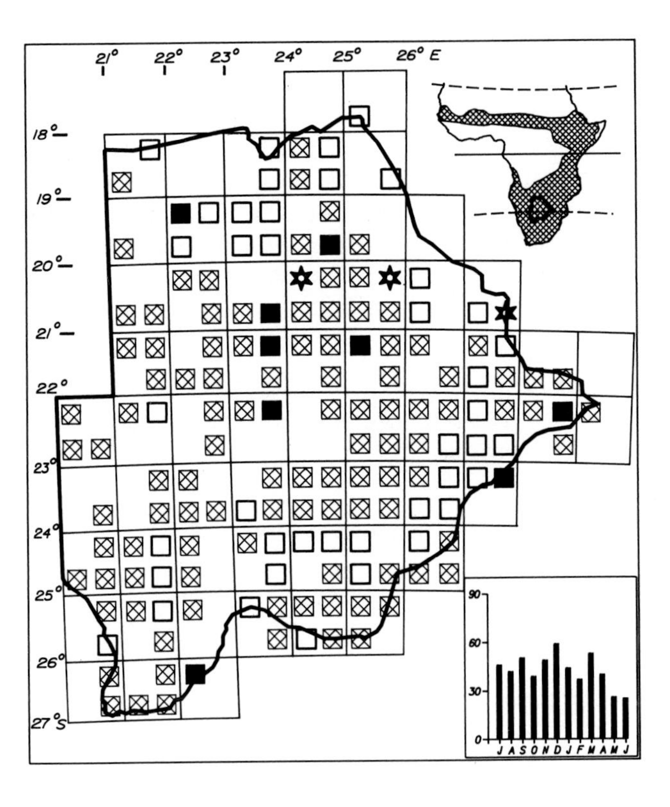

Status

Fairly common to very common resident throughout the country. Sparse in the river systems of the north. Sometimes underrecorded because of its nocturnal habits. Egglaying October, November, December. Solitary or in pairs.

Habitat

Short or sparse grassland in bush and tree savanna. semi-desert scrub, base of rocky hills, fossil valleys, dry grass pans. Rarely near water. Also occurs in cultivation, clearings in dry woodland, overgrazed areas near habitation and in towns. Shades under bushes, trees or woody shrubs during the day.

Analysis

157 squares (68%) Total count 598 (0,33%)

WATER DIKKOP

298

Burhinus vermiculatus

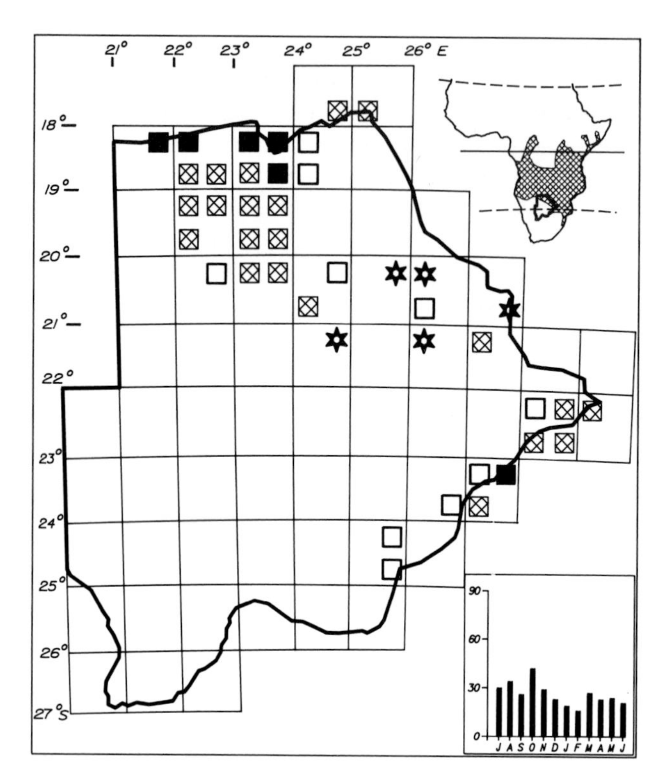

Status

Common to very common resident of the Okavango, Linyanti and Chobe river systems. Sparse to fairly common along the Boteti River to Sukwane and Nhabe River to Lake Ngami. In the east sparse to locally common along the Limpopo River and sparse along the Ngotwane River near Gaborone. Egglaying November to January. Solitary or in pairs, occasionally gathers into flocks of 10–30 birds.

Habitat

Banks and beaches of rivers, edges of lakes and dams, margins of lagoons and waterways.

Analysis

42 squares (18%) Total count 325 (0,18%)

BURCHELL'S COURSER

299

Cursorius rufus

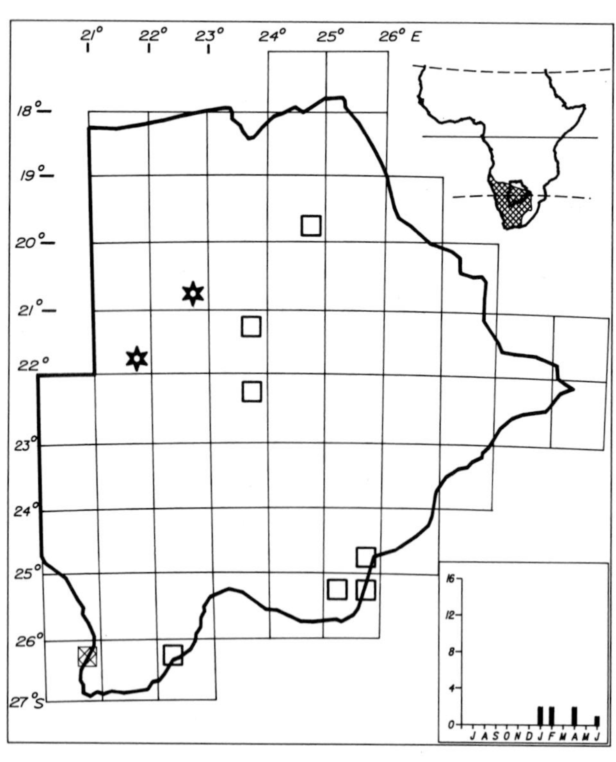

Status

A nomadic species of uncertain status. Known from 11 records in the past decade in widely scattered localities. Its status has changed considerably since 1963 when Smithers (1964) wrote that it was 'widespread on open plains and around pans in central and southwest Kalahari'. Maclean (1985) comments on a decline in numbers in the southern part of its range. It has been recorded more than once in three squares only. Poorly known in Botswana now and warrants special study.

Habitat

Open short grassland, bare or lightly-grassed pans, semi-desert scrub savanna.

Analysis

10 squares (4%) Total count 13 (0,007%)

TEMMINCK'S COURSER 300

Cursorius temminckii

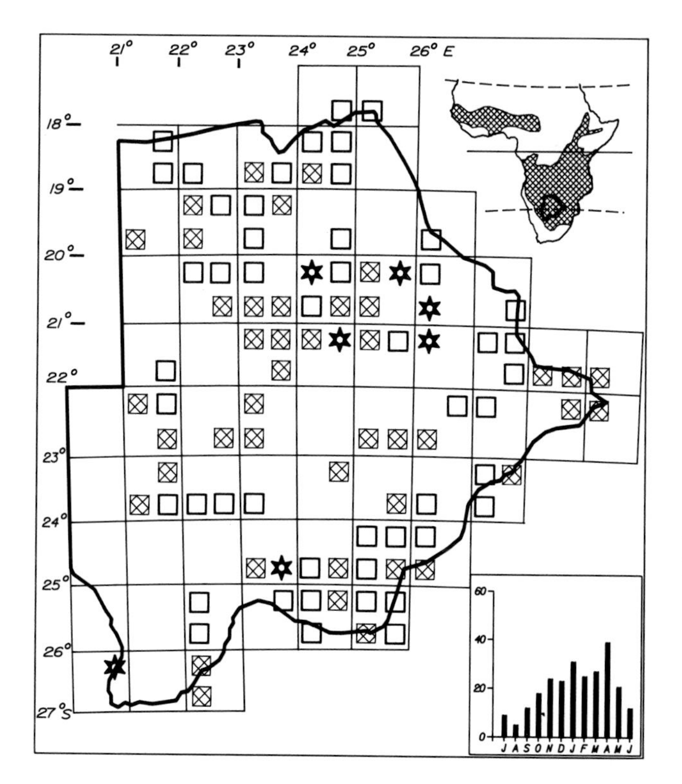

Status

Sparse to fairly common resident and partial migrant occurring throughout the country. It is highly nomadic. Absences in some areas are attributed to poor field coverage at appropriate times of the year. Present mainly from October to early May. Some birds occur throughout the year. The type specimen of the subspecies *C. t. damarensis* was collected at Sekoma in 1961. Egglaying all months July to December. Usually in pairs or family groups.

Habitat

Short grass in open areas, dry lightly-grassed pans, overgrazed ground, football fields, airfields. Also in open woodland, tree and bush savanna with recently burnt ground and in semidesert scrub savanna.

Analysis

99 squares (43%) Total count 274 (0,15%)

DOUBLEBANDED COURSER 301

Rhinoptilus africanus

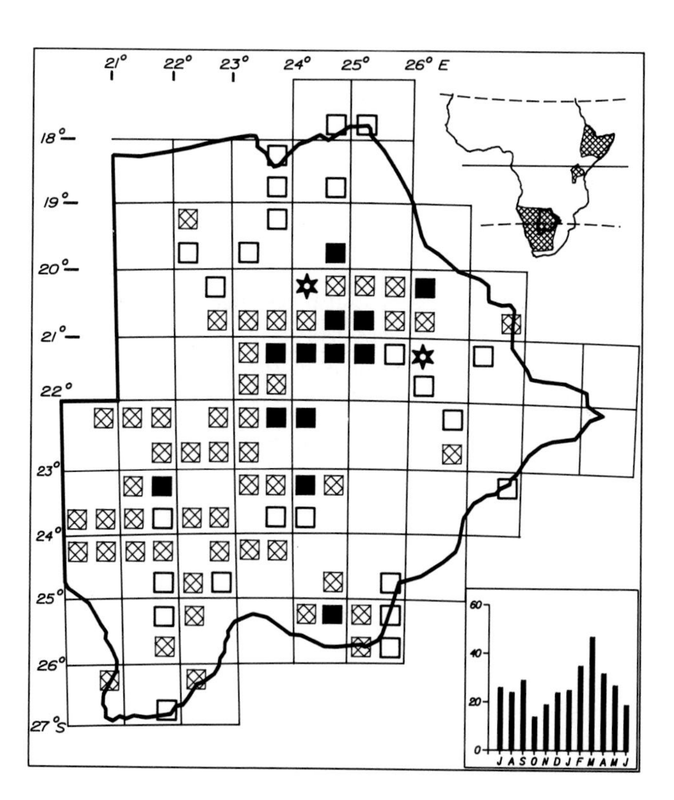

Status

Common to very common resident in central and western regions south of 20°S. Very common at Nxai Pan and in the Makgadikgadi Depression south through Deception Valley to the Okwa Valley. Sparse and thinly distributed in the north and east. Common to very common on plains and pans in the southeast and west. The commonest and most predictable courser in Botswana. Egglaying all months December to March. Usually solitary or in pairs.

Habitat

Bare or short-grassed pans, stony or sandy ground on ecotone of pans, open grassland plains with short grass or bare ground between grass tussocks, open scrub savanna. Its main range reflects the distribution of Kalahari pans.

Analysis

89 squares (39%) Total count 369 (0,21%)

THREEBANDED COURSER 302

Rhinoptilus cinctus

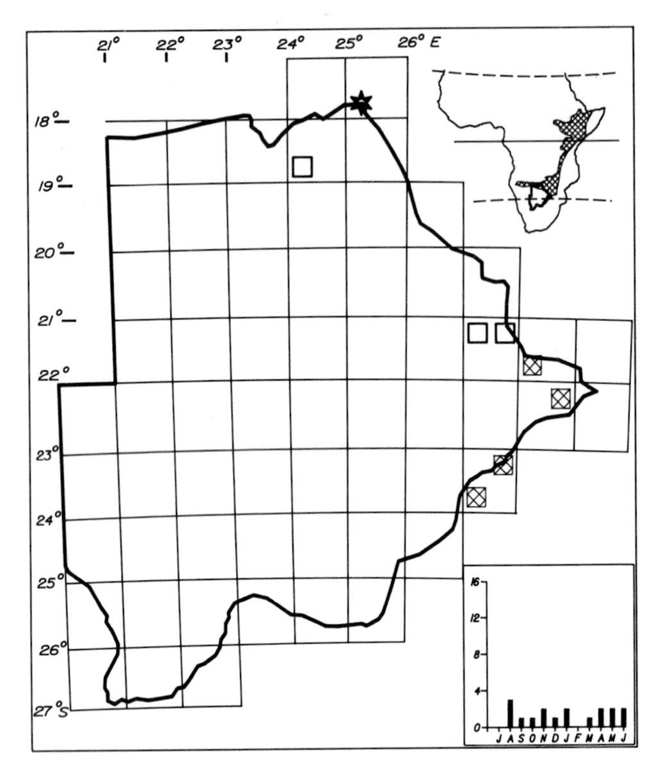

Status

Sparse to uncommon resident in the east along the Shashe and Limpopo valleys. Recorded once by Smithers (1964) at Kasane and there is one recent record from Savuti. Most regularly reported from Martin's Drift. It is an elusive species, mainly crepuscular, which is easily overlooked unless specific search is made for it. Occurs throughout the year. Usually solitary or in pairs.

Habitat

Wooded glades in river valleys, particularly in areas with river terraces of short grass or bare ground. Also at the base of wooded rocky hills or rock outcrops.

Analysis

8 squares (3%) Total count 20 (0,01%)

BRONZEWINGED COURSER 303

Rhinoptilus chalcopterus

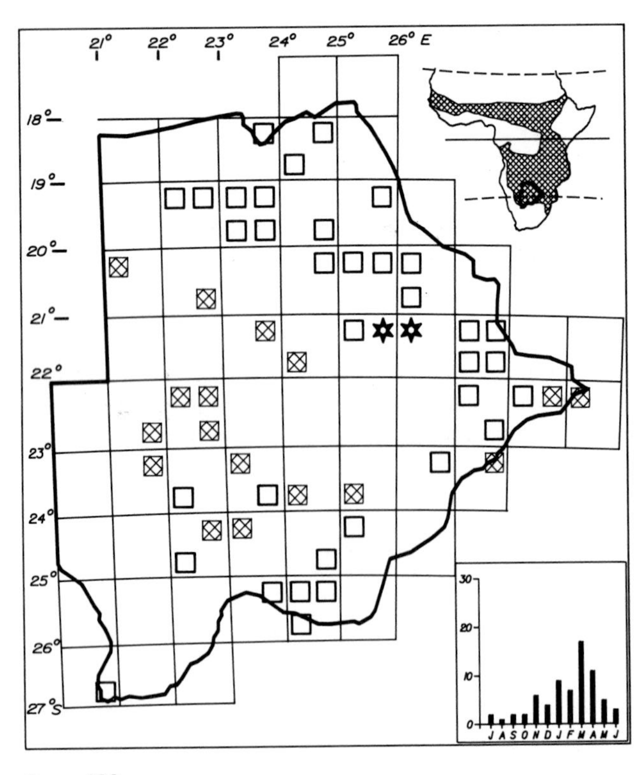

Status

Sparse to fairly common resident throughout the country. Part of the population is migratory and present mainly from November to April. Crepuscular and often overlooked. Usually solitary, but loose congregations—such as 12 birds along 100 m of woodland track—have been recorded.

Habitat

Any woodland and tree and bush savanna where it rests in shade during the day. Forages in open areas, such as clearings and roads at night. Occasionally in towns and villages.

Analysis

54 squares (23%) Total count 79 (0,04%)

REDWINGED PRATINCOLE 304

Glareola pratincola

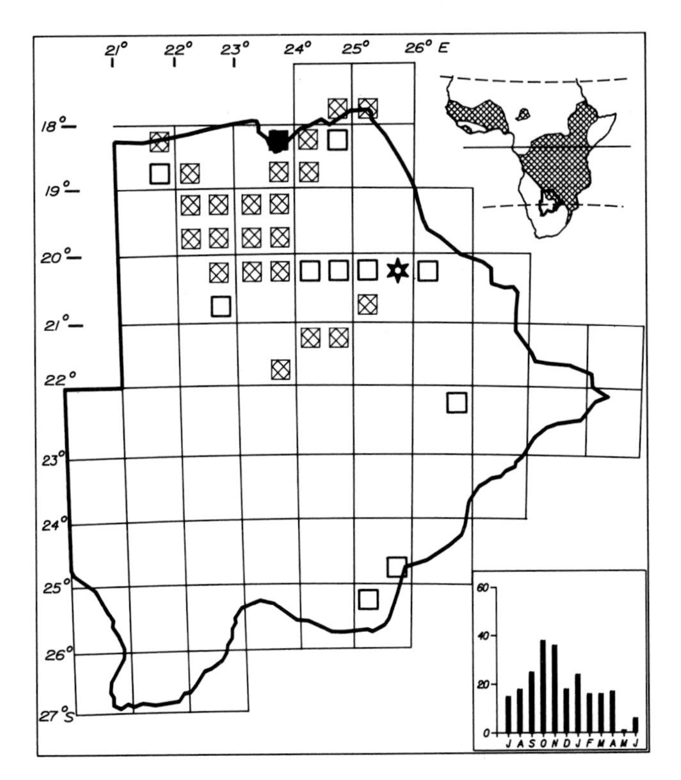

Status
Common breeding intra-African migrant to the Okavango, Linyanti, Chobe and Boteti river systems. Occurs commonly as far south as Lake Ngami, Mopipi, Rakops and Deception Valley. It is less common in the Makgadikgadi region. Recorded very sparsely in the east and southeast as far south as Mmathethe. Present throughout the year but there are fewer records for May and June. Egglaying September, October and November. Usually in flocks of 20–500 birds, but many flocks may gather together at breeding grounds so that thousands of birds are present.

Habitat
Open areas of flat bare sandy or stony ground or short grass near water—shores of lakes, lagoons, dams and on floodplains. Nests on the ground in shallow scrapes, in cattle hoof prints or at the edge of cow dung.

Analysis
34 squares (14%) Total count 242 (0,13%)

BLACKWINGED PRATINCOLE 305

Glareola nordmanni

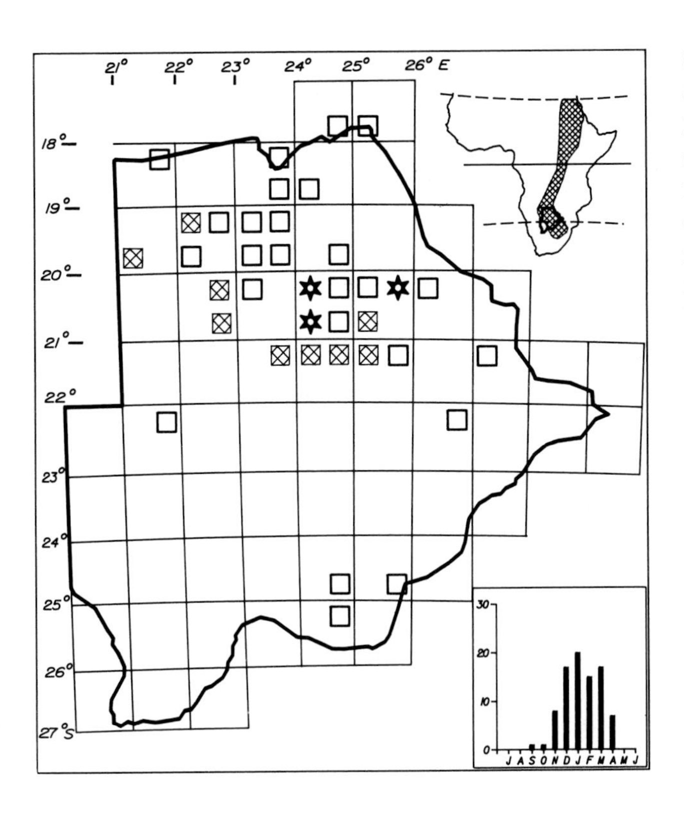

Status
Sparse to fairly common Palaearctic migrant to areas north of 22°S. Very sparse in the east and sparse but regular in the southeast. It is less common than the Redwinged Pratincole in the Okavango, Linyanti and Chobe regions but more common than it in the Makgadikgadi region. Most frequently recorded between Deception Valley and Orapa. Present mainly from November to April. Usually in flocks of 10–100 birds, seldom solitary.

Habitat
Flat open grassland—plains, pans, floodplains, fossil valleys and cultivation. Usually near water but also in dry conditions. Recorded once in March in arid grassland in the Okwa Valley at 2221C.

Analysis
37 squares (16%) Total count 96 (0,05%)

GREYHEADED GULL

315

Larus cirrocephalus

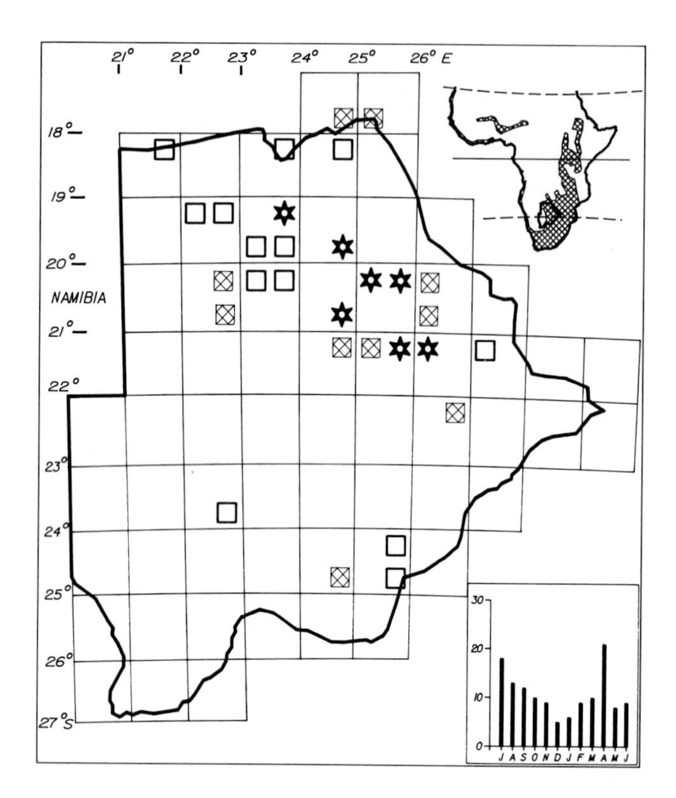

Status

Sparse to locally fairly common resident of the north and east. It may be more widespread in high rainfall years and extend, for example, to pans in the west. Present throughout the year. Local movements occur with changing habitat conditions. Most commonly recorded at permanent water as at Kasane, Serondella, Lake Ngami, Nata River mouth, Mopipi Dam and Jwaneng. Egglaying July and August. May breed only in some years. Usually in small groups or solitary, flocks of up to 50 birds when breeding.

Habitat

Rivers, estuaries, lakes, dams, lagoons, flooded pans and sewage ponds.

Analysis

30 squares (13%) Total count 149 (0,08%)

CASPIAN TERN

322

Hydroprogne caspia

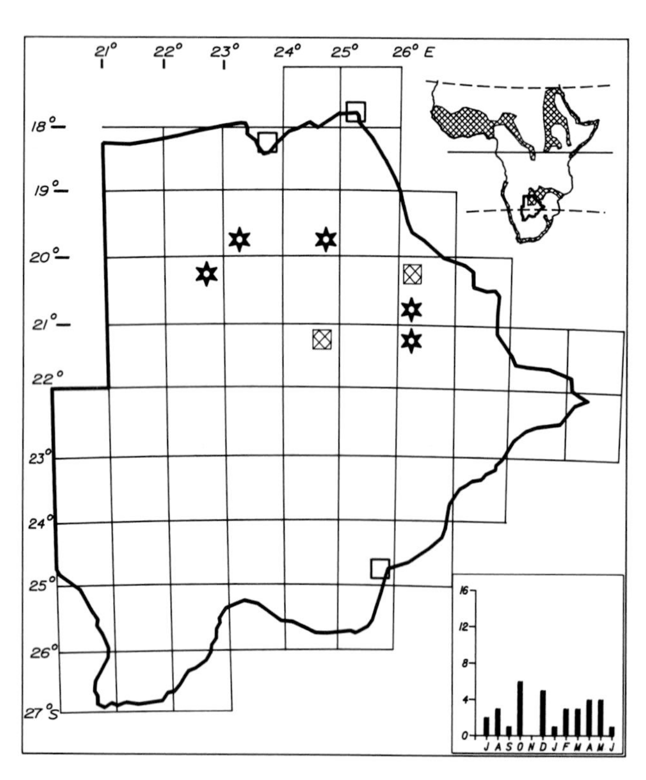

Status

Sparse to uncommon resident at isolated localities in the north. Recorded mainly at Nata and Mopipi but has been recorded more widely in the Makgadikgadi Pans in the past. There are also old records from Lake Ngami, Nxai Pan and Maun where it is likely to occur again when suitable conditions prevail. Also recorded once during this survey at Kasane and Linyanti. Egglaying March and April. Usually in small flocks.

Habitat

Estuaries, flooded pans, lakes, dams.

Analysis

10 squares (4%) Total count 38 (0,02%)

WHISKERED TERN

338

Chlidonias hybridus

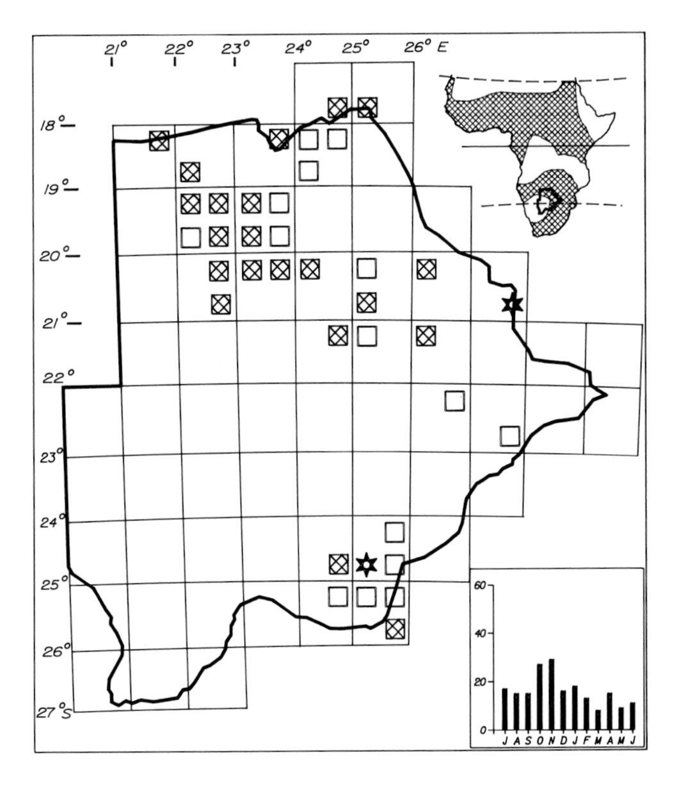

Status

Fairly common to common breeding local migrant to areas north of 22°S. Sparse in the east and sparse to fairly common in the southeast where it also breeds. Expected to be more widespread in high rainfall years. Occurs throughout the year —in breeding dress from November to April. Egglaying February and March. Usually in small flocks.

Habitat

Rivers, lakes, dams, flooded pans, marshes, sewage ponds: seasonally inundated depressions with tall aquatic vegetation are also used for breeding.

Analysis

38 squares (17%) Total count 210 (0,12%)

WHITEWINGED TERN

339

Chlidonias leucopterus

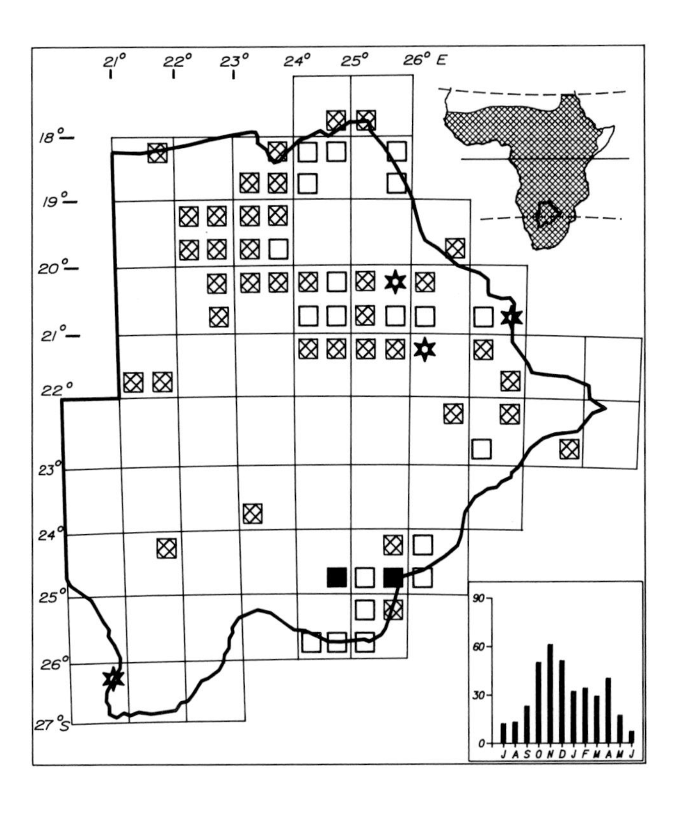

Status

Fairly common to common Palaearctic migrant recorded in all regions of the country. It is widespread north of 22°S, very localised in the east and west and regularly recorded in the southeast where it is locally very common at Jwaneng and Gaborone. Present mainly from October to April with some birds arriving in mid-August (15th) and some departing in mid-May (10th). A few birds remain throughout June and July in some years. Birds in breeding dress have been recorded in August, September, March, April and May. Part of the population occurs only on passage. Usually in flocks of 10–200 birds.

Habitat

Lakes, dams, rivers, lagoons, flooded grassland and pans, sewage ponds. Also occurs over dry habitat such as farmland, cultivation, open grassland in river valleys.

Analysis

63 squares (27%) Total count 413 (0,23%)

AFRICAN SKIMMER

343

Rynchops flavirostris

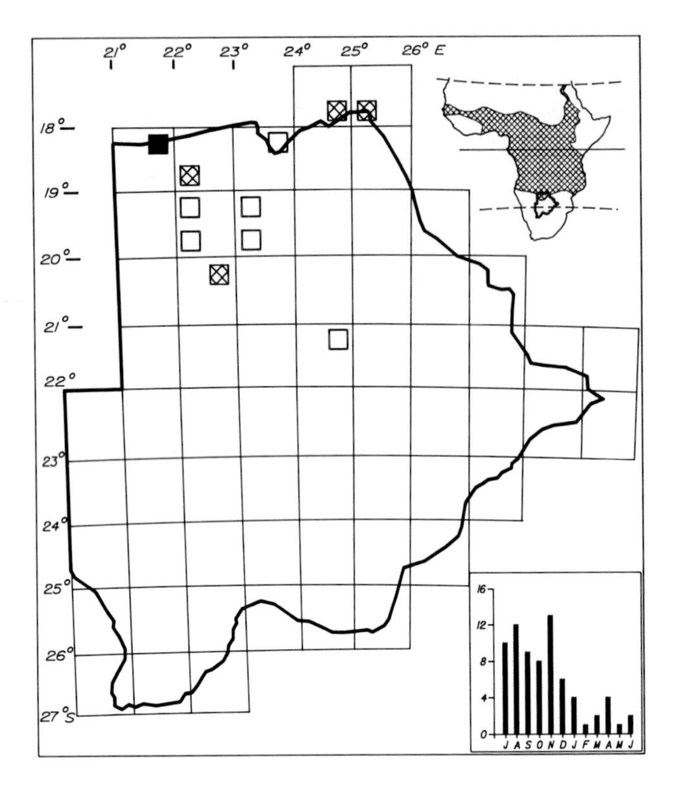

Status

Sparse to locally common resident and partial intra-African migrant in the Okavango, Linyanti and Chobe river systems to as far south as Lake Ngami. Recorded only once during this survey at Mopipi but it may be more common there and along the Boteti River in high rainfall years. Most records fall between July and November but it is present in all months. Its numbers are declining in southern Africa due to the damming and silting of rivers. Its vulnerability needs special investigation. Egglaying August, September and October—all from Shakawe. Bred at Mopipi and Lake Ngami in the past. May breed in some years only. Usually in flocks of 5–30 birds.

Habitat

Large rivers, lakes, lagoons and large dams. Nests on sandbanks.

Analysis

11 squares (5%) Total count 80 (0,04%)

NAMAQUA SANDGROUSE

344

Pterocles namaqua

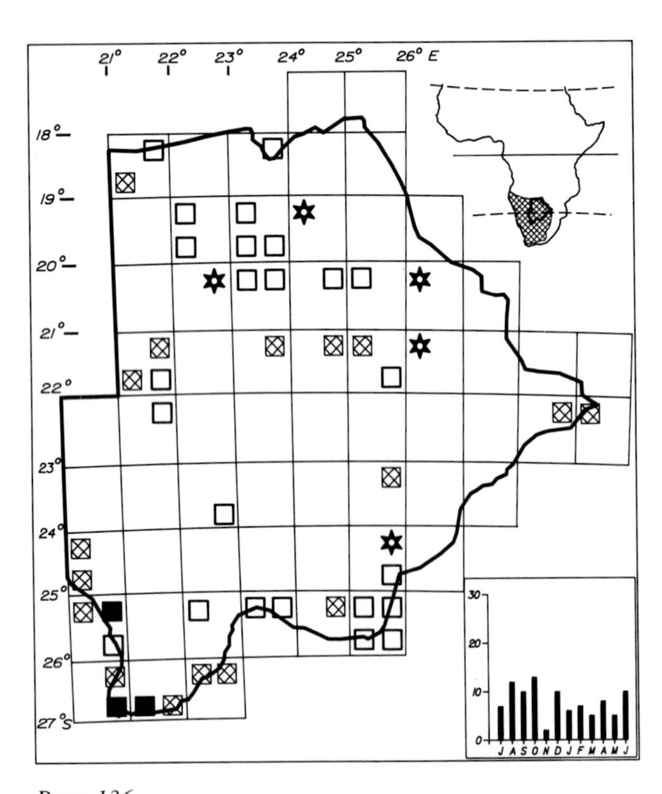

Status

Common to very common resident in the southwest. Elsewhere throughout the country it is sparse to uncommon and thinly distributed. It occurs in all regions throughout the year. Egglaying August and September. Usually in pairs or small flocks of 5–10 birds, sometimes in larger flocks of up to 100 birds.

Habitat

Semidesert scrub, bare or sparsely covered dunes, bare rocky or stony ground in river valleys and at pans, arid grassland, open areas in bush savanna.

Analysis

49 squares (21%) Total count 117 (0,06%)

BURCHELL'S SANDGROUSE

345

Pterocles burchelli

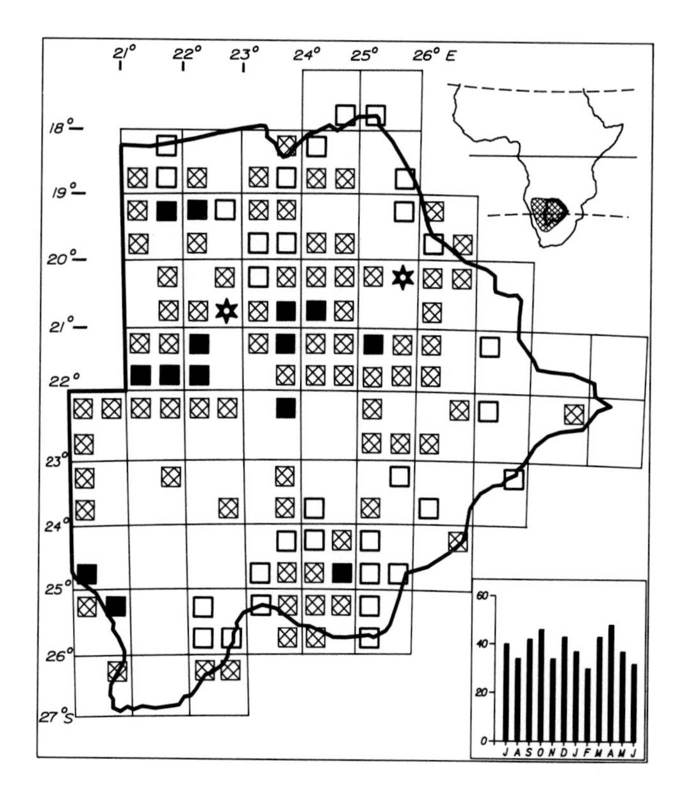

Alternative name

Spotted Sandgrouse

Status

Fairly common to very common resident throughout the country. It is the commonest sandgrouse. Uncommon east of 27°E and seldom recorded in the western woodlands where further study is required to ascertain its true status. Movements occur irregularly—dependent on the availability of water and breeding status. Egglaying April. Usually in pairs or flocks of 5–50 birds, but larger flocks may occur.

Habitat

Grassland in semidesert tree and bush savanna, edges of pans, grassland plains. Mostly in grass 0,3 to 1 m high with bare ground between tussocks.

Analysis

121 squares (53%) Total count 516 (0,29%)

YELLOWTHROATED SANDGROUSE

346

Pterocles gutturalis

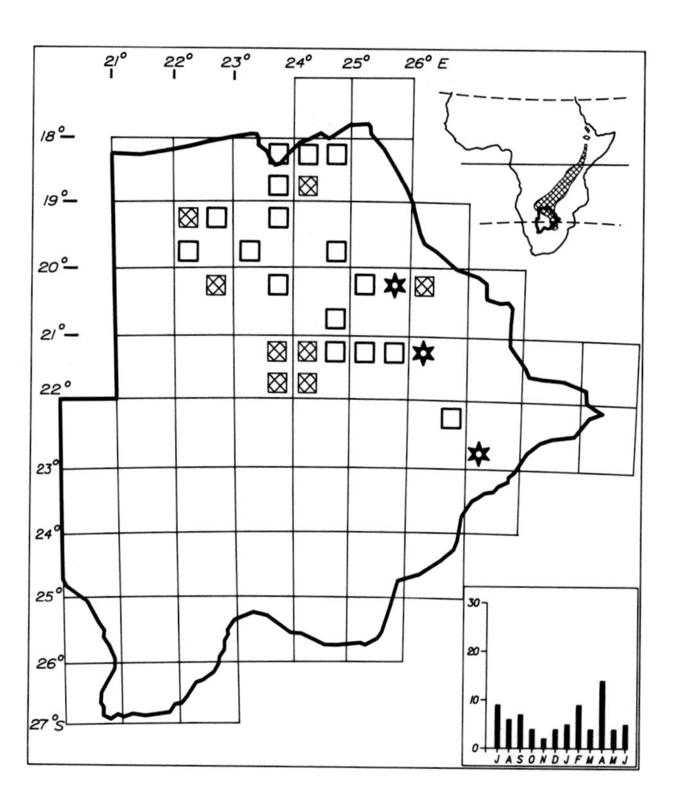

Status

Sparse to fairly common resident and partial migrant. Occurs mainly north of 22°S. It may extend further south in some years—it is known to occur between 24°S and 26°S in adjacent western Transvaal. Recorded as common at Gomare, Lake Ngami and Nata. Although it is present in every month, part of the population is thought to move away in winter to breed further north in Zambia. Its movements are not well understood and the current records are too few to make a conclusive analysis. Usually in pairs or flocks of 5–50 birds.

Habitat

Floodplains and short grassland near swamps, lakes, pans and rivers. Also on old cultivation and open semiarid grassland.

Analysis

27 squares (12%) Total count 81 (0,045%)

DOUBLEBANDED SANDGROUSE

Pterocles bicinctus

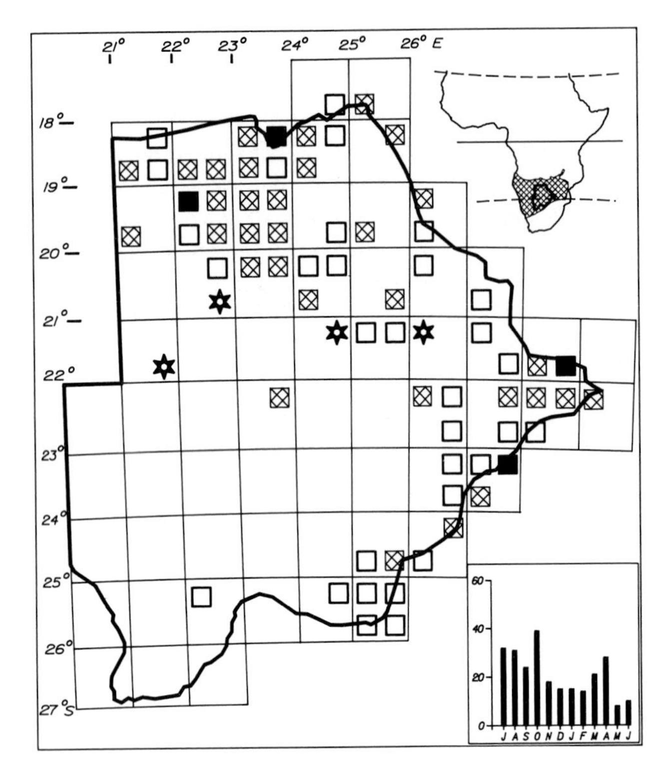

Status

Fairly common to very common resident of the north and east. Sparse to uncommon in the southeast. It is absent from areas without trees such as much of the Makgadikgadi Pans. Occurs mainly in high rainfall areas but has been recorded in Kalahari woodland in the eastern Okwa Valley and near Mabuasehube. Egglaying April, May and August. Usually in pairs or small flocks, but large flocks of several hundred may fly together to drinking points.

Habitat

Tree and bush savanna and edges of woodland. Usually close to woodland or clumps of trees and bushes. Of the four sandgrouse species occurring in Botswana this species utilises the least open habitat.

Analysis

71 squares (31%) Total count 320 (0,18%)

ROCK PIGEON

Columba guinea

Alternative name

Speckled Pigeon

Status

Fairly common to very common resident of the east and southeast where it occurs at the western edge of its range in the Transvaal and Zimbabwe. Isolated populations occur at some human settlements over most of the country from Nossob Camp to Kasane. Deliberately introduced into Ghanzi in the 1960s. Its range in Botswana is expanding with the construction of buildings. Egglaying May, June, July, October and February. Usually in pairs or small flocks of 4–20 birds.

Habitat

Cliffs and gorges in hilly and mountainous areas. Now very well adapted to breeding on buildings and manmade ledges on bridges, dam walls, construction beams etc. Buildings in new towns or new developments near current distribution are rapidly colonised e.g. Jwaneng in 1982 and Morupule in 1986.

Analysis

42 squares (18%) Total count 418 (0,23%)

REDEYED DOVE

352

Streptopelia semitorquata

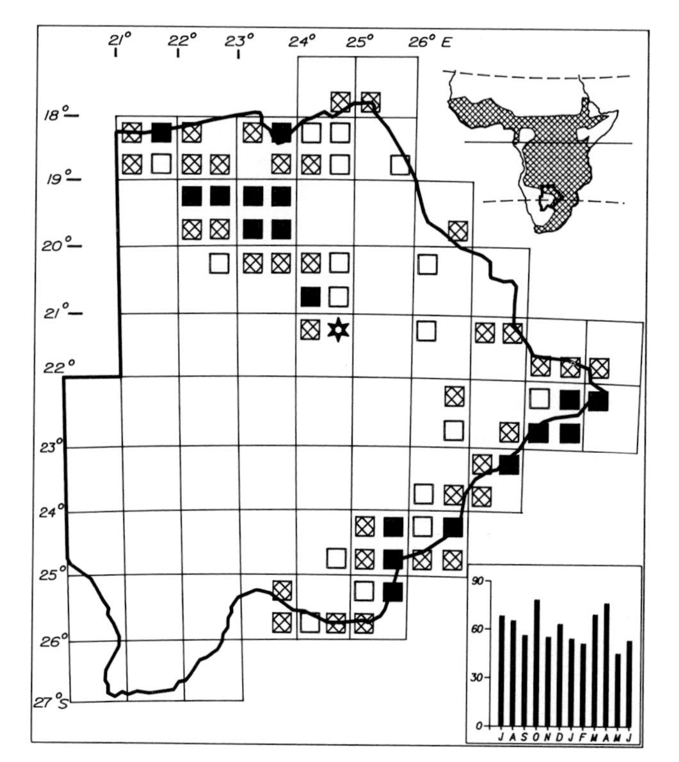

Status

Common to very common resident in the river systems of the north, east and southeast. Extends along the Molopo River to as far as Bray. Recorded along the Boteti River south to Rakops but not recently reported from Mopipi. Its status in the northeast needs further research as it may occur more commonly than shown in the Nata River drainage at least in high rainfall years. Egglaying March, June and August. Usually solitary, in pairs or flocks of 10–50 birds.

Habitat

Tall trees along rivers and drainage lines, mature woodland, tree and bush savanna in high rainfall regions. Often on farmland, edges of cultivation and occasionally in gardens.

Analysis

71 squares (31%) Total count 778 (0,43%)

MOURNING DOVE

353

Streptopelia decipiens

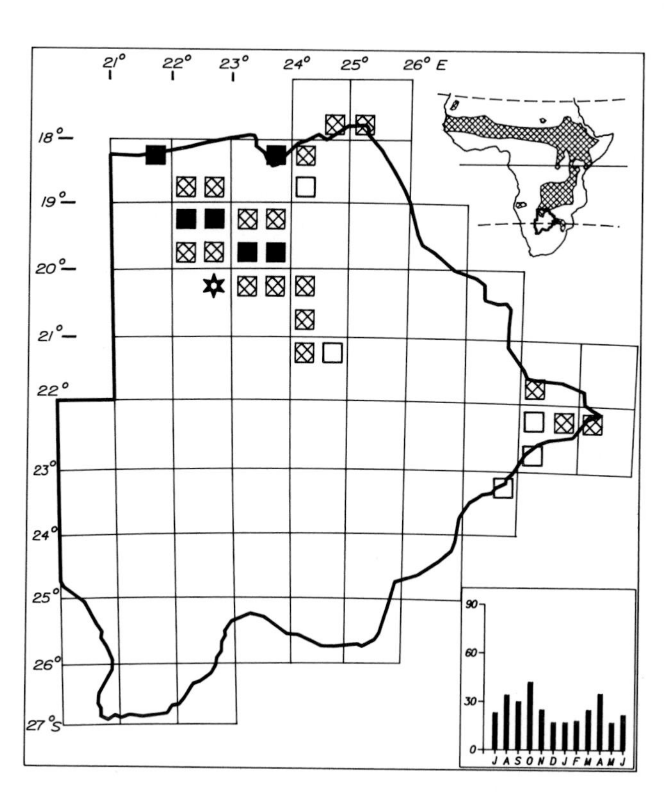

Status

Common to very common resident of the Okavango, Linyanti and Chobe river systems. Extends south along the Boteti River to Mopipi where it is sparse. Reappears in the extreme east in the region of the Shashe/Limpopo confluence at the western limit of its range along the Limpopo Valley from Mozambique. In the east it is markedly less common than the Redeyed Dove. It is locally as common and usually more numerous than the Redeyed Dove in the northern river systems. Usually in pairs, or flocks of up to 50 birds at waterholes.

Habitat

Tall riverine trees and riparian fringing forest, riverine *Acacia* particularly *Acacia albida.* Confined to river valleys and associated vegetation where it is frequently found in gardens and around habitation e.g. game lodges, fishing camps and villages.

Analysis

29 squares (13%) Total count 323 (0,18%)

CAPE TURTLE DOVE

354

Streptopelia capicola

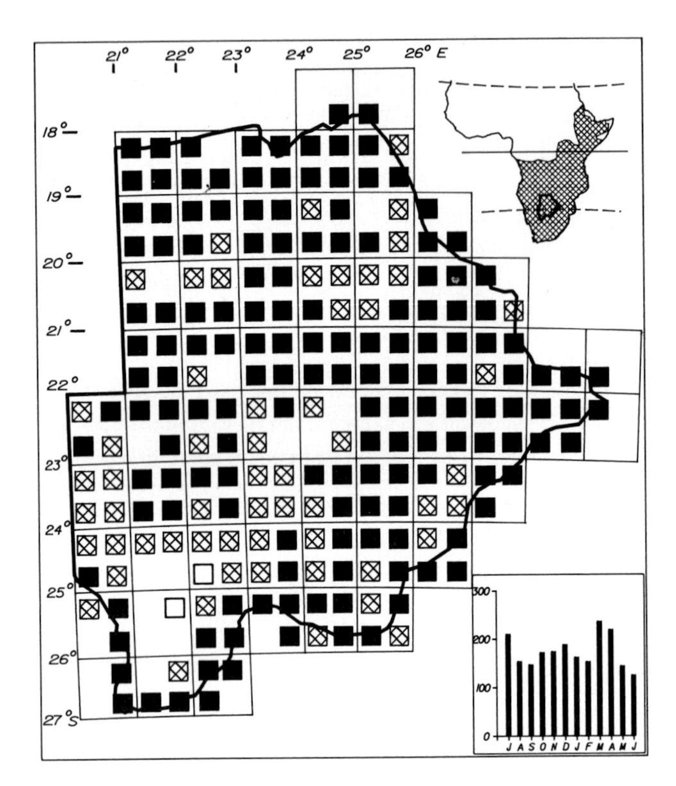

Status

Common to very common resident in all regions. One of the most common and widespread species in Botswana. Sometimes as common in semidesert savanna as in the high rainfall regions. Local and seasonal movements occur depending on the availability of surface water and, like the Laughing Dove, it is sometimes found only near boreholes and cattle posts. Egglaying all months October to May but probably throughout the year in high rainfall years. Solitary, in pairs or flocks.

Habitat

All types of woodland and savanna, cultivation, gardens and riparian vegetation—a wide range of habitats associated with the presence of trees usually in more closed conditions than the Laughing Dove. Its presence everywhere is dependant on the availability of surface water. It can be abundant at isolated drinking sites.

Analysis

216 squares (94%) Total count 2247 (1,25%)

LAUGHING DOVE

355

Streptopelia senegalensis

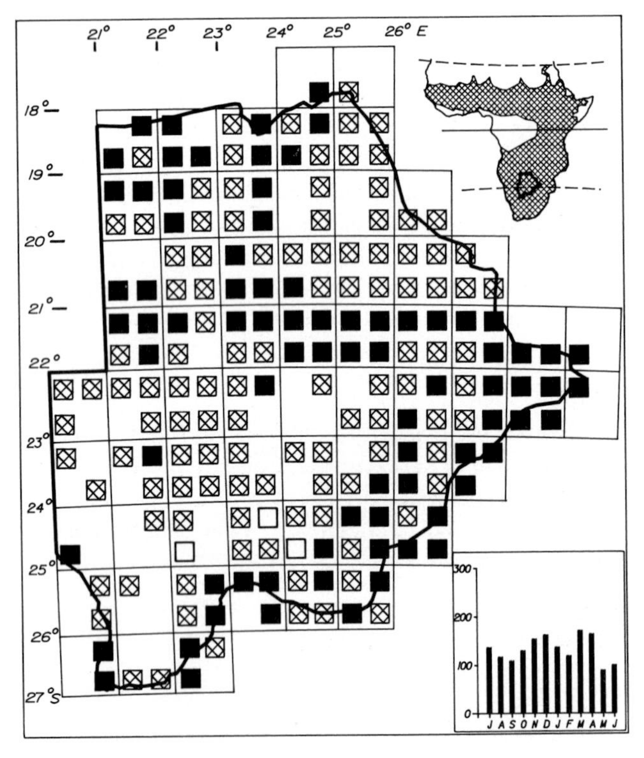

Status

Common to very common resident north of 22°S and in the Limpopo, Ngotwane, Molopo and Nossob river drainages. In western and central regions it is uncommon to locally common and possibly more restricted to areas with surface water than the Cape Turtle Dove. In these latter regions it may be absent over large areas only to appear in huge numbers at watering points. Movements occur depending on local conditions and these may be part of a more widespread seasonal migration. Egglaying October to December and February to May. Solitary, in pairs or in flocks.

Habitat

Open woodland, tree and bush savanna, cultivation, towns and villages, cattle posts particularly those on cattle trek routes, Kalahari pans with water, boreholes.

Analysis

194 squares (84%) Total count 1745 (0,97%)

NAMAQUA DOVE 356

Oena capensis

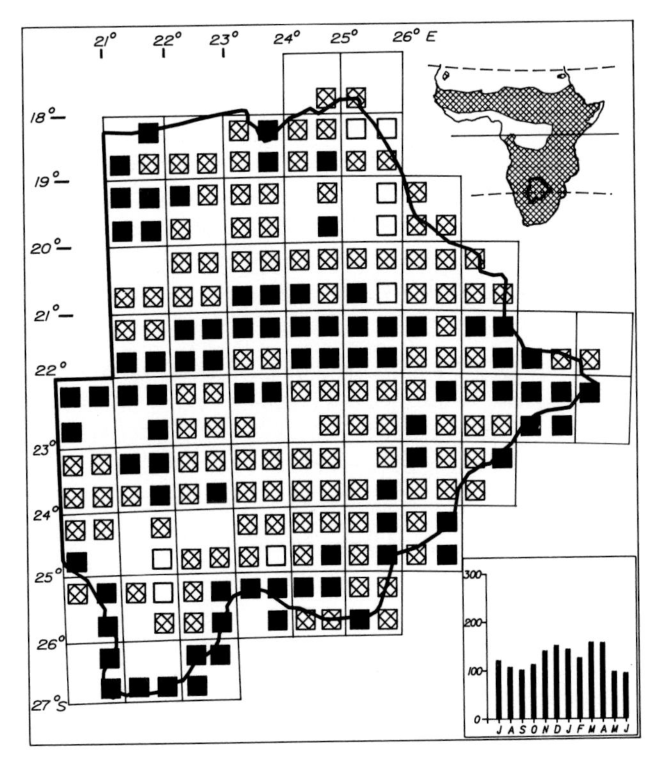

Status

Fairly common to very common resident in all regions. Less common in the woodland areas of the northeast than elsewhere. Local and seasonal movements occur as well as more widespread population movements into neighbouring countries. In Zambia it is a partial intra-African migrant and birds with long range movements are likely to arrive in Botswana, at least in the north. Southern and northern birds may behave differently in their movements. Egglaying March. Usually solitary or in pairs except at drinking spots where large numbers may gather.

Habitat

Acacia tree and bush savanna, semidesert scrub savanna, clearings in woodland and riverine vegetation, cultivation, bare sandy or stony ground in arid or overgrazed regions.

Analysis

208 squares (90%) Total count 1658 (0,92%)

GREENSPOTTED DOVE 358

Turtur chalcospilos

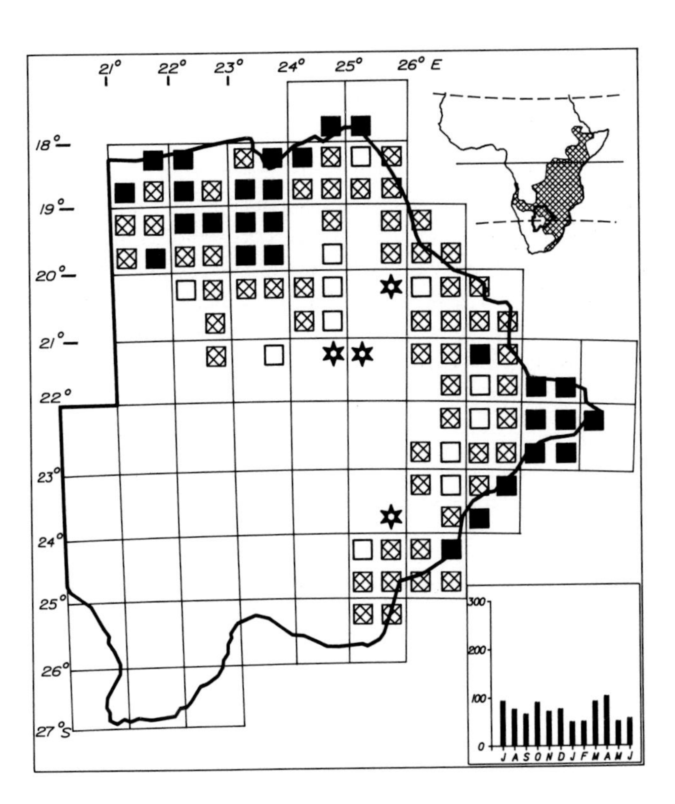

Status

Fairly common to locally very common resident of the north and east to as far south as Lobatse. From the northern wetlands it reaches as far south as the Kuke fence in the west but no further. South of 22°S in the east it is confined to the drainage of the Limpopo and Ngotwane rivers. Egglaying April. Solitary or in pairs.

Habitat

Thicket and dense cover in any woodland in the high rainfall regions, riparian forest and riverine bush. Uncommon in many areas of miombo due to the lack of understorey and absent from mopane unless mixed with thicket.

Analysis

98 squares (43%) Total count 926 (0,51%)

GREEN PIGEON

361

Treron calva

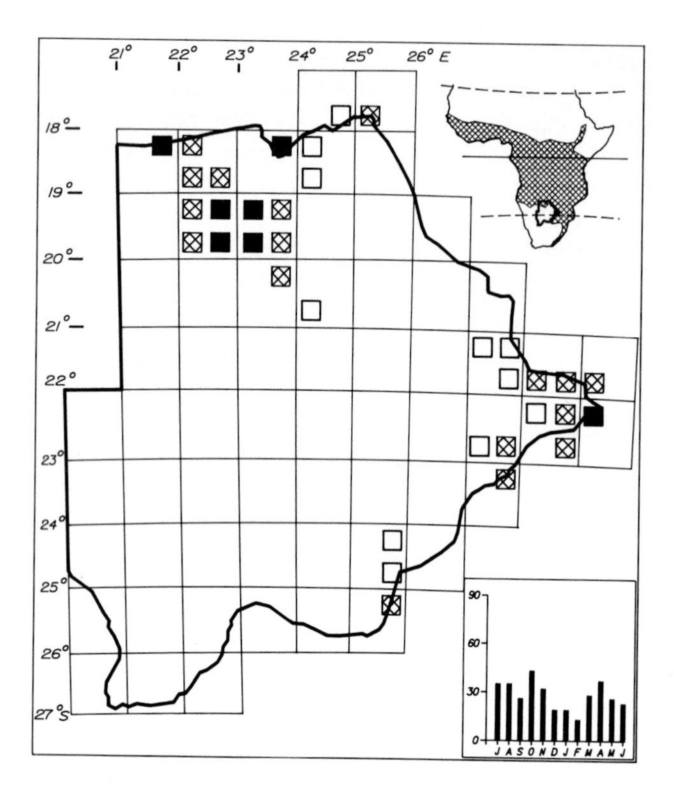

Status

Common to very common resident in the Okavango Delta and at Linyanti Camp. Sparse to fairly common in the Chobe region and in the east from Francistown to Martin's Drift. Locally very common at Pont's Drift. Reappears very locally in the southeast at Gaborone and Lobatse and recorded once near Molepolole. Its occurence in these latter localities may represent a recent extension of range. Egglaying December. Usually in small groups of 3–10 birds.

Habitat

Riparian forest and adjacent woodland in areas with fruiting trees, especially wild figs. It sometimes occurs in isolated patches of fig or fruit trees as at Kgale Hill near Gaborone. Virtually restricted to river valleys in Botswana.

Analysis

35 squares (15%) Total count 352 (0,19%)

CAPE PARROT

362

Poicephalus robustus

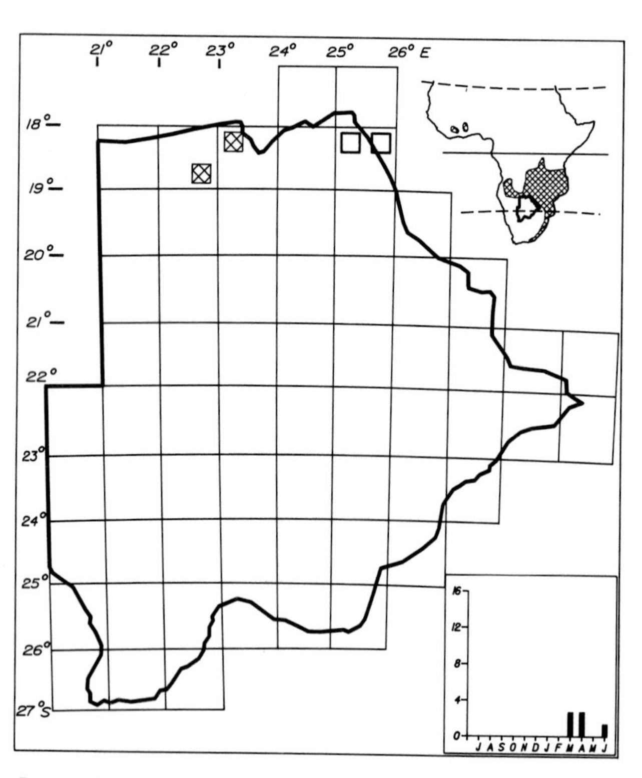

Alternative name

Brownnecked Parrot

Status

Sparse to uncommon resident of the extreme northern woodlands. Recorded once at Mashatu (2229A—not shown on map) in June 1988 . It is very poorly known and only during this survey has it been definitely established as a resident of this region. Occurs at the southern limit of its continental range at these longitudes. Known to wander long distances (up to 80 km) to feed. Solitary and in pairs, but may occur in flocks.

Habitat

Mature and undisturbed *Baikiaea* woodland and riparian forest. Also well-developed broadleafed woodland.

Analysis

4 squares (2%) Total count 5 (0,003%)

MEYER'S PARROT

364

Poicephalus meyeri

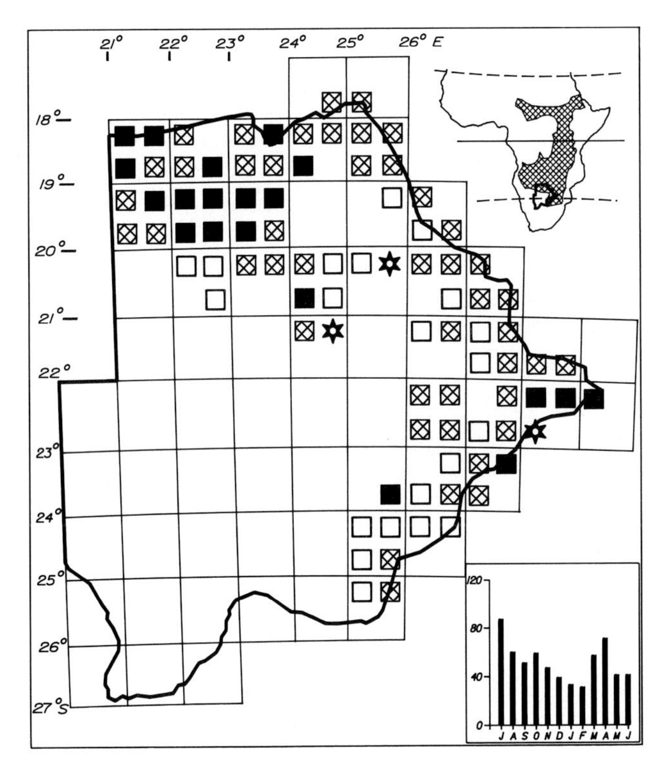

Status

Common to very common resident in the north to as far south as Tale Pan and Rakops. Sparse to fairly common in the southeast as far south as Lobatse. In the east it is very common in the eastern Tuli Block and sparse to common elswhere within the Shashe, Limpopo and Ngotwane river drainage. Local movements occur. Egglaying April and May. Usually in small groups of 3–10 birds.

Habitat

Deciduous woodland and tree savanna, bush savanna with some tall trees, riverine *Acacia* and riparian forest. Forages in the canopy of trees bearing fruit, nuts or berries.

Analysis

89 squares (39%) Total count 644 (0,36%)

GREY LOURIE

373

Corythaixoides concolor

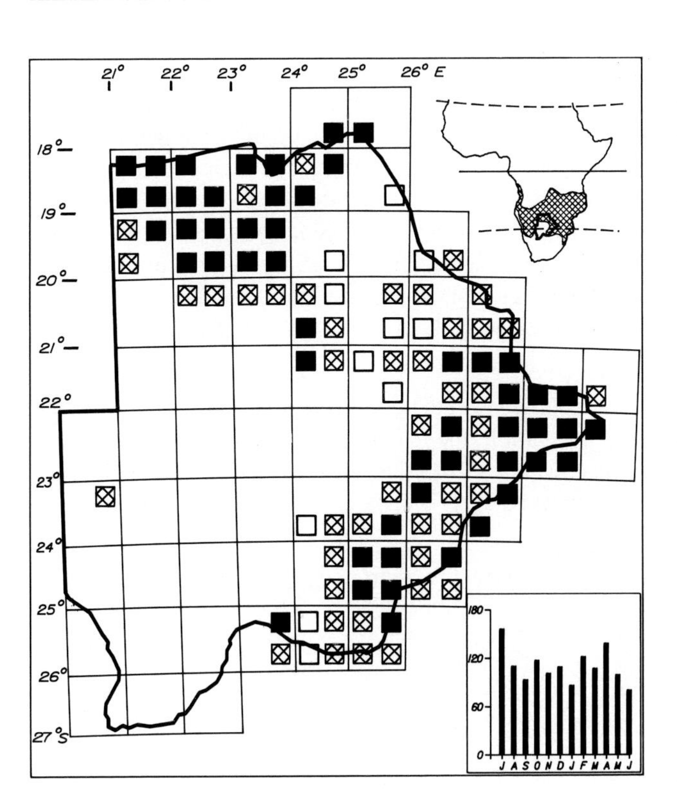

Status

Common to very common resident in the north, east and southeast extending west along the Molopo River to as far as Bray. Sparse and uncommon in the northeastern woodlands. Recorded once at the Ncojane Ranches (2320B) where it occurs at the eastern limit of its Namibian range. Further search is required west of Ghanzi and in the far west to ascertain its status in this region. Egglaying July, August and October. Usually in pairs, occasionally in small groups.

Habitat

Tree and bush savanna particularly in river valleys, open *Acacia* and mixed woodland. More common near water or along watercourses even when seasonally dry.

Analysis

108 squares (47%) Total count 1353 (0,75%)

EUROPEAN CUCKOO **374**

Cuculus canorus

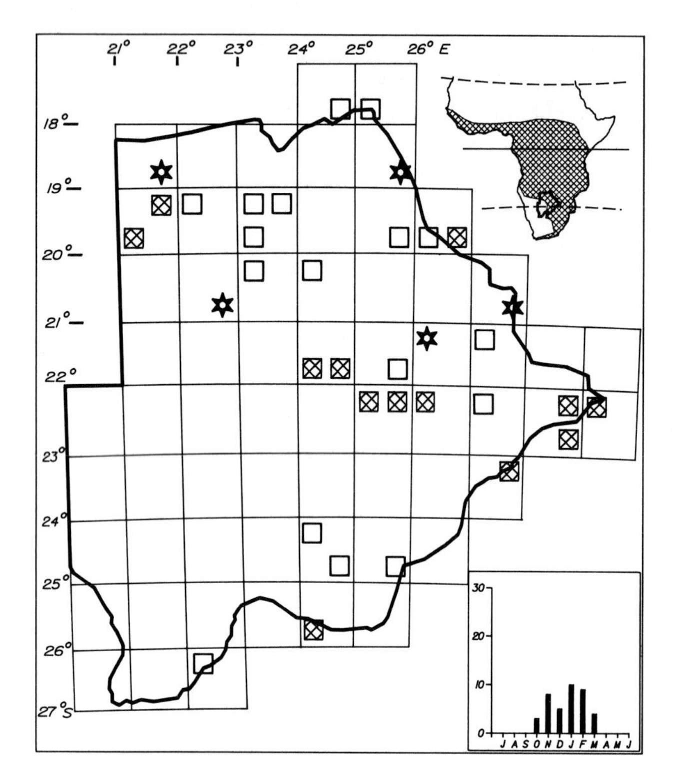

Status

Sparse to uncommon Palaearctic migrant mainly to the north and east. Very sparse in the southeast. Recorded twice at Boshoek and once at Tshabong along the Molopo River. Present from October to March with most records in January and February. Although it can be mistaken for the African Cuckoo, it is not as common as the latter in Botswana whereas the reverse is the situation reported in Zimbabwe. Differentiated by absence of vocalisation, morphology of bill and tail and by its more confiding and conspicuous behaviour. Solitary.

Habitat

Broadleafed and mixed *Acacia* woodland, tree savanna, semidesert with scattered trees. Infrequently in riparian trees and riverine bush.

Analysis

35 squares (15%) Total count 45 (0,02%)

AFRICAN CUCKOO **375**

Cuculus gularis

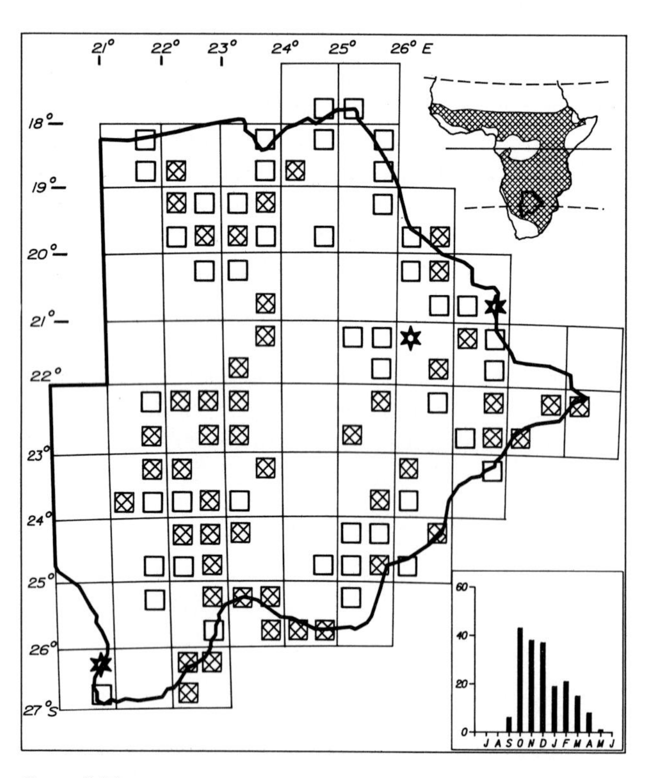

Status

Sparse to fairly common breeding intra-African migrant recorded throughout the country. Present from late September (28th) to the end of April with one record in early May. Most records fall during October to December when the adults are calling in the early breeding season. Vocalisations drop off dramatically after December. Egglaying October and November. Parasitizes the Forktailed Drongo and no other host species is known in Botswana. Solitary.

Habitat

All types of woodland, tree and bush savanna. Appears to be more common in the dry *Acacia* savannas of the west and along the Molopo River than in the broadleafed woodlands of the north and east.

Analysis

96 squares (42%) Total count 205 (0,12%)

REDCHESTED CUCKOO

Cuculus solitarius

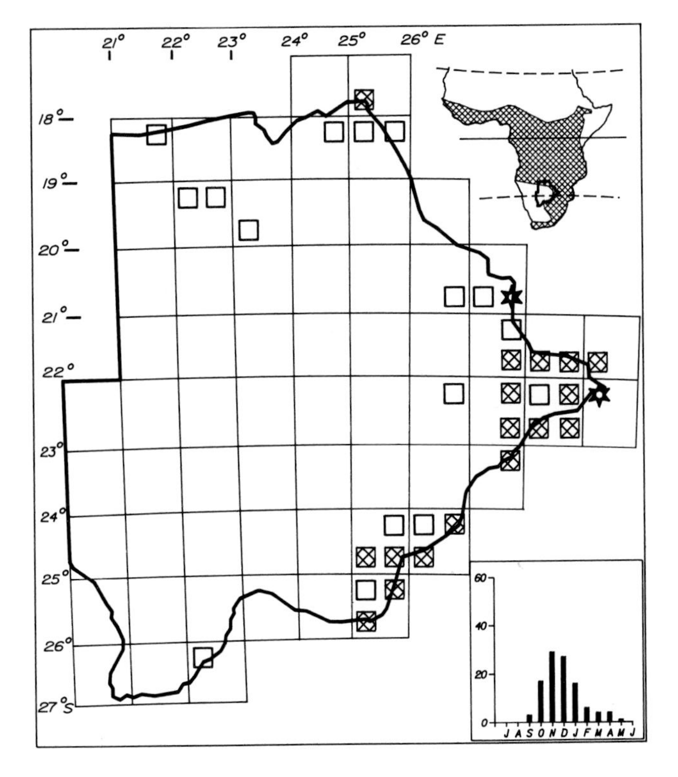

Status

Sparse to fairly common breeding intra-African migrant to the east and southeast. Very sparse in the north except at Kasane. Recorded once at Tshabong. The distribution is confined to the river systems. Arrives in late September but mainly in October and departs in April (one record in early May). Most commonly recorded between October and January. Its hosts are likely to include robins, thrushes, chats and Cape Wagtail but there are no known breeding records in Botswana. Usually solitary.

Habitat

Tree and bush savanna in river valleys, riparian forest, broadleafed and mixed *Acacia* woodland in the high rainfall areas and in the major river systems.

Analysis

35 squares (15%) Total count 123 (0,07%)

BLACK CUCKOO

Cuculus clamosus

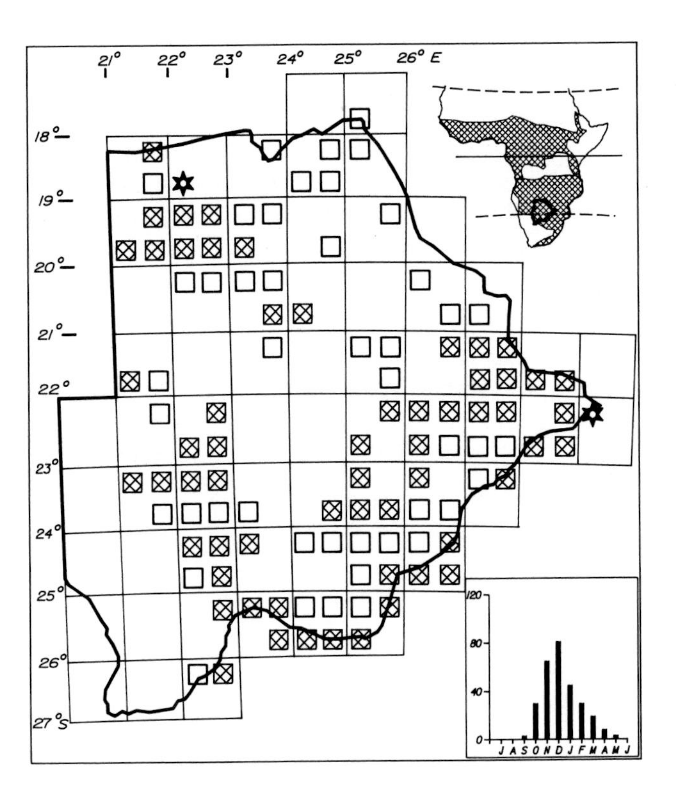

Status

Sparse to common breeding intra-African migrant to all regions except the far west and southwest. Its status in the central Kalahari needs further study. Present mainly from October to March with first arrivals in late September and latest departures in early May. Egglaying December, January and March. Three records confirm the Crimsonbreasted Shrike as the usual host and the March record is from a Swamp Boubou—the first ever record of this latter host. Usually solitary, sometimes seen or called up in pairs.

Habitat

Acacia and other deciduous tree savanna, edges of broadleafed and mixed woodlands, bush savanna, riverine bush, gardens.

Analysis

105 squares (46%) Total count 297 (0,17%)

GREAT SPOTTED CUCKOO

380

Clamator glandarius

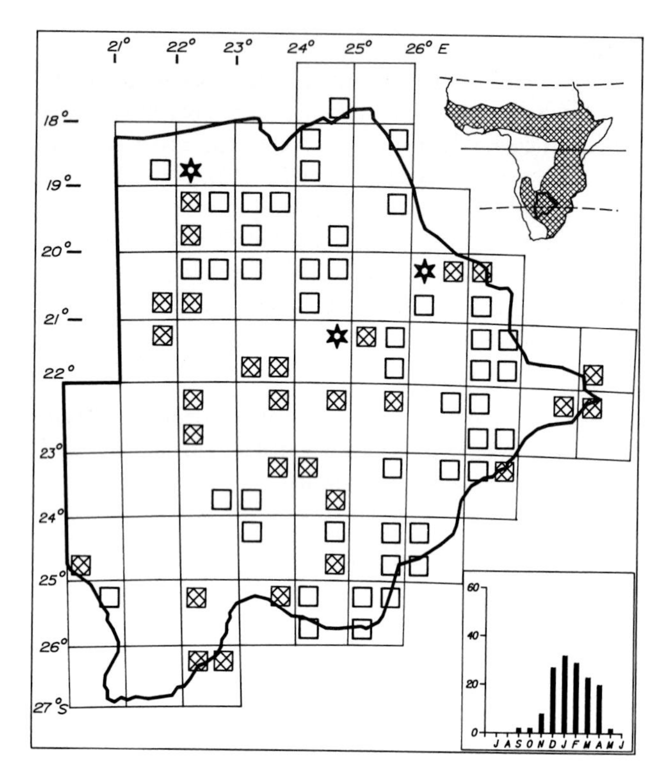

Status

Sparse to uncommon breeding intra-African migrant. Recorded in scattered localities throughout the country and nowhere common. Numbers may fluctuate from year to year. Most records fall between December and April but earliest arrivals are reported in late September and last departures in early May. Egglaying December and March. Crows and Longtailed Starling are the confirmed hosts. Earliest breeding behaviour 23 November. Juveniles are seen from February onwards. Usually solitary, occasionally in pairs.

Habitat

Acacia woodland and tree and bush savanna. Also in open broadleafed and mixed woodlands. Occurs mainly in mid-stratum of leafy trees or in tall thicket and dense vegetation.

Analysis

77 squares (33%) Total count 163 (0,09%)

STRIPED CUCKOO

381

Clamator levaillantii

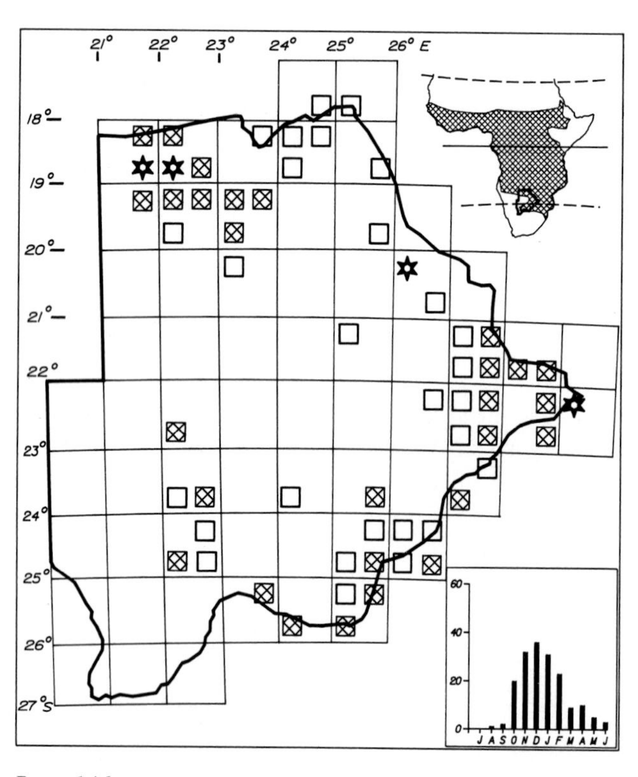

Status

Sparse to fairly common intra-African migrant mainly to the north and east but extending into the western woodlands between Kang, Khakhea and Mabuasehube. Present mainly from October to April. Some birds may be present throughout the year. Records are fewest between May and September. Egglaying February. Arrowmarked Babbler is the confirmed host. Usually solitary.

Habitat

Broadleafed and *Acacia* woodland in areas with dense cover of thicket and bushes. Also riverine bush and luxuriant gardens.

Analysis

59 squares (26%) Total count 182 (0,1%)

JACOBIN CUCKOO

382

Clamator jacobinus

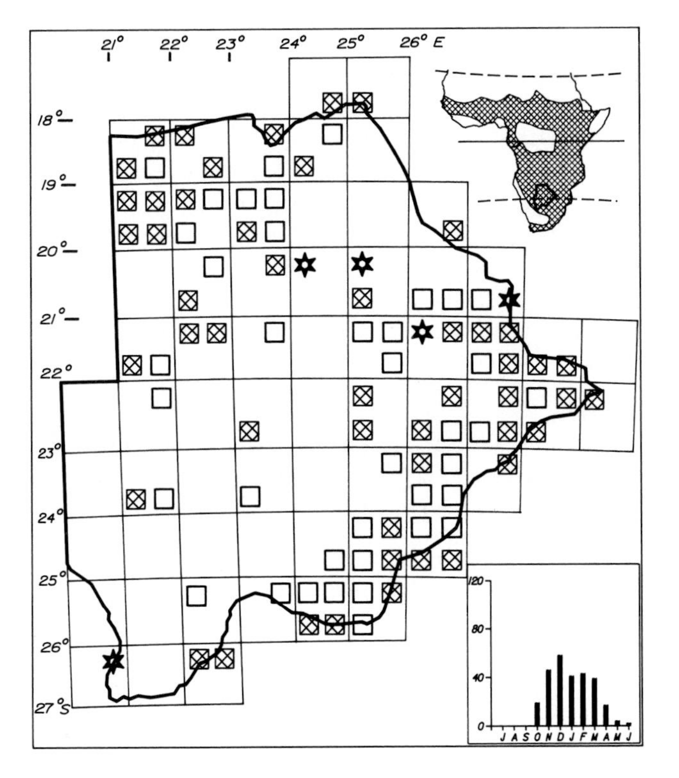

Status

Sparse to locally common migrant mainly to the north and east but with scattered records throughout the country. Present from October to April with few birds remaining until May and June. Three subspecies may occur: *C.j. pica* as a nonbreeding intra-African migrant; *C.j. jacobinus* as a nonbreeding migrant from the Indian subcontinent; *C.j. serratus* as a breeding intra-African migrant. Egglaying January and February. Parasitizes the Redeyed and Black-eyed Bulbuls. Usually solitary, occasionally 2 or 3 birds occur together.

Habitat

Broadleafed and *Acacia* woodland, tree and bush savanna. On dry floodplain with clumps of thicket where seen alongside the Striped Cuckoo at Serondella and Savuti Marsh. Also in gardens.

Analysis

93 squares (40%) Total count 294 (0,16%)

KLAAS'S CUCKOO

385

Chrysococcyx klaas

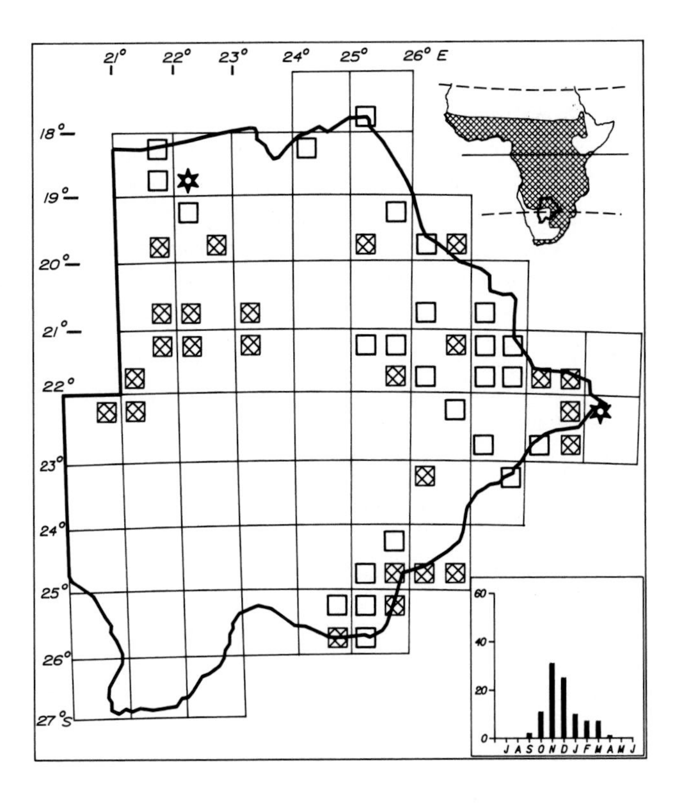

Status

Sparse to fairly common intra-African migrant to the northwest and east. Reported very sparsely in most of the northern woodlands. Nowhere common. Present from September to April with most records from October to March when it is calling. Egglaying January and February. Two known hosts are Longbilled Crombec and Marico Sunbird. Usually solitary.

Habitat

Tree and bush savanna with dense vegetation, thickets, riverine bush and edges of riparian forest.

Analysis

52 squares (23%) Total count 103 (0,06%)

DIEDERIK CUCKOO

386

Chrysococcyx caprius

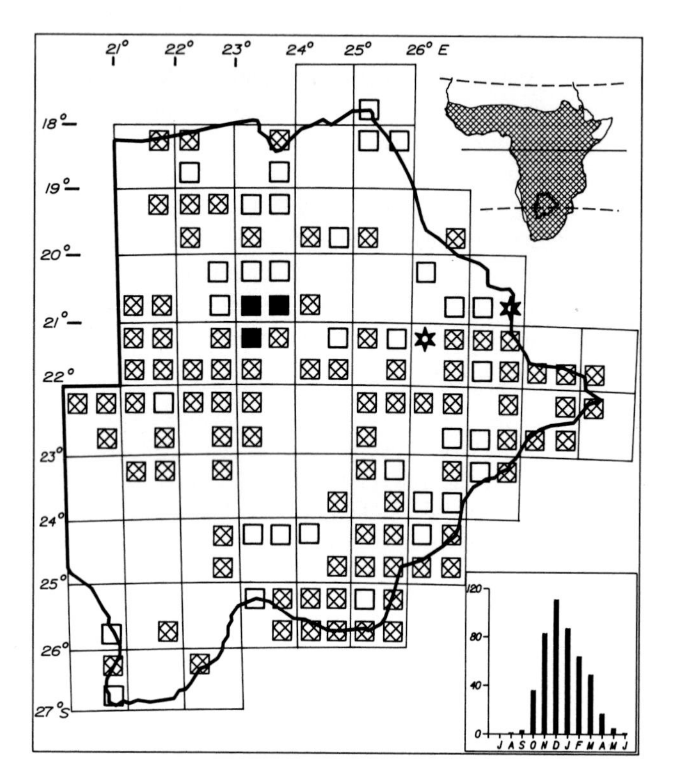

Status

Sparse to common breeding intra-African migrant to all regions of the country. Its status in central and western areas requires further study. Present mainly between October and April, earliest arrival 28 September. Some birds remain until May and June and few may stay throughout the year in some years. Egglaying December to February. All records with Cape Sparrow as host. Other likely hosts are the Ploceid weavers especially the widespread Masked Weaver. Usually solitary, but 2 or 3 territorial males may sometimes be seen together.

Habitat

Acacia tree and bush savanna is the typical habitat in high and low rainfall areas. Also occurs in mixed tree savanna, edges of woodland, riparian growth, gardens and cultivation.

Analysis

125 squares (54%) Total count 485 (0,27%)

BLACK COUCAL

388

Centropus bengalensis

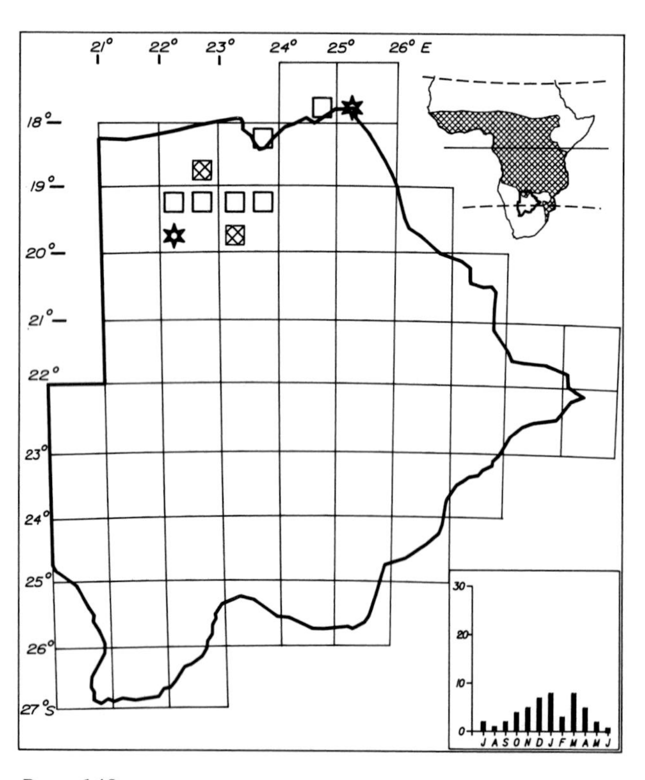

Status

Sparse to locally fairly common breeding intra-African migrant to the Okavango, Linyanti and Chobe river systems. Most records fall between October and April. Some birds remain throughout the year in the Okavango Delta. Probably overlooked to some extent but numbers may fluctuate from year to year. Poorly known in Botswana and is a good candidate for special study. Egglaying February. Usually in pairs or solitary.

Habitat

Tall rank grass on floodplains, the edges of marshland and on well-vegetated seasonal pans.

Analysis

10 squares (4%) Total count 49 (0,03%)

COPPERYTAILED COUCAL 389

Centropus cupreicaudus

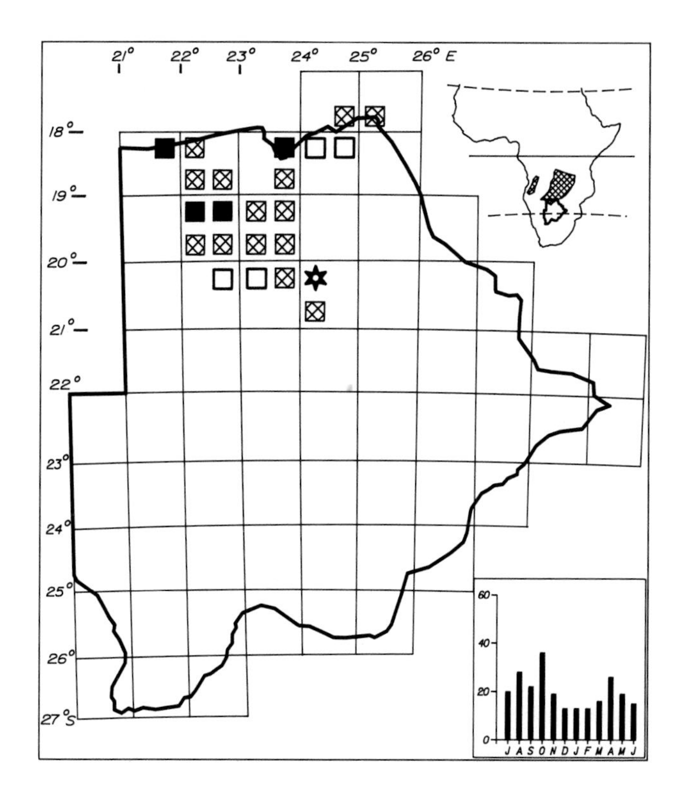

Status

Common to very common resident of the Okavango, Linyanti and Chobe river systems to which it is confined. Recorded twice from Lake Ngami during the drought period of this survey. May expand its range to adjacent areas in higher rainfall years and extend, for example, further south along the Boteti River. Solitary or in pairs.

Habitat

Marshes and swamps with tall rank vegetation including papyrus and reed beds. Also along well-vegetated drainage lines, on the edges of riparian forest and on wet floodplains.

Analysis

23 squares (10%) Total count 251 (0,14%)

SENEGAL COUCAL 390

Centropus senegalensis

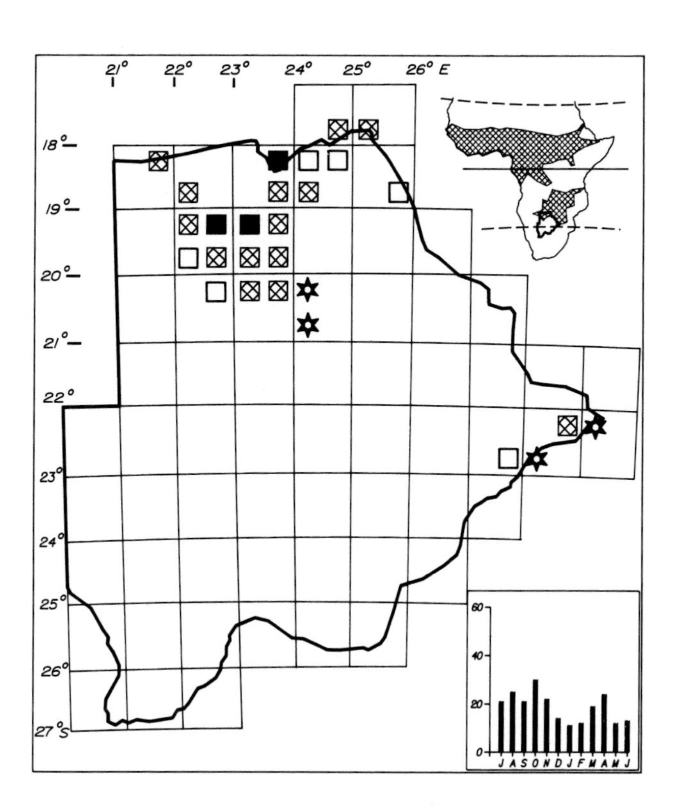

Status

Uncommon to very common resident of the Okavango Delta, Savuti, Linyanti and Chobe regions to as far south as Lake Ngami and Makalamabedi. There are past records from the upper Boteti River and in the east along the Limpopo River from where there are three recent records. Also recorded once in the Sibuyu Forest Reserve in the northeast. Egglaying January, February and March. Usually solitary.

Habitat

Long grass and thickets in tree savanna, dense riverine bush, edges of woodland with thick cover. Sometimes occurs in dry reed beds and tall grass on floodplains.

Analysis

27 squares (12%) Total count 232 (0,13%)

BURCHELL'S COUCAL

391

Centropus superciliosus

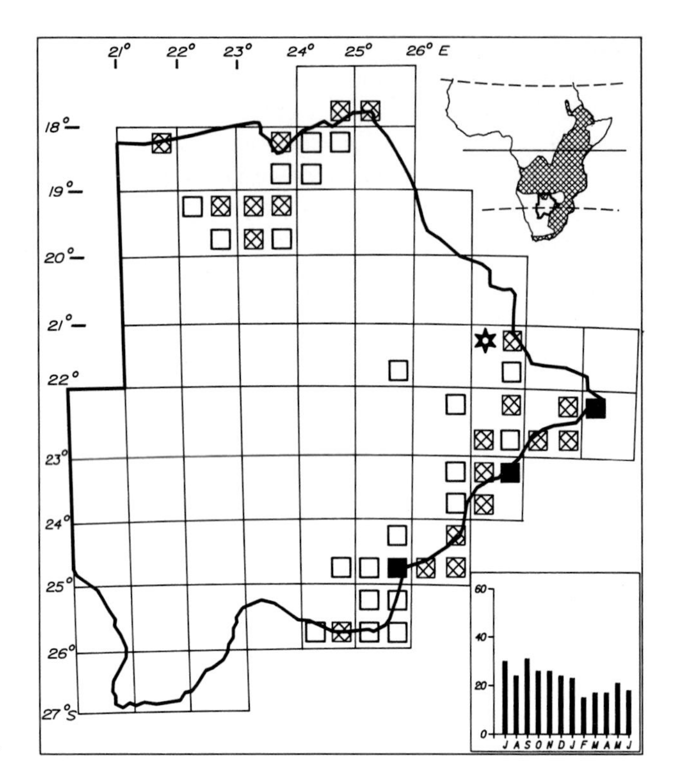

Alternative name

Whitebrowed Coucal

Status

Sparse to fairly common resident of the Okavango, Linyanti and Chobe river systems. Sparse to locally very common in the east and southeast along the Limpopo River drainage to Ramatlabama. Two subspecies occur which are morphologically distinguishable in the field: *C. b. loandae* in the north; and *C. b. burchelli* in the east. Egglaying December. Solitary or in pairs.

Habitat

Rank growth, thicket and dense vegetation along wooded streams and rivers and around dams and ponds. Occasionally wanders into adjacent riparian forest. Also in gardens with lush plants and creepers or other dense cover.

Analysis

45 squares (20%) Total count 299 (0,17%)

BARN OWL

392

Tyto alba

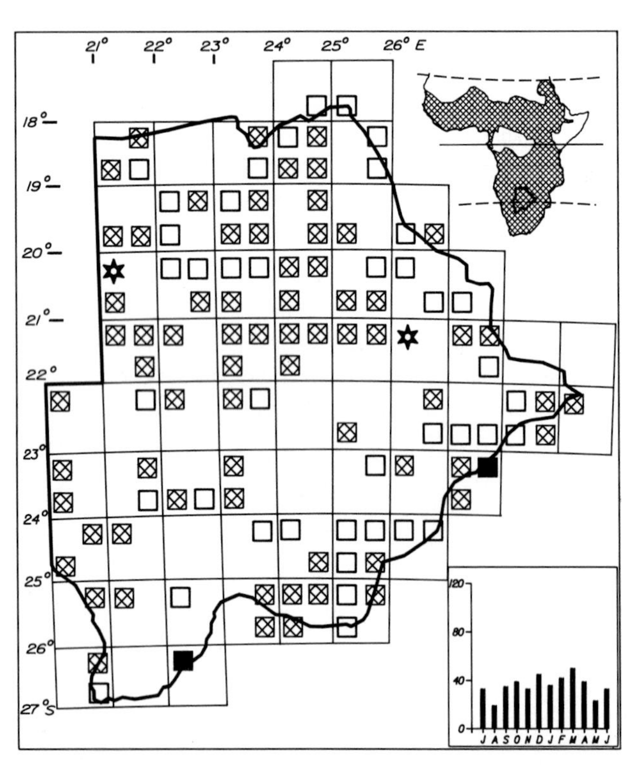

Status

Sparse to locally common resident throughout the country. It probably occurs in all squares but is easily overlooked in rural areas. Egglaying March, April, May. Usually solitary.

Habitat

Very varied, ranging from urban to desert. It is locally common at some farms, game lodges, villages, towns, cattle posts and boreholes. In semidesert it is found near large trees, patches of woodland, small hills and in fossil valleys with steep sides such as some areas of the Okwa Valley. Its presence depends on the availability of rodent prey and suitable nest sites. Twice found drowned—in an oil drum in semidesert and in a water tank on a remote cattle ranch.

Analysis

114 squares (50%) Total count 475 (0,27%)

WOOD OWL

Strix woodfordii

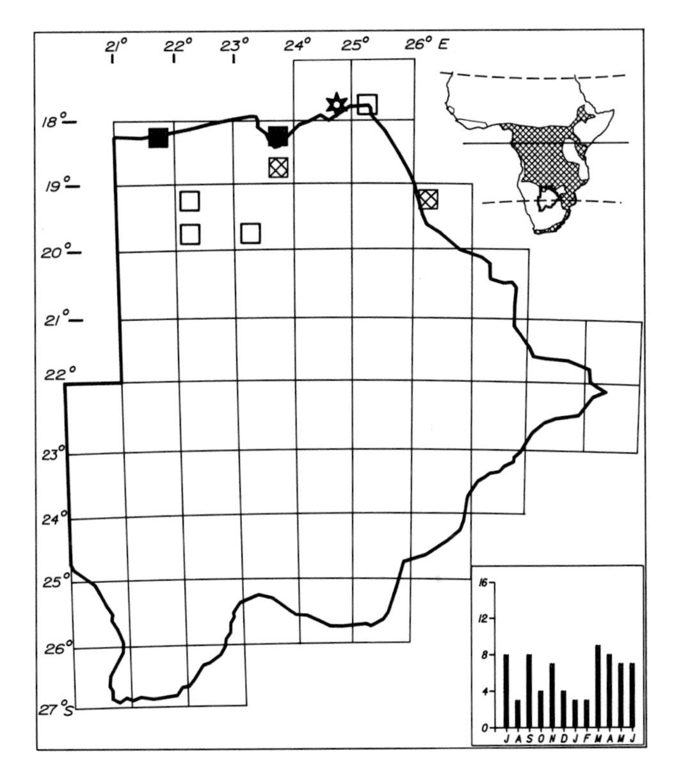

Status

Local resident at Shakawe and Linyanti Camp. Elsewhere in the north it is very sparse and thinly distributed. Easily overlooked and probably more common than indicated. Egglaying September and October. Usually in pairs.

Habitat

Riparian forest including isolated clumps. Rich woodland such as well-developed miombo and *Baikiaea* with dense understorey and thickets as found, for example, in 1926A.

Analysis

9 squares (4%) Total count 72 (0,04%)

MARSH OWL

Asio capensis

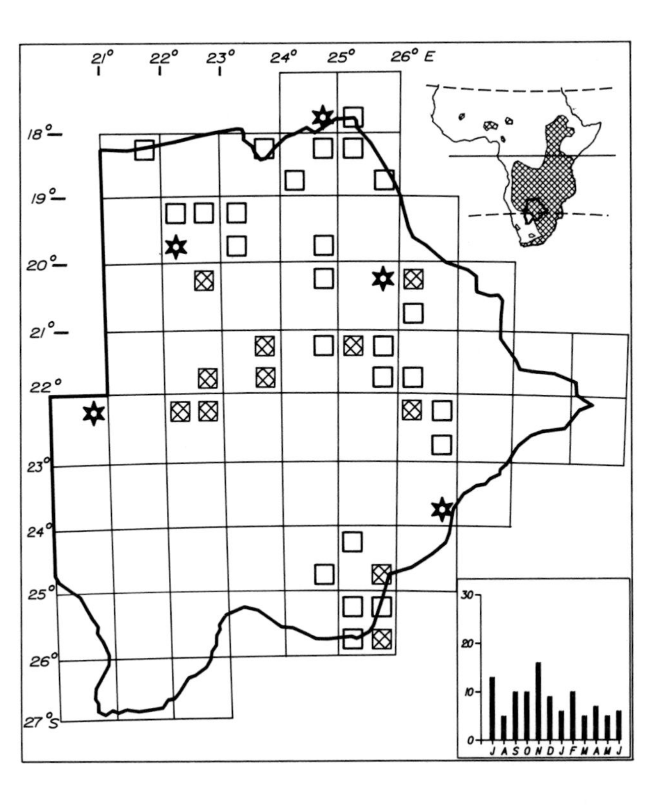

Status

Sparse to fairly common resident in the north, central and southeastern regions. Movements into the grasslands of the central Kalahari are suspected in high rainfall cycles and possibly seasonally. Poorly known and is a good candidate for special study. Egglaying October and May. Usually in small groups of 3–6 birds, solitary or in pairs.

Habitat

In the northern wetlands it is found near marshes and swamps and in rank grass along drainage lines. Occurs throughout the Makgadikgadi basin in seasonally moist grassland even when dry. At Orapa, Mopipi and in the southeast it is found mainly in grassland near pans and dams and on marshland. In semidesert it occurs in rank grass along fossil valleys such as Deception, Okwa and Naledi and in open grassland in tree and bush savanna during summer rains.

Analysis

41 squares (18%) Total count 115 (0,06%)

SCOPS OWL

396

Otus senegalensis

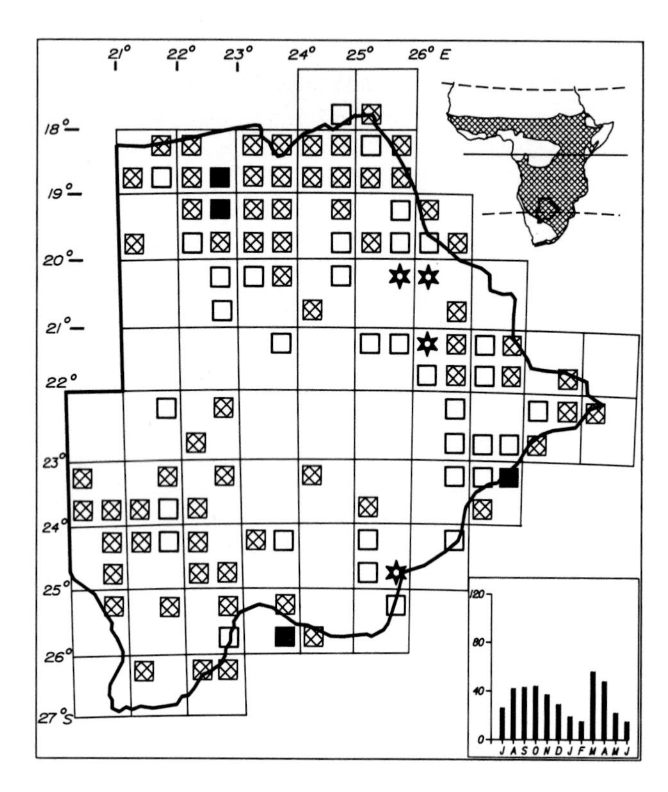

Status
Fairly common to locally very common resident of woodlands in the north, east and west. Sparse on the periphery of these areas and in the southeast. Mainly absent from central Kalahari savannas. Egglaying September and October. Solitary or in pairs.

Habitat
Tall mature broadleafed and *Acacia* woodland in areas where there are many trees and not in isolated clumps. Also in degraded woodland along the Molopo River.

Analysis
106 squares (46%) Total count 407 (0,23%)

WHITEFACED OWL

397

Otus leucotis

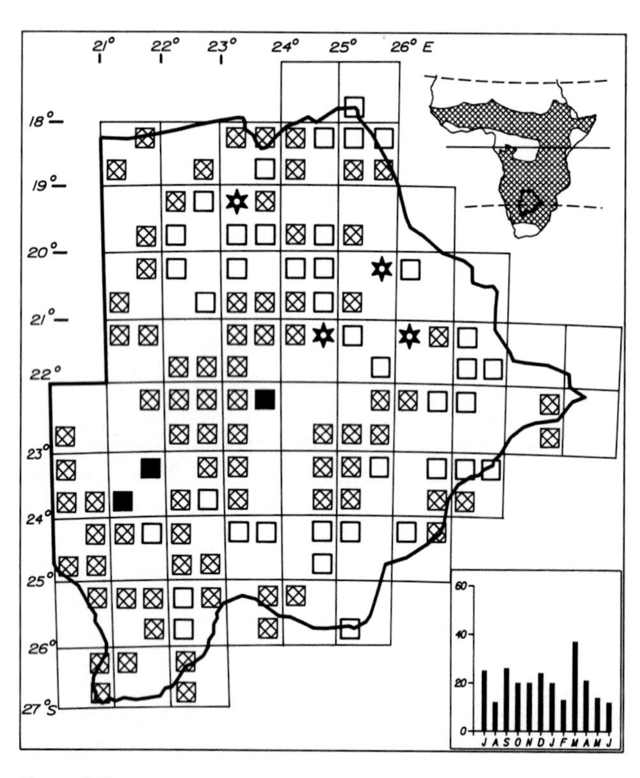

Status
Fairly common to common resident throughout the country. Probably occurs in all squares but is under-recorded because its call does not carry far. Local movements occur with the availability of rodent prey. Egglaying September and November. Solitary, in pairs or family groups.

Habitat
Acacia tree and bush savanna, dry woodland, riparian *Acacia* and riverine bush. Its habitat varies, for example, from rich riverine bush at Nata to scattered thorn trees in arid savanna in the Nossob Valley.

Analysis
124 squares (54%) Total count 256 (0,14%)

PEARLSPOTTED OWL 398

Glaucidium perlatum

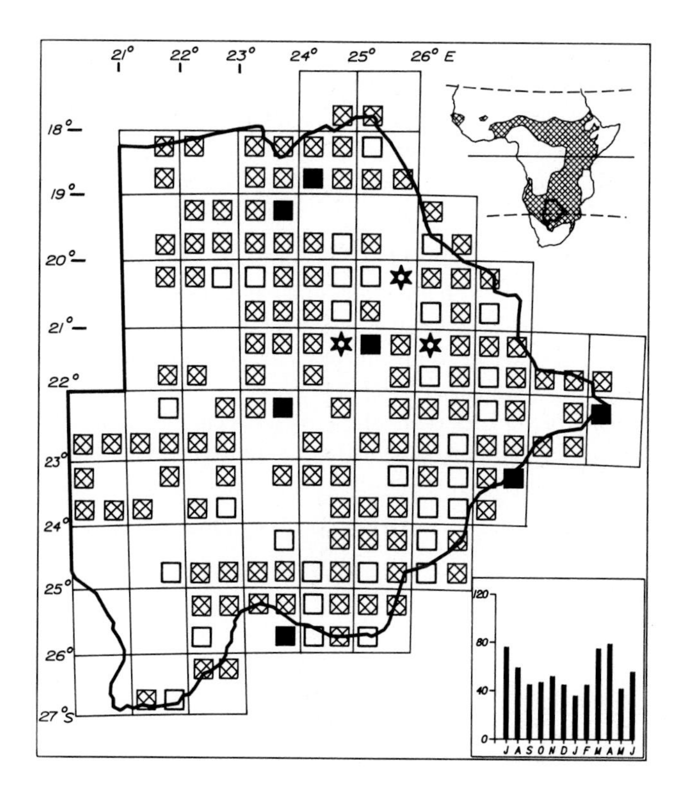

Status

Common to locally very common resident in all regions except the southwest dunelands where it is mainly absent even from apparently suitable habitat. It is the most frequently recorded owl in Botswana. Its call is frequently heard at dusk but it also calls during the day. There are no known breeding records. Often in pairs.

Habitat

Open areas of woodland, tree and bush savanna in high and low rainfall regions. Also on farmland, cultivation, gardens and riverine bush in areas with tall trees and scattered bushes.

Analysis

156 squares (68%) Total count 682 (0,38%)

BARRED OWL 399

Glaucidium capense

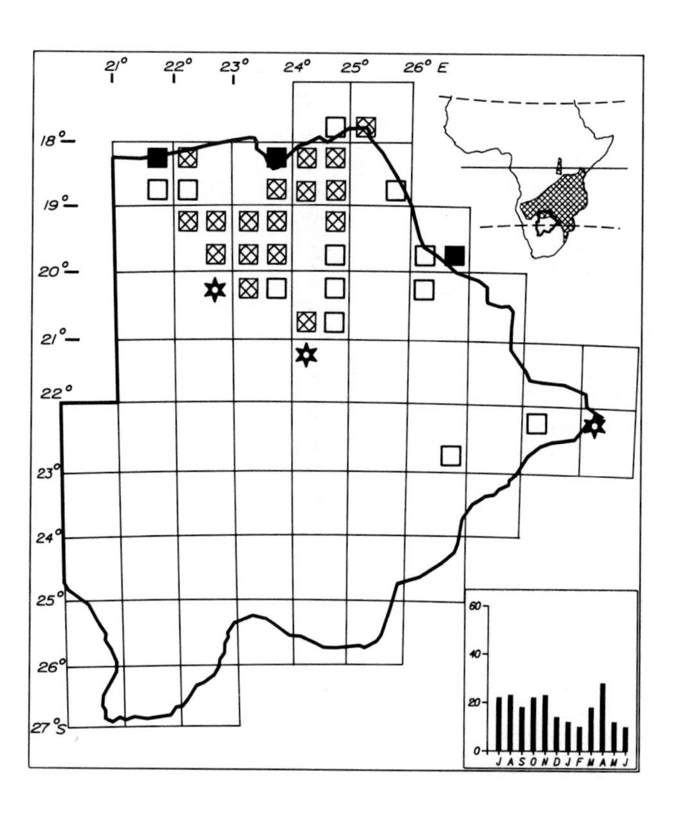

Status

Fairly common to common resident of the Okavango, Linyanti, Chobe, Boteti and Nata river systems. Very common at Shakawe, Linyanti Camp and Tsebanana. Sparse in the east where recorded once on the Lotsane and Thune river drainage. Not recently reported from the Limpopo River where suitable habitat exists but where it is known to be 'generally absent in the western Limpopo Valley' in Zimbabwe (Irwin, 1981). Usually solitary.

Habitat

Riparian forest and riverine bush in the high rainfall areas. Extends into adjacent rich miombo, *Baikiaea* and *Acacia* woodland but nearly always near water.

Analysis

35 squares (15%) Total count 225 (0,13%)

SPOTTED EAGLE OWL

401

Bubo africanus

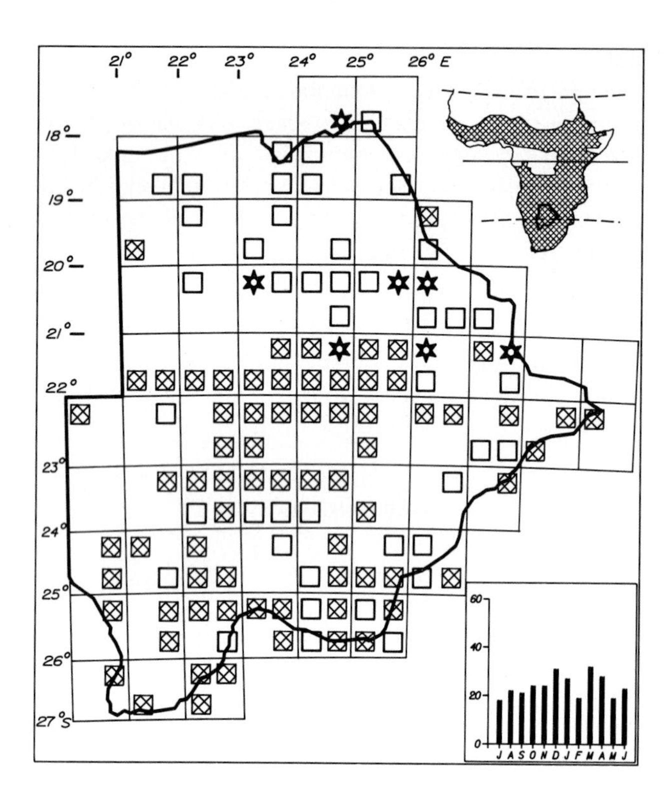

Status
Sparse to uncommon resident north of 21°S. Sparse to fairly common in all other regions. Regularly recorded from Deception Valley, Jwaneng and Mabuasehube. Egglaying all months from July to October. Usually solitary.

Habitat
More commonly recorded in semidesert and arid woodlands in this survey than in high rainfall areas and towns. Usually found near trees but it also occurs in semidesert bush savanna where it rests and possibly breeds on the ground even in areas without rocks. Also in fossil valleys. It is quite frequently a casualty of vehicles on roads at night. Occurs in towns and villages, at cattle posts, game lodges, farms and boreholes.

Analysis
121 squares (53%)   Total count 337 (0,19%)

GIANT EAGLE OWL

402

Bubo lacteus

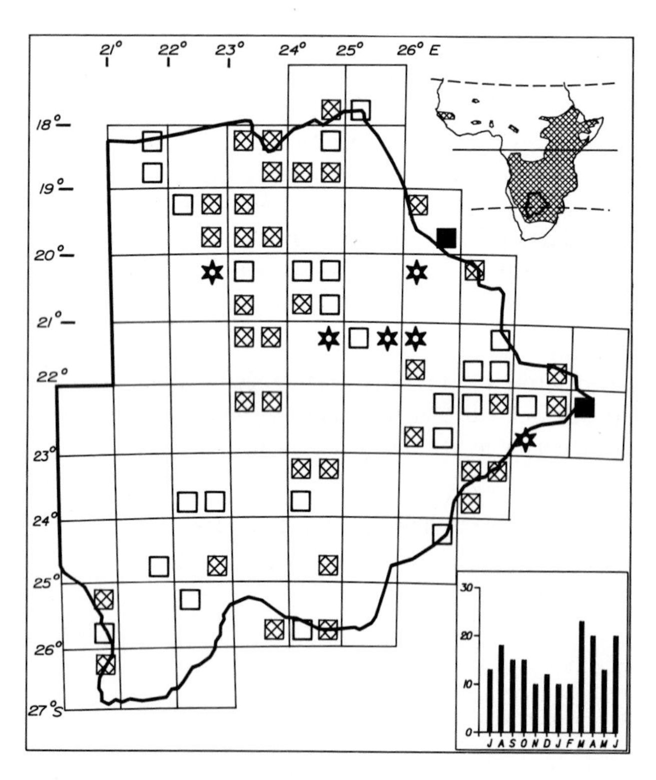

Status
Distributed sparsely throughout the country but is commonest in the river systems of the north and east. Locally common to very common at Linyanti, Savuti, Maun, Tsebanana, Mathathane, Pont's Drift and Martin's Drift. Uncommon along the Molopo and Nossob valleys. Egglaying July, August and September. Usually solitary, occasionally in pairs.

Habitat
Riparian forest and large trees along water courses. Also in mature woodland in the central Kalahari and in the western woodlands.

Analysis
70 squares (30%)   Total count 204 (0,11%)

PEL'S FISHING OWL

Scotopelia peli

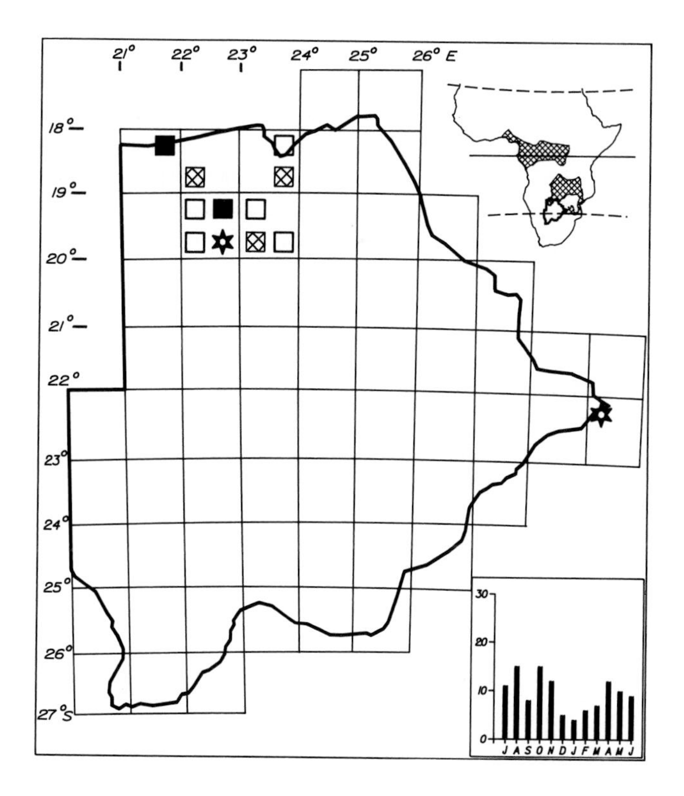

Status

Sparse to fairly common in restricted habitat in the Okavango and Linyanti river systems. Locally very common at Shakawe and the western part of the Moremi Wildlife Reserve. Not recorded in the Limpopo Valley in this survey but there is an earlier record from this area at the western limit of its range in Zimbabwe. It is not likely that there are more than 100 pairs in the country and it must be considered as a vulnerable species. Egglaying March and April. Solitary or in pairs.

Habitat

Riparian forest and large trees along perennial waterways. Feeds mainly on fish and crustaceans.

Analysis

11 squares (5%) Total count 126 (0,07%)

EUROPEAN NIGHTJAR

Caprimulgus europaeus

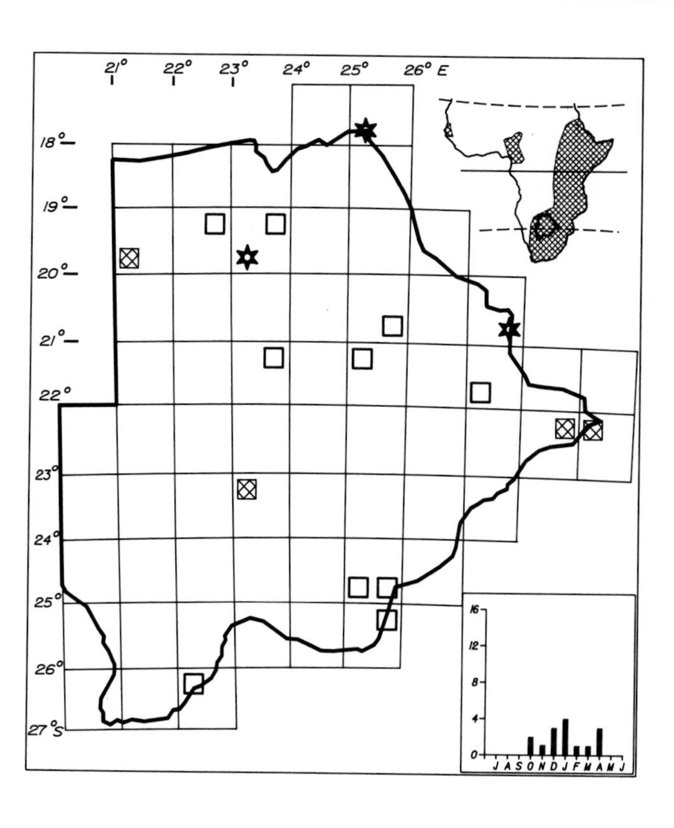

Status

A rarely recorded Palaearctic migrant which may occur regularly throughout the country in small numbers. It has only been recorded once at each of the localities shown except Lobatse and Gaborone from each of which there are two records. Identified from road killed specimens as well as sight records. Very poorly known. Present between October and April. Usually solitary.

Habitat

Open areas of tree and bush savanna, semidesert scrub savanna, cleared areas such as roads and veterinary cordon fences. Also in river valleys and fossil valleys.

Analysis

17 squares (7%) Total count 20 (0,01%)

FIERYNECKED NIGHTJAR

405

Caprimulgus pectoralis

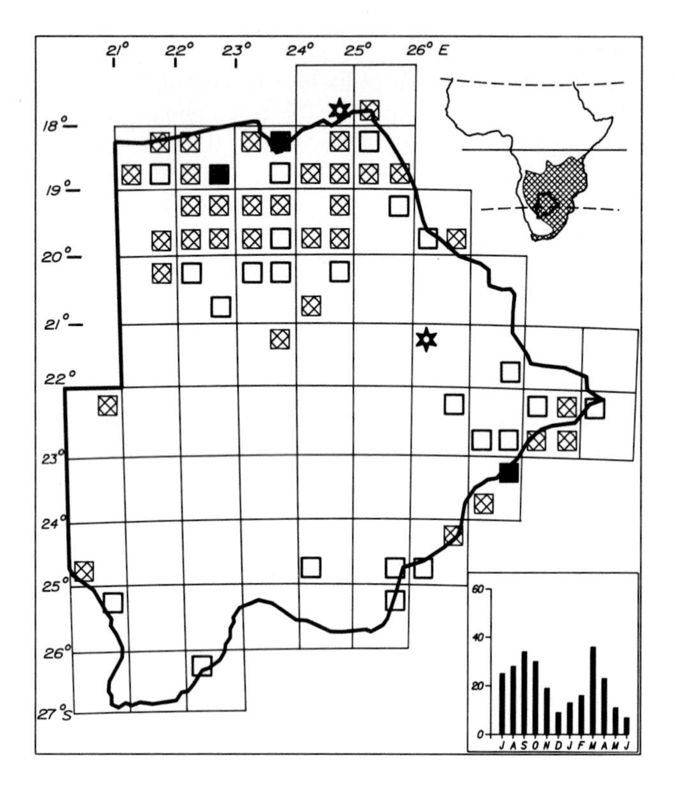

Status

Fairly common to very common resident and breeding intra-African migrant mainly to areas north of 21°S. Sparse to common in the east. In the southeast and southwest it is very sparse and may only occur in some years of good rainfall. Present in all months but most commonly recorded between July and October and again in March and April. Egglaying October but the breeding season is probably from August to December. Usually solitary or in pairs.

Habitat

Well-developed broadleafed woodland. Lays eggs on leaf litter under trees. Should be regarded as a tropical woodland savanna species and all records from dry woodland and Kalahari savanna are of special interest.

Analysis

60 squares (26%) Total count 263 (0,15%)

RUFOUSCHEEKED NIGHTJAR

406

Caprimulgus rufigena

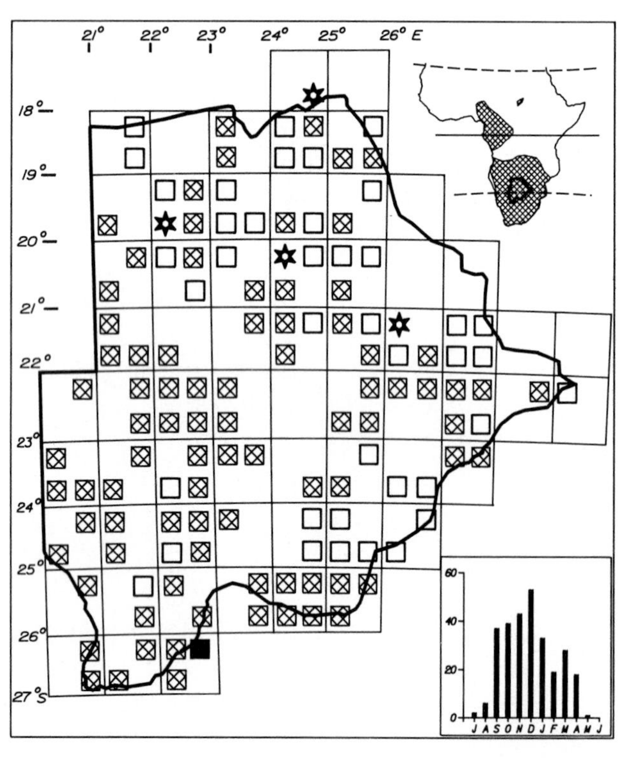

Status

Sparse to common breeding intra-African migrant to all regions of the country. It is the most commonly recorded and widespread nightjar in Botswana. Present mainly from September to April with some early arrivals in July and August and one record in May. Egglaying September, October and December. Usually in pairs.

Habitat

Any woodland where it occurs near the edges and in clearings. Also in tree and bush savanna in high and low rainfall regions, semidesert bush and scrub savanna. Lays eggs on the ground or in leaf litter under bushes and trees.

Analysis

130 squares (56%) Total count 317 (0,17%)

NATAL NIGHTJAR

407

Caprimulgus natalensis

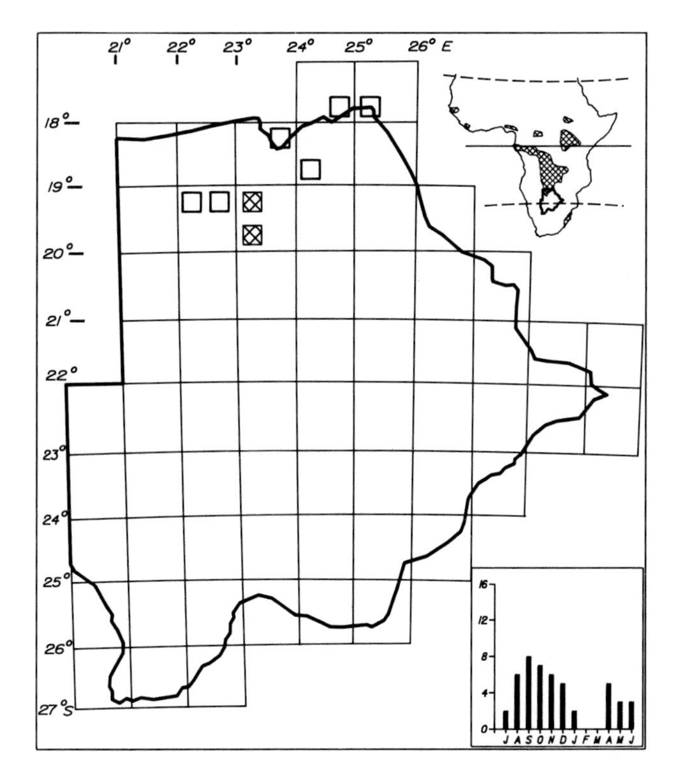

Status

Sparse to locally fairly common resident of the Okavango Delta and in the Chobe region. Its presence in the delta has only recently been established although it was previously known from Chobe. An alternative and better name is Swamp Nightjar. It has been recorded mainly from the southern and central parts of the Moremi Wildlife Reserve. Recorded more often at Serondella than at Kasane. It is not well known and warrants special study. Usually solitary.

Habitat

Floodplains and edges of marshes and swamps. Usually found on the periphery of open areas near woodland or savanna.

Analysis

8 squares (4%) Total count 50 (0,03%)

FRECKLED NIGHTJAR

408

Caprimulgus tristigma

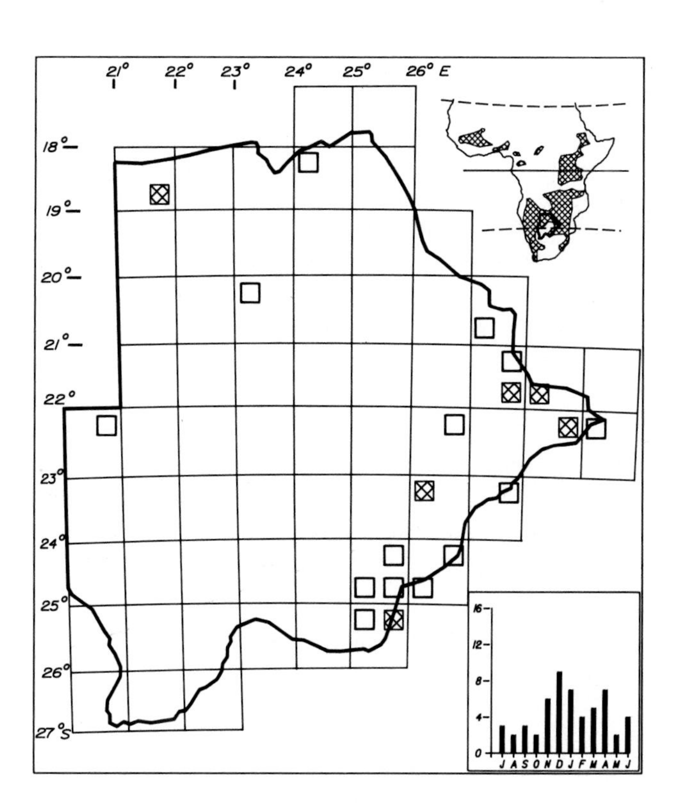

Status

Sparse to uncommon resident of hilly areas in the southeast and east. Very sparse and unpredictable elswhere except at the Tsodilo Hills where it is fairly common. Recorded mainly from granitic hills at Ootse, Kanye and Gaborone in the southeast. In the east recorded from Serowe and the Shoshong Hills and from rocky outcrops near Bobonong through Selebi Phikwe to north of Francistown. Usually solitary or in pairs.

Habitat

Rock outcrops and ledges on hills and mountains in open or lightly vegetated areas with scattered trees. Also on granite boulders in wooded habitat at Selebi Phikwe and adjacent areas.

Analysis

20 squares (9%) Total count 62 (0,03%)

MOZAMBIQUE NIGHTJAR 409

Caprimulgus fossii

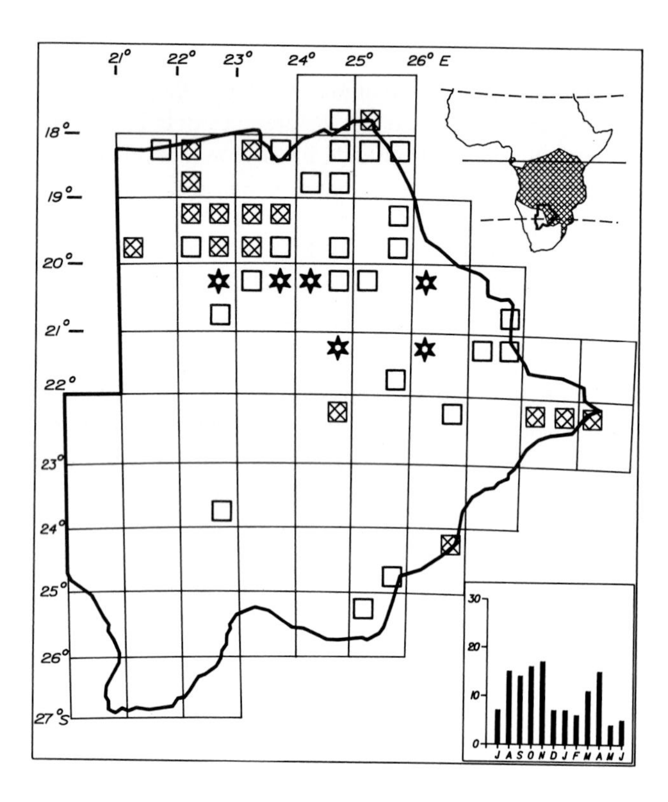

Status

Sparse to common resident of the north. South of 21°S it is sparse to uncommon and thinly distributed mainly in the east and southeast. It may be more common in these latter regions in high rainfall years. Recorded once at Kang. Some local movements occur. Found in Botswana at the western and southern limits of its continental range. Egglaying September, October and November. Solitary or in pairs.

Habitat

Open woodland, tree and bush savanna in areas with bare sandy ground. Also in river valleys where it occurs in areas with scrub and scattered low bushes on riparian terraces.

Analysis

47 squares (20%) Total count 142 (0,08%)

PENNANTWINGED NIGHTJAR 410

Macrodipteryx vexillaria

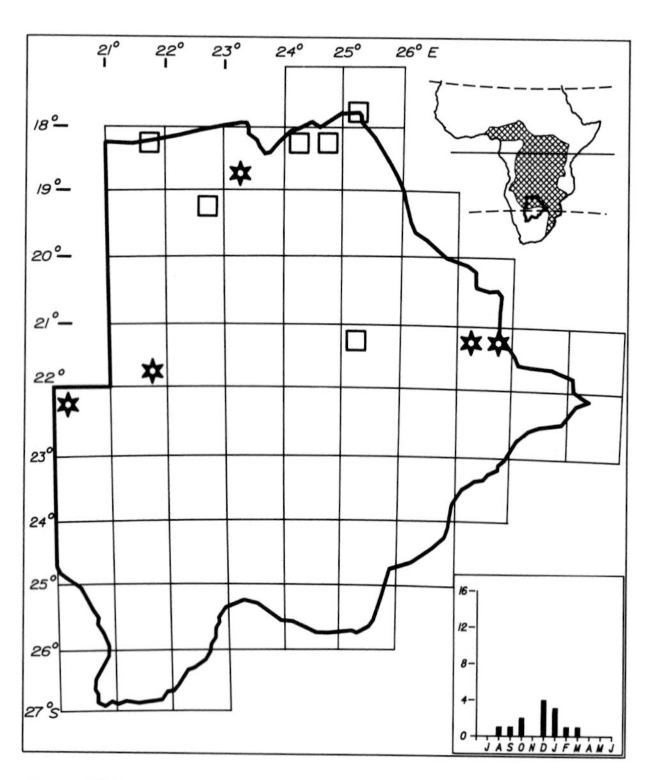

Status

A very sparse breeding intra-African migrant to areas north of 22°S. During the drought between 1980 and 1987 this species was recorded only at Kasane (1725C). From 1988 onwards it has been recorded on more than 10 occasions in different localities. Likely to occur mainly in the woodlands of the northeast and northwest but will extend its range further south, for example to Francistown, Orapa and Ghanzi, in some years. Recorded between August and March. Usually reported as solitary males in breeding plumage.

Habitat

Mature broadleafed woodland in high rainfall regions. Forages in open areas such as clearings, tracks and roads; is sometimes a casualty of vehicles in this latter habitat. It also occurs near hills; usually on bare sandy or stony ground.

Analysis

11 squares (4%) Total count 23 (0,01%)

EUROPEAN SWIFT

411

Apus apus

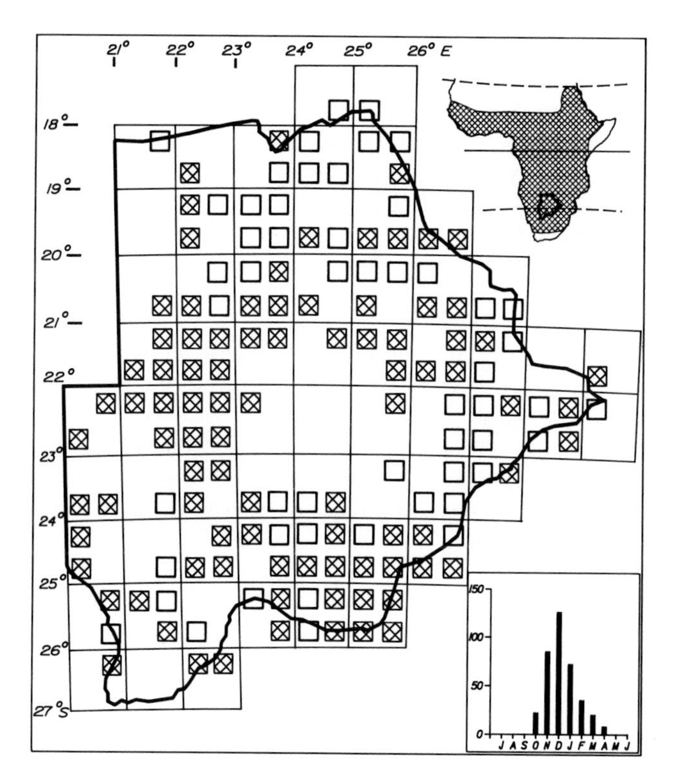

Status

Sparse to common Palaearctic migrant to all regions of the country. Likely to occur in the central Kalahari more often than reported there during this survey. It is present from late October to March with a few birds still on northward passage in April. Often seen near rain clouds and the first annual sightings frequently occur with the first thunderstorms of the year. Constantly moving from one area to another and rarely seen at the same locality on consecutive days. Usually in flocks of 20–200 birds, flocks of several thousands occur less commonly.

Habitat

Migrates over all types of habitat usually flying very high just within binocular vision and easily overlooked. Forages over large open areas such as plains, floodplains, dams, lakes, pans, swamps and rivers; particularly over wet habitat where insects are plentiful. Also over woodland and savanna descending to feed over the ground or tree tops at times.

Analysis

142 squares (62%) Total count 384 (0,22%)

BLACK SWIFT

412

Apus barbatus

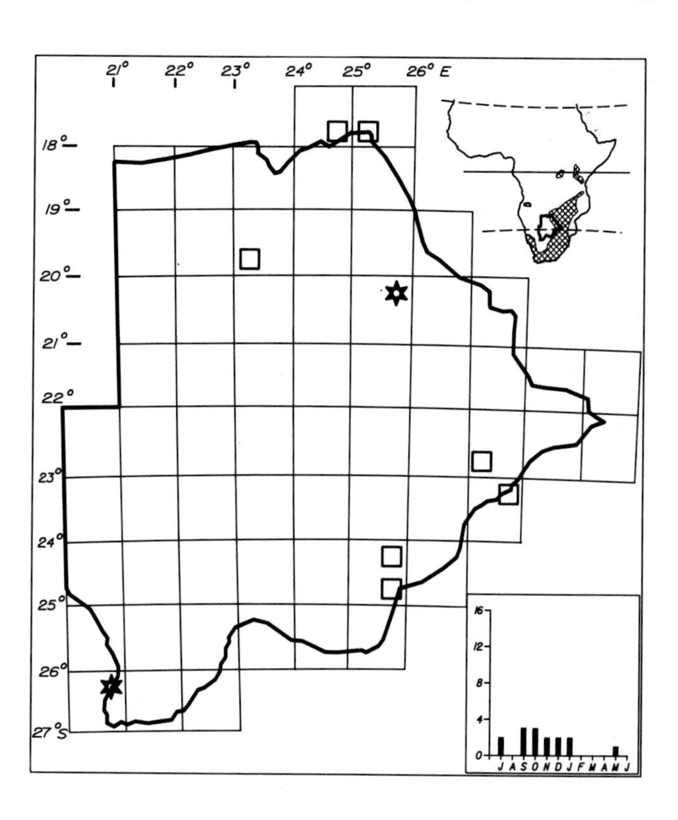

Status

Status uncertain and poorly known. Recorded very sparsely from hilly regions of the east and southeast. Records from the north may represent wanderers from the Victoria Falls gorges where there is a well-known breeding colony. There is no evidence yet that it is a regular intra-African migrant to Botswana as is the case in South Africa although most of the few current records fall between September and January. It is difficult to differentiate it in the field from the European Swift.

Habitat

Hills and gorges in the east and southeast. Forages over any habitat but mainly near hills or along river valleys.

Analysis

9 squares (4%) Total count 19 (0,01%)

WHITERUMPED SWIFT

415

Apus caffer

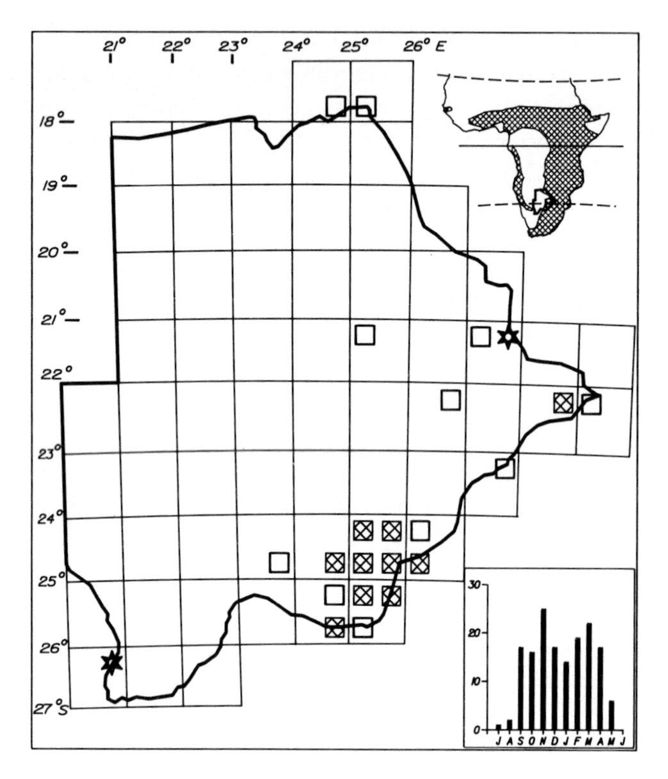

Status

Fairly common to common breeding intra-African migrant in the southeast to as far west as Khakhea. Sparse to uncommon in the east to as far west as Orapa. Uncommon at Kasane and Serondella. Present mainly from September to April with some early arrivals in July and August and late departures in May. Egglaying January. Usually in pairs, occasionally forages in small groups.

Habitat

Manmade structures such as buildings, bridges, dam walls and culverts on which it breeds in the nests of swallows—mainly Lesser and Greater Striped Swallows. Thus mostly seen in urban areas where it forages over any open area and over water at dams, ponds and swimming pools. Occasionally near hills and gorges where it may breed.

Analysis

23 squares (10%) Total count 169 (0,09%)

HORUS SWIFT

416

Apus horus

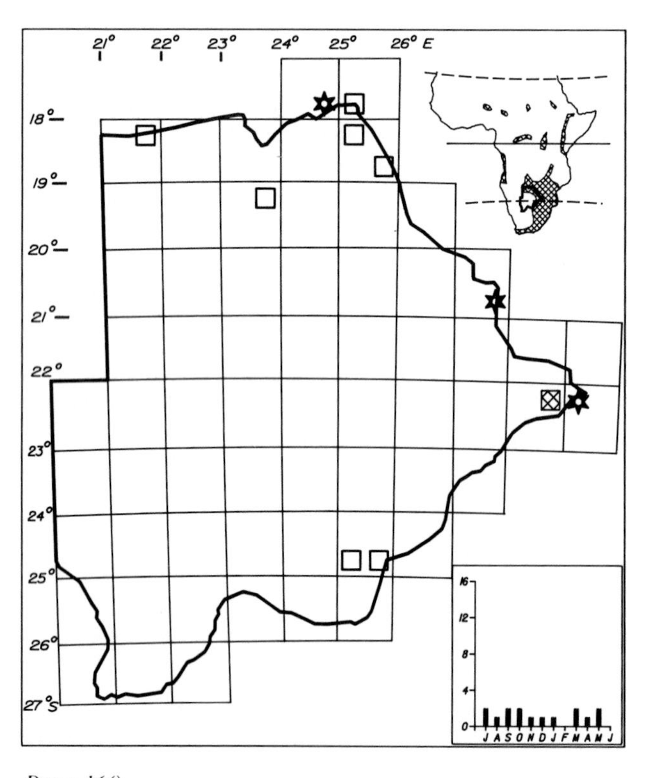

Status

A very sparse intra-African summer migrant in the north and east. Recorded also in July and August at Kasane and in May at Mpandamatenga and Gaborone. Most of the very few records fall between September and April but it is possibly resident and present throughout the year in the extreme northeast. Poorly known and requires more detailed study.

Habitat

Along major rivers and adjacent habitats. Also at dams in Gaborone and Kanye. Requires sandbanks for breeding.

Analysis

11 squares (5%) Total count 17 (0,01%)

LITTLE SWIFT — 417

Apus affinis

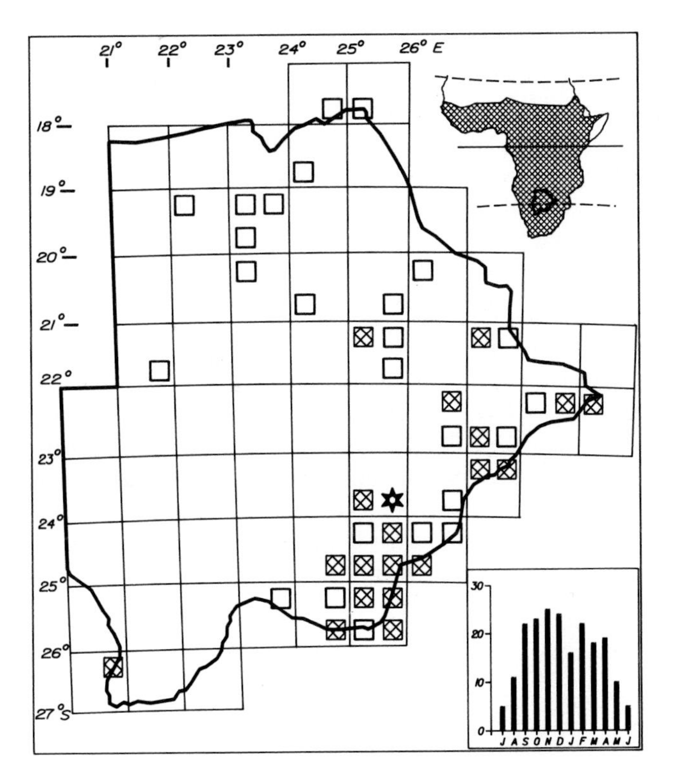

Status

Fairly common to locally common resident in the east and southeast to as far west as Orapa and Bray. Sparse in the north. Isolated colonies occur at Ghanzi and Twee Rivieren. Part of the population is migratory and present only between September and April. Its status and distribution benefits from and reflects the density of tall buildings. Egglaying September, October and February. Breeds colonially and forages in noisy flocks, usually of 20–100 birds.

Habitat

Nests on urban and suburban buildings, industrial sites, tall mine constructions (Jwaneng, Morupule, Selebi Phikwe, Orapa, Letlhakane), cooling towers, grain silos, dam walls and bridges. Also on cliffs and gorges. Forages over gardens, parks, farmland and savanna but especially over open water. Its distribution is restricted by the availability of water for drinking.

Analysis

45 squares (20%) Total count 212 (0,12%)

ALPINE SWIFT — 418

Apus melba

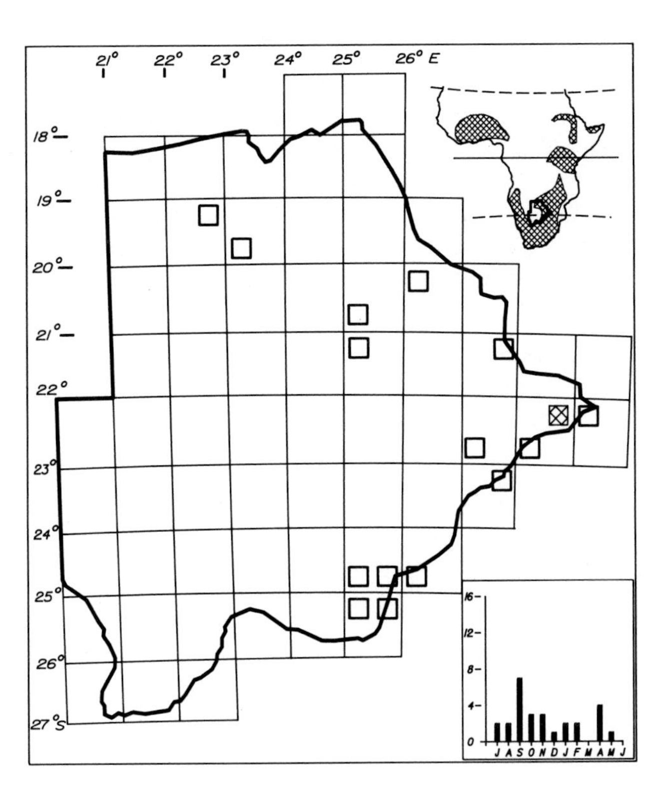

Status

A sparse to uncommon passage migrant in the north, east and southeast. It is not well known and requires more detailed study. Breeding in the hills of the southeast and east must be considered but no evidence of this has yet been ascertained. Breeds in the Transvaal. Based on the few available records, peaks of passage are in September and April. Sometimes solitary, more often in small groups with or without other swifts.

Habitat

Mountains and hills in the southeast. Over woodland and savanna in the Limpopo Valley. Over open pans and grass plains in the Makgadikgadi region. Over floodplains and woodland in the north.

Analysis

16 squares (7%) Total count 27 (0,02%)

PALM SWIFT

Cypsiurus parvus

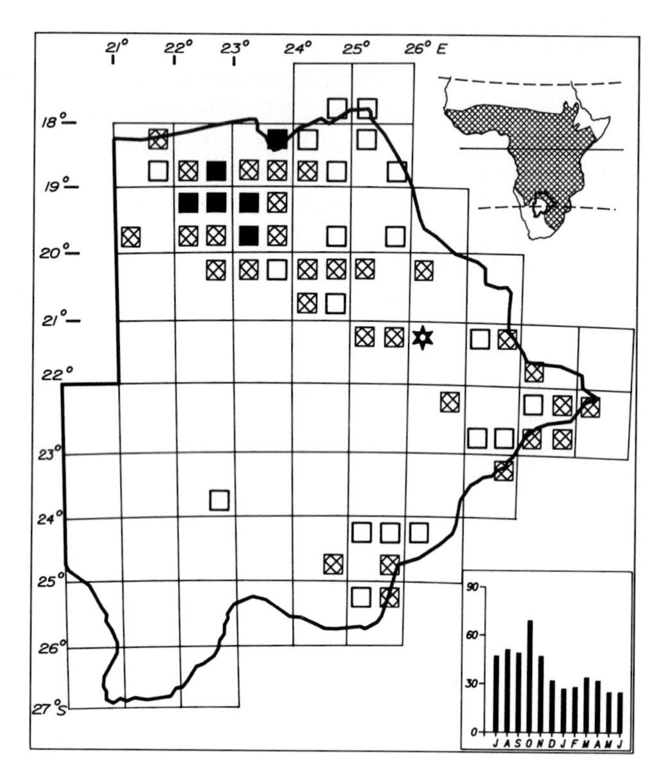

Status

Common to very common resident in the Okavango Delta and Linyanti region south to Nxai Pan, the Boteti River and Lake Ngami. Sparse in the northeast. Sparse to locally common in the east and southeast. Recorded twice at Kang. Egglaying September, October and January. Usually in small groups of 4–15 birds. Several pairs usually nest in the same palm tree.

Habitat

Any habitat with palm trees whether indigenous or exotic. Indigenous palms are found commonly in the Okavango Delta and at Nxai Pan and sparsely along the Boteti River to Orapa and in the Tuli Block. Exotics are found in towns such as Lobatse, Gaborone and Francistown and in villages such as Kang, Serowe and Nata. Its range may increase with continued planting of palms near habitation as has occured in the southwestern Transvaal.

Analysis

57 squares (25%)   Total count 471 (0,26%)

SPECKLED MOUSEBIRD

Colius striatus

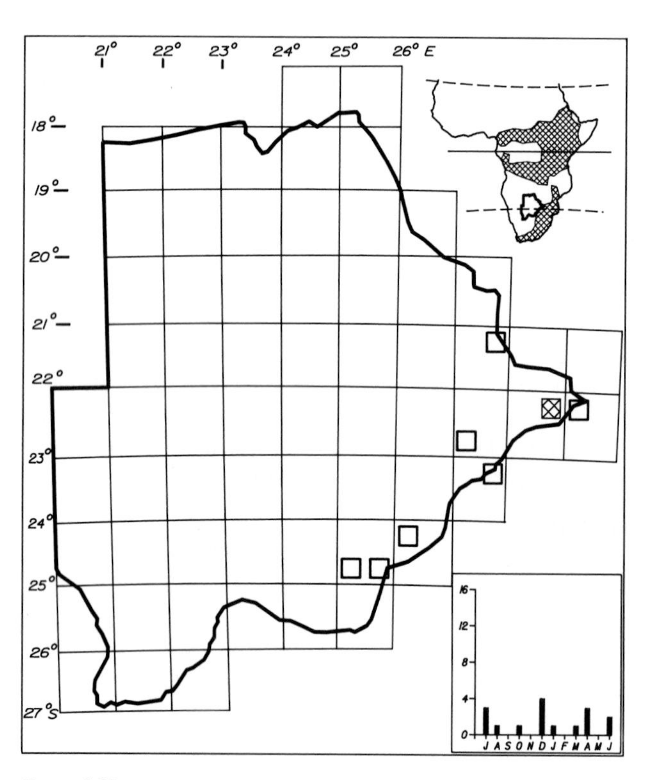

Status

Very scarce resident whose range in southern Africa just reaches across the east and southeast borders of Botswana from the Transvaal. Its status is not certain as its presence in the country has only recently been established. It is unlikely to extend west of the Limpopo drainage. Its numbers may be on the increase as a result of increased cultivation and gardens in this area. Usually in small flocks of 4–10 birds.

Habitat

Dense vegetation in tree and bush savanna and in riverine bush. Occasionally in gardens. Often at or near trees and bushes with fruit or berries.

Analysis

8 squares (3%)   Total count 15 (0,008%)

WHITEBACKED MOUSEBIRD

425

Colius colius

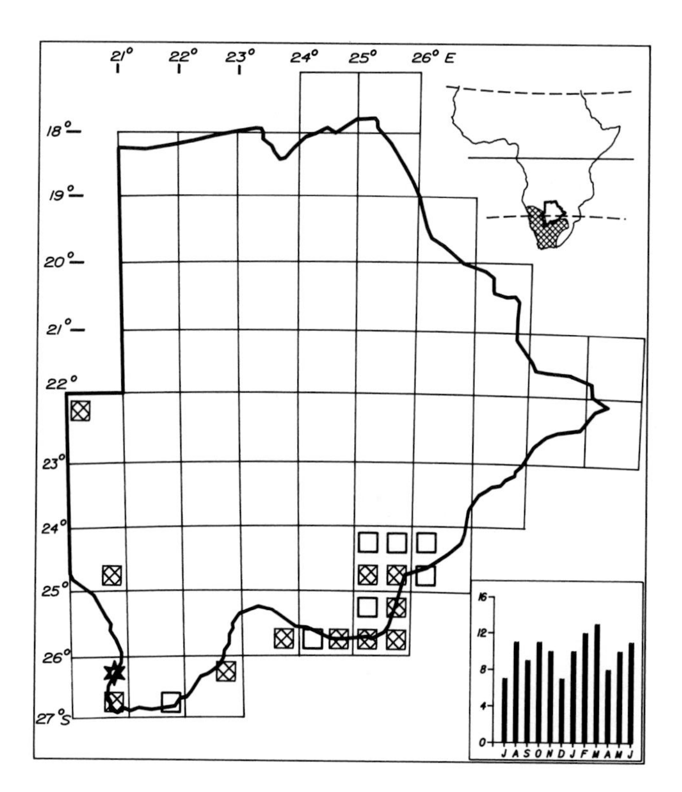

Status

Sparse to common resident of the southeast. Sparse to uncommon elswhere in its range. Extends along the whole length of the Molopo River to Bokspits. Recorded once in the northwestern sector of the Gemsbok National Park and once at Mamuno. In the west it is at the eastern limit and in the south at the northern limit of its southern African range. Egglaying September. Usually in small flocks of 4–10 birds.

Habitat

Tree and bush savanna in low rainfall regions, riverine bush and thickets, gardens, arid scrub savanna with scattered trees and bushes.

Analysis

19 squares (8%) Total count 131 (0,07%)

REDFACED MOUSEBIRD

426

Urocolius indicus

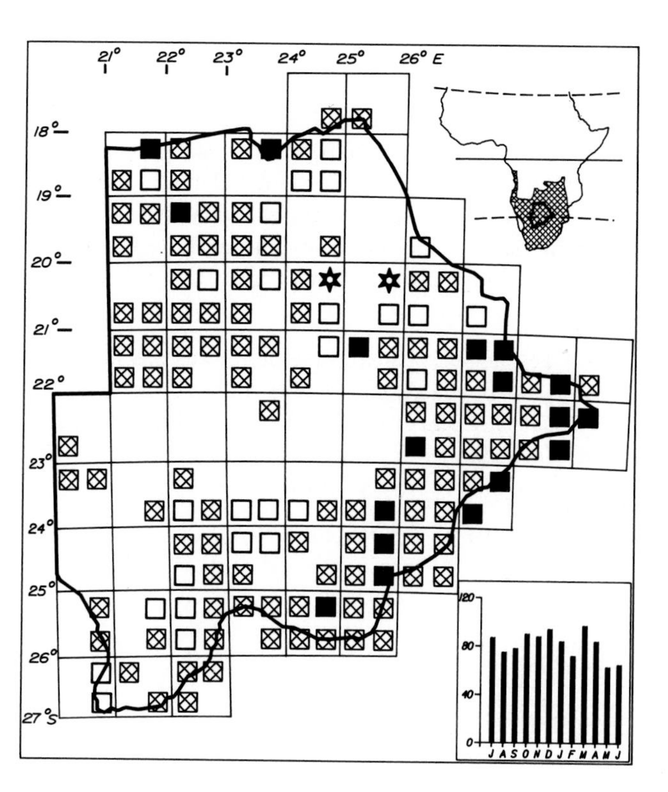

Status

Common to very common resident in the southeast, east and northwest. Sparse to fairly common in the northeast and southwest. Mainly absent in the central and western Kalahari. It is by far the most common and widespread mousebird in the country. Egglaying all months September to February. Usually in small flocks of 4–10 birds.

Habitat

Tree and bush savanna in high and low rainfall regions. Often in areas with thicket or dense tall bushes of *Acacia mellifera* or *Acacia tortilis*. Also in open mixed woodland, gardens, riparian vegetation and riverine *Acacia* in areas with flowers or fruit. Not in the richer (70% canopy) woodlands such as miombo except occasionally on the edge.

Analysis

150 squares (65%) Total count 1017 (0,57%)

NARINA TROGON

Apaloderma narina

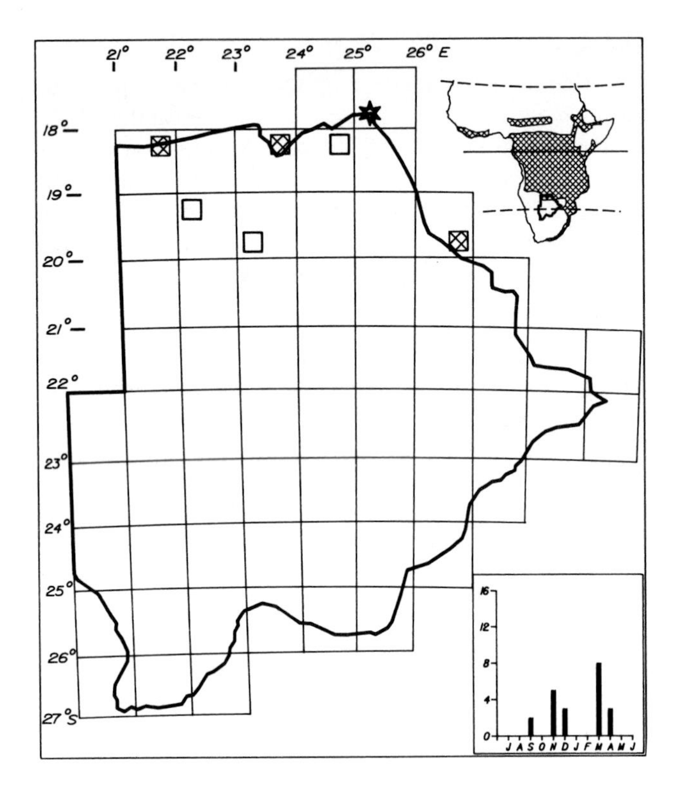

Status

Sparse to locally uncommon breeding intra-African migrant patchily recorded in the north, mainly at Shakawe and Linyanti Camp. Present from September to April, mostly mid-November to March at Linyanti. Breeding behaviour and nest site December. Some birds may be resident throughout the year and overlooked when not calling. Males respond well to imitation of call in the breeding season. Poorly known and merits special study. Usually solitary.

Habitat

Riparian forest, dense broadleafed woodland and riverine thicket.

Analysis

7 squares (3%) Total count 23 (0,01%)

PIED KINGFISHER

Ceryle rudis

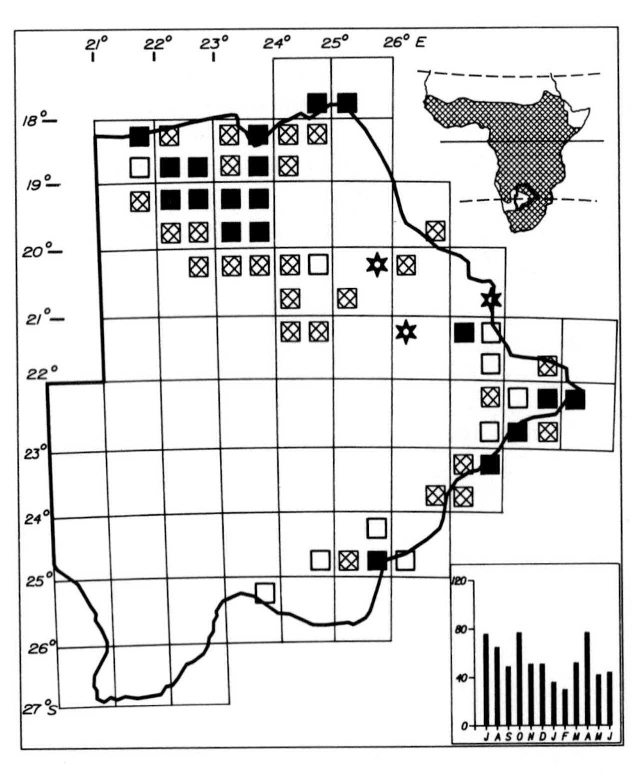

Status

Common to very common resident of the Okavango, Linyanti, Chobe, Nhabe, Boteti, Nata, Shashe, and Limpopo river systems and at Lake Ngami, Shashe, Mopipi, Gaborone and Kanye dams. Mainly sedentary—it will move to adjacent wetlands during wet periods and be confined to more permanent water in dry spells. Egglaying July. Occurs in pairs, occasionally in small groups.

Habitat

Rivers, lakes, lagoons, dams, ponds and sewage ponds. Less frequently on flooded pans in river valleys. On seasonal rivers it usually occurs only near remnant pools or where the river is dammed.

Analysis

58 squares (25%) Total count 674 (0,37%)

GIANT KINGFISHER

Megaceryle maxima

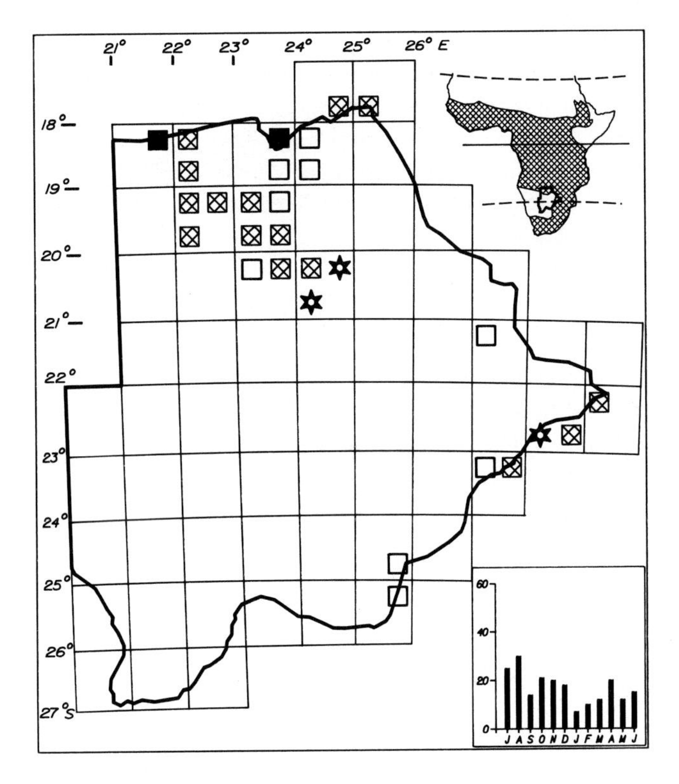

Status

Sparse to fairly common resident of the northern wetlands to as far south as the upper reaches of the Boteti River. Very common at Shakawe and Linyanti Camp. Uncommon to fairly common on the Limpopo River; sparse in Gaborone and Lobatse; recorded only once at Shashe Dam. There is much suitable habitat along several rivers into which this species may spread during wetter cycles. Solitary or in pairs.

Habitat

Perennial rivers. Tree-lined waterways, dams and swamps.

Analysis

29 squares (13%)   Total count 211 (0,12%)

HALFCOLLARED KINGFISHER

Alcedo semitorquata

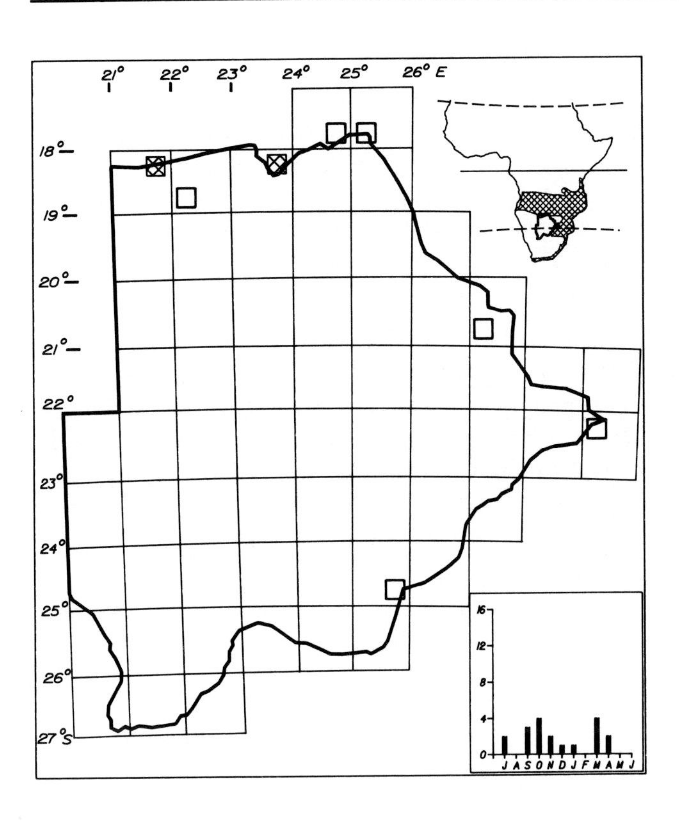

Status

Uncommon resident at Shakawe and Linyanti Camp. Very sparse at few other sites on the Okavango and Chobe rivers. Recorded once on each of the Shashe, Limpopo and Ngotwane rivers. Normally sedentary; secretive in habits. Occurs in Botswana at the extreme southern and western limits of its range in Africa. Egglaying April. Solitary or in pairs.

Habitat

Streams and rivers in areas with riparian forest or dense overhanging riverine vegetation; particularly places where the vegetation almost encloses the waterway.

Analysis

8 squares (3%)   Total count 21 (0,01%)

MALACHITE KINGFISHER

431

Alcedo cristata

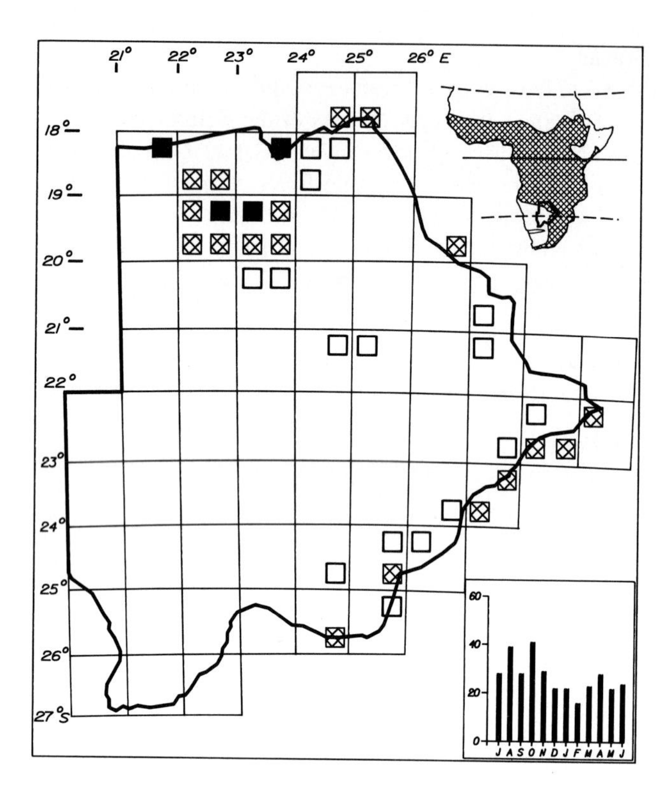

Status

Common to very common resident of the Okavango, Linyanti and Chobe river systems. Sparse to fairly common along the Shashe, Limpopo, Ngotwane and the eastern Molopo rivers. May turn up at any dam or pond in the east when suitable conditions are available such as at Tsebanana Pan, Mopipi Dam and Jwaneng Golf Club Dam. Movements occur seasonally but it is mainly sedentary. Solitary or in pairs.

Habitat

In reeds, grass or sedges at marshes, swamps, rivers, dams, lagoons, flooded pans and lakes. Perches on low branches over water or on the margins. Usually flies just above the surface of the water.

Analysis

38 squares (17%) Total count 329 (0,18%)

PYGMY KINGFISHER

432

Ipsidina picta

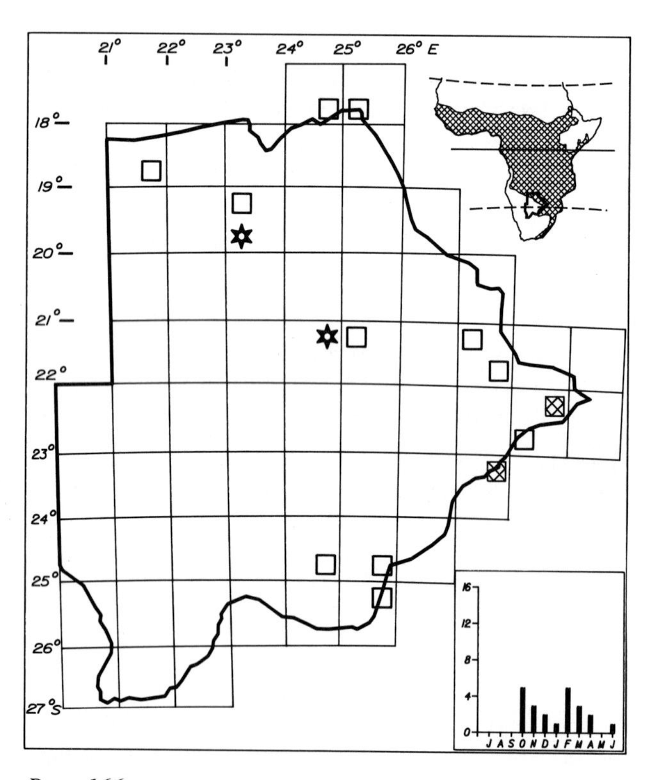

Status

Scarce to uncommon intra-African migrant from October to April. Only recorded patchily in the north and east along the major river systems and twice at Jwaneng. Breeding is unknown in Botswana where it occurs at the western limits of its continental range at this latitude. Breeding is known from adjacent Zimbabwe and Transvaal. Solitary.

Habitat

Mature woodland and riparian forest in the high rainfall areas. Probably occurs over savanna on passage as suggested by its occurrence at Jwaneng. These two records were of birds which had flown into the town at night possibly disorientated by the very bright lights of the mine. This likely theory may also apply to the three Orapa records and to those from Francistown and Gaborone.

Analysis

15 squares (7%) Total count 26 (0,01%)

WOODLAND KINGFISHER

Halcyon senegalensis

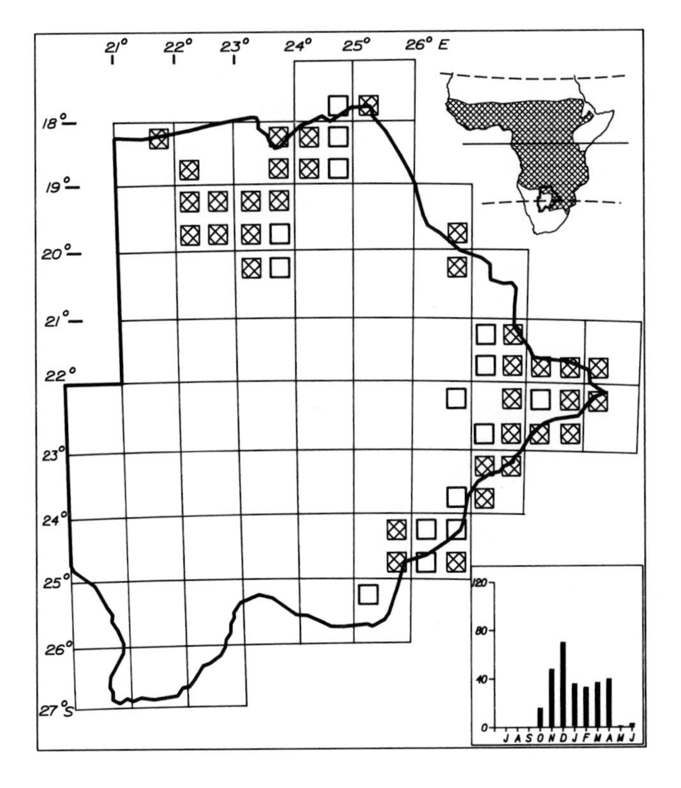

Status

Common breeding intra-African migrant to the north and east. Present from October to April. Few birds remain until June and some may remain throughout the year. Its distribution is limited to the drainage of the major river systems. Usually in pairs.

Habitat

Riparian forest, broadleafed and mixed woodlands in the high rainfall regions. Also riverine *Acacia*. Occurs mainly in river valleys.

Analysis

49 squares (21%) Total count 302 (0,17%)

BROWNHOODED KINGFISHER

Halcyon albiventris

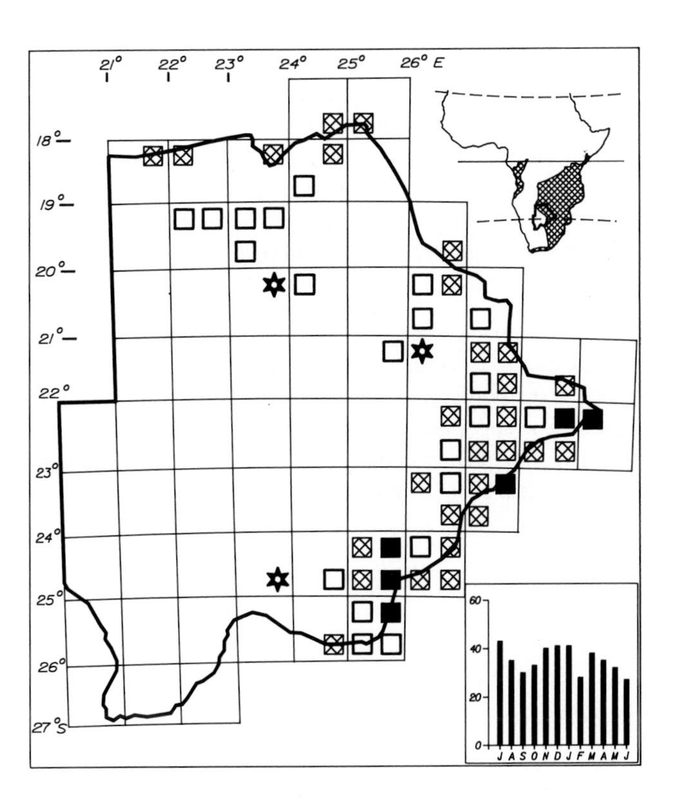

Status

Fairly common to very common resident of the east from Francistown to Ramatlabama. Sparse to fairly common but patchily recorded in the north and mainly absent in the northeast except around Nata, Tsebanana and the eastern margins of the Makgadikgadi Pans. Does not occur west of the Limpopo watershed and is absent on Kalahari sand except in river valleys. Solitary or in pairs.

Habitat

Any woodland in the high rainfall regions. Riparian forest, riverine bush and dense growth along drainage lines in tree and bush savanna.

Analysis

58 squares (25%) Total count 441 (0,25%)

GREYHOODED KINGFISHER

436

Halcyon leucocephala

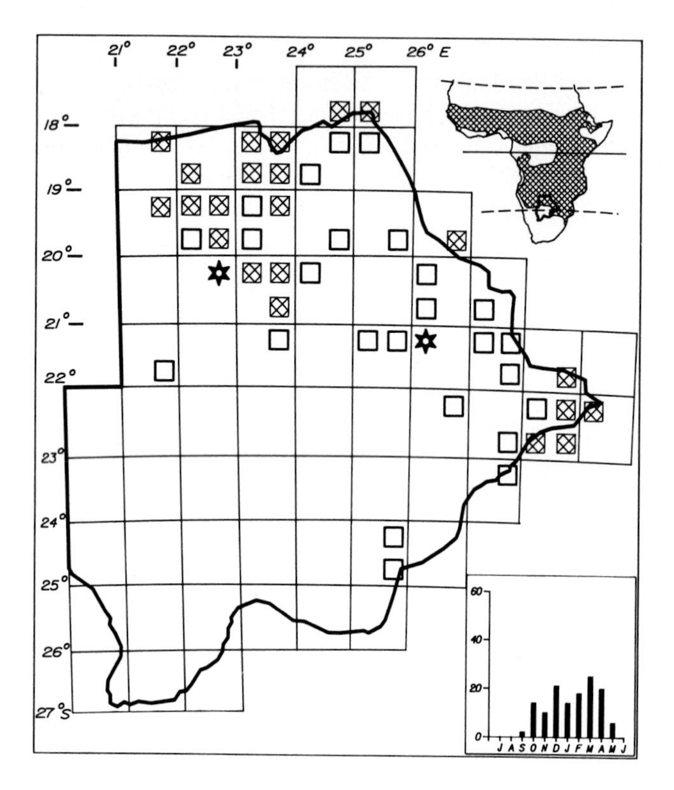

Status

Sparse to fairly common breeding intra-African migrant to the north and east, mainly between October and April with a few September and May records. Rare south of 23°S. Commonest in the Okavango, Linyanti and Chobe river systems. Estimated to be more widespread in high rainfall years when it extends as far south as the Kuke fence and Ghanzi in the west and Molepolole and Gaborone in the southeast. No confirmed breeding. Usually solitary or in pairs.

Habitat

Mature deciduous woodlands in the high rainfall regions, usually near water or in river valleys. In broadleafed and mixed mopane woodlands in the north and in mixed *Acacia* woodlands in the north and east. In dry *Acacia* and *Combretum* woodland near the Kuke fence.

Analysis

49 squares (21%) Total count 143 (0,08%)

STRIPED KINGFISHER

437

Halcyon chelicuti

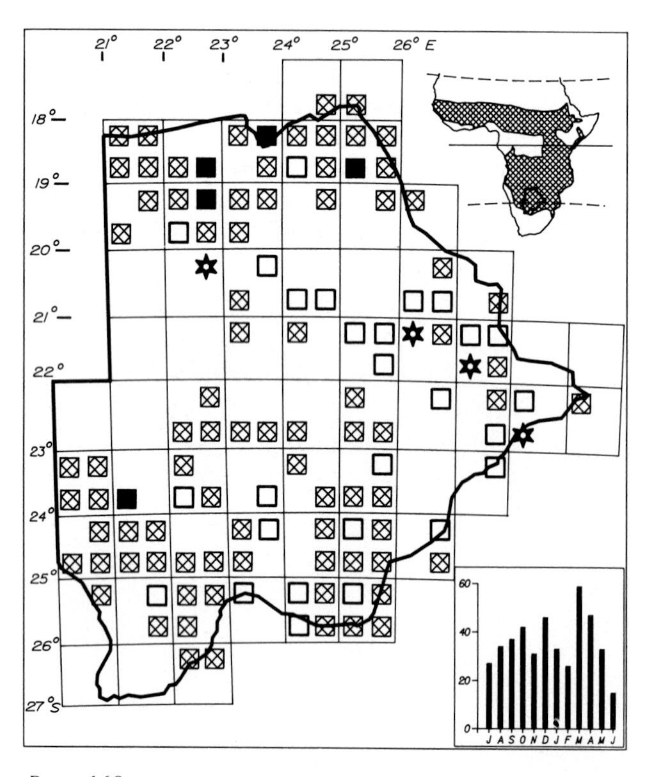

Status

Fairly common to locally very common resident occurring in all regions of the country. Present throughout the central Kalahari and the dry woodlands of the west. Absent in areas without trees. Probably overlooked in some areas. In the south it occurs at the southern limit of its continental range. Egglaying September. Solitary or in pairs.

Habitat

Widespread in semiarid woodland and tree savanna. Commonest in broadleafed woodland in the high rainfall regions. Usually found in continuous areas of woodland and rarely in isolated woodland patches. Its distribution is not dependant on river systems as is the case with all other kingfishers occurring in Botswana.

Analysis

117 squares (51%) Total count 453 (0,25%)

EUROPEAN BEE-EATER

Merops apiaster

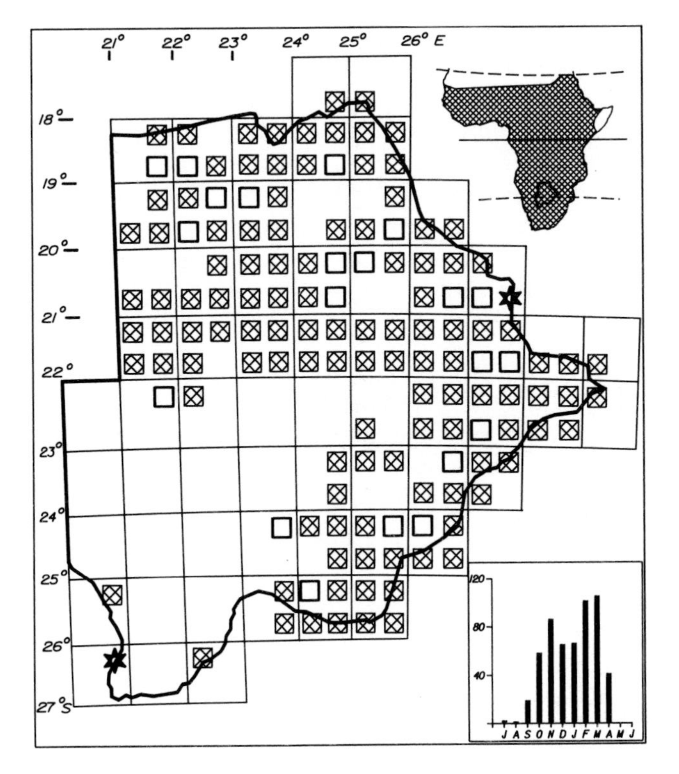

Status

A common Palaearctic migrant to the northern and eastern sectors of the country. It may be more widespread in years of high rainfall. There is no evidence that the southern African breeding population breeds in Botswana but it does so in Transvaal and as close as Aansluit (2622C) on the Kuruman River in Cape Province. Present from September to April with peaks on passage in February and March. Usually in flocks of 5–50 birds.

Habitat

Woodland and tree savanna. May occur over any habitat on passage including the central Kalahari. Often seen perched on roadside telephone wires or overhead power lines.

Analysis

139 squares (60%) Total count 575 (0,32%)

BLUECHEEKED BEE-EATER

Merops persicus

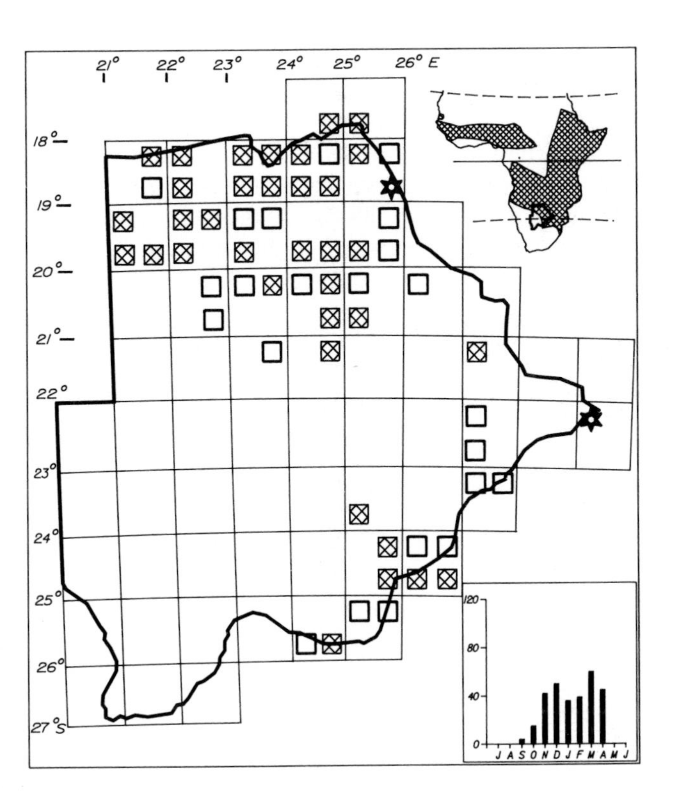

Status

Common Palaearctic migrant to the northern third of the country. Sparse to locally fairly common in the east and southeast mainly in river valleys. Its distribution in Botswana reflects its greater association with water habitats than the European Bee-eater. Present from late September to early April, mainly between November and March. Occurs in flocks of 5–50 birds.

Habitat

Rivers, dams, marshes, lakes and floodplains, particularly those with fringing woodland. It occurs in a wide variety of habitats on passage but rarely far from water. Migration probably takes place mainly along river valleys. Perches on roadside telephone wires and on any exposed perch for hawking insects.

Analysis

60 squares (26%) Total count 303 (0,17%)

CARMINE BEE-EATER

441

Merops nubicoides

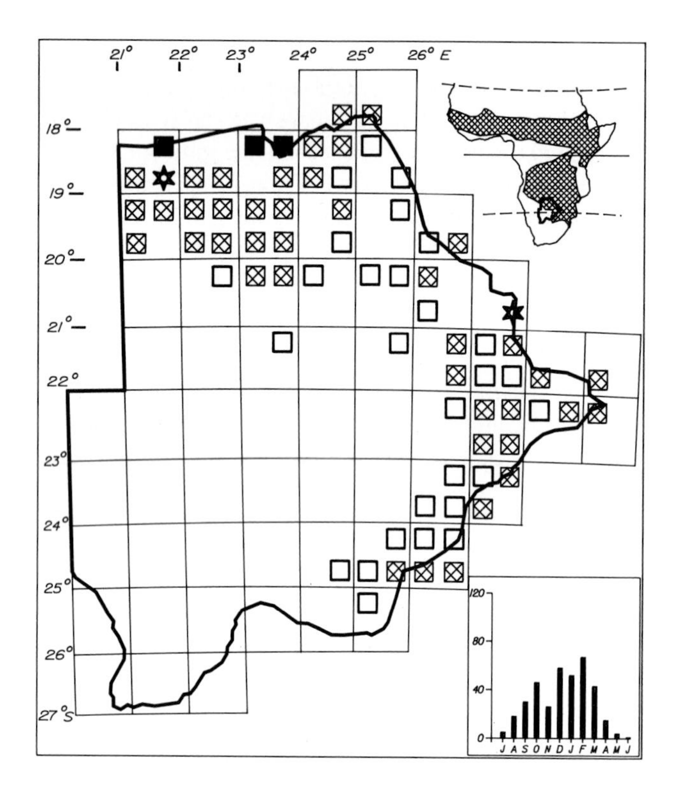

Status

Locally common to very common breeding intra-African migrant in the north. Sparse to fairly common passage migrant in the east and southeast. Most birds on passage probably originate from the Botswana breeding population but some may be from Zambia, Angola and Zimbabwe. Present mainly from September to March for breeding but recorded in every month. Occurs in small groups and flocks on passage, breeding colonies vary from about 10–100 birds.

Habitat

Breeds in steep or vertical river banks along the major rivers and forages over water and floodplain or from perches on the edges of woodland and tree savanna. Migrates mainly over woodland and tree savanna, also over semidesert.

Analysis

74 squares (32%) Total count 379 (0,21%)

WHITEFRONTED BEE-EATER

443

Merops bullockoides

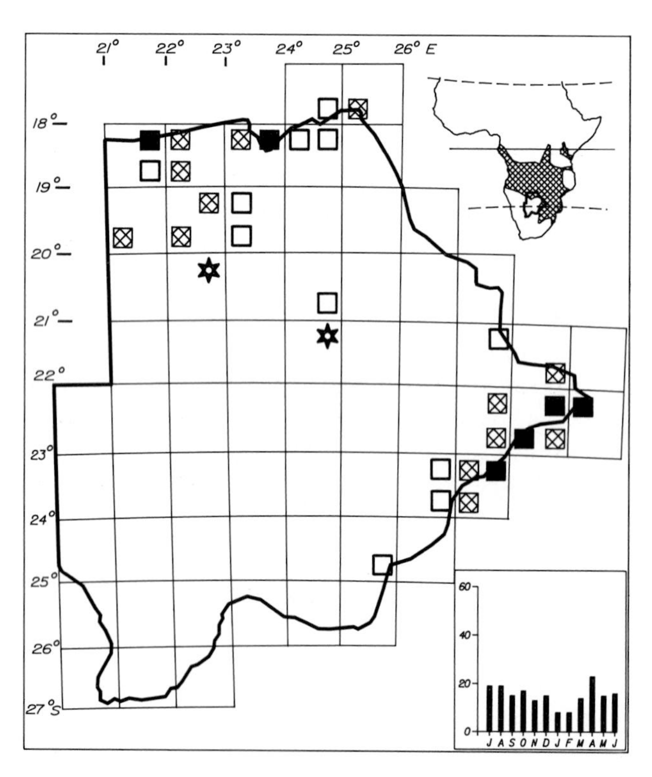

Status

Very common resident on major rivers at Shakawe, Linyanti and the northern Limpopo. Elsewhere in the north and east it is sparse to uncommon in scattered localities in river valleys. Occurs in Botswana at the southern and western limits of its African range at these latitudes. Egglaying September. Found mainly in small groups, less often solitary or in pairs.

Habitat

Riparian forest and mature woodland in river valleys. Requires steep banks for breeding. Occurs mainly along the course of large rivers and will remain at suitable nesting sites possibly for several seasons after the last flowing of the river.

Analysis

32 squares (14%) Total count 188 (0,10%)

LITTLE BEE-EATER 444

Merops pusillus

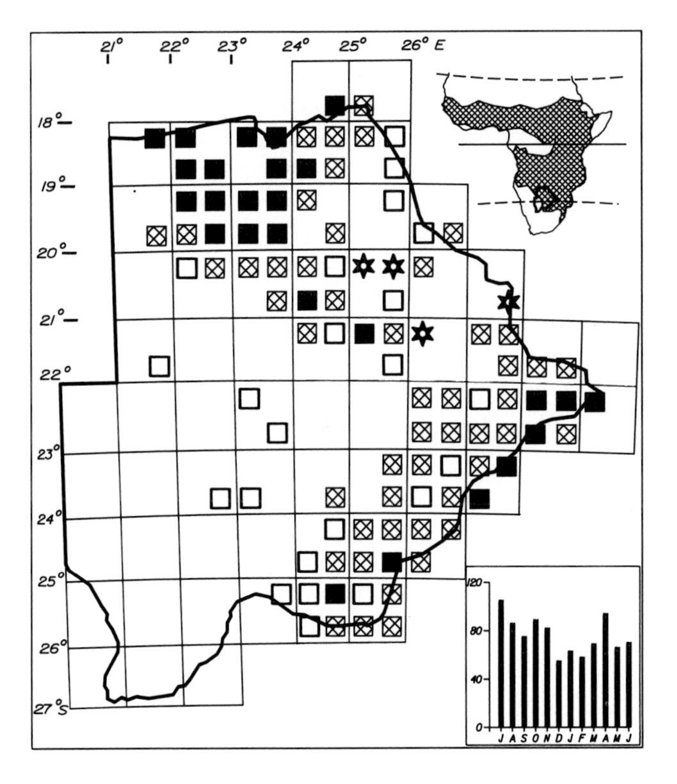

Status

Common to very common resident in the north and east being commonest in moist areas with mature and rich vegetation. Very scarce at scattered localities in arid savanna into which it may be spreading as a result of manmade water impoundments at cattle posts and villages. Mainly sedentary. Egglaying October and November. Solitary, in pairs or small groups.

Habitat

Edges of woodland and in tree and bush savanna in the higher rainfall regions, favouring well-developed trees either near water or in river valleys. Very uncommon in pure mopane. Often occurs in riparian trees and bushes and in reedbeds on the edge of water.

Analysis

102 squares (44%) Total count 937 (0,52%)

SWALLOWTAILED BEE-EATER 445

Merops hirundineus

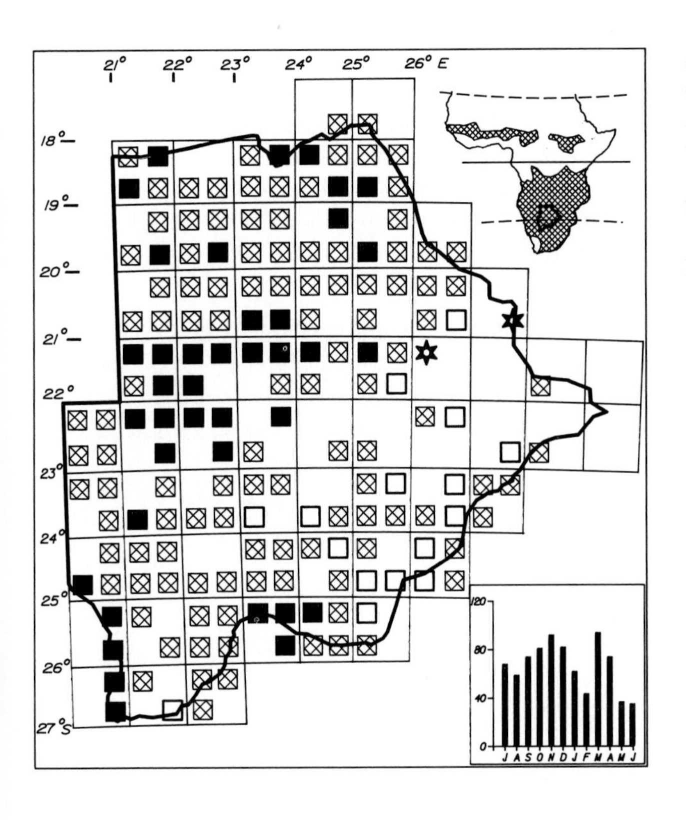

Status

Common to very common resident throughout the country except in the east where it is largely replaced by the Little Bee-eater. Present in good numbers in all months. Part of the population is thought to be migratory and present only in the breeding season from September to March, moving to Zimbabwe and the Transvaal between April and August. Egglaying September and October. Usually in small groups.

Habitat

Typically occurs in semiarid tree or bush savanna on Kalahari sand. Also on the edge of any type of woodland in the north including miombo, *Baikiaea* and mopane. May occur in any wooded habitat on seasonal migration including gardens and riparian vegetation.

Analysis

169 squares (73%) Total count 855 (0,48%)

EUROPEAN ROLLER

446

Coracias garrulus

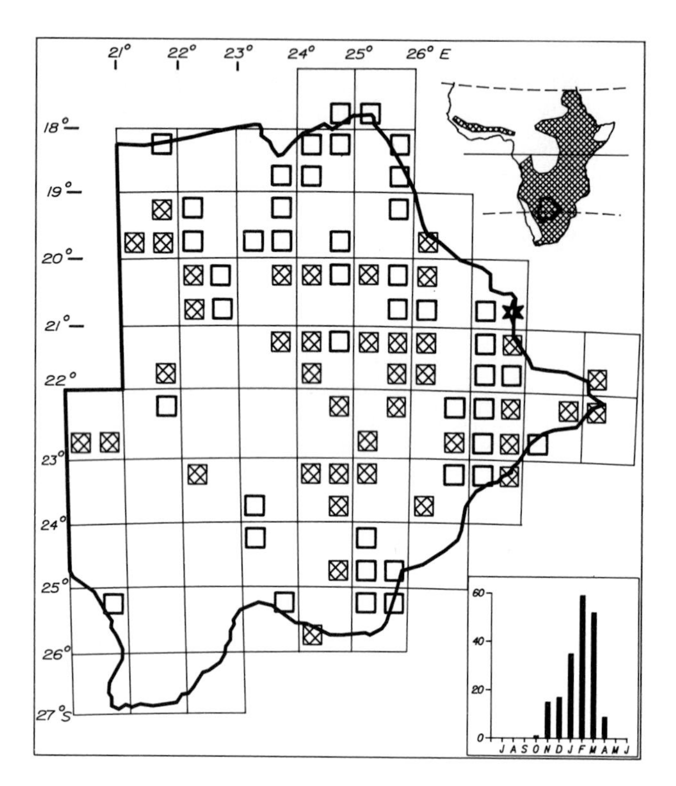

Status

Sparse to uncommon Palaearctic migrant between late October and April, commonest on northward passage in January to March. Peak southward passage occurs in November. There is some evidence that it congregates just south of the Makgadikgadi Pans over a short period on northward migration but further detailed study is required. 77% of all records are from east of 24°E. Usually solitary, but migrates in flocks of 10–50 birds at peaks in the northern part of the country.

Habitat

Migration occurs mainly over woodland and savanna in the east. Some temporarily resident birds are found in semidesert savanna and Kalahari woodland patches as well as the high rainfall woodlands and savannas.

Analysis

84 squares (37%) Total count 209 (0,12%)

LILACBREASTED ROLLER

447

Coracias caudata

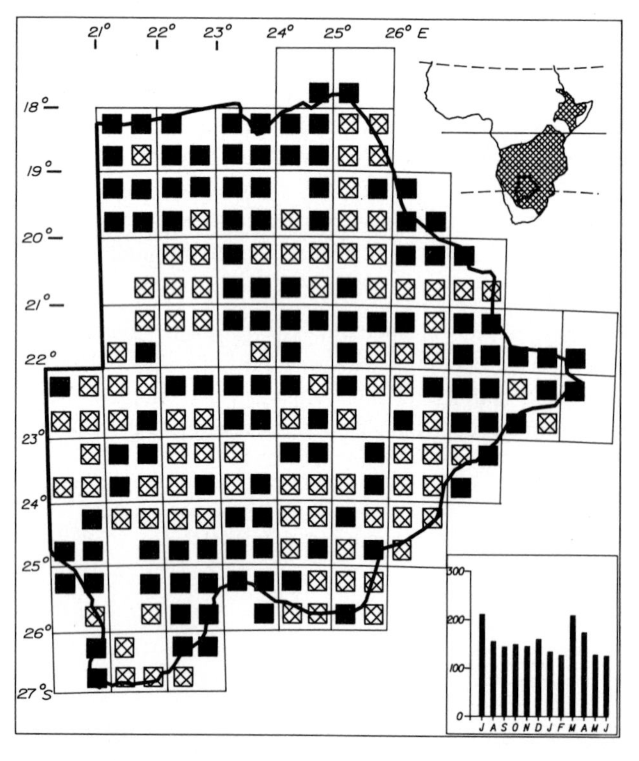

Status

Common to very common resident throughout the country. It has been recorded as very common in 50% of all squares. Local and more widespread movements may occur but these have not been studied. Its distribution in Botswana is close to the southern limit of its range on the continent. Egglaying October. Usually solitary or in pairs.

Habitat

Any woodland in moist or arid conditions where it occurs on the edges or in open areas. Also in tree and bush savanna and open arid semidesert but not where there are no trees for perching. Pounces from perch onto prey on the ground .

Analysis

210 squares (91%) Total count 1894 (1,06%)

RACKETTAILED ROLLER

Coracias spatulata

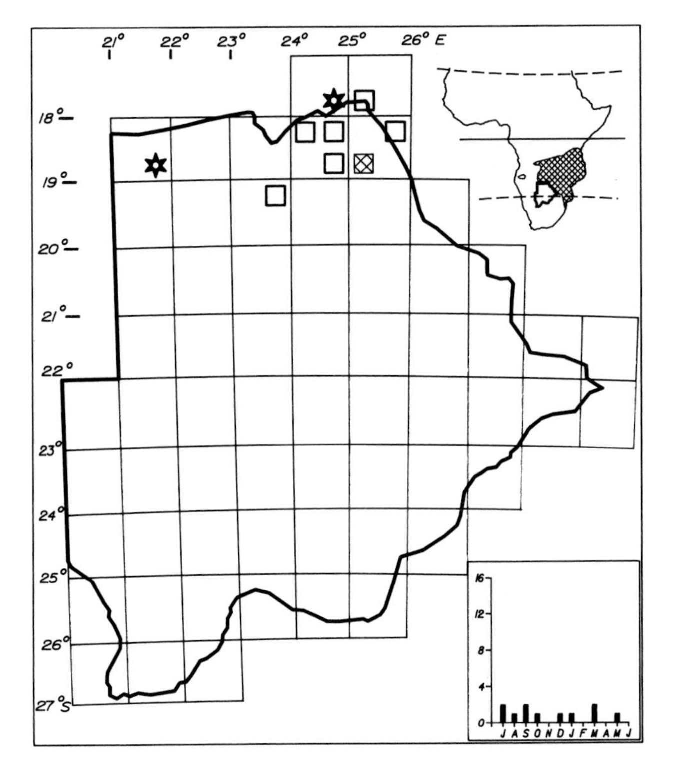

Status

Sparse and uncommon in the northeastern woodlands to which region it may now be confined. Recorded once near the Tsodilo Hills but not again in the last 20 years. Occurs at the extreme southwestern limit of its continental range. Poorly studied in Botswana. Some seasonal movements may occur into adjacent areas of Zimbabwe but it is likely to be a breeding resident and not a regular migrant. In pairs and solitary.

Habitat

Well-developed miombo and *Baikiaea* woodland within which it resides unobtrusively. Not reported from mopane in Botswana but it can be expected to occur in tall mature stands at least when mixed with other broadleafed trees.

Analysis

9 squares (4%) Total count 12 (0,006%)

PURPLE ROLLER

Coracias naevia

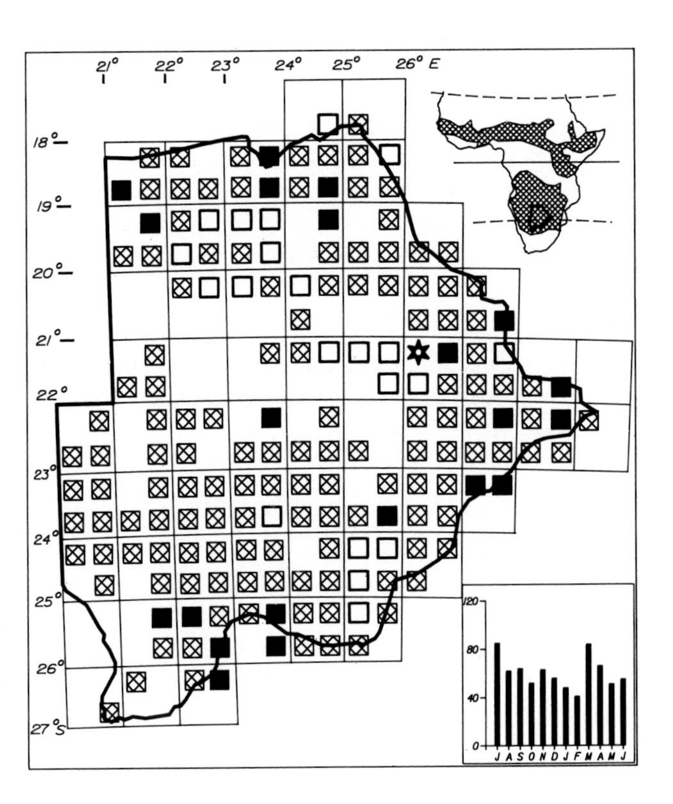

Status

Fairly common to common resident throughout the country. Recorded as very common in 9% of squares. Its status in the northwest Kalahari needs to be ascertained by further fieldwork. Present throughout the year but local and international movements may occur—it is mainly an intra-African migrant in Zambia and seasonal movements are found in Zimbabwe. Egglaying October. Usually in pairs or solitary.

Habitat

Tree savanna particularly *Acacia*, and mixed broadleafed woodlands, on Kalahari sands and hardveld. Also in semiarid areas where there are good stands of trees. Its niche appears to lie between the enclosed woodland of the Rackettailed Roller and the open savannas of the Lilacbreasted Roller.

Analysis

173 squares (75%) Total count 750 (0,42%)

BROADBILLED ROLLER

450

Eurystomus glaucurus

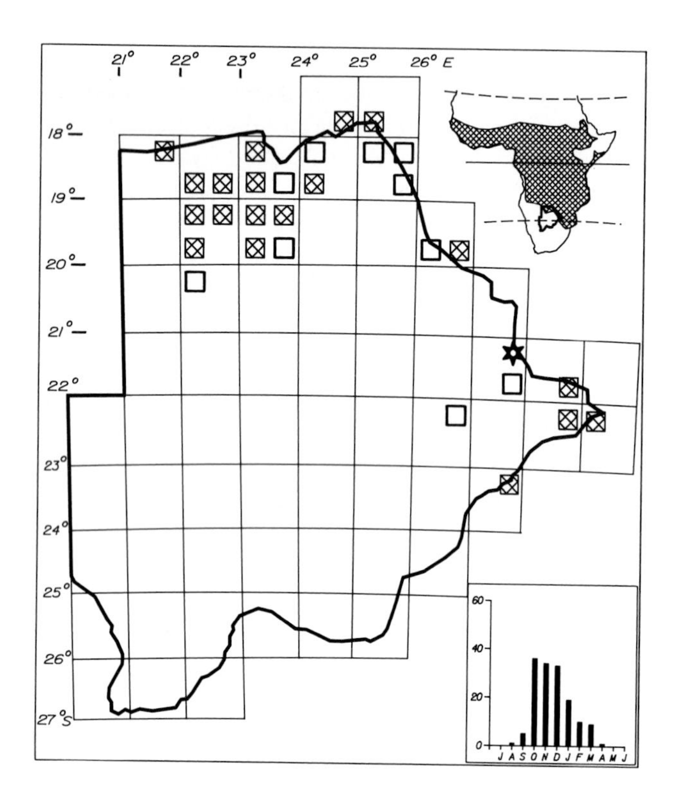

Status

Fairly common to common breeding intra-African migrant in the north and east from September to April. Occurs at the southern limit of its continental distribution. Numbers fluctuate in different years. Current analysis shows that it is more common between October and January which is the main breeding period. Conspicuous when displaying otherwise unobtrusive. Active nest reported in October. Occurs in pairs.

Habitat

Riparian forest and adjacent tall mature woodland in major river valleys such as the Okavango, Linyanti, Chobe, Nata, Shashe and Limpopo. Also in mature broadleafed woodland. Usually perches near the tops of tall trees.

Analysis

31 squares (13%) Total count 160 (0,09%)

HOOPOE

451

Upupa epops

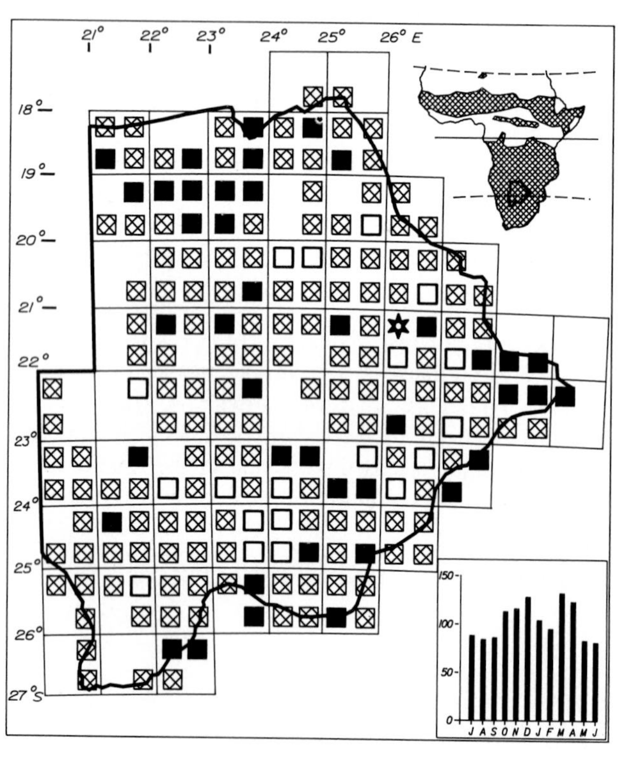

Status

Common resident throughout the country; very common in 18% of squares. Commonest in the moist areas of the north and east. Present in all months but records are fewer in the winter months and highest between October and April. Breeds August to December. Local and international movements occur. Usually solitary, in pairs, less commonly in small groups.

Habitat

Any woodland, tree and bush savanna, farmland, gardens and river valleys. Rarely in arid scrub with widely scattered trees possibly only on migration.

Analysis

200 squares (87%) Total count 1269 (0,71%)

REDBILLED WOODHOOPOE

452

Phoeniculus purpureus

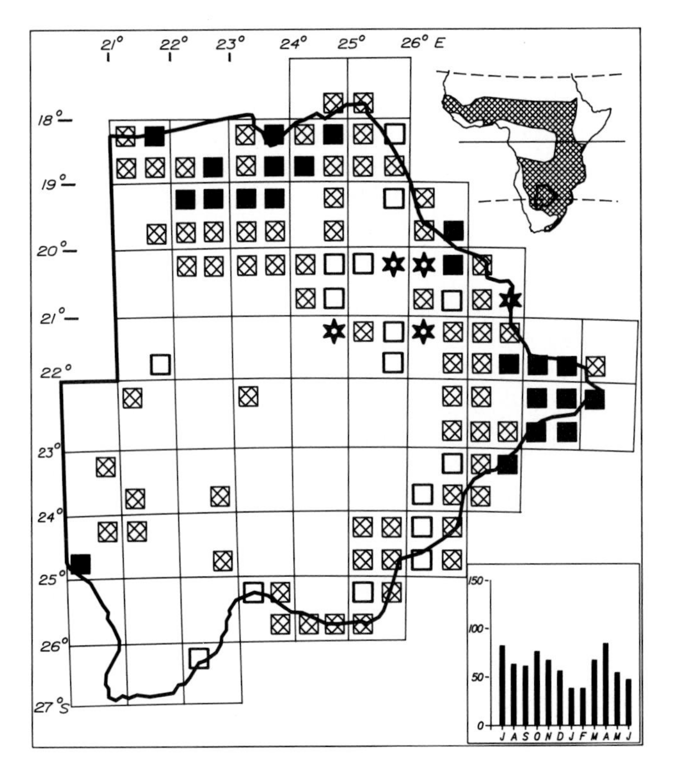

Status

Common to very common resident of the north and east. Its range extends westwards along the Molopo River to join a population in the arid camelthorn woodlands of the west. Absent from most of the central Kalahari. Present in all months. Breeds mainly in September and October. Occurs in groups of 5–15 birds.

Habitat

Mixed deciduous and *Acacia* woodlands where there are tall trees and particularly in river valleys. In the west it occurs in parklike areas of *Acacia erioloba* as well as other mature arid woodland.

Analysis

109 squares (47%) Total count 767 (0,43%)

SCIMITARBILLED WOODHOOPOE

454

Rhinopomastus cyanomelas

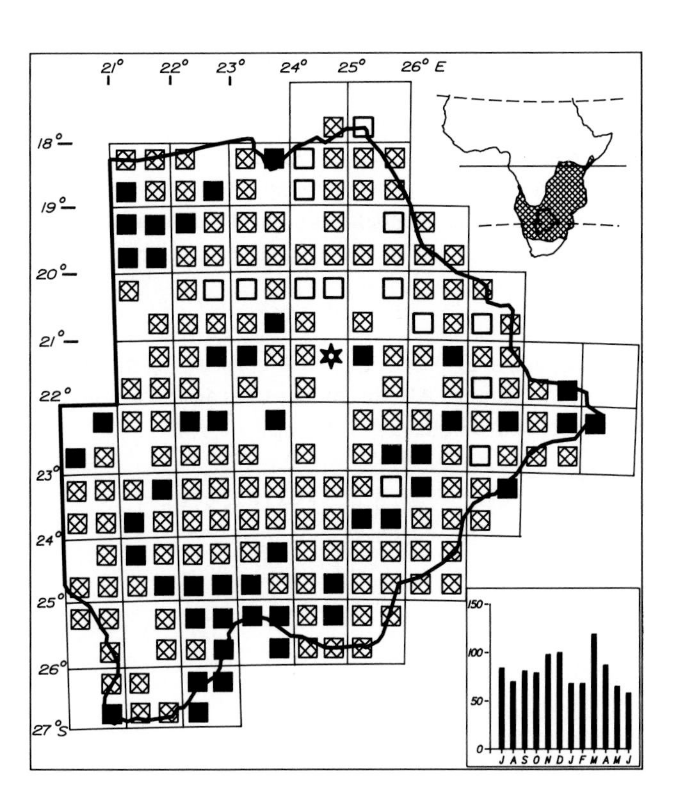

Status

Common resident throughout the country; very common in 21% of squares. It is mainly sedentary and present in all areas throughout the year. Breeds mainly in September and October. Solitary, in pairs when breeding, occasionally in small loose groups.

Habitat

Acacia tree and bush savanna in moist to semiarid conditions, mixed deciduous and broadleafed woodland, riverine trees and bush, gardens and parks.

Analysis

202 squares (88%) Total count 1019 (0,57%)

TRUMPETER HORNBILL

Bycanistes bucinator

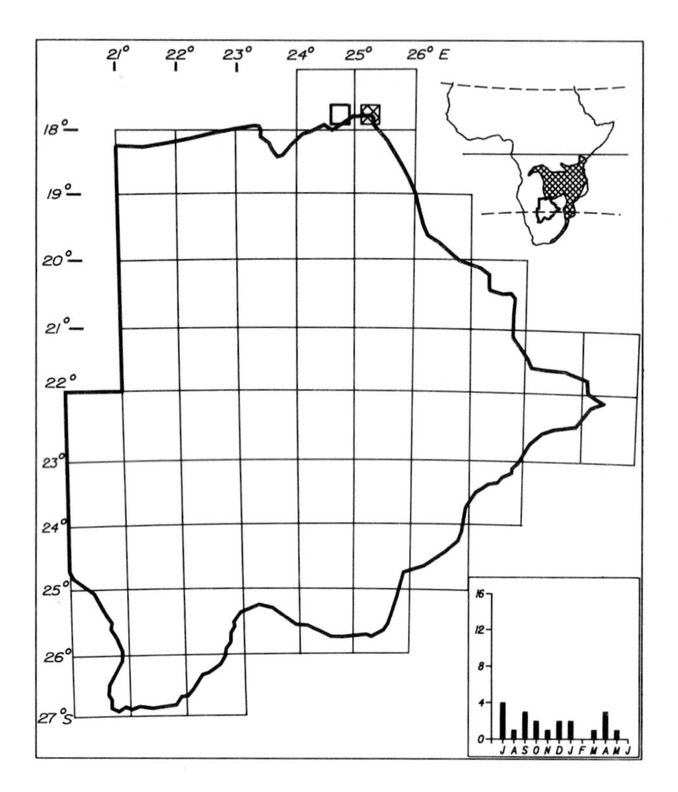

Status

Sparse to uncommon resident of the eastern Chobe and Zambezi river valleys around Serondela, Kasane and Kazungula. It occurs at the western and southern extremity of its continental range in this area and is unlikely to be found more extensively because of its habitat requirements. Very little study has been made of this species in Botswana.

Habitat

Primarily a forest species feeding on fruit in the canopy and extending short distances into adjacent rich woodlands when ripe food is available. It is thus confined to the Zambezi riparian forest and is unlikely to extend westwards along the Chobe River where such habitat does not exist. Occurs on the Limpopo River east of the Tuli region and may eventually be found in Botswana in this area.

Analysis

2 squares (1%) Total count 23 (0,013%)

GREY HORNBILL

Tockus nasutus

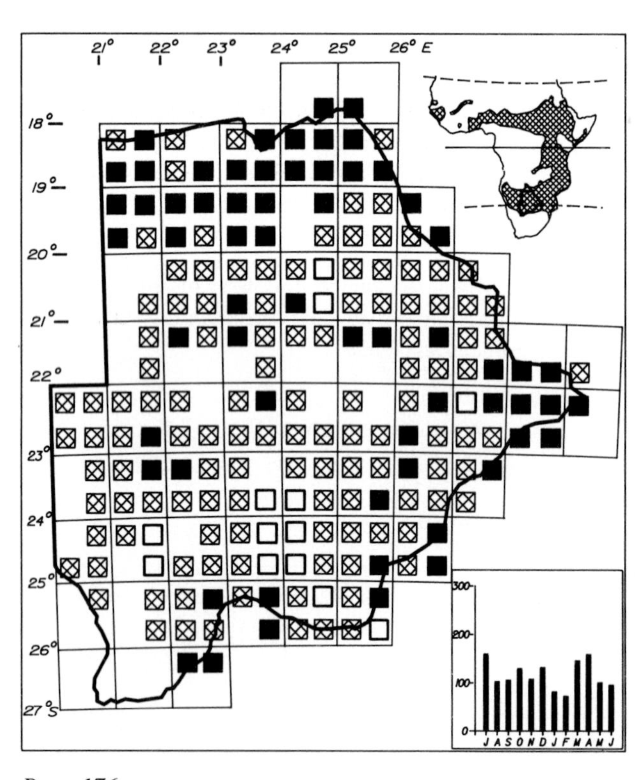

Status

Fairly common to very common resident throughout the country. Extensively very common in the richer woodlands of the north and east and less common in semiarid savanna. Some local movements may occur. Present in all months. Its status in the southwest dunelands merits more detailed study. Egglaying October. Occurs in small groups, in pairs or solitary.

Habitat

Any woodland and tree and bush savanna in moist and semiarid conditions. Also in riverine trees and bushes and on well-vegetated duneland.

Analysis

195 squares (85%) Total count 1434 (0,81%)

REDBILLED HORNBILL 458

Tockus erythrorhynchus

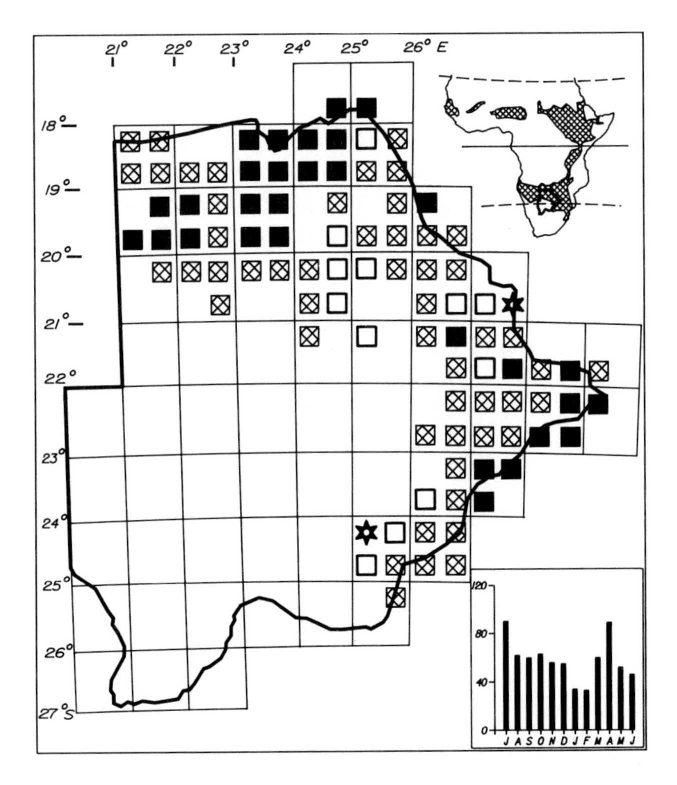

Status

Fairly common to very common resident of the north and east extending south along the Limpopo and Ngotwane drainage to as far as Lobatse. Mainly very common in areas of mopane woodland. Local movements may occur. Usually in small flocks.

Habitat

Broadleafed woodland with open spaces of bare ground or short grass cover particularly in pure or mixed mopane. Feeds on the ground. Also in tree and bush savanna in the higher rainfall areas, in riverine bush and *Acacia* on alluvial soils.

Analysis

97 squares (42%) Total count 717 (0,40%)

YELLOWBILLED HORNBILL 459

Tockus leucomelas

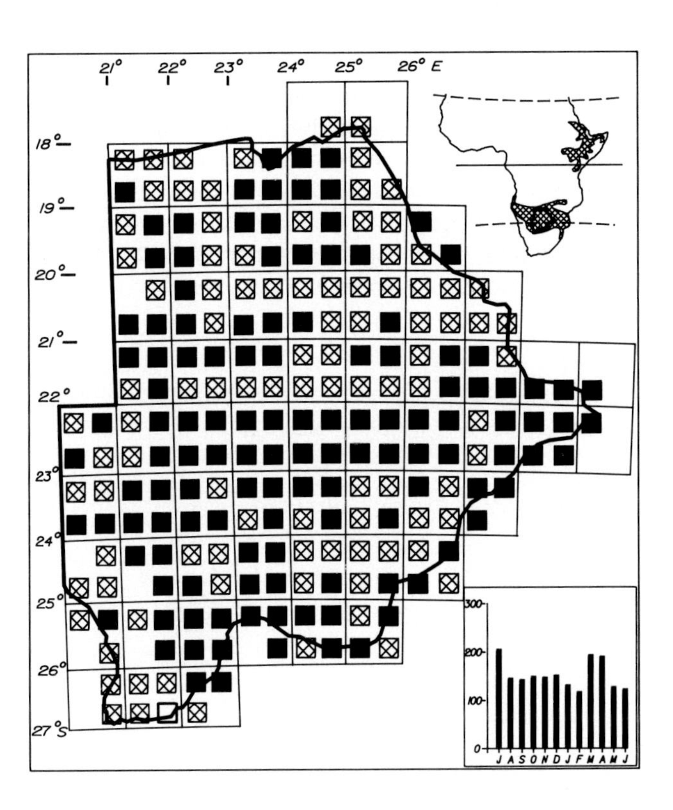

Status

Common to very common resident throughout the country. It is the commonest hornbill and one of the commonest species in the country. It is more widespread in Botswana than in any other country in which it occurs. Least predictable in the southwest. Breeds mainly from October to December. Occurs in small groups, solitary or in pairs.

Habitat

Savanna and woodland of all types except arid treeless scrub savanna. Feeds mainly on the ground.

Analysis

224 squares (97%) Total count 1876 (1,05%)

BRADFIELD'S HORNBILL

461

Tockus bradfieldi

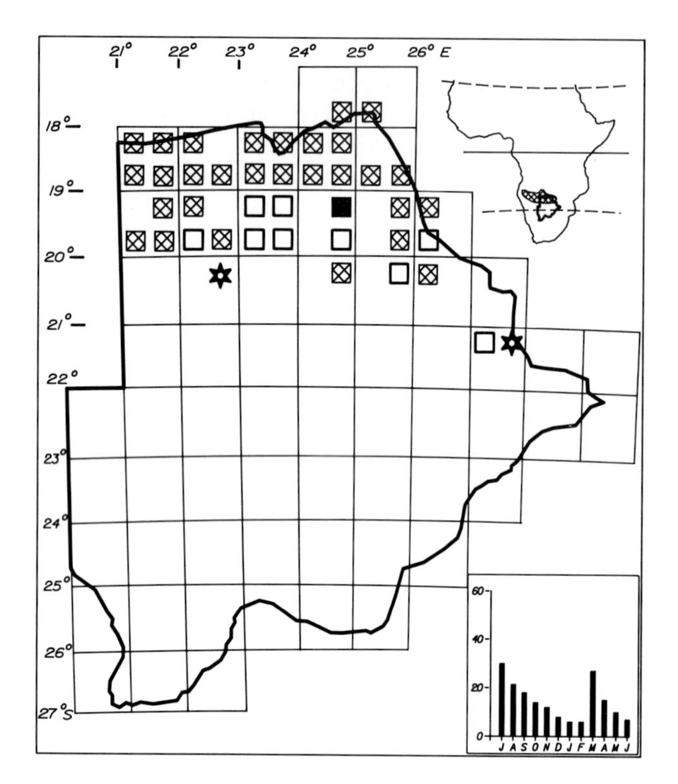

Status

Sparse to fairly common resident of the north to as far south as Nata in the northeast. Recorded once recently and in the past at Francistown. Common at Shakawe and very common just north of Nxai Pan. It is not well known and is a good candidate for special study in Botswana. Local movements occur but no clear pattern has emerged. Solitary, in pairs or small groups.

Habitat

Broadleafed woodlands particularly *Baikiaea* and mixed mopane. Feeds on the ground and in trees and is omnivorous. Recorded in croton oil bushes at Savuti Marsh during drought conditions.

Analysis

41 squares (18%) Total count 179 (0,10%)

SOUTHERN GROUND HORNBILL

463

Bucorvus leadbeateri

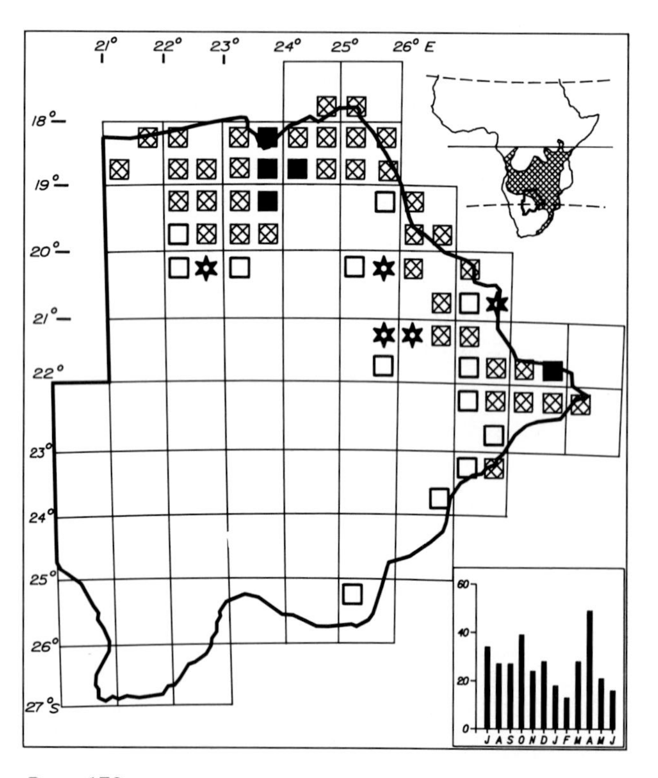

Status

Common resident of the northern woodlands. Very common at Linyanti, Savuti and eastern Moremi. In the northeast and east it is sparse to common to as far south as Buffels Drift. Recorded once near Mmathethe in the southeast. It does not extend into the dry Kalahari savannas. Occurs in Botswana at the western limit of its range in southern Africa. Egglaying October and November. Usually in pairs or small groups of 5–10 birds.

Habitat

Any woodland in the high rainfall areas. Also tree and bush savanna, cultivation, farmland. Nests in tree hollows.

Analysis

60 squares (26%) Total count 345 (0,19%)

BLACKCOLLARED BARBET

464

Lybius torquatus

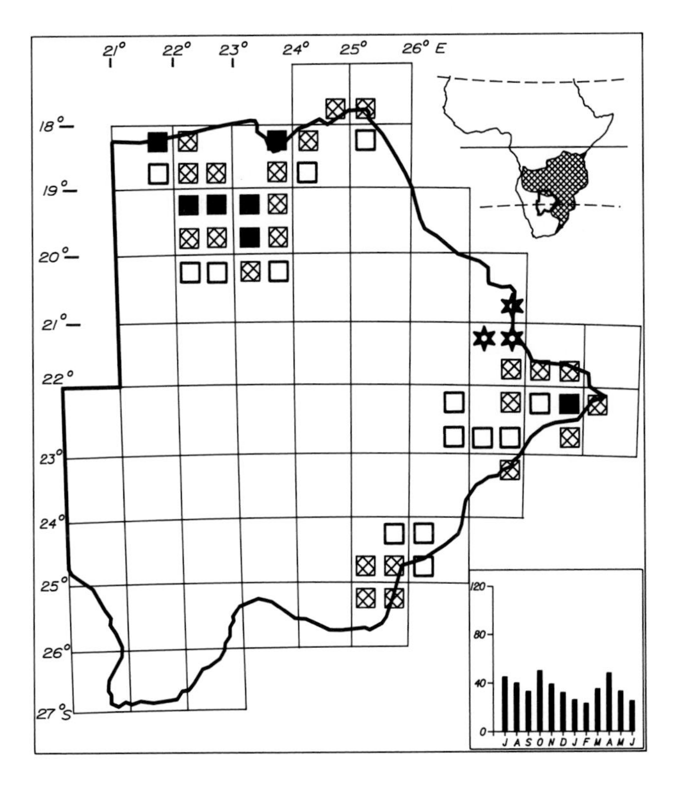

Status

Common to very common resident in the Okavango Delta and in the east south of Francistown. Sparse to fairly common in the southeast and mostly absent in the northeast. There may be some seasonal movements up and down the river valleys. Egglaying December. Usually in pairs and has a duetting call.

Habitat

Mature broadleafed woodlands, tree and bush savanna in river valleys. Less frequently in riparian forest and gardens when fruit is available.

Analysis

47 squares (20%) Total count 451 (0,25%)

ACACIA PIED BARBET

465

Tricholaema leucomelas

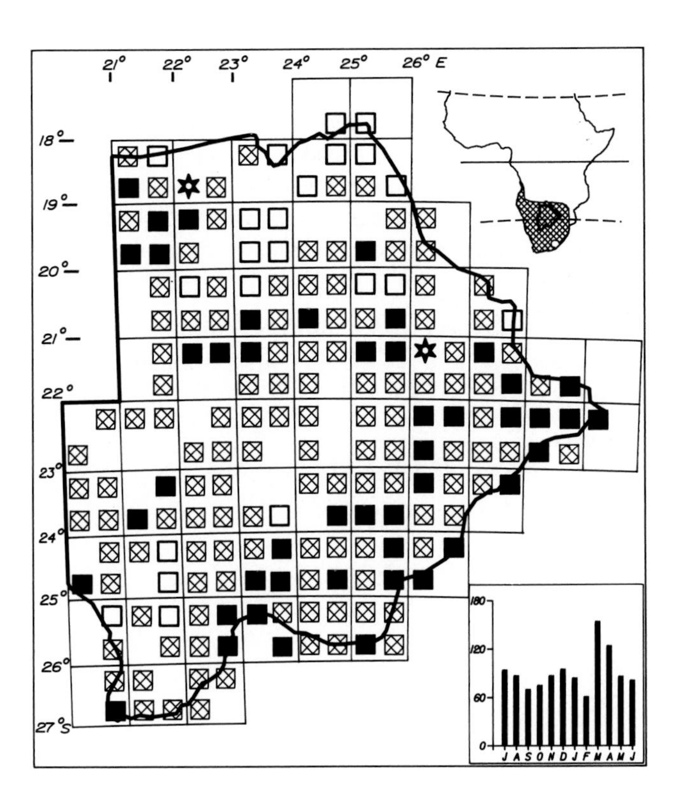

Status

Common to very common resident throughout except in the northeast where it is generally sparse. Reported as very common in 20% of squares. The most common and widespread barbet. Mainly sedentary. Egglaying October to December. Usually occurs in pairs or solitary.

Habitat

Thorn tree and bush savanna, particularly *Acacia*. Also occurs in open areas of richer woodlands, in riparian *Acacia*, in scrub savanna with scattered stunted trees and in gardens and cultivation.

Analysis

192 squares (83%) Total count 1158 (0,65%)

YELLOWFRONTED TINKER BARBET 470

Pogoniulus chrysoconus

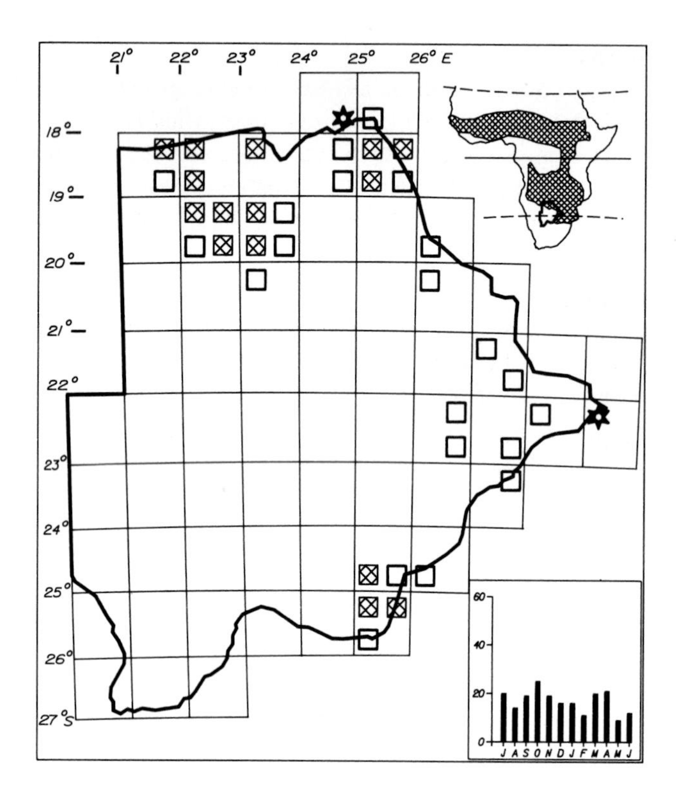

Status

Sparse to uncommon resident of the north, east and southeast. Only reported as common at Shakawe, Maun and Lobatse. Occurs in Botswana at the western limits of its continental range in these latitudes. Local seasonal movements may occur to and from river valleys. Egglaying January. Usually solitary or in pairs.

Habitat

Broadleafed woodland in river valleys extending to adjacent well-wooded savanna, miombo and *Baikiaea* woodland, particularly where there are fruiting trees. Also in riparian forest and gardens of residences near rivers.

Analysis

38 squares (17%) Total count 210 (0,12%)

CRESTED BARBET 473

Trachyphonus vaillantii

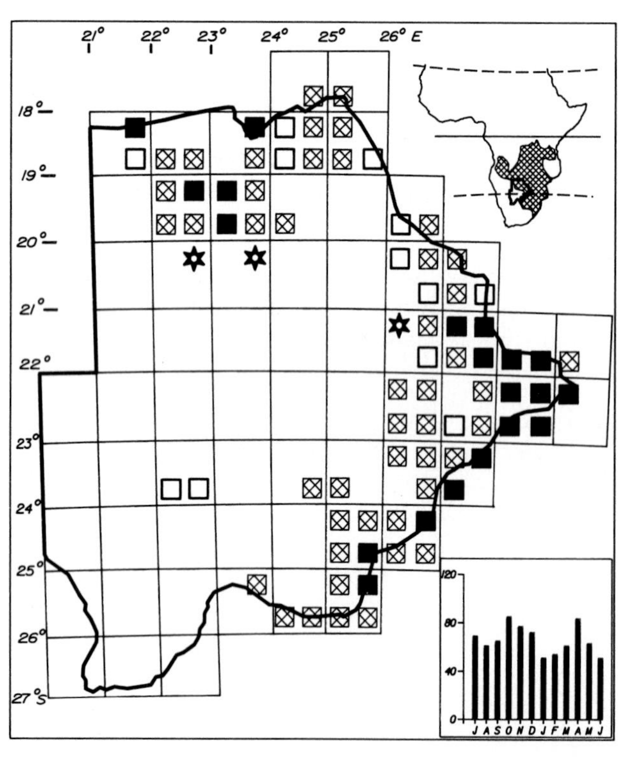

Status

Common to very common resident of the north and east. Extends west along the Molopo River to as far as Bray. Recorded on two occasions at and near Kang in the western woodland. Further study is required to ascertain its true status in this area. Mainly sedentary. Egglaying November and December. Occurs in pairs or solitary.

Habitat

Dense vegetation and thicket in woodland, tree and bush savanna and riverine areas. Also in gardens in urban and rural settings.

Analysis

81 squares (35%) Total count 793 (0,44%)

GREATER HONEYGUIDE

Indicator indicator

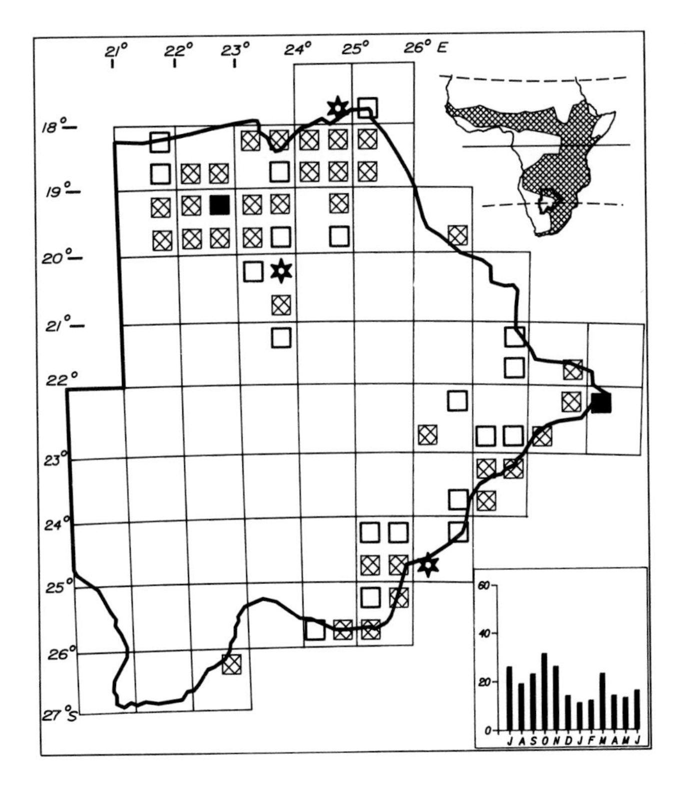

Status

Sparse to fairly common resident of the north and east. Common to very common in the Okavango Delta. Extends along the Molopo River to as far west as McCarthysrus. Known to breed regularly in African Hoopoe nests and possibly several other hole nesting species. Egglaying June (one record). Usually solitary.

Habitat

Deciduous woodland, riparian *Acacia*, edges of forest, tree savanna.

Analysis

58 squares (25%) Total count 240 (0,13%)

LESSER HONEYGUIDE

Indicator minor

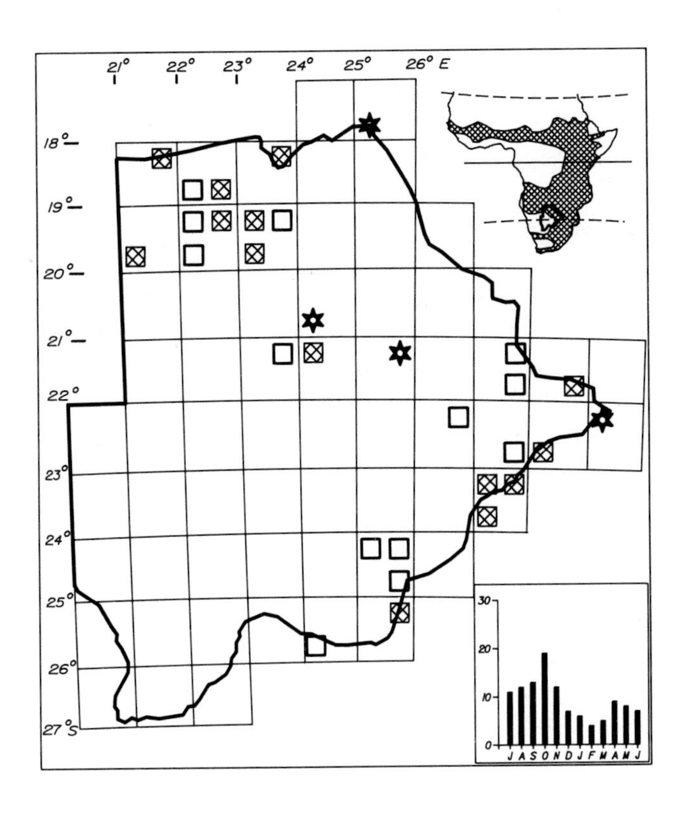

Status

A sparse or uncommon resident irregularly recorded in the northern and eastern river systems and along the Molopo River to as far west as Boshoek. Recorded as common only at Maun. Mainly sedentary. There are no breeding records for this species in Botswana. Usually solitary.

Habitat

Riparian fringing forest and deciduous woodland in river valleys. Absent from *Acacia* savanna on Kalahari sand and from mopane. Most commonly seen in riverine trees and bush along major rivers.

Analysis

31 squares (13%) Total count 126 (0,07%)

SHARPBILLED HONEYGUIDE

478

Prodotiscus regulus

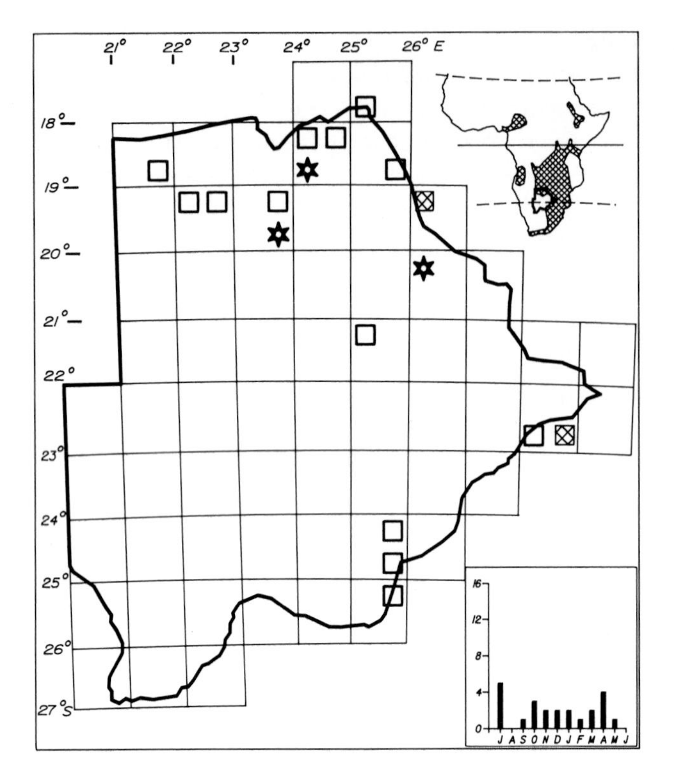

Alternative name

Brownbacked Honeyguide

Status

A sparse and uncommonly recorded resident of the north and east to as far south as Lobatse. It is easily overlooked. Mainly sedentary and possibly localised in occurrence. Usually solitary.

Habitat

Tree savanna, in the midstratum of open deciduous woodlands, usually where there are scattered bushes or thicket. Also in riverine trees and bush. Hawks from a perch like a flycatcher and gleans like a warbler.

Analysis

18 squares (8%) Total count 27 (0,02%)

BENNETT'S WOODPECKER

481

Campethera bennettii

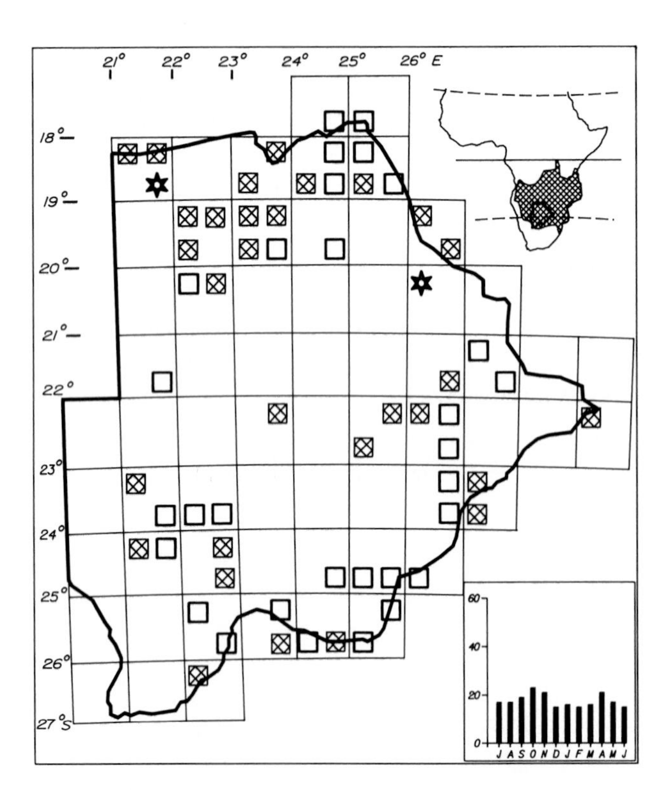

Status

Sparse to fairly common resident of woodland throughout the country. Absent from most of the central Kalahari and the southwest duneland. There is no evidence of movement. Often in pairs or solitary.

Habitat

Any type of mature woodland including woodland patches in semiarid areas of the eastern Kalahari and the more extensive areas of woodland of the west. Also in broadleafed and *Acacia* tree savanna. Less common in woodland in poor rainfall regions than the Goldentailed Woodpecker.

Analysis

62 squares (27%) Total count 221 (0,12%)

GOLDENTAILED WOODPECKER 483

Campethera abingoni

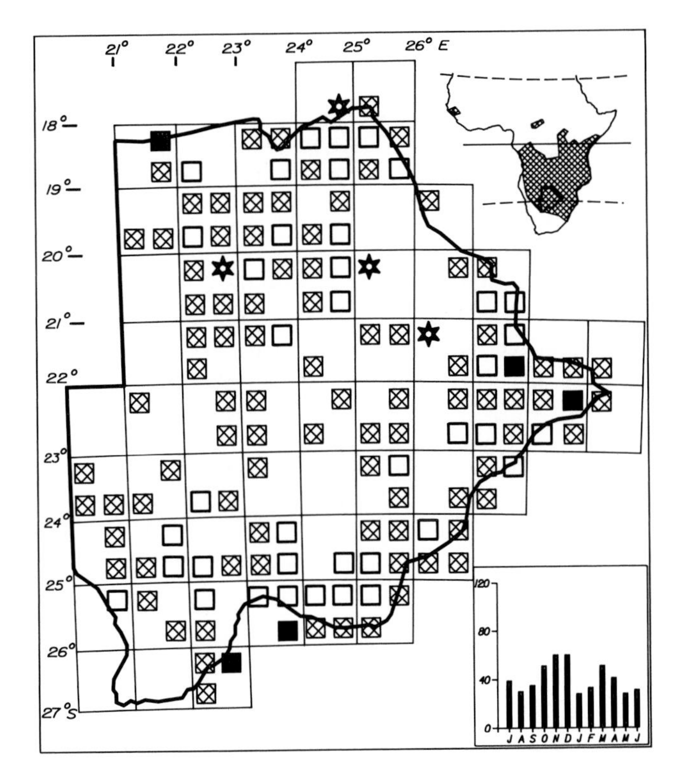

Status

Sparse to common resident throughout the country. Sparse in the mainly grassland areas of the southeast and apparently absent in the southwest corner. Occurs even in semidesert scrub with very few trees. Very common in some river valleys. Solitary or in pairs.

Habitat

Any woodland, deciduous tree and bush savanna, scrub savanna, riverine forest and bush including fringing evergreen forest, gardens and farmland.

Analysis

134 squares (58%) Total count 508 (0,28%)

CARDINAL WOODPECKER 486

Dendropicos fuscescens

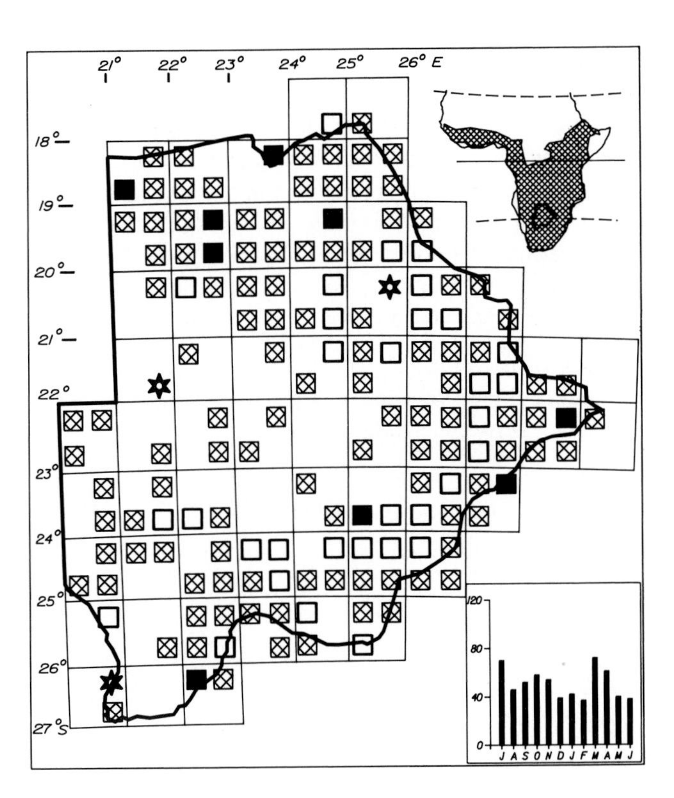

Status

Common to very common resident of the north and east. Sparse to fairly common elsewhere except at Tshabong where it has been recorded as very common. It is the most commonly recorded woodpecker in Botswana. Solitary or in pairs. Often in woodland parties with other birds.

Habitat

Any woodland, tree or bush savanna, semiarid scrub with scattered stunted trees, riverine bush, riparian forest, gardens.

Analysis

152 squares (66%) Total count 624 (0,35%)

BEARDED WOODPECKER

Thripias namaquus

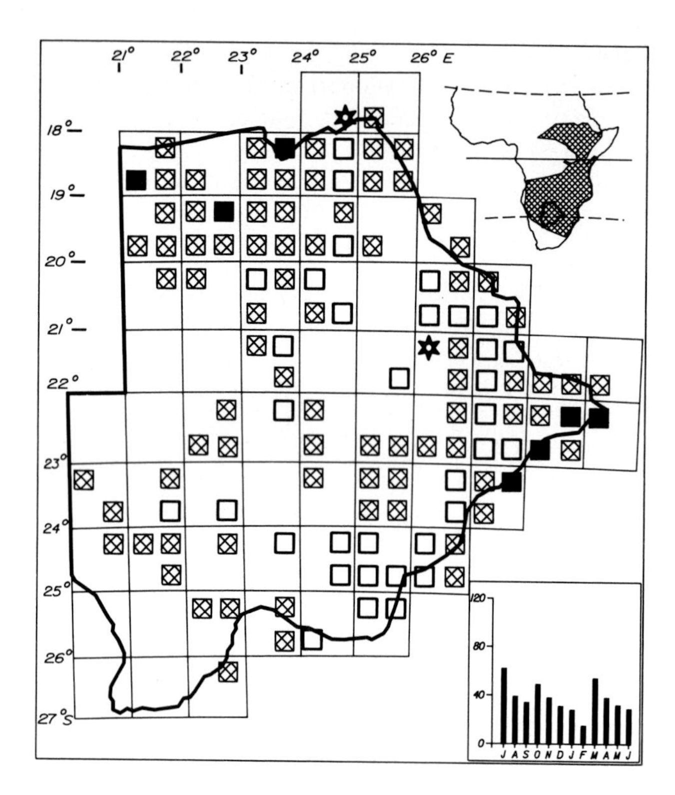

Status

Common to very common resident of the Okavango and Tuli regions becoming less common southward and westward. Sparse in the southeast, uncommon in the west, and apparently absent in the southwest duneland. Usually in pairs.

Habitat

Well-developed woodland with tall trees, tolerates dry savanna and woodland even in the central Kalahari but is more common in moist situations. Also in riparian forest with tall trees such as *Acacia albida*.

Analysis

120 squares (52%) Total count 464 (0,26%)

MONOTONOUS LARK

Mirafra passerina

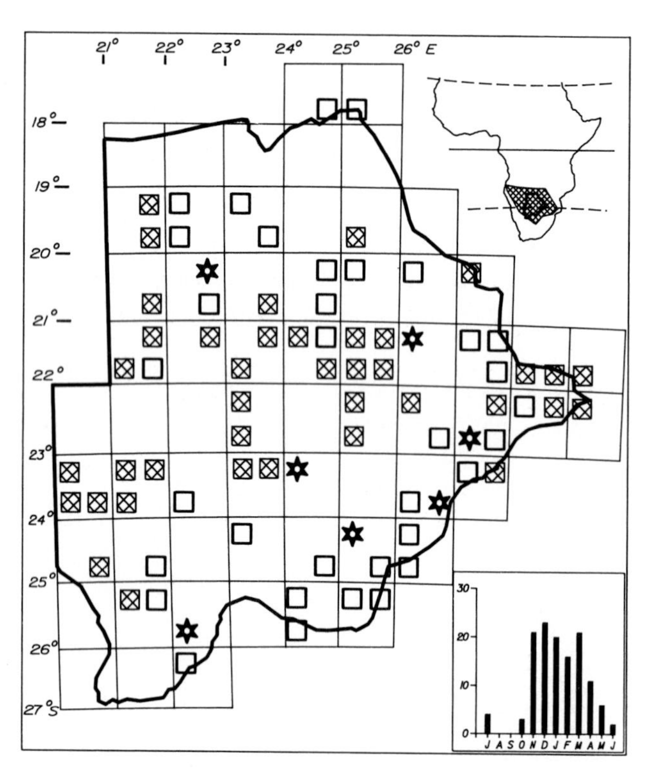

Status

Sparse to locally common breeding intra-African migrant. Patchily distributed throughout the country. More common in some years than others. It can be very common in one area in one year but absent from that area in the following year. Present mainly from November to April—earliest date 29 October, latest 19 May. Some birds remain in June and July. Egglaying January and February. In pairs or solitary, usually several pairs occur together at a locality.

Habitat

Open tree and bush savanna with patchy grass cover; open woodland, cultivation and fallow lands.

Analysis

80 squares (35%) Total count 141 (0,08%)

RUFOUSNAPED LARK

494

Mirafra africana

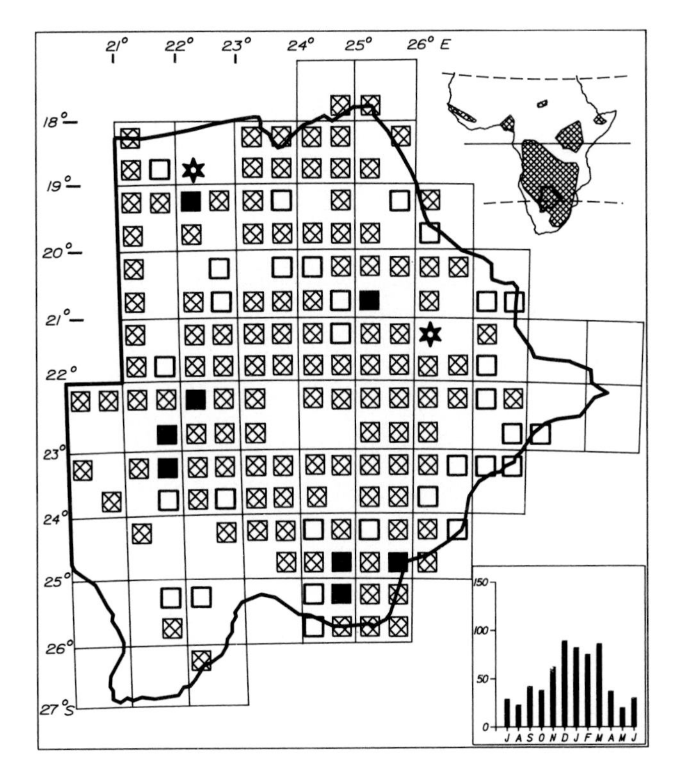

Status

Fairly common to locally very common resident in all regions except the extreme southwest where it is mainly absent. Present throughout the year but most frequently recorded when in song from October to April. Egglaying February. Usually solitary.

Habitat

Grassland in open tree and bush savanna, plains, dry grass pans, farmland, fallow lands and floodplains. Perches conspicuously on bushes, termite mounds and fence posts.

Analysis

154 squares (67%) Total count 654 (0,37%)

CLAPPER LARK

495

Mirafra apiata

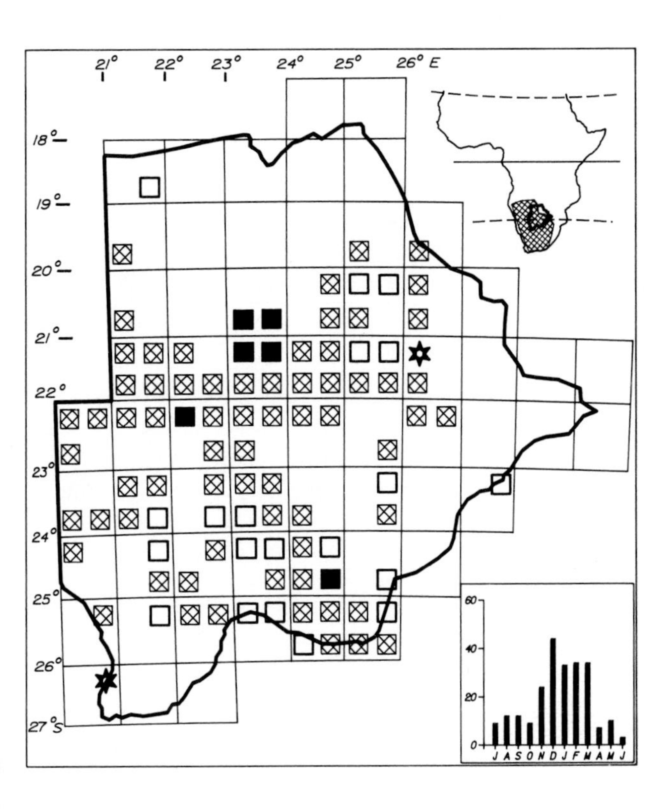

Status

Fairly common to common resident south of 20°S and west of 27°E. Very common on open plains along the eastern Kuke fence, near the Okwa Valley in the western Kalahari and on plains around Jwaneng. Mainly absent in the north, east and southwest. Overlaps with the Flappet Lark on the northern side of the Makgadikgadi. Colour variations suggest that different races occur in the country. Egglaying December. Solitary or in pairs.

Habitat

Open bush savanna in semi-arid regions, grass plains, edges of Kalahari pans with scattered bushes and long grass, scrub savanna, old cultivation.

Analysis

95 squares (41%) Total count 252 (0,14%)

FLAPPET LARK

496

Mirafra rufocinnamomea

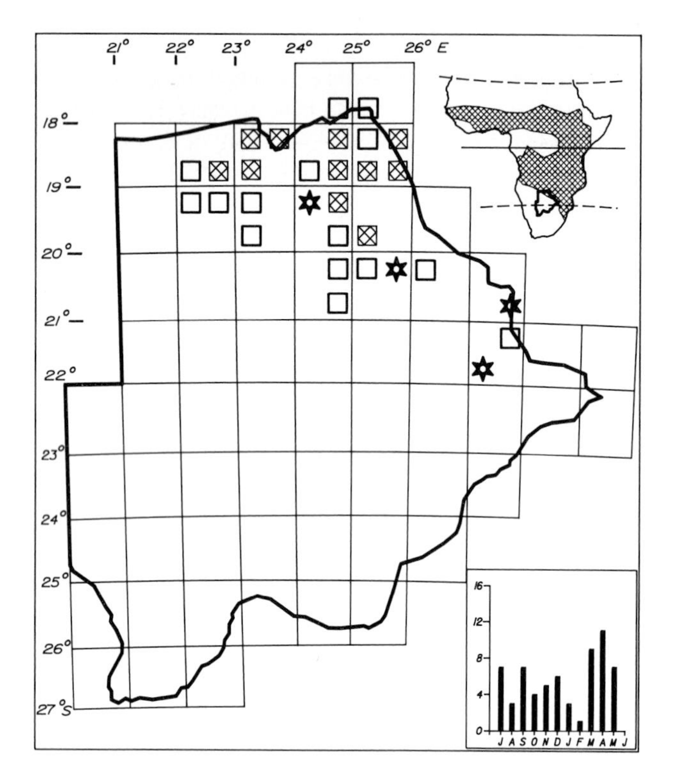

Status

Sparse to fairly common resident of the north and northeast, rarely reaching as far south as Francistown. Commonest in the well-wooded areas of the northeast. Extends westwards mainly to the north of the Okavango Delta. Solitary or in pairs.

Habitat

Edges of miombo and *Baikiaea* woodland, open areas along drainage lines in bush and tree savanna, edges of mopane woodland with secondary growth and open spaces of short grass, seasonal pans. Uncommonly in mopane scrub with poor groundcover.

Analysis

30 squares (13%) Total count 71 (0,04%)

FAWNCOLOURED LARK

497

Mirafra africanoides

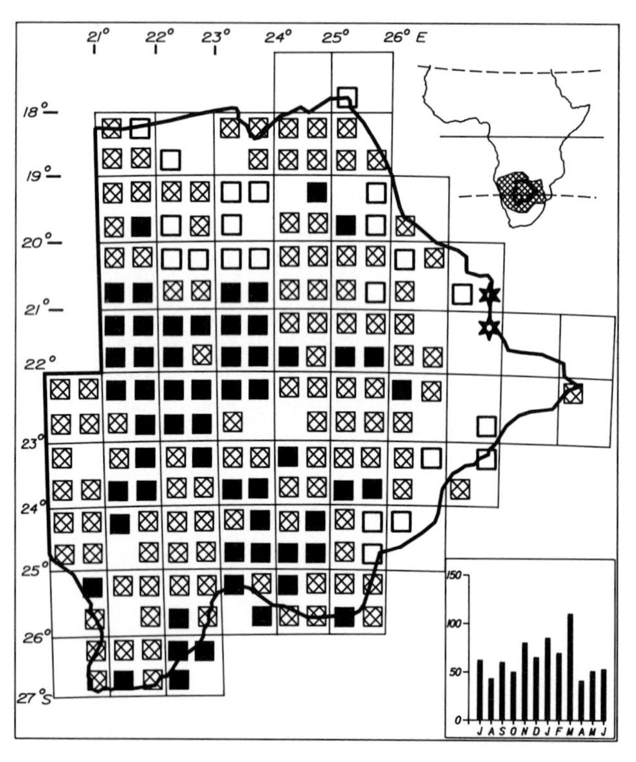

Status

Common to very common resident throughout Botswana west of 27°E. Sparse or uncommon in the Okavango Delta and in the higher rainfall areas of the east. Generally very common in the central, western and southern Kalahari. The most typical and widespread lark of the Kalahari semidesert. Egglaying October. Solitary, or in pairs holding territories so that 10–20 birds along a kilometre of track is not uncommon when breeding.

Habitat

Open bush and scrub savanna on Kalahari sands. Commonest in open spaces with low woody bushes, dead trees, or other bare woody perches. Often on telephone wires or fence wires. Also perches on isolated leafy bushes and thorn trees.

Analysis

192 squares (83%) Total count 789 (0,44%)

SABOTA LARK 498

Mirafra sabota

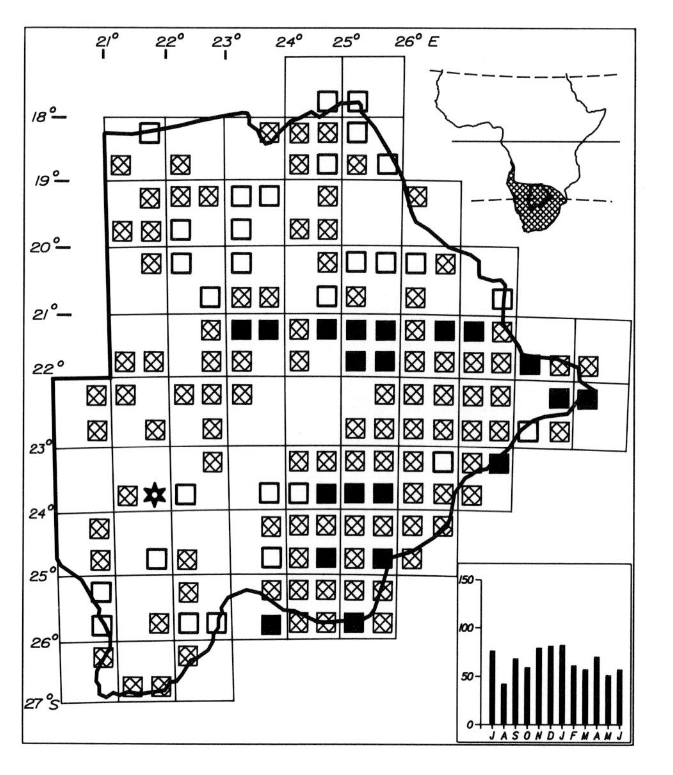

Status

Fairly common to very common resident south of 21°S and east of 24°E. Also very common in the Kalahari wooded savannas along the eastern Kuke fence and around Deception Valley. In the north, west and southwest it is sparse or uncommon, patchily distributed and unpredictable. Birds in the western Molopo Valley have heavy bills and appear larger. These latter may be a separate subspecies and some authors suggest a separate species *M. naevia*. Egglaying all months October to March. Solitary or in pairs.

Habitat

Acacia and thorn savanna in semidesert, broadleafed and mixed tree savanna in the higher rainfall areas. Absent from open Kalahari savanna and from the interior of woodland although it does occur on the edges particularly at patches on rocky ridges. It is also found in dry river valleys with scattered rocks and stones with little vegetation as in the western Molopo.

Analysis

146 squares (63%) Total count 828 (0,46%)

SHORTCLAWED LARK 501

Mirafra chuana

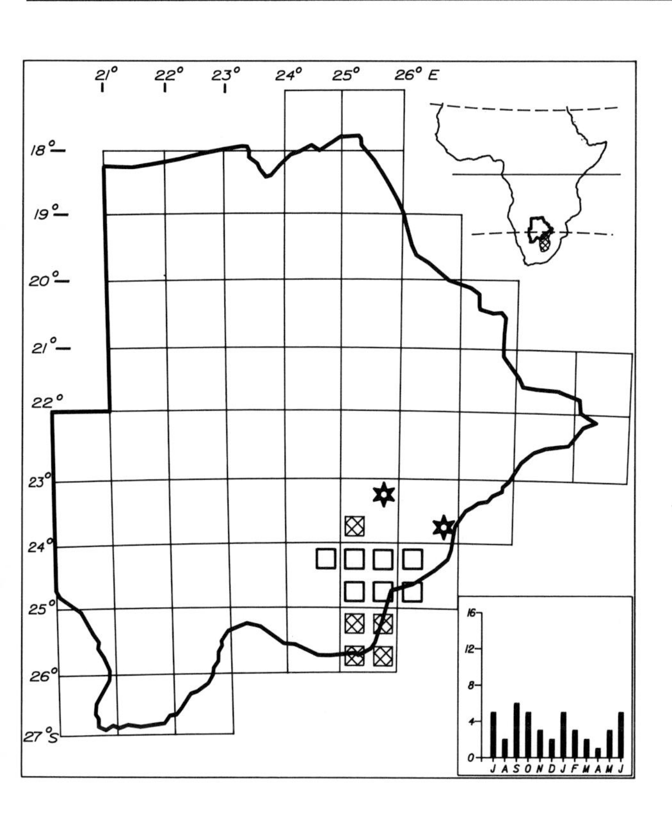

Status

Sparse to locally fairly common resident of the southeast. It has been regularly recorded at Ramatlabama, Good Hope and near Mmathethe. Occurs as far north as Lephepe and may occur sparsely in the Limpopo Valley. Its distribution in Botswana is in the northern portion of its small range in southern Africa. It is not well known. Active nest abandoned before egglaying in September. Usually solitary.

Habitat

Short grassland with scattered bushes on shallow soils, open sparsely grassed areas in bush or tree savanna (usually *Acacia*), old cultivation and fallow lands. Sometimes in grassed areas with stony ground or with rock ridges near the surface.

Analysis

14 squares (6%) Total count 51 (0,03%)

DUSKY LARK

Pinarocorys nigricans

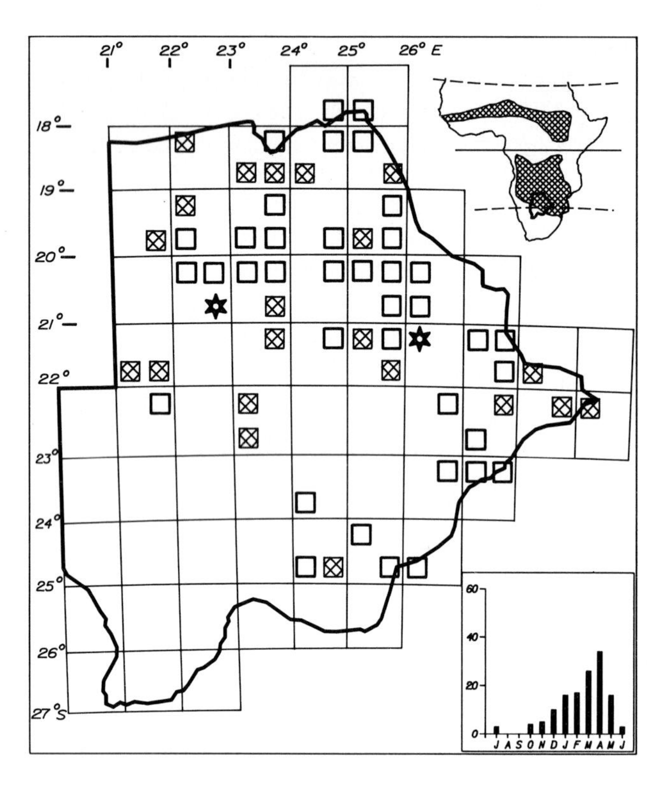

Status

Sparse to uncommon nonbreeding intra-African migrant mainly to areas north of 23°S. Arrives in early October and remains until May with peak northward movement in April. A few birds stay in June and July. Known to reside at some localities from January and February until departure in April or May. Some nomadic movements also occur. Movements are most noticeable on northward passage and in northern areas. Usually solitary, but occurs in loose congregations on passage at times.

Habitat

Open short grass in tree and bush savanna. Also on the edges of woodland and in glades and other open spaces in woodland. Found on untarred roads especially at dusk and on burnt ground and airfields.

Analysis

60 squares (26%) Total count 147 (0,08%)

SPIKEHEELED LARK

Chersomanes albofasciata

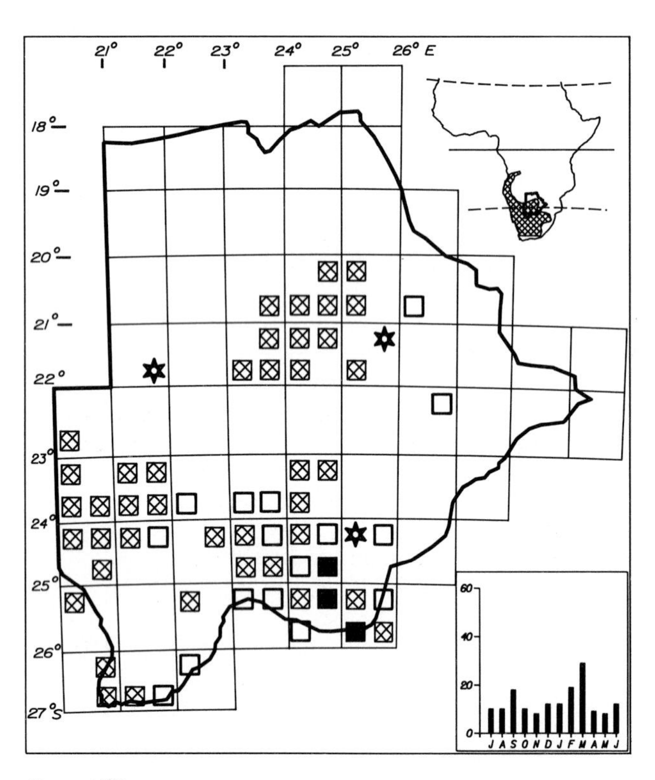

Status

Fairly common to common resident of the south and west and in the western part of the Makgadikgadi Depression. Its status in the southwest dunelands and in the central Kalahari needs further investigation. Sparse east of 26°E and absent north of 20°S. Mainly sedentary but local seasonal movements may occur. Egglaying October and February. Solitary, in pairs when breeding.

Habitat

Short grass in semidesert and bush savanna, especially clumps of grass with intervening bare ground. Common on vegetated pans in the south and west and may occur in similar habitat in the central Kalahari. Also occurs in fossil valleys and on the edges of open plains. Will breed in small areas of suitable habitat such as on the verges of roads, cutlines and tracks.

Analysis

63 squares (27%) Total count 173 (0,09%)

REDCAPPED LARK

Calandrella cinerea

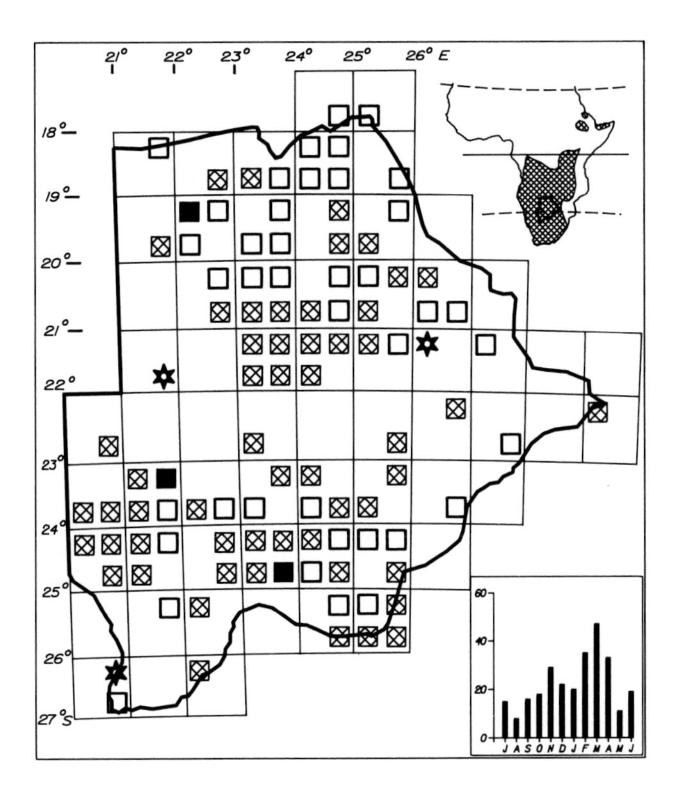

Status

Sparse to common resident and local migrant throughout the country. Most commonly recorded between November and April. In March and April it congregates in flocks on migration but the extent and nature of this movement is unknown. It may be a local postbreeding dispersal or an intra-African migration. It is a winter visitor (mainly April to October) in Zambia. Several races occur which behave in different ways. Egglaying November and April. Usually in groups of 5–30 birds, but several hundreds may congregate together on migration.

Habitat

Flat dry ground with sparse grass cover typically on pans, floodplains, burnt grassland, airfields, football pitches. Occasionally on fallow lands.

Analysis

101 squares (44%) Total count 296 (0,17%)

PINKBILLED LARK

Spizocorys conirostris

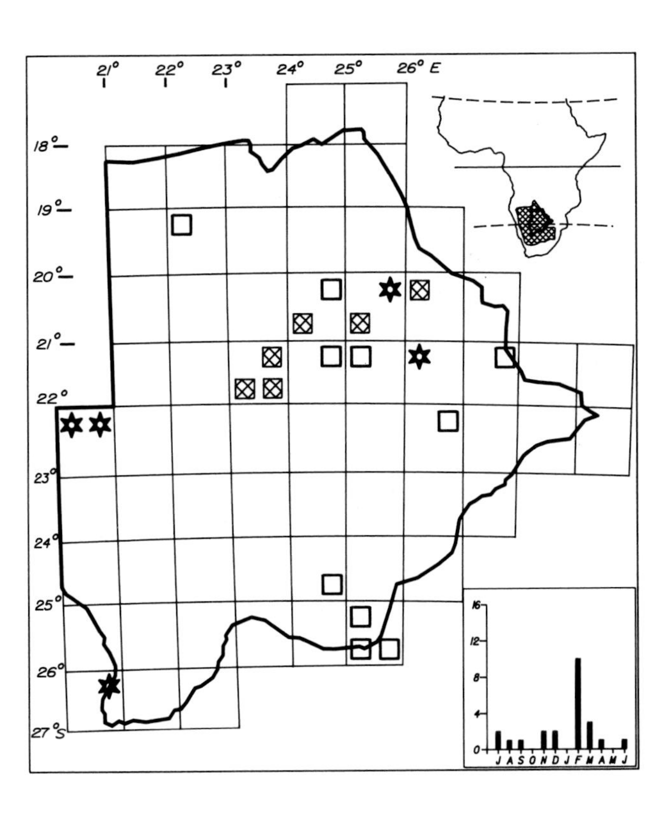

Status

Sparse to uncommon resident. Occurs in two main areas—the southeast and the Makgadikgadi region west to Deception Valley. It is easily overlooked and may be present regularly in other areas. It may be more numerous in years of high rainfall. There are old records from west of Ghanzi and the Nossob Valley. Local movements occur and possibly international movements to and from the Transvaal. Egg-laying all months from January to April. Usually in pairs or small groups.

Habitat

Short grass on plains, dry floodplains and as found in fossil valleys such as Deception and Naledi and the floor of the ancient Makgadikgadi lake. Also on stony ground with grass cover on small hills and undulating terrain in river valleys.

Analysis

21 squares (9%) Total count 32 (0,18%)

CHESTNUTBACKED FINCHLARK

515

Eremopterix leucotis

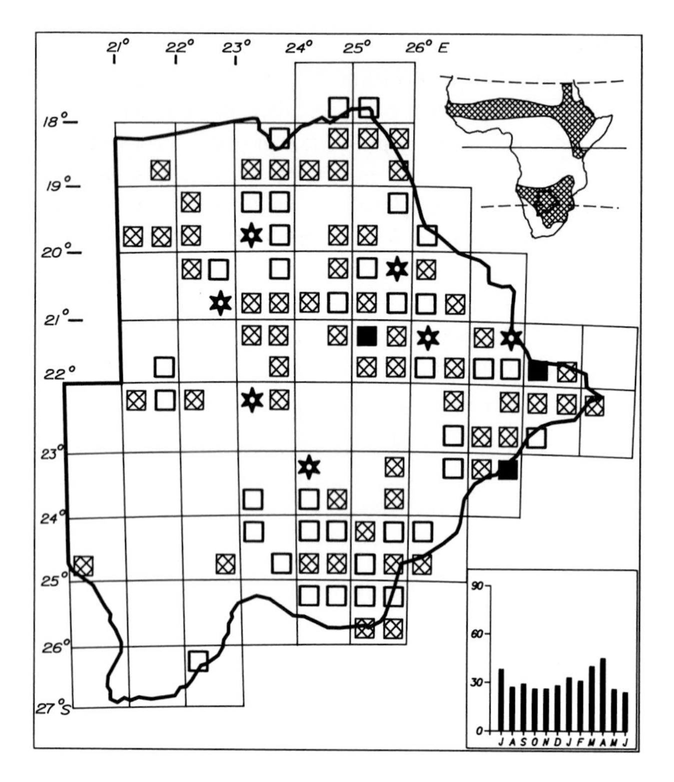

Status

Sparse to locally common resident in all regions except the west and southwest where it is rare and mainly absent. It is commonest in the east. There is no clearly defined separation between it and the Greybacked Finchlark. Both sometimes occur together on the same ground and their ranges overlap extensively in the central and southern regions. Highly nomadic. Egglaying January and February. Usually in groups of 5–30 birds.

Habitat

Flat open grass areas and bare ground between clumps of grass, on plains, vegetated pans, open areas in bush and tree savanna and in old cultivation.

Analysis

102 squares (44%) Total count 400 (0,22%)

GREYBACKED FINCHLARK

516

Eremopterix verticalis

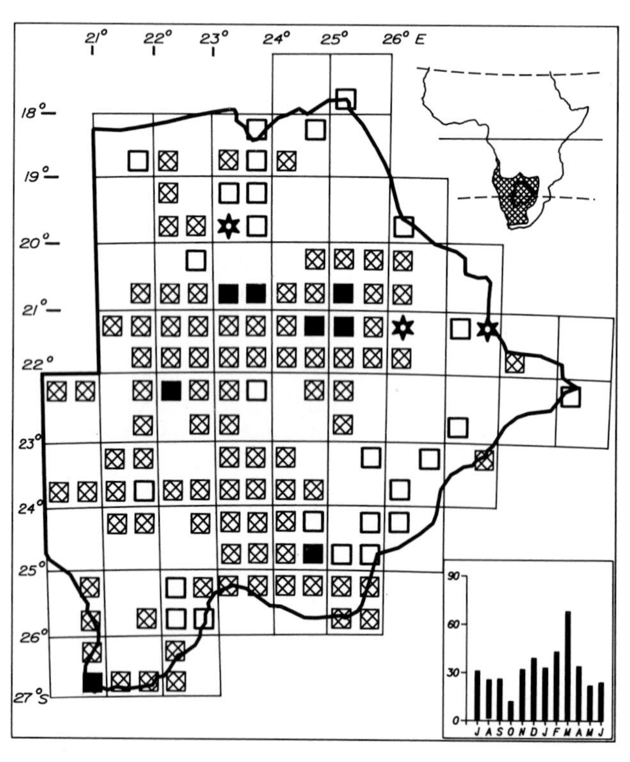

Status

Fairly common to common resident mainly west of 26°E and south of 20°S. Sparse to uncommon in the north and east. Generally more numerous and widespread than the Chestnut-backed Finchlark. Nomadic. Egglaying December and April. In groups of 5–50 birds, but congregations of several hundreds to a thousand birds can occur in one area. Known for eruptions of several thousands of birds but this has not been witnessed in this survey.

Habitat

Open grassland in semidesert and arid regions, plains, pans, open *Acacia* bush and tree savanna. On stony or sandy ground in dry river valleys, fossil valleys and duneland provided there are some grass clumps. Occurs in drier conditions than the Chestnutbacked Finchlark.

Analysis

125 squares (54%) Total count 429 (0,24%)

EUROPEAN SWALLOW

518

Hirundo rustica

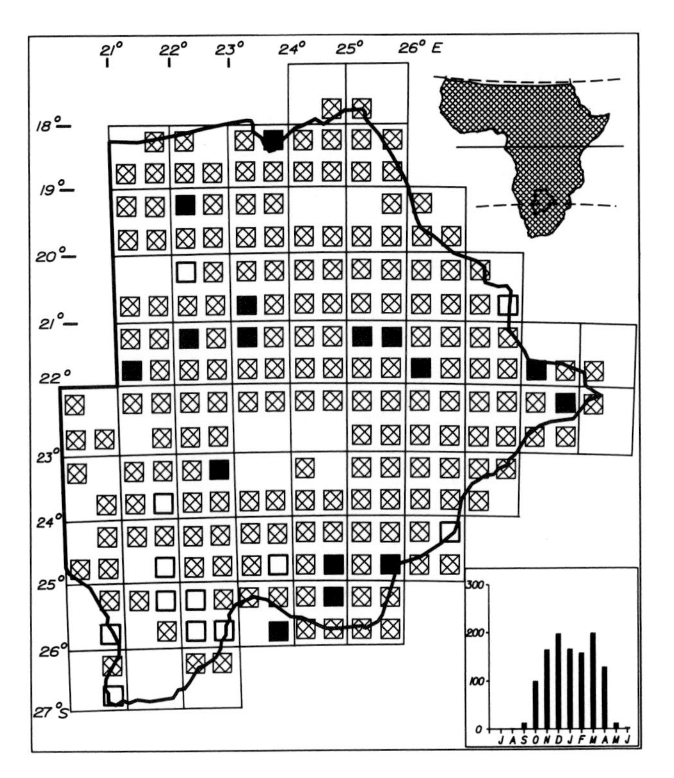

Status

Fairly common to very common Palaearctic migrant throughout the country. It is the commonest hirundine occurring in Botswana and in many areas of the west it is the only recorded member of the family. Present mainly from October to April with a few birds arriving in September and some remaining until May. Overwintering birds are rare. Flocking prior to northward migration starts in February. Passage movement varies from year to year with more birds remaining in Botswana in some years than others. Immature birds are often seen. Usually in flocks of 10–200 birds. Roosts of several thousands of birds are known.

Habitat

Occurs over any habitat but particularly large numbers are seen near water or over grassland after rain. Roosts in reed beds or maize fields. Resting birds are commonly found on telephone wires in the east and on dead trees in Kalahari savanna.

Analysis

203 squares (88%) Total count 1177 (0,65%)

WHITETHROATED SWALLOW

520

Hirundo albigularis

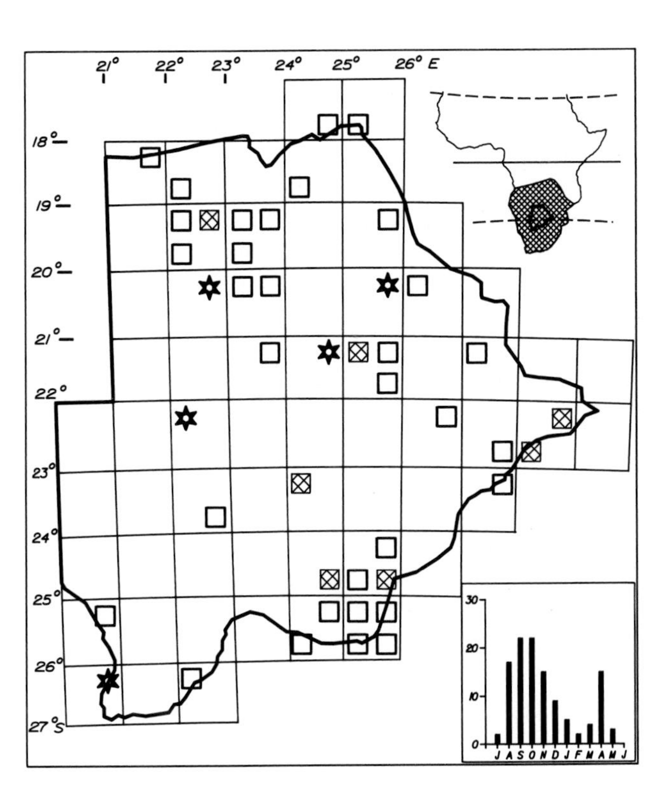

Status

Sparse to uncommon intra-African migrant occurring in all regions but mainly in the north, east and southeast. Present from August to April. Mainly recorded on southward passage between August and November (peak in September and October). The fewer records between December and March suggests that most birds have passed through to breeding grounds further south. Usually solitary, in pairs or in small loose flocks of 3–8 birds.

Habitat

Grassland and open savanna usually near water or in river valleys. Also in towns and near manmade structures e.g. buildings and bridges on which it nests in adjacent Transvaal.

Analysis

43 squares (19%) Total count 138 (0,075%)

WIRETAILED SWALLOW

522

Hirundo smithii

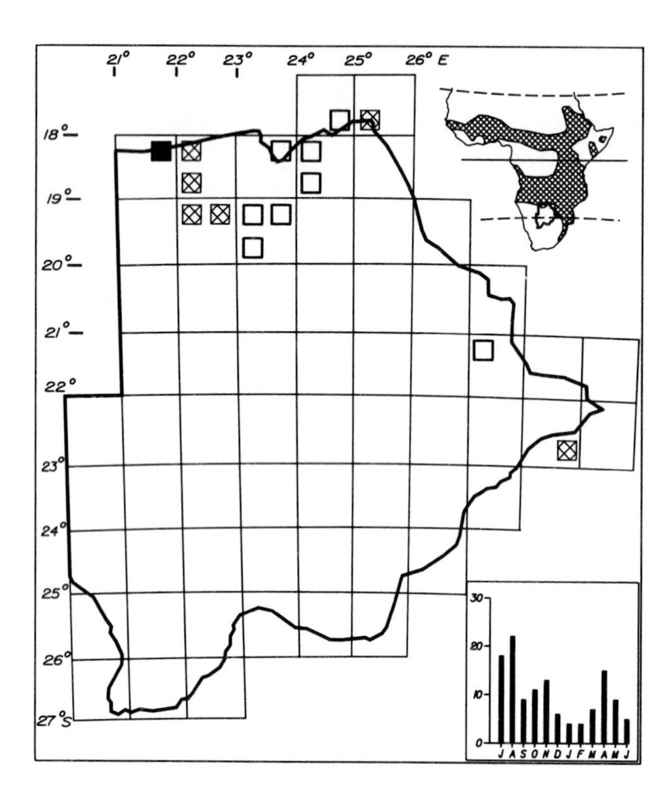

Status

Common to very common resident of the Okavango river system. Sparse on the periphery of this region at Linyanti, Savuti and along the Chobe River. Very common at Shakawe and common at Kasane. In the east it has been recorded once at Shashe Dam and once on the Limpopo River near Baines Drift. Well-known unusual nest site in the bar at Shakawe Fishing Camp. Egglaying August, October and March. Solitary or in pairs.

Habitat

Permanent water and adjacent woodland and habitation. Rivers, waterways, lagoons and dams.

Analysis

15 squares (7%)   Total count 127 (0,07%)

PEARLBREASTED SWALLOW

523

Hirundo dimidiata

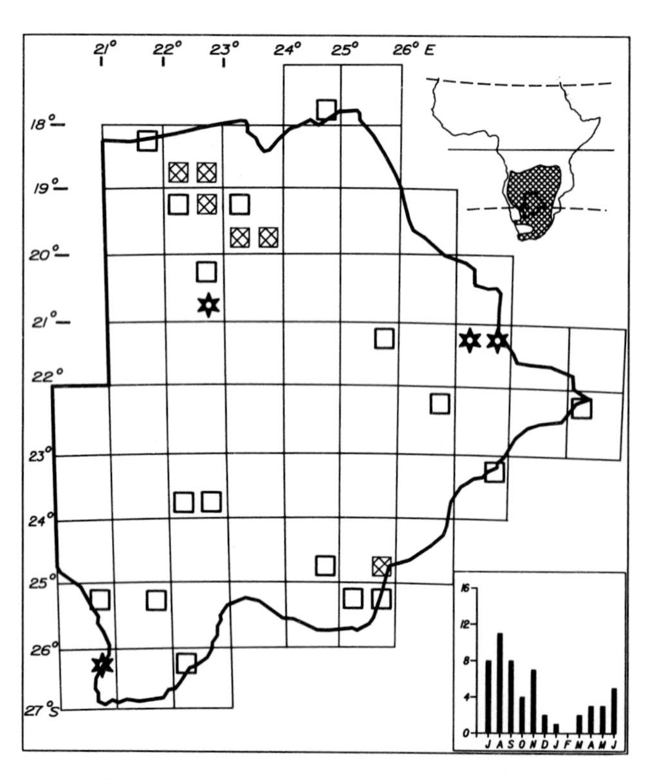

Status

Sparse to uncommon intra-African migrant recorded in all regions of the country. Poorly known. Most records are from the Okavango Delta. The few current records suggest it is mainly a passage migrant in winter months from June to November with a peak in August. Its movements are not well understood and may be complex in the southern African region. There are no records from the northeastern woodland where it might be expected to breed if it were resident as is part of the population in Zimbabwe and Transvaal. Usually in small loose groups of 3–10 birds or solitary.

Habitat

Broadleafed and riverine woodland in the north. Dry grassed pans with surrounding woodland in the west, open areas of woodland and tree savanna in the east and southeast.

Analysis

26 squares (11%)   Total count 63 (0,035%)

REDBREASTED SWALLOW 524

Hirundo semirufa

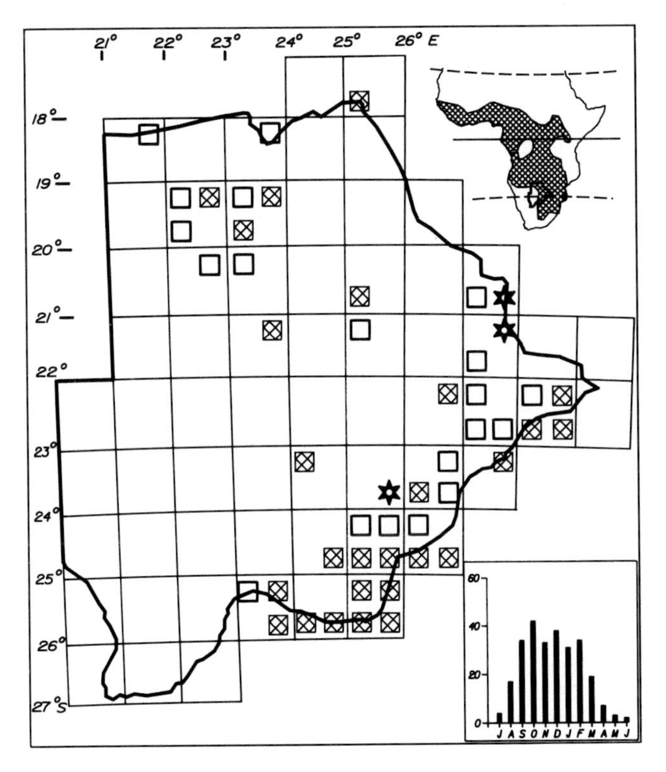

Status

Sparse to locally common breeding intra-African migrant to the north, east and southeast. Commonest in the southeast and east where it regularly nests in road culverts and where the road system is most prolific. Present mainly from August to March with some birds arriving in July and departing in April and May. Very few birds have been recorded in June in the north. Egglaying September. Usually in pairs or solitary.

Habitat

Open areas of tree and bush savanna, edges of woodland, grassland plains, cultivation, cleared areas along roads and in towns. Frequently perches on telephone or fence wires near nest sites in road culverts.

Analysis

48 squares (21%) Total count 285 (0,16%)

MOSQUE SWALLOW 525

Hirundo senegalensis

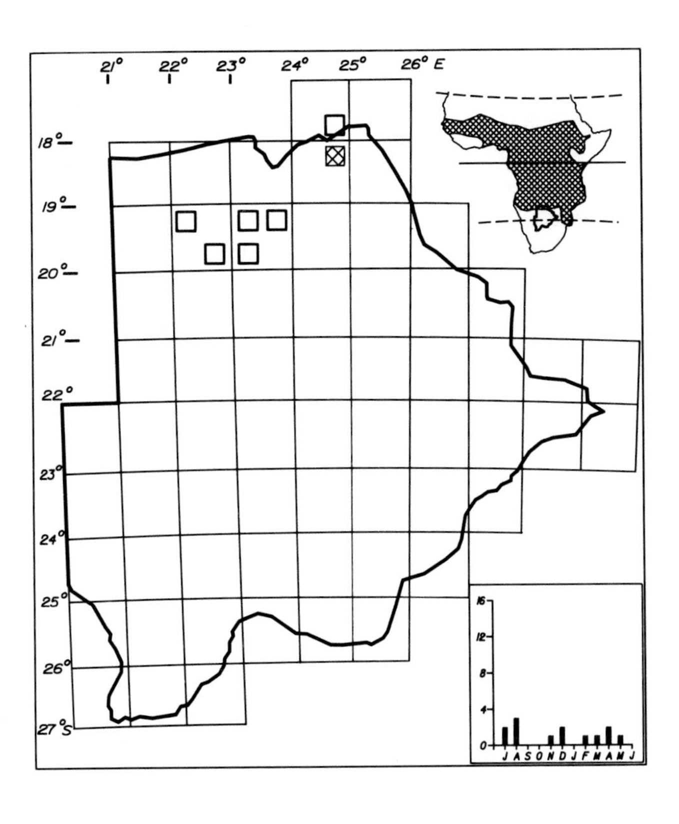

Status

A very sparse and irregular resident in the Okavango Delta and the Chobe region. It is poorly known. It occurs at the southern limit of its continental range at these longitudes. Recorded in most months of the year. An active nest site in a tree hollow was found in April. Usually in pairs or solitary.

Habitat

Tall mature woodland in high rainfall areas and in river valleys, edges of riverine forest and over adjacent flood-plains.

Analysis

7 squares (3%) Total count 14 (0,007%)

GREATER STRIPED SWALLOW
526

Hirundo cucullata

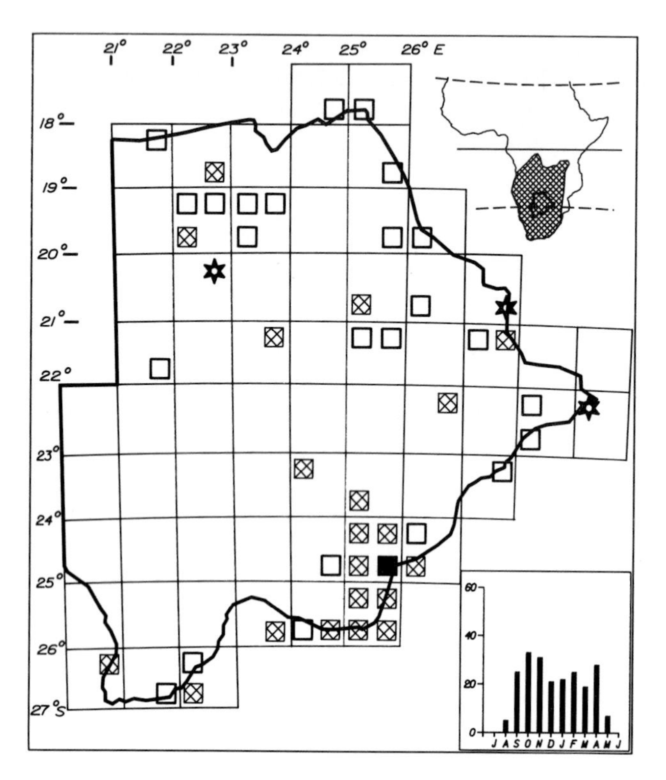

Status
Fairly common to very common breeding intra-African migrant in the southeast where except along the Molopo River it has not been recorded west of 24°E. Sparse to uncommon in the north, east and southwest where records are mainly of passage birds. Present from late August to April, latest date 10 May. Probably breeds only south of the Tropic of Capricorn. Egglaying November. Usually in pairs or solitary, postbreeding flocks of 20–100 birds congregate in March and April prior to northern migration.

Habitat
Nesting occurs mainly on manmade structures such as buildings, bridges and culverts. Occurs in open areas in towns, villages and farms, and at roadsides. Also over lightly-wooded savanna, cultivation, grassland plains and marshland. Wanderers and passage birds may occur over almost any habitat.

Analysis
48 squares (21%) Total count 235 (0,13%)

LESSER STRIPED SWALLOW
527

Hirundo abyssinica

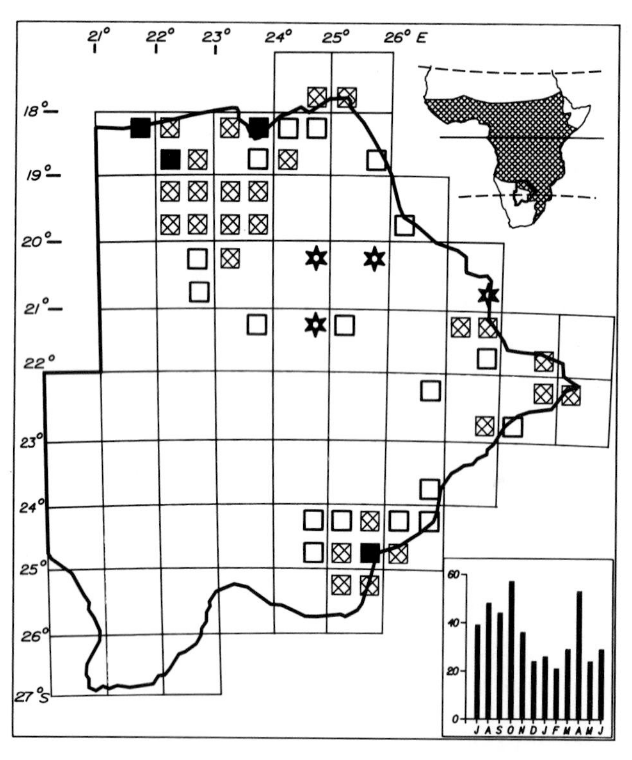

Status
Fairly common to very common breeding intra-African migrant to the Okavango, Linyanti and Chobe river systems. Sparse elswhere in the north. In the east it is sparse to uncommon mainly in the Shashe and Limpopo drainage. In the southeast it is sparse to locally very common. In the north it is mainly present between July and April but about 10% of records are for May and June and this proportion may be resident. In the southeast there is an influx in August to October and again in March and April. Some birds remain in this southern area to breed but most are on passage. Egglaying November and December. Usually in pairs or flocks of 10–30 birds.

Habitat
Woodland, tree and bush savanna in high rainfall areas and in river valleys. Also in towns and villages, at farms and rural dwellings. Nests on buildings and bridges.

Analysis
52 squares (23%) Total count 447 (0,25%)

SOUTH AFRICAN CLIFF SWALLOW 528

Hirundo spilodera

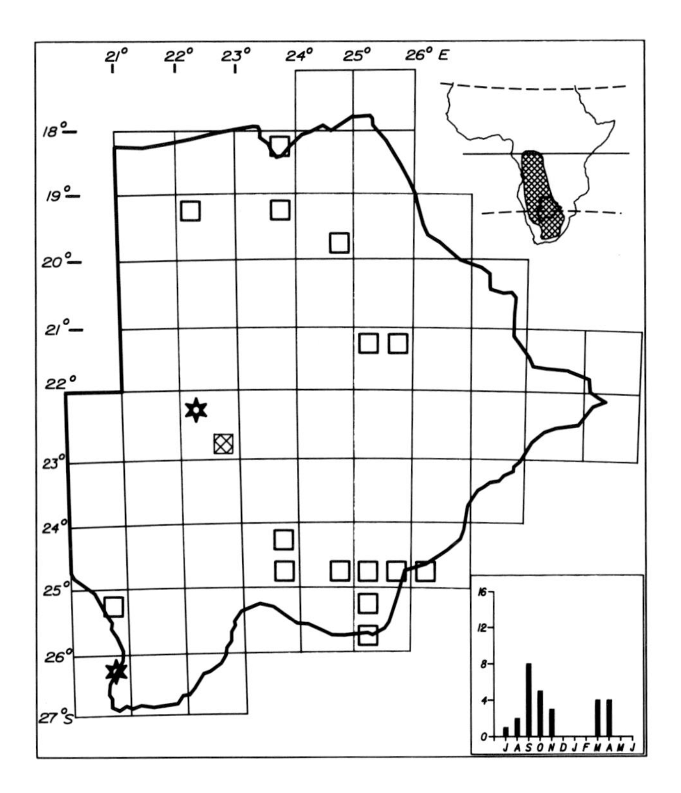

Status

Sparse intra-African passage migrant. Mainly recorded in the southeast where it occurs near the northern limit of its breeding range in South Africa. Elsewhere the very few records are of passage birds or wanderers. It is not clear why there are so few records for Botswana which is geographically on the migration route and so close to the breeding grounds—the nearest breeding record is within 20 km of the southeastern border. Occurs between late July and November and again in March and April. In small numbers of 1–5 birds.

Habitat

Most records are of birds near human habitation—towns, airfields, bridges, buildings. Also woodland edge, dry Kalahari pan, fossil valley, open grassland and bush savanna. Probably over any habitat on passage.

Analysis

18 squares (8%) Total count 32 (0,017%)

ROCK MARTIN 529

Hirundo fuligula

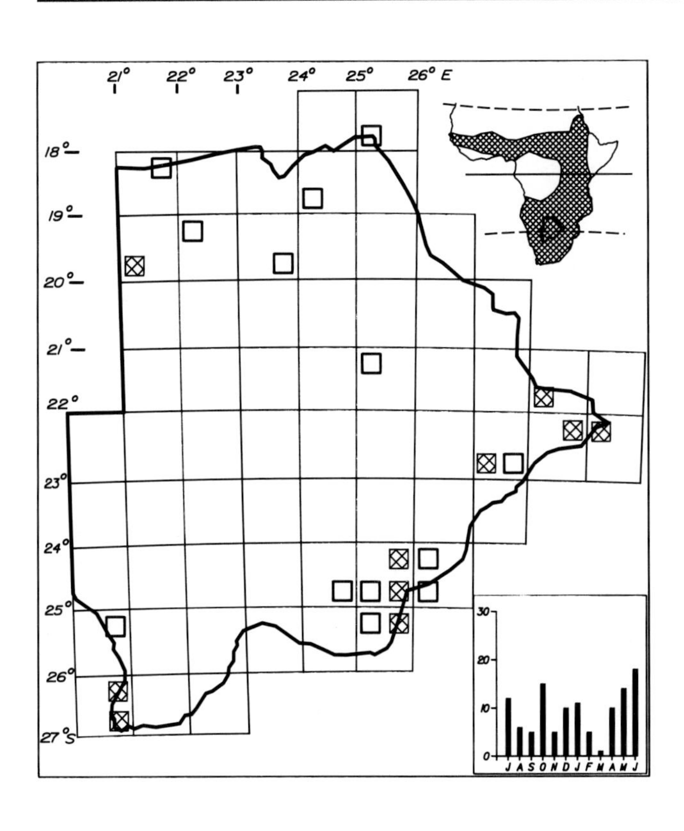

Status

Sparse to fairly common resident in the east and southeast. Locally common in the Nossob Valley. Elsewhere it is sparse in isolated localities. Mainly sedentary but local movements may occur as there are only scant records for some months in well observed areas e.g. Lobatse and Gaborone in February and March. Usually solitary or in pairs, occasionally in groups of 5–10 birds.

Habitat

The distribution is principally in the hilly regions of the southeast and east and along the Nossob Valley escarpment in the southwest. Elsewhere it occurs at isolated hills. Rocky hills, cliffs, gorges, escarpments. Occasionally at tall buildings and even on smaller buildings in remote areas as at Kasane and Twee Rivieren. Also at dam walls and mine rock dumps.

Analysis

23 squares (10%) Total count 123 (0,07%)

HOUSE MARTIN

530

Delichon urbica

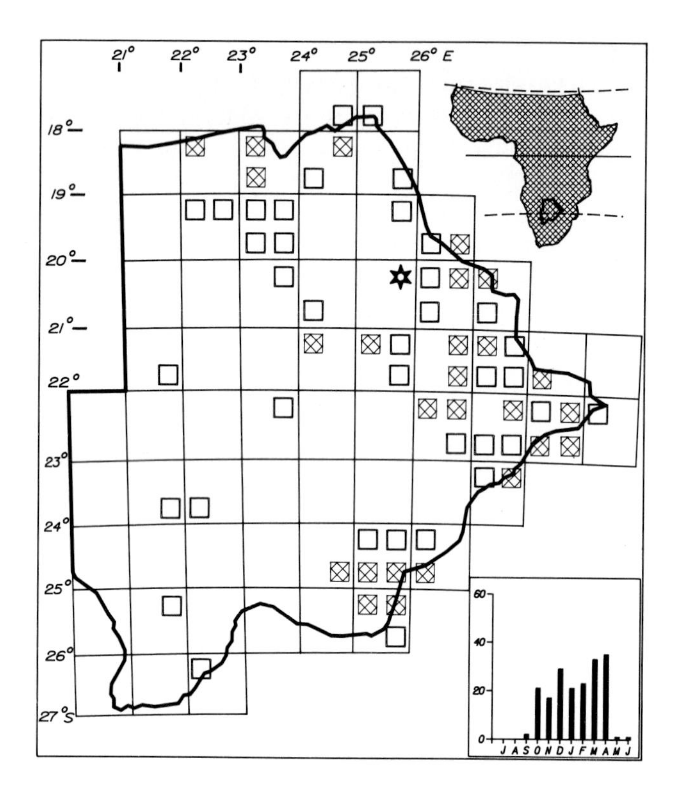

Status

Sparse to fairly common Palaearctic migrant to all regions of the country. Recorded mainly in the north, east and southeast. It flies high and often with other hirundines and is easily overlooked. Present from October to late April, rarely in late September and early May. Flocking prior to northward migration starts in late February. Usually in small groups of 4–20 birds, but larger flocks occur on northward migration.

Habitat

Over any habitat from riparian forest and woodland to semidesert scrub. Regularly over water and towns. Rests on roofs of buildings, bare trees and telephone wires, often with European Swallows.

Analysis

65 squares (28%) Total count 196 (0,11%)

GREYRUMPED SWALLOW

531

Pseudhirundo griseopyga

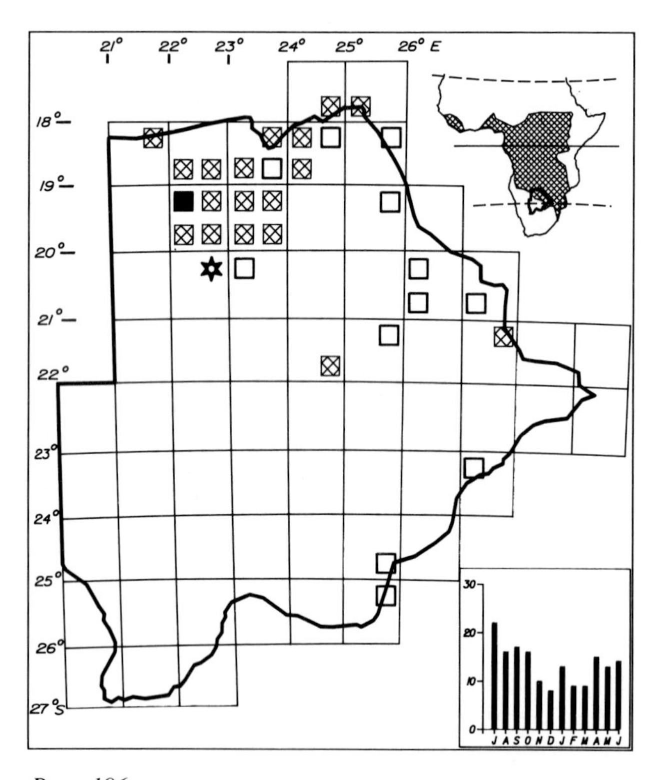

Status

Fairly common to common resident and partial migrant in the Okavango, Linyanti and Chobe river systems. Sparse to uncommon elsewhere north of 22°S. Very sparse along the Limpopo and Ngotwane valleys—most records from the Gaborone area are from dams. Occurs mainly between April and November but some birds are present throughout the year. Regular movements are likely but the pattern is clouded by the irregular movements of part of the population. Egglaying September. Usually in small flocks of 10–30 birds, but flocks of several hundreds of birds occur on migration.

Habitat

Grass plains, floodplains, open woodland and tree savanna, clearings and cultivation. Nests in burrows and is usually found in areas of short grass or recently burnt ground. Also occurs in river valleys and over open areas of water. On migration it has been recorded in large flocks over semidesert scrub.

Analysis

32 squares (14%) Total count 174 (0,09%)

SAND MARTIN

532

Riparia riparia

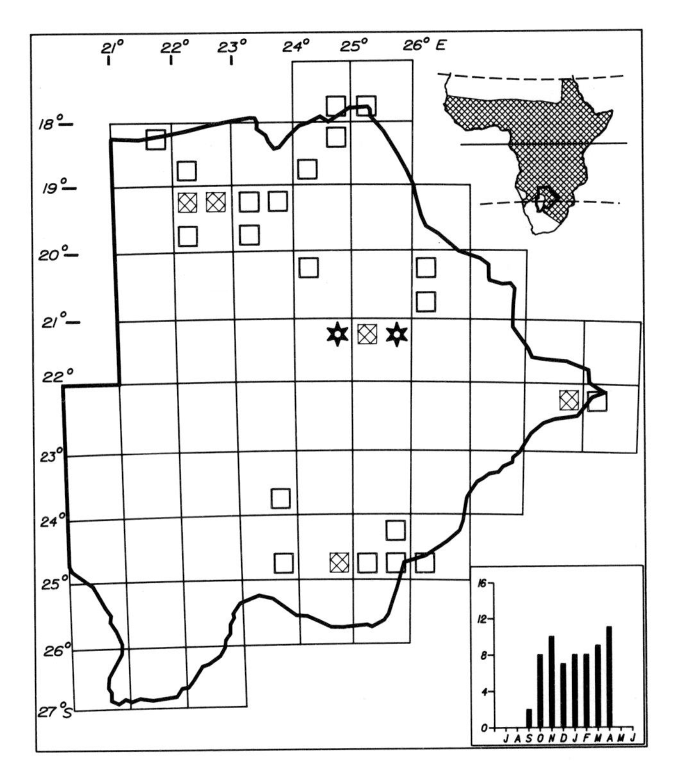

Status

Sparse to uncommon Palaearctic migrant to the north, east and southeast. It appears to be more common in some years than others. It is never common but may occur in large flocks (several hundred birds) near water at one locality for a few days before moving elswhere. Present mainly from mid-October to mid-April. It is unlikely to occur except as a rarity over most of the central and western Kalahari. Usually solitary or small groups, but large flocks of 100–500 birds may congregate on northward migration.

Habitat

Floodplains, river valleys, marshes, dams, pans, sewage ponds, grass plains and airfields. Usually near water. Perches with other hirundines on telephone wires on migration.

Analysis

27 squares (12%) Total count 71 (0,04%)

BROWNTHROATED MARTIN

533

Riparia paludicola

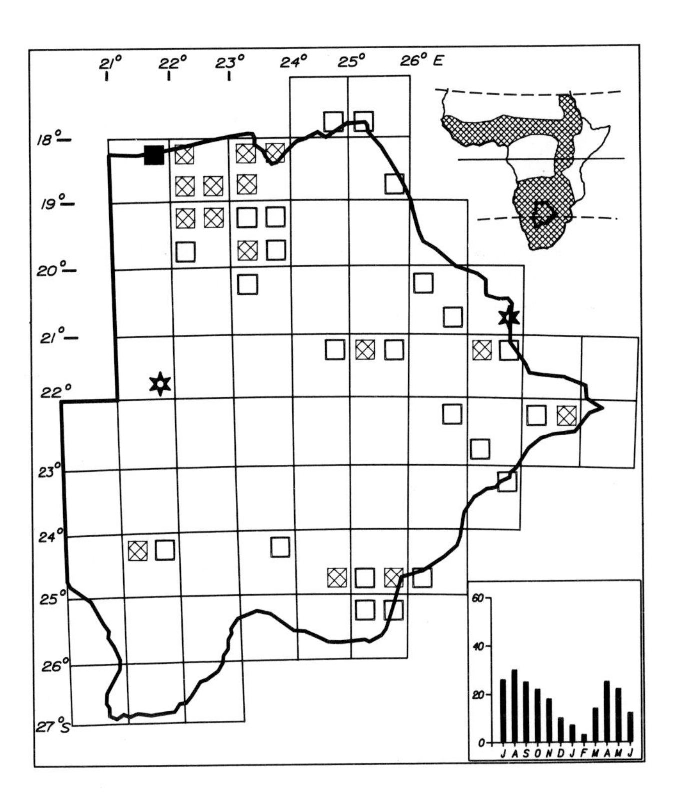

Alternative name

African Sand Martin

Status

Fairly common to locally common resident in the Okavango, Linyanti and Chobe river systems. Elsewhere it is sparse to uncommon mainly in the east and southeast particularly at open areas with water such as Mopipi, Orapa, Nata, Francistown, Jwaneng and Gaborone. Seasonal movements occur and most records fall between April and November. Breeding is likely along the major rivers and at large inland waters in winter months. Usually in small flocks of 5–20 birds, but larger flocks of several hundred birds occur.

Habitat

River valleys, lagoons, waterways, lakes, dams, floodplains and pans. Nests in sand banks. Twice seen flying over woodland and bush savanna in the west in March, probably on migration.

Analysis

41 squares (18%) Total count 221 (0,12%)

BANDED MARTIN

534

Riparia cincta

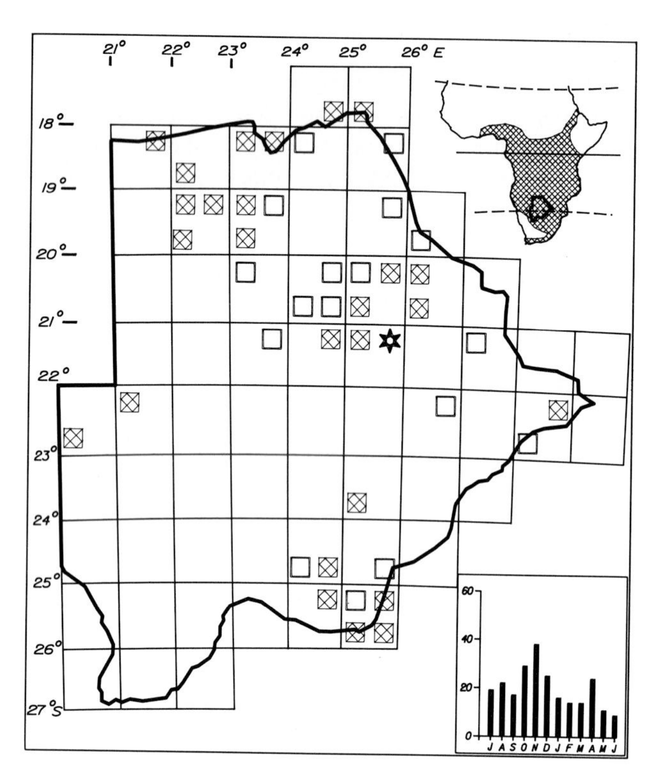

Status

Sparse to fairly common intra-African migrant recorded mainly in the north and southeast. Elsewhere it is sparse to uncommon. Its status is not well understood and is probably made up of a summer breeding population as well birds on passage at various times of the year. Most frequently recorded between July and December and again in April. There are records for all months of the year. Breeding records are required to clarify the position. Usually solitary or in pairs, occasionally in small loose groups of up to 10 birds.

Habitat

Open grass plains, floodplains, fossil valleys, airfields, pans, semidesert scrub. Usually over large areas of flat dry grassland. Its distribution is mainly over the northern floodplains, the Makgadikgadi Pans grassland and the open grass plains of the southeast.

Analysis

44 squares (19%) Total count 253 (0,14%)

BLACK CUCKOOSHRIKE

538

Campephaga flava

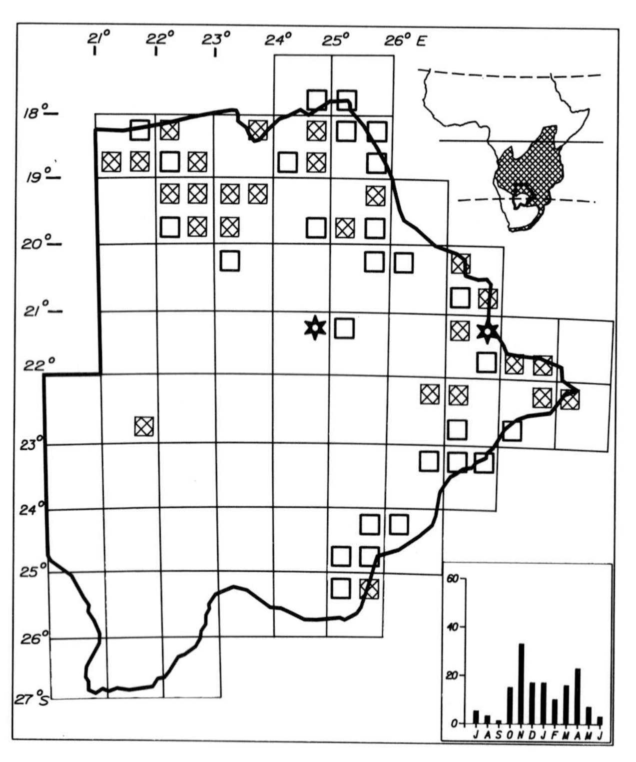

Status

Sparse to fairly common resident and partial migrant in the north, east and southeast. Occurs mainly between October and April as a breeding summer migrant but some birds are present in all months in the river valleys throughout its range. A pair was recorded once near Takatshwane (2221D) in December. Further investigation is needed to ascertain its status in the western woodlands. Occurs in Botswana at the western limit of its southern African range. Usually in pairs or solitary.

Habitat

Broadleafed and mixed woodland, edges of riparian forest, riverine bush. Occurs mainly in the drainage of the major rivers.

Analysis

55 squares (24%) Total count 163 (0,09%)

WHITEBREASTED CUCKOOSHRIKE

539

Coracina pectoralis

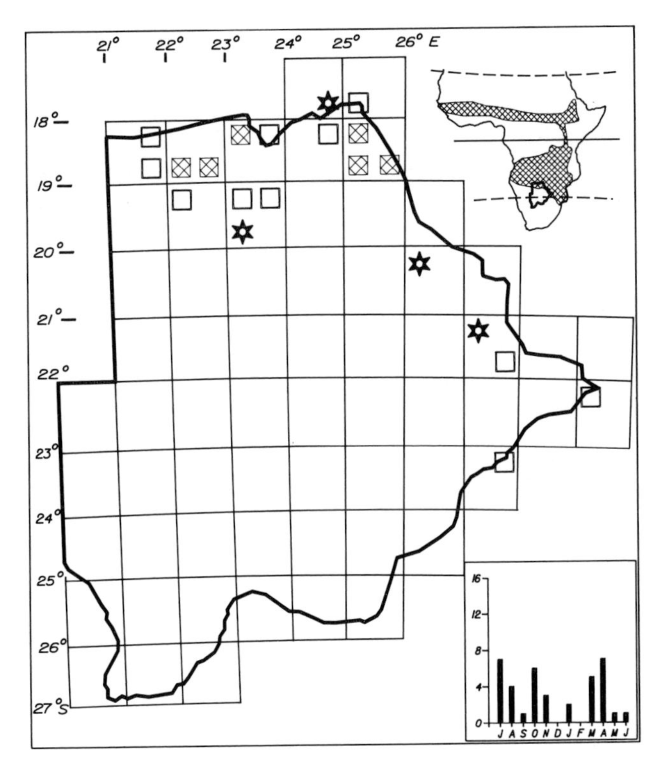

Status

Sparse to uncommon resident in the northern broadleafed woodlands. In the east it is very sparse and has been recorded only once in each of the localities shown. Old records at Maun, Nata and Francistown may indicate that its range has undergone a recent contraction, possibly as a result of increased human population in those areas. It is poorly known in Botswana and warrants special study. It occurs at the western limit of its southern African range. Usually solitary or in pairs. Occasionally in mixed bird parties.

Habitat

Mature broadleafed woodland where it is found mainly in the canopy. Also in tall mixed riverine woodland but not in riparian forest.

Analysis

22 squares (10%) Total count 45 (0,025%)

FORKTAILED DRONGO

541

Dicrurus adsimilis

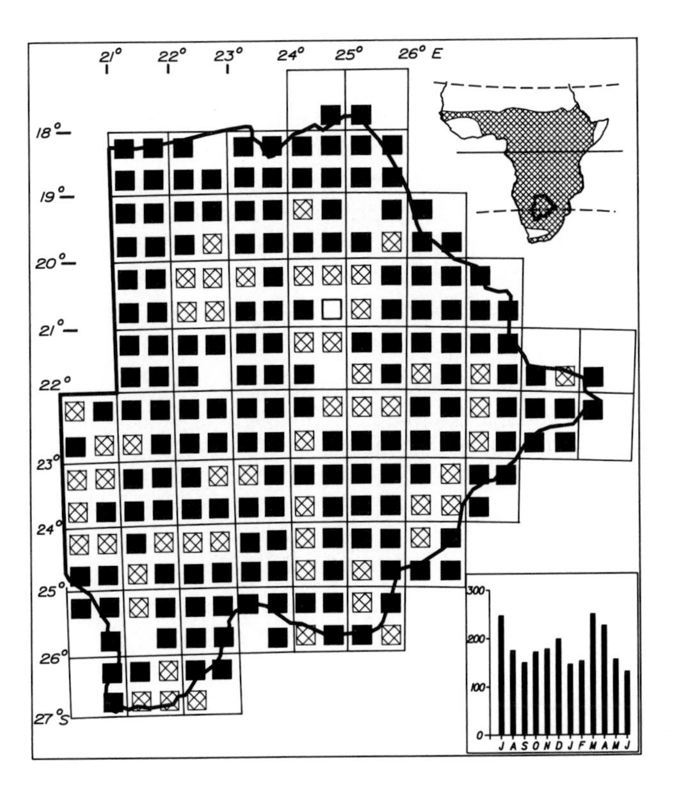

Status

Very common resident throughout. The most commonly recorded species in Botswana. Host to the African Cuckoo. Egglaying October, November and December. Usually solitary or in pairs, but congregates in flocks at bush fires or at alate emergences. Often with or near woodland bird parties but less of an indicator of these in Botswana than in countries to the north.

Habitat

Any type of woodland, tree and bush savanna, riverine *Acacia*, patches of woodland in semidesert, gardens, plantations and edges of riparian forest.

Analysis

225 squares (98%) Total count 2256 (1,27%)

EUROPEAN GOLDEN ORIOLE

543

Oriolus oriolus

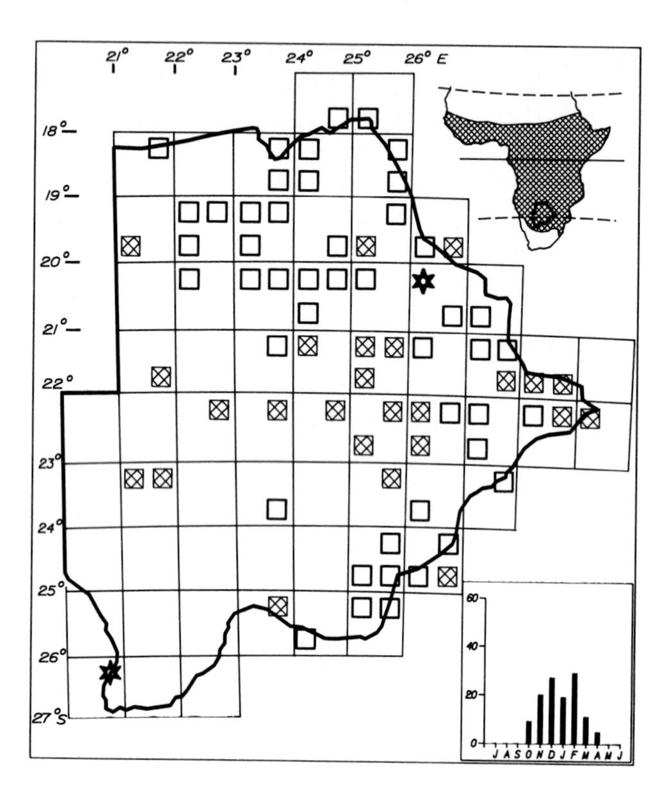

Status

Sparse to fairly common Palaearctic migrant. Present from mid-October (17th) to mid-April (15th) and commonest between November and February. Northward passage is particularly evident in February when loose groups of up to 20 birds can be seen together. Locally nomadic in their wintering quarters. Solitary birds are mostly males; females and juveniles are usually in groups with or without males.

Habitat

Any woodland and tree savanna including isolated patches of trees in semidesert in the central Kalahari. Also in gardens, plantations and edges of riparian forest.

Analysis

73 squares (32%) Total count 132 (0,07%)

AFRICAN GOLDEN ORIOLE

544

Oriolus auratus

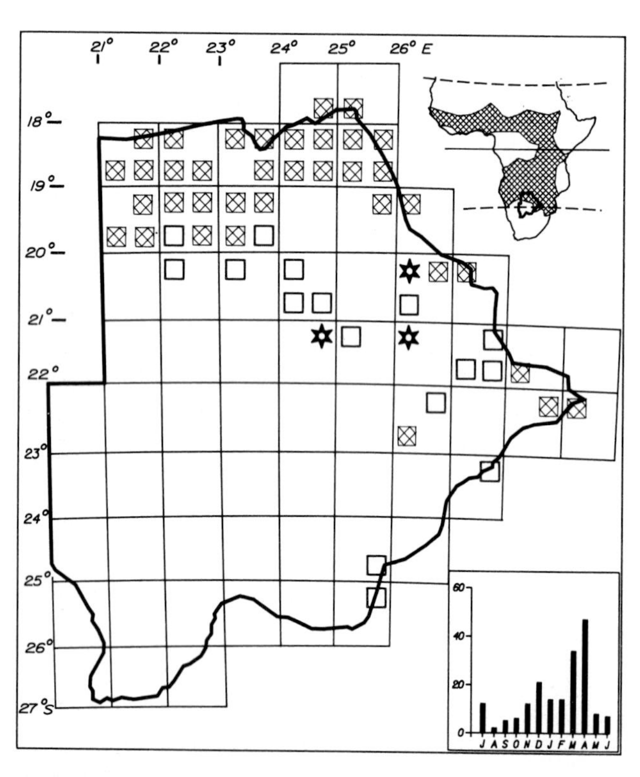

Status

Sparse to fairly common resident and intra-African migrant in the north and east. Commonest in the northern woodlands. Rare in the southeast at Gaborone and Lobatse. Present throughout the year but numbers are augmented by migrants between November and April. Its movements in Botswana and adjacent territories is not well understood. Egglaying March (one record) and probably in earlier summer months. Solitary or in pairs.

Habitat

Mature miombo, *Baikiaea* and mopane woodlands, riverine *Acacia* and riparian forest. Commonest in river valleys. Often at fruiting trees. Mainly in the canopy and upper branches of trees.

Analysis

55 squares (24%) Total count 192 (0,11%)

BLACKHEADED ORIOLE

Oriolus larvatus

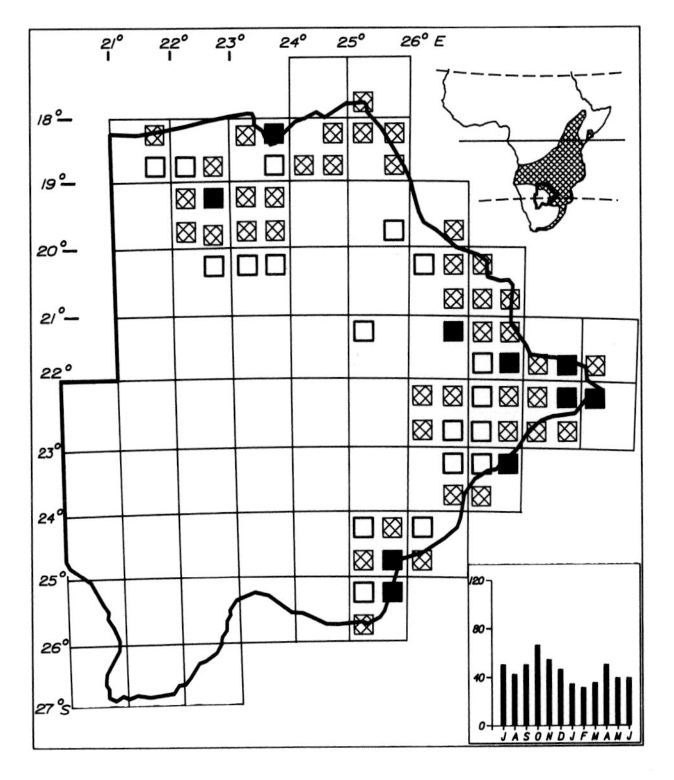

Status

Fairly common to locally very common resident of the north and east. In the southeast it is common to as far west as Kanye but does not occur further west. Mainly sedentary but some short distance local movements may occur. Egglaying November and December. Solitary or in pairs.

Habitat

Well-developed broadleafed woodland usually in proximity to rivers or dams. Always found in areas with surface water. Mostly in the canopy of larger trees. Also occurs in large exotic trees in gardens and its range may expand with urban development in the east.

Analysis

69 squares (30%)   Total count 556 (0,31%)

BLACK CROW

Corvus capensis

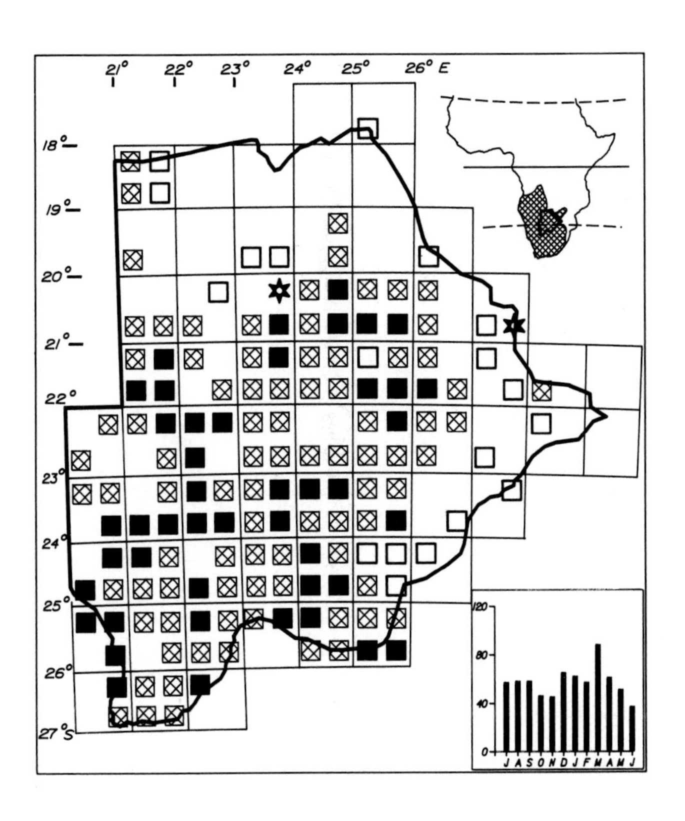

Status

Common to very common resident south of 20°S and west of 27°E. Sparse to uncommon in the north and east. Ecologically separate from the Pied Crow with which its range overlaps considerably in the eastern half of the country. Egglaying October, November and December. Its old nests are frequently used by Greater Kestrel and the two species are often seen together. Usually in pairs and often in threes. Postbreeding congregations of 20–80 birds together may be found in March and April.

Habitat

Open Kalahari tree and bush savanna, semidesert bush and scrub savanna, Kalahari pans and fossil valleys. Perches readily in any bush or tree including dead or stunted trees but frequently found on the ground.

Analysis

154 squares (67%)   Total count 722 (0,41%)

PIED CROW

Corvus albus

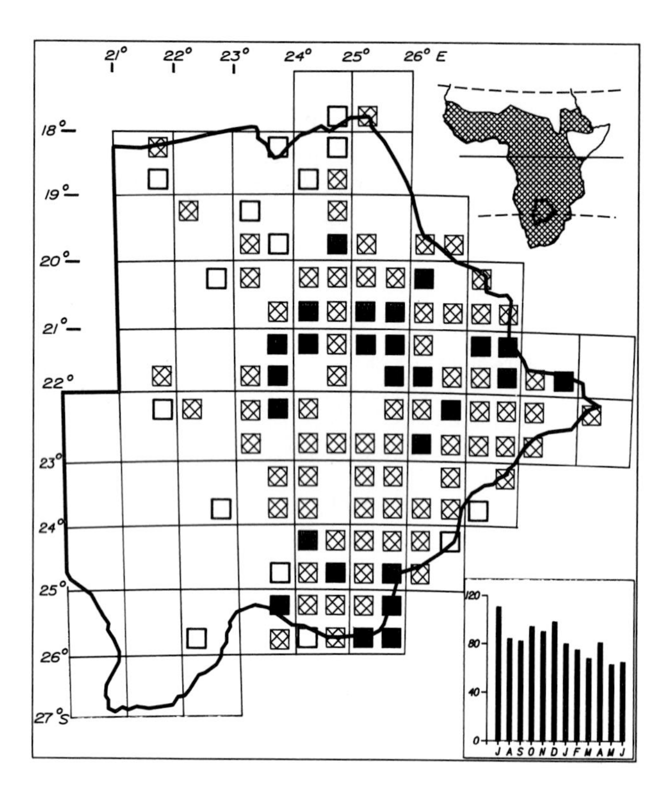

Status

Fairly common to locally very common resident east of 23°E. In the west it is largely restricted to the vicinity of towns such as Shakawe, Gomare, Ghanzi, Kang and Tshabong. Uncommon north of Nxai Pan. Expected to increase numerically and in range with increasing human population. Egglaying October and November. Usually solitary or in pairs. Post-breeding flocks occur between February and April.

Habitat

Urban areas on buildings, in gardens, parks, clearings, at rubbish dumps, sewage ponds, farms, villages and cattle posts. In rural areas it occurs in tree and bush savanna mainly in the high rainfall regions. In semidesert at boreholes and cattle posts. Nests quite commonly on electricity pylons and perches on telegraph poles and fences.

Analysis

111 squares (48%) Total count 1010 (0,57%)

ASHY TIT

Parus cinerascens

Alternative name

Acacia Grey Tit

Status

Common resident south of 20°S and west of 27°E. Sparse to fairly common in the northwest and rare in the east and northeast. It is generally very common along the Molopo River. Its range overlaps very little with that of the Southern Black Tit. Usually solitary or in pairs. Frequently a member of bird parties.

Habitat

Acacia savanna in the drier regions. Occurs in open tree savanna including parklike areas of *Acacia erioloba* as well as fairly dense areas of mixed trees and bushes. Also in Kalahari woodland and riverine bush.

Analysis

154 squares (67%) Total count 538 (0,3%)

SOUTHERN BLACK TIT

554

Parus niger

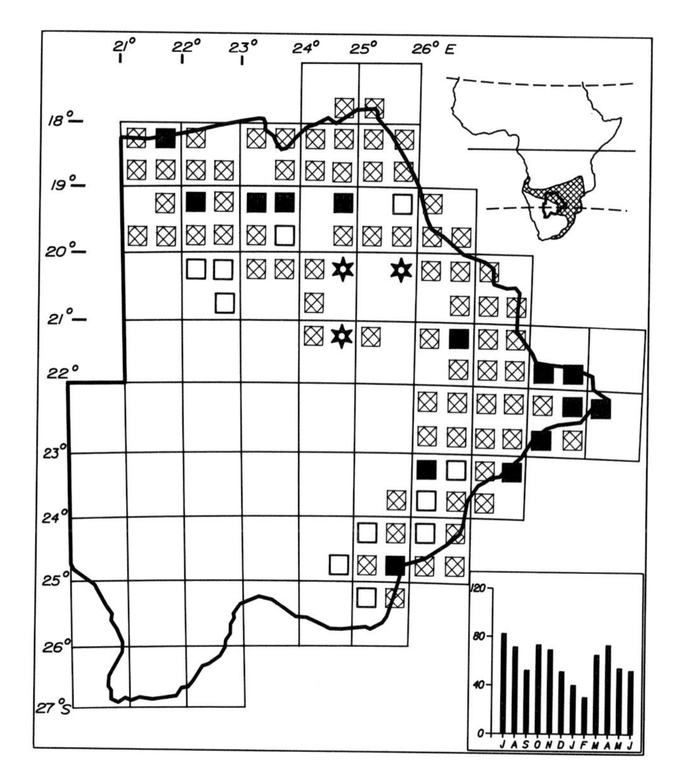

Status

Fairly common to very common resident of the north and east to as far south as Lobatse. Occurs at the western limit of its range in southern Africa. Very common in certain areas throughout its range such as near the Limpopo/Shashe River confluence, the Okavango Delta, at Nxai Pan, Shakawe, Gaborone and west of Mahalapye. Usually in pairs. Occurs often in mixed bird parties.

Habitat

Any woodland in the higher rainfall areas. Favours *Acacia* bushes and trees along drainage lines. Bush and tree savanna.

Analysis

98 squares (43%) Total count 731 (0,41%)

CAPE PENDULINE TIT

557

Anthoscopus minutus

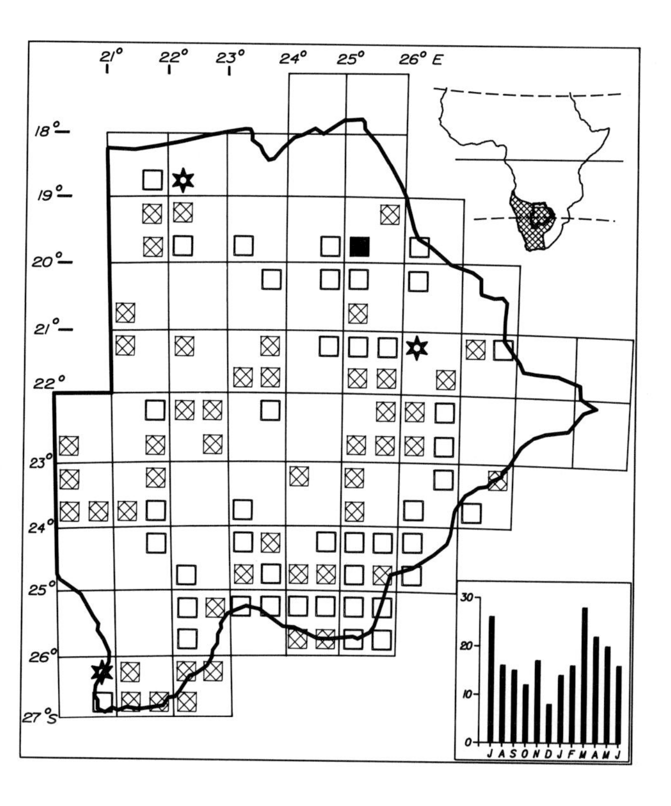

Status

Sparse to fairly common resident. Rare north of 19°S and not recorded east of 28°E. Patchily distributed but easily overlooked. Its status in the central Kalahari and southwest duneland needs more detailed investigation. Replaced by the Grey Penduline Tit in the more richly vegetated areas of the north and east. Egglaying November to February. Usually in pairs or small groups.

Habitat

Bush savanna in the drier regions particularly in *Terminalia sericea*. Occurs on the edges of *Acacia* woodland and in Kalahari tree and scrub savanna. Conspicuous nests are suspended from the side of trees and bushes two to four meters from the ground.

Analysis

95 squares (41%) Total count 227 (0,13%)

GREY PENDULINE TIT

Anthoscopus caroli

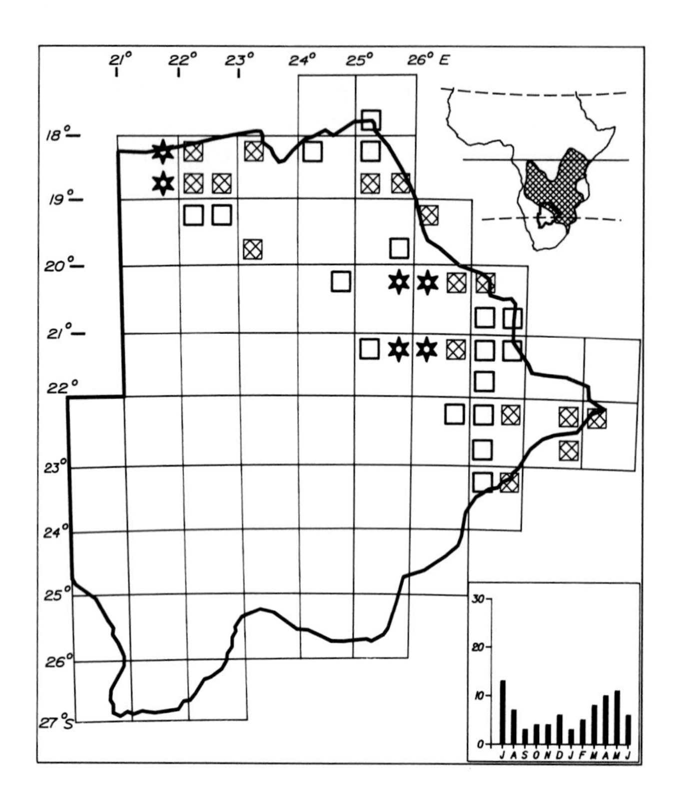

Status

Sparse to fairly common resident of the north and east mainly north of 20°S and east of 26°E. Extends southward in the east to as far as the Tropic of Capricorn. May extend further west in high rainfall years at least as far as Orapa but not into semidesert. Not well known. Overlaps with the Cape Penduline Tit in the drier parts of its range. Usually in pairs or small groups.

Habitat

Broadleafed woodland, riverine *Acacia*, mixed *Acacia* tree and bush savanna. Often along drainage lines or in river valleys where there are tall trees and underlying grass or secondary growth.

Analysis

39 squares (17%) Total count 90 (0,05%)

ARROWMARKED BABBLER

Turdoides jardineii

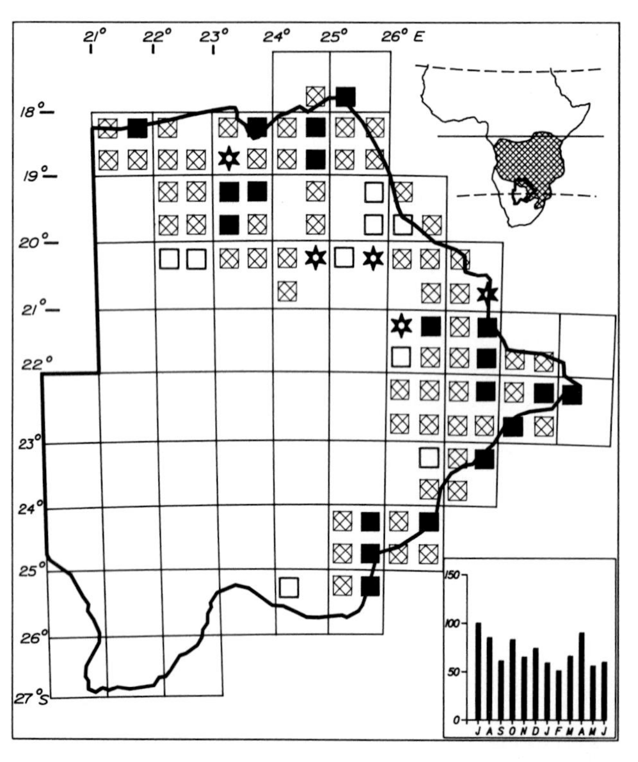

Status

Fairly common to locally very common resident of the north and east to as far south as Lobatse. Its range in Botswana represents the western limit of its range in southern Africa. Overlaps considerably with the Pied Babbler and although they may occur in the same habitat, they occupy different niches. Its range stops abruptly west and south of seasonal rivers e.g. west of Kanye and south of the Boteti/Mopipi drainage. Egglaying October to March. Usually in groups of 5–15 birds.

Habitat

Dense vegetation and thickets in tree and bush savanna in high rainfall areas. Also in riparian forest, riverine bush and any woodland with thick cover.

Analysis

87 squares (38%) Total count 878 (0,49%)

BLACKFACED BABBLER

Turdoides melanops

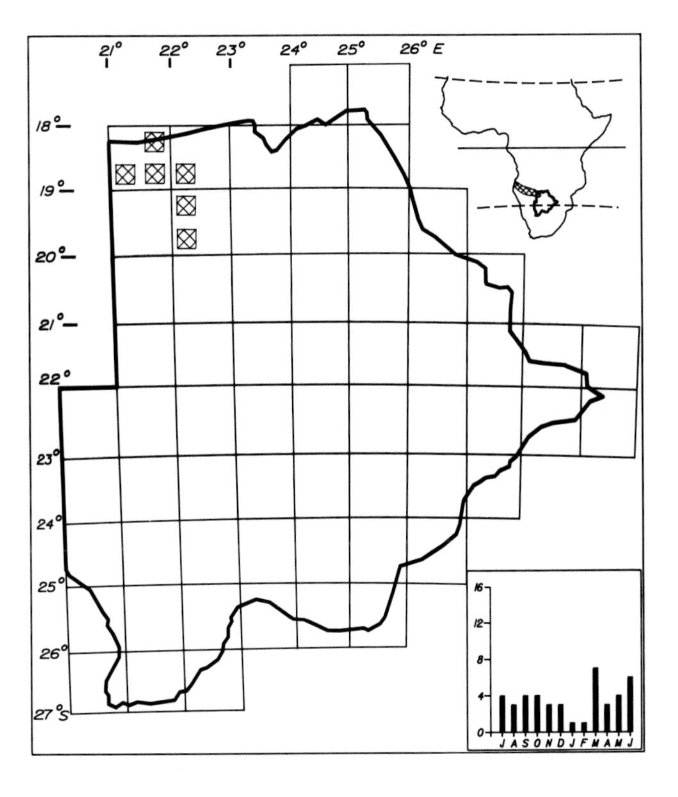

Status

Uncommon to locally common resident of the northwest—west of the Okavango Delta—where it occurs at the eastern and southern limit of its range in Namibia and Angola. It does not occur in the Okavango Delta although it reaches the western edge at Gomare and Nokaneng. Occupies a different niche from the three other babbler species occurring in the same area. Poorly known. Usually in groups of 5–15 birds.

Habitat

Long grass, thicket, creepers and secondary growth in the understorey of broadleafed and mixed *Acacia* woodland. Usually in areas where the trees are tall and often where the groundstratum is well separated from the canopy.

Analysis

6 squares (3%) Total count 44 (0,02%)

WHITERUMPED BABBLER

Turdoides leucopygius

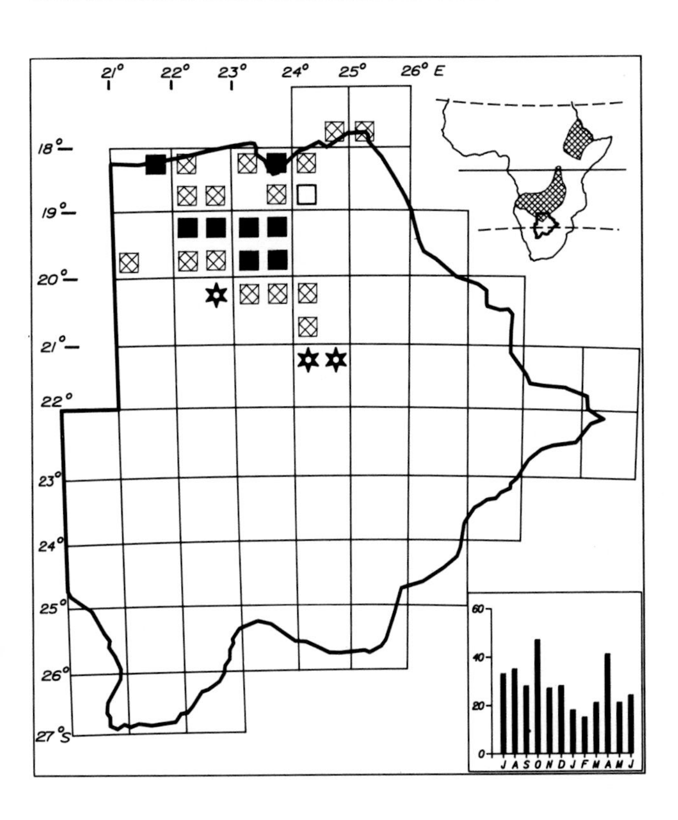

Status

Common to very common resident of the Okavango, Nhabe, Boteti, Linyanti and Chobe river systems. It has not been recorded in the southern parts of the Boteti River nor at Lake Ngami during this survey probably as a result of the drought. Its range is likely to expand into adjacent areas in high rainfall cycles. Usually in groups of 5–15 birds.

Habitat

Bushes, thickets and dense vegetation on floodplains and in perennial river valleys, usually between water and woodland. Thicket termite mounds along drainage lines, edges of riparian forest, tall dense vegetation on marshes and swamps, occasionally in reedbeds and papyrus.

Analysis

27 squares (12%) Total count 347 (0,19%)

PIED BABBLER

563

Turdoides bicolor

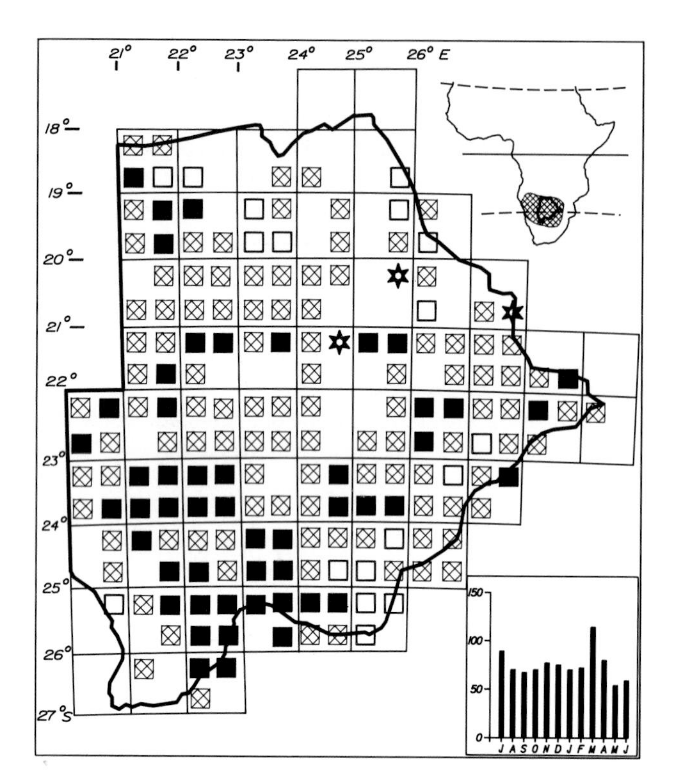

Status

Common to very common resident throughout the country except in the northeastern woodlands where it is mainly absent. It is the typical babbler of the Kalahari savannas and is very common in parts of the south and west. Often found alongside the Crimsonbreasted Shrike although no symbiosis is known nor suspected. Egglaying October to February. Usually in groups of 5–15 birds.

Habitat

Tree and bush savanna particularly in *Acacia* and other thorn vegetation. Usually where the trees and bushes are close together but will occur in open areas if there are intermittent thickets or patches of dense growth as on large anthills. Not found in scrub savanna nor in rich woodland in the north. Often in *Acacia mellifera* bushes in the west.

Analysis

176 squares (77%) Total count 949 (0,53%)

REDEYED BULBUL

567

Pycnonotus nigricans

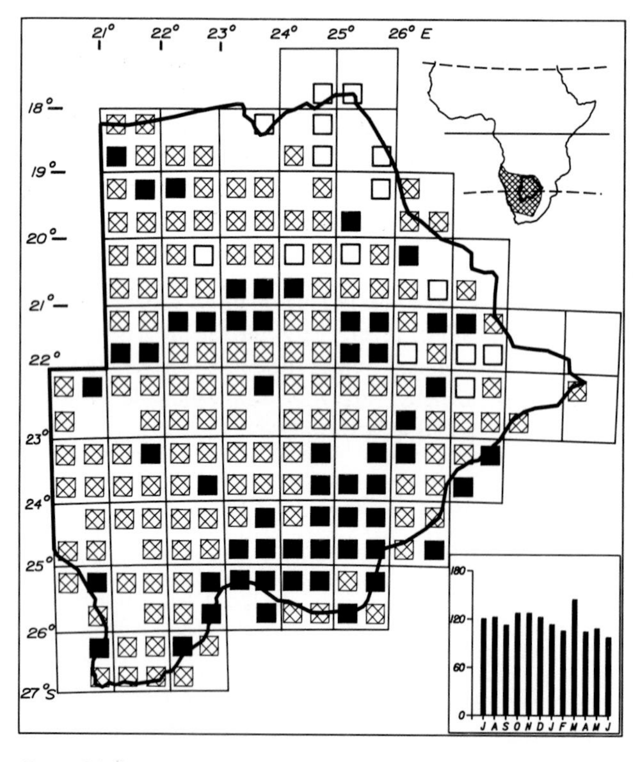

Status

Common to very common resident throughout the country. In the north and east it is partly replaced by the Blackeyed Bulbul which is predominant in these areas. Direct competition is likely but the two species occur alongside each other at times. Seasonal movements occur. Egglaying in all months from October to March. Usually solitary or in pairs.

Habitat

Tree and bush savanna in semidesert and arid regions but also in the high rainfall areas. Where it is not in competition with the Blackeyed Bulbul it can occupy lush habitats such as gardens, riparian vegetation, farmland and plantations but appears to be absent from these habitats within the range of the Blackeyed Bulbul.

Analysis

202 squares (88%) Total count 1460 (0,82%)

BLACKEYED BULBUL

Pycnonotus barbatus

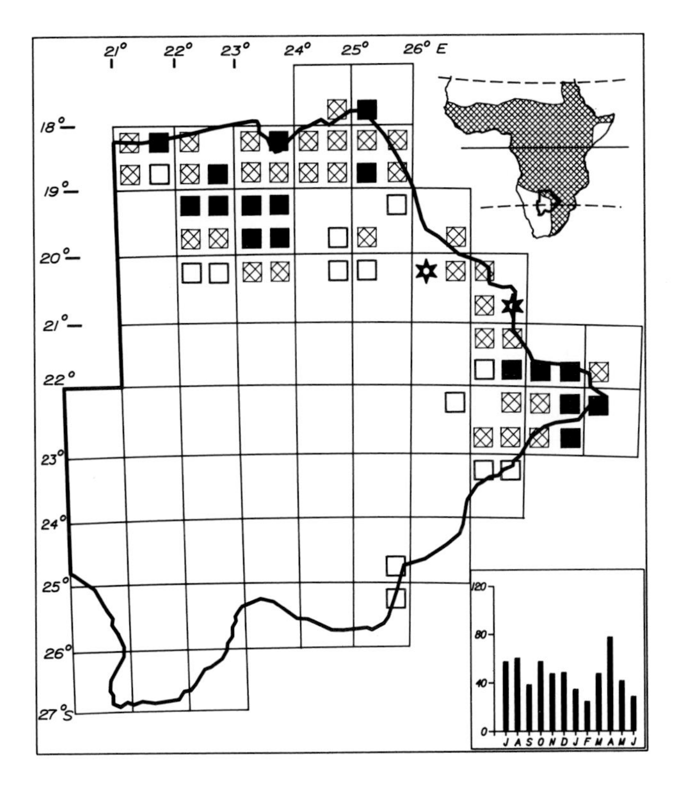

Status

Common to very common throughout its range in the north and east and sparse on the periphery of this range. Normally occurs as far south as the Tropic of Capricorn in the east. Rarely found in the southeast at Gaborone and Lobatse. Seasonal movements occur but these are poorly understood and need special study to record absences as well as presences. It outnumbers the Redeyed Bulbul in the high rainfall areas at all times of the year. Usually solitary or in pairs.

Habitat

Any woodland in high rainfall regions, tree and bush savanna, riparian forest, riverine bush, gardens, parks and cultivation.

Analysis

63 squares (27%) Total count 575 (0,32%)

TERRESTRIAL BULBUL

Phyllastrephus terrestris

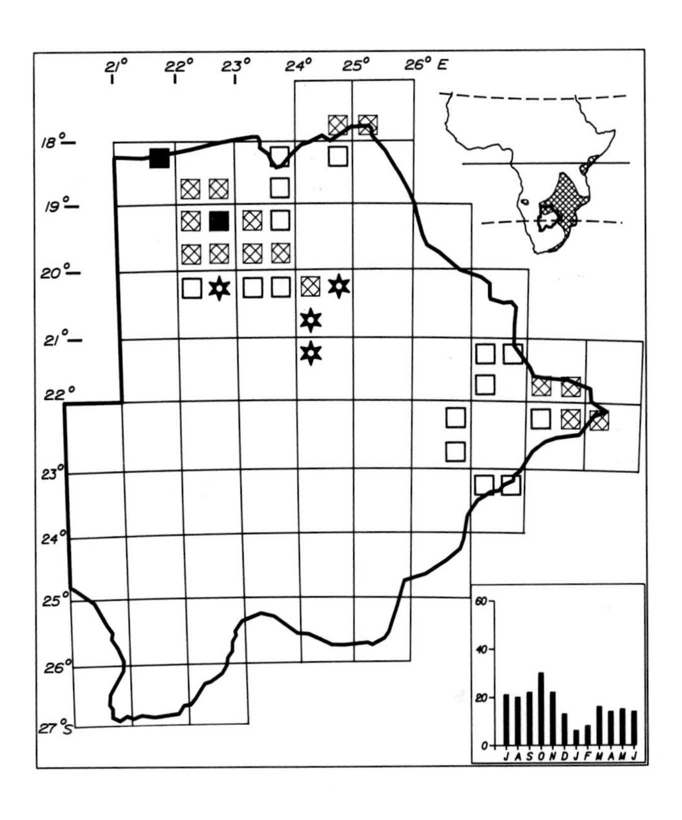

Status

Sparse to locally very common resident of the Okavango, Linyanti and Chobe river systems. Not recorded from the southern reaches of the Boteti River in this survey. Sparse along the smaller rivers of the east but common near the Shashe/Limpopo confluence. Reaches as far south as Martin's Drift. Usually in small parties of 4–10 birds.

Habitat

Groundstratum and midstratum of riparian forest or dense riverine vegetation, usually where there is leaf litter.

Analysis

36 squares (16%) Total count 209 (0,12%)

YELLOWBELLIED BULBUL

574

Chlorocichla flaviventris

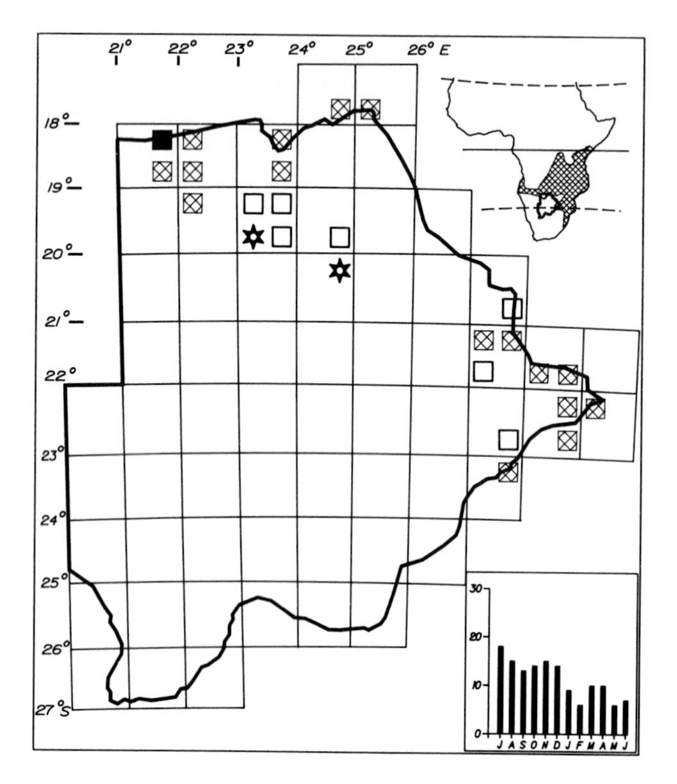

Status

Sparse to fairly common resident of the Okavango, Linyanti and Chobe river systems in the north and the Shashe, Tati, Motloutse and Limpopo rivers in the east. Its range may be contracting with the loss of riparian vegetation due to human interference in the east. It should be considered as a vulnerable species in Botswana. Locally common at Shakawe, Linyanti Camp and Kasane. Seasonal movements may occur. Usually in small groups of 4–10 birds.

Habitat

Rich and tall stands of riparian evergreen forest even when this habitat is present along seasonal rivers and may be only 10 m wide. In areas of permanent water it utilises bushes, thickets and secondary growth on the edges of forest and on moist floodplains.

Analysis

26 squares (11%) Total count 145 (0,08%)

KURRICHANE THRUSH

576

Turdus libonyana

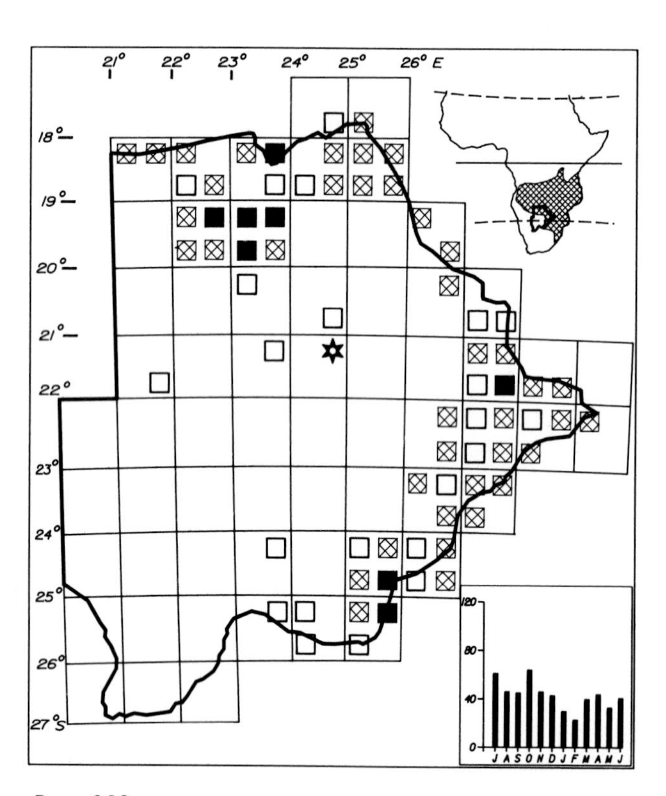

Status

Fairly common to very common resident of the north. Extends sparsely south along the Boteti River and recorded once from Deception Valley and Ghanzi. In the east it is sparse to locally very common to as far south as Phitsane Molopo. Rare along the Molopo River to Bray and recorded once near Mabutsane (2423B). Egglaying November, December and January. Usually solitary or in pairs.

Habitat

Broadleafed and mixed *Acacia* woodland including tall mopane but absent from mopane scrub. Usually in woodland with ground cover and bare patches of soil. In the north it is common in miombo and *Baikiaea* woodland. In the east it occurs commonly in *Acacia/Commiphora* woodland in valleys.

Analysis

72 squares (31%) Total count 540 (0,3%)

OLIVE THRUSH

577

Turdus olivaceus

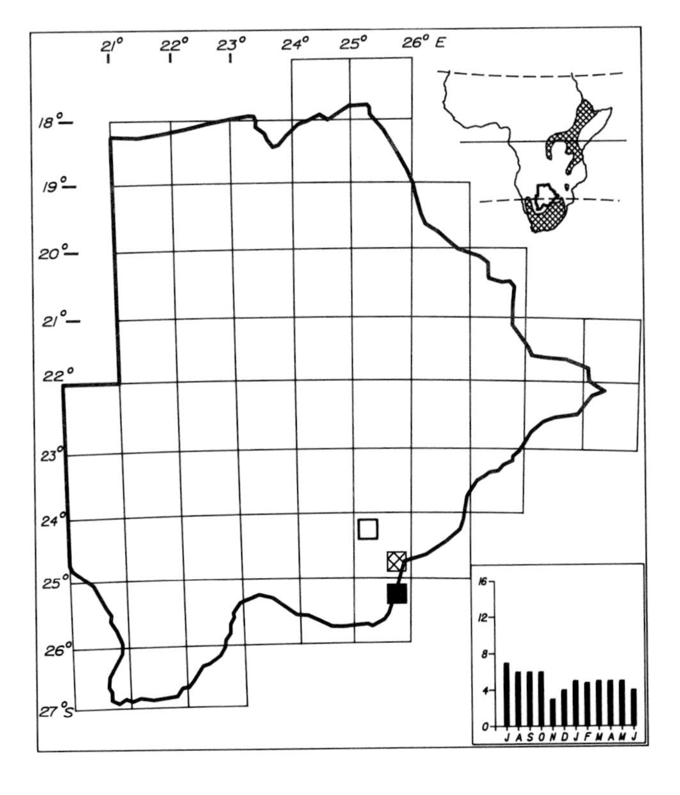

Status

Common resident of the southeast in the Gaborone and Lobatse areas. Recorded once northwest of Molepolole. Occurs in Botswana at the northern limit of its South African range. Appears to have extended its range since 1976—not recorded by Irvine and Beesley (1976)—as a result of the development of gardens with tall trees and exotic vegetation. Likely to extend as far west as Kanye. Egglaying October and November. Usually in pairs.

Habitat

Dense vegetation and shade in gardens, woodland and riverine bush and trees.

Analysis

3 squares (1%) Total count 61 (0,03%)

GROUNDSCRAPER THRUSH

580

Turdus litsitsirupa

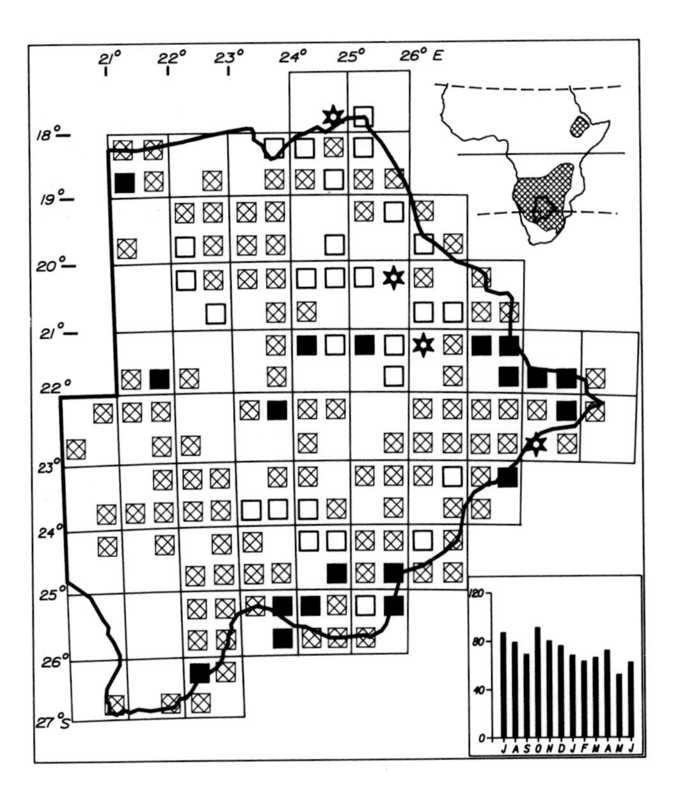

Status

Fairly common to locally very common resident throughout the country except in the southwest dunelands. Its status there and north of Ghanzi needs further study. Egglaying November to February. Seasonal movements occur which are not clearly defined. Usually solitary or in pairs.

Habitat

Any woodland where there is short grass groundcover or bare ground. Tree and bush savanna. Usually in open areas with scattered clumps of bushes and tall trees.

Analysis

150 squares (65%) Total count 898 (0,5%)

SHORTTOED ROCK THRUSH

583

Monticola brevipes

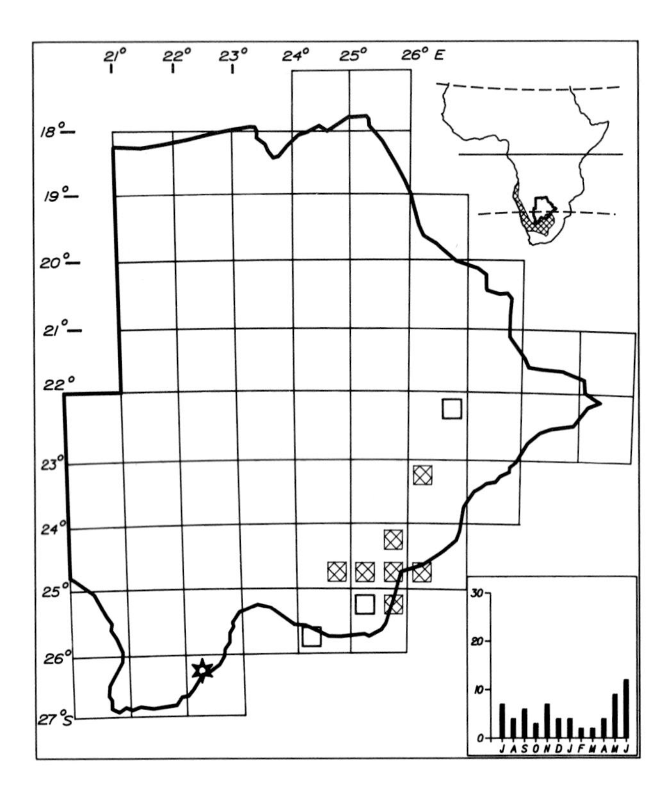

Status

Sparse to fairly common resident of the southeast occurring at the northeastern limit of its southern African range. Regular as far north as Molepolole and west to hills east of Jwaneng. Its status at Serowe and the Shoshong Hills needs further research as both areas have suitable rocky hills but only one record each. Smithers (1964) recorded it regularly at Tshabong but it has not been recorded there during this survey. Seasonal movements occur. Egglaying September. Usually solitary, occasionally in pairs.

Habitat

Rocky ridges, rock outcrops, stony ground in hilly country, on escarpments and in valleys, usually in the vicinity of light woodland or open tree savanna. Often perches on telephone and fence poles and on telephone wires.

Analysis

11 squares (5%) Total count 72 (0,04%)

MIOMBO ROCK THRUSH

584

Monticola angolensis

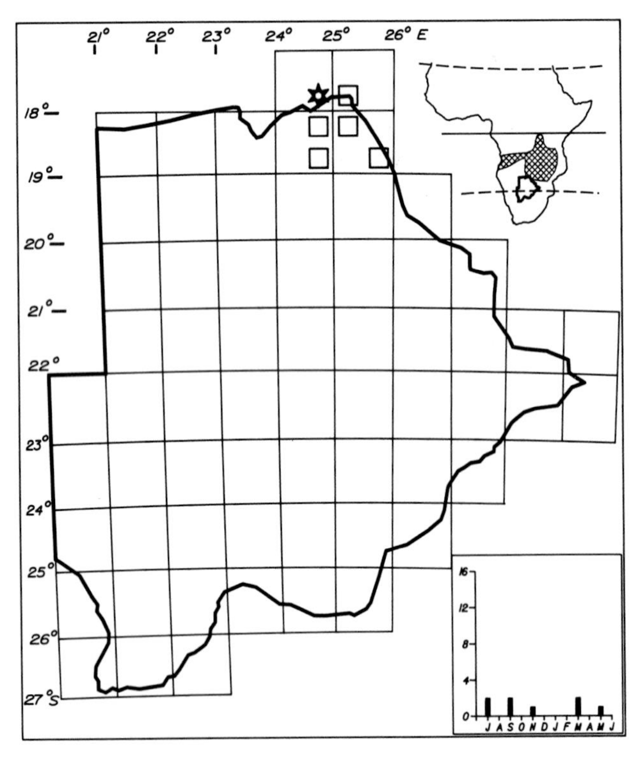

Status

Sparse resident of the northeastern woodlands. Easily overlooked and poorly known. The few available records suggest that it is resident throughout the year. Occurs in Botswana as a western extension of its range in Zimbabwe and at the southern limit of its continental range. Usually solitary or in pairs.

Habitat

Miombo and *Baikiaea* woodland particularly near hills and rock ridges. Also in light woodland on shallow soils.

Analysis

6 squares (3%) Total count 8 (0,004%)

CAPPED WHEATEAR

Oenanthe pileata

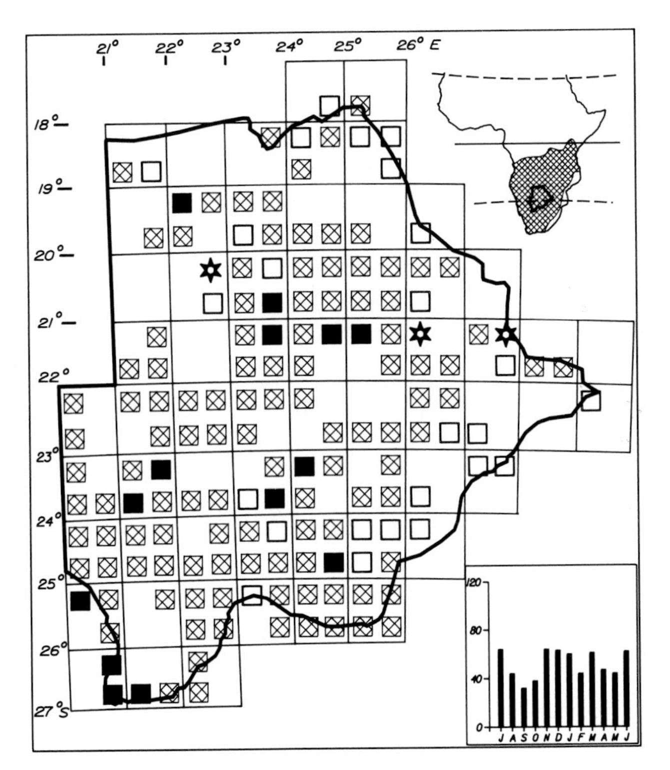

Status

Fairly common to locally very common intra-African migrant throughout the country. Two international population movements are suspected. The first arrives from the north in November and remains until April or May. The second, which peaks in June and July, is a winter migration from the southwest of the continent to Zimbabwe to breed. However, some birds may be resident and all populations may breed. Present in all months. Solitary or in pairs.

Habitat

Open bare ground or short grass on pans, dry floodplains, plains, airfields, cultivation, open tree and bush savanna. Occasionally on recently burnt ground. Predictable on most Kalahari pans when dry.

Analysis

152 squares (66%) Total count 665 (0,37%)

FAMILIAR CHAT

Cercomela familiaris

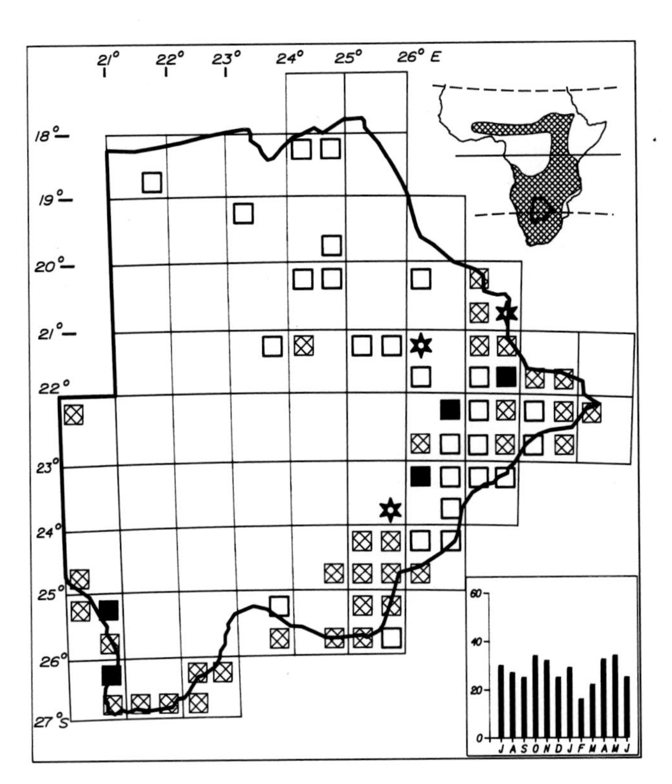

Status

Fairly common to locally very common resident of the hilly regions of the east and southeast from Maitengwe to Ramatlabama. Extends westwards along the Molopo River to the Nossob Valley where it is very common. Elsewhere it is sparse or uncommon in scattered localities such as Nxai Pan, Orapa, Letlhakeng, Deception Valley, Mamuno, Tsodilo Hills and Gcoha Hills. The subspecies *C.f. galtoni* replaces the usual subspecies *C.f. hellmayri* in the Nossob Valley. Egglaying September, October, December. Solitary or in pairs.

Habitat

Rocky or stony ground in hilly regions, lightly wooded rock or boulder outcrops, escarpments of dry river valleys, wooded gorges, small cliff faces, rock dumps (mines), towns and villages.

Analysis

67 squares (29%) Total count 360 (0,2%)

MOCKING CHAT 593

Thamnolaea cinnamomeiventris

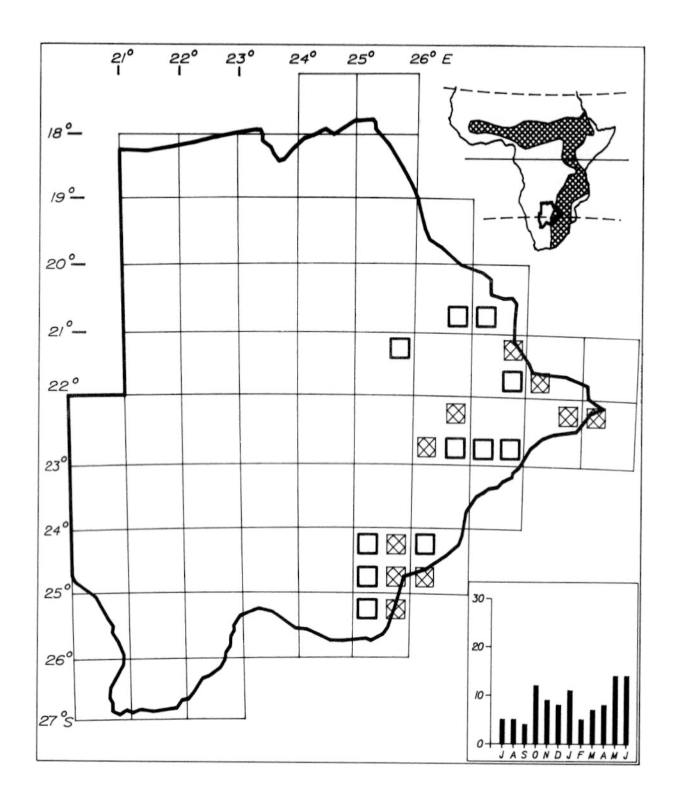

Status

Sparse to fairly common resident of the hilly areas of the east and southeast to as far north as Mosetse. Recorded once at Orapa. Occurs mainly in the granitic hills from Lobatse to Thamaga, Molepolole and Gaborone. Also on the Shoshong Hills, Tswapong Hills and boulder outcrops further north and east around Selebi Phikwe, Bobonong, and Francistown. Occurs at the western limit of its southern African range. Egglaying November. Usually solitary or in pairs, but 5 or 6 birds often occur together at the same site.

Habitat

Cliffs, gorges, rock faces and boulder outcrops usually near or in wooded areas.

Analysis

20 squares (9%) Total count 113 (0,06%)

ARNOT'S CHAT 594

Thamnolaea arnoti

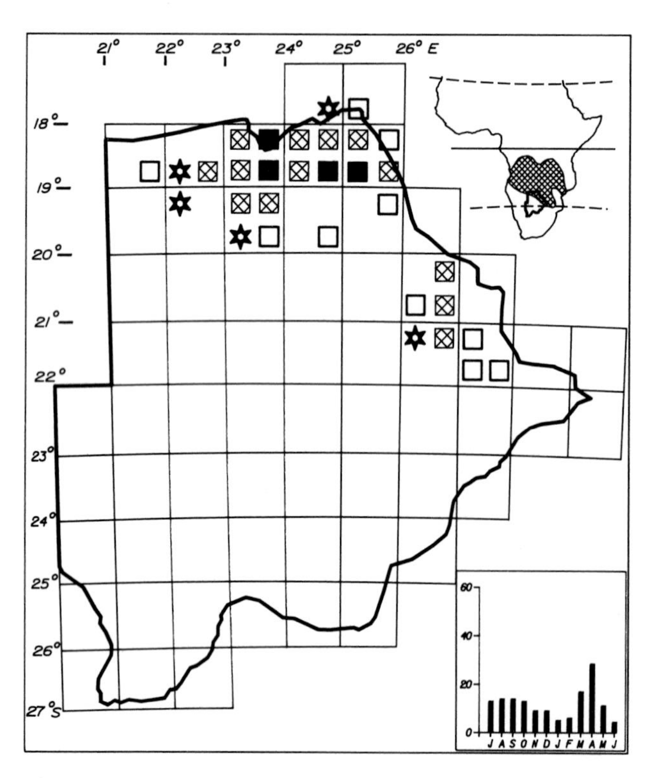

Status

Sparse to locally very common resident of the north and northeast to as far south as Selebi Phikwe. Extends westwards through the Okavango Delta to Nxamaseri where it is sparse. During this survey it was absent from Gomare and Sepopa from where it had previously been recorded. Its apparent absence between Nata and Mpandamatenga requires further investigation. Egglaying October. Usually in pairs or small groups.

Habitat

Mature broadleafed woodland, such as miombo, *Baikiaea* and mopane, where it occurs in midstratum. Its distribution appears to follow that of the mopane belts ending abruptly south of Maun and Selebi Phikwe, but it is very sparse in the poorer mopane as around Nxai Pan.

Analysis

32 squares (14%) Total count 149 (0,08%)

ANTEATING CHAT

Myrmecocichla formicivora

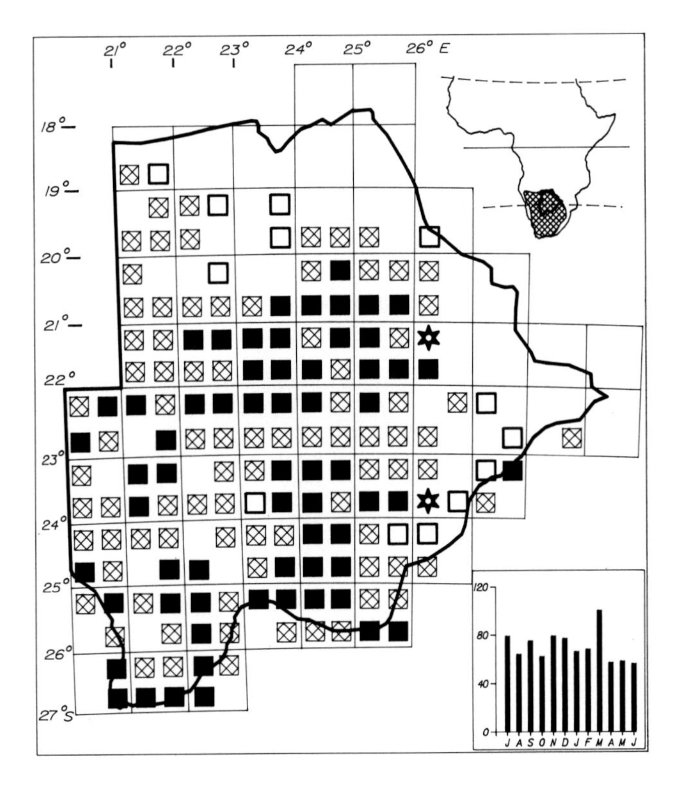

Status

Common to very common resident throughout the country south of 20°S and west of 28°E. Sparse to uncommon to the north and east of these parameters. It is one of the commonest and most predictable species of the Kalahari semidesert. Egglaying November and December. Usually in pairs on adjacent territories or in family groups of 5–10 birds after breeding.

Habitat

Open grassy areas in semidesert tree and bush savanna, semidesert scrub savanna, grassland plains, dry sparsely-grassed pans and low shrubs on ecotone of Kalahari pans. Nests in antbear burrows or in natural or manmade excavations in sandy or stony ground.

Analysis

163 squares (71%) Total count 909 (0,51%)

STONECHAT

Saxicola torquata

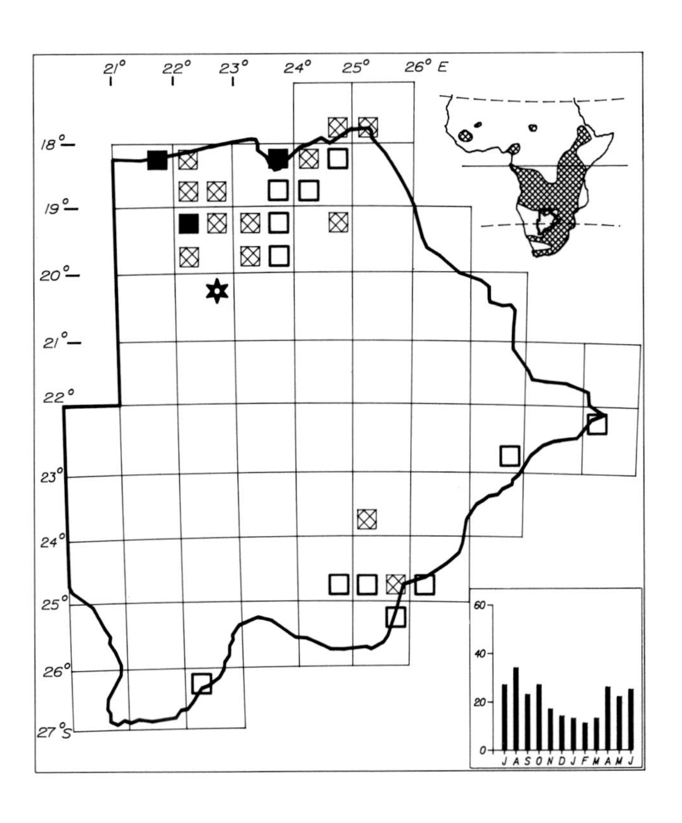

Status

Fairly common to locally very common resident of the Okavango, Linyanti and Chobe river systems. Very sparse along the Limpopo River in the east. Sparse to uncommon in the southeast to as far west as Jwaneng. Recorded once at Tshabong in the southwest. Local and seasonal movements occur with higher numbers in winter months. Egglaying September. Usually solitary, in pairs or family groups. Juveniles are sometimes mistaken for Whinchat.

Habitat

Rank grass and sedges on the margins of rivers, lakes, lagoons and dams. Long grass, bushes and shrubs on floodplains, pans, marshes and seasonally inundated grassland.

Analysis

29 squares (13%) Total count 261 (0,15%)

HEUGLIN'S ROBIN 599

Cossypha heuglini

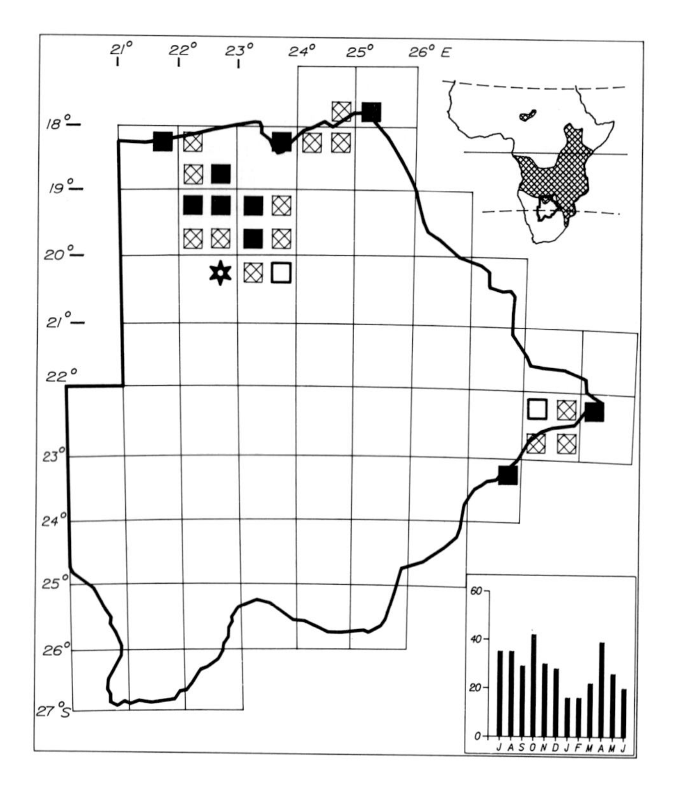

Status

Common to very common resident of the Okavango, Linyanti and Chobe river systems to as far south as Makalamabedi. It has not been recorded south along the Boteti River where suitable habitat exists. Reappears in the east along the Limpopo River to as far south as Martin's Drift. Occurs in Botswana at the southern and western limits of its continental range at these longitudes. Egglaying in all months from October to January. Solitary or in pairs.

Habitat

Dense undergrowth in riparian forest, bushes and thickets in riverine vegetation and adjacent woodland and in gardens near perennial rivers.

Analysis

25 squares (11%) Total count 343 (0,19%)

WHITETHROATED ROBIN 602

Cossypha humeralis

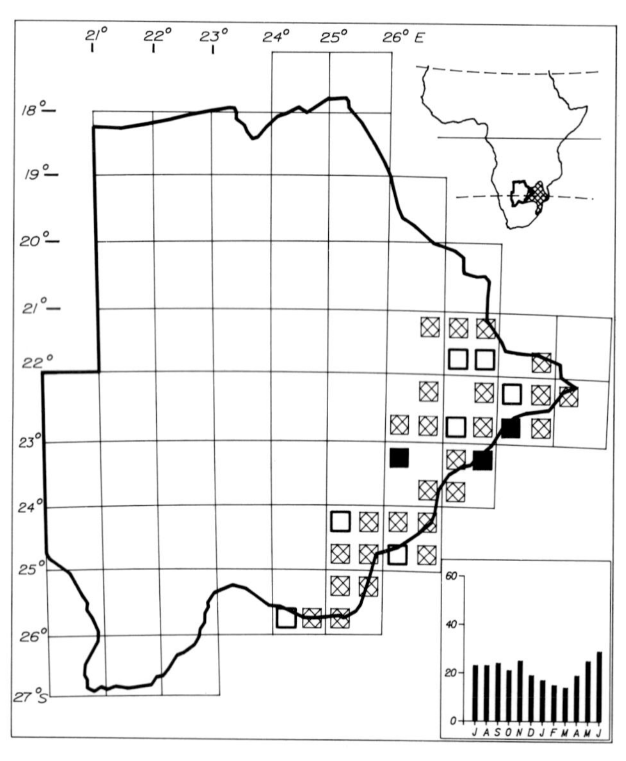

Status

Fairly common to locally very common resident of the east and southeast from Francistown to Phitsane Molopo. Extends west along the Molopo River to as far as Boshoek. Not recorded north of Francistown. Occurs in Botswana at the western limit of its range in southern Africa. Egglaying December. Solitary or in pairs.

Habitat

Acacia bushes and thorn thicket in river valleys, dense scrub vegetation on valley slopes, riverine thicket, riparian forest, gardens.

Analysis

34 squares (15%) Total count 266 (0,15%)

THRUSH NIGHTINGALE

609

Luscinia luscinia

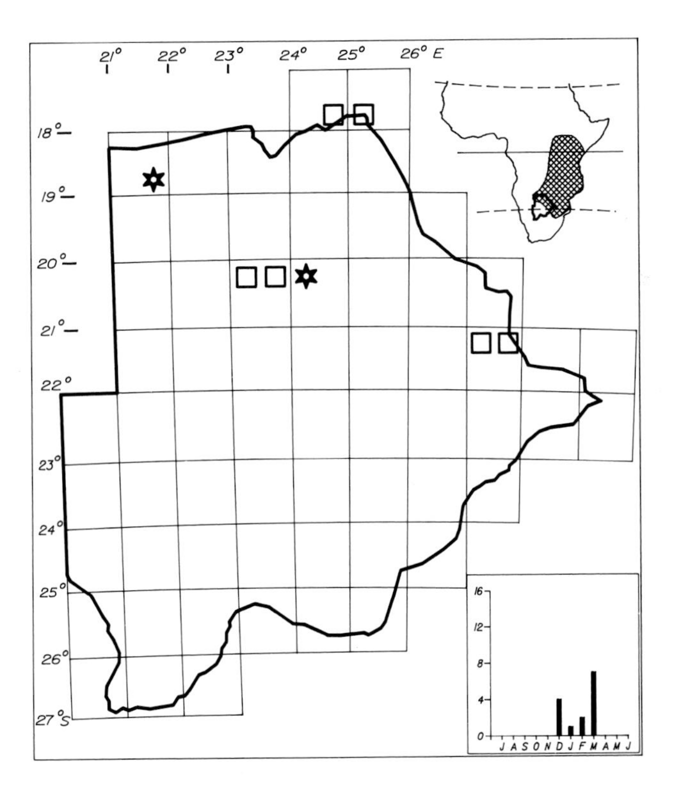

Status

Very sparse Palaearctic migrant recorded at isolated localities in the north to as far south as Francistown. Recorded between December (16th) and March (16th) with most records in March. Poorly known. All the recent records are from vegetation in river valleys—along the Chobe River at Kasane and Serondella, along the Thamalakhane River south of Maun, along the Boteti River near Makalamabedi and along the Shashe and Tati rivers at Francistown. Solitary.

Habitat

Riverine bush, thicket and shrubs in river valleys.

Analysis

8 squares (3%) Total count 17 (0,009%)

BOULDER CHAT

610

Pinarornis plumosus

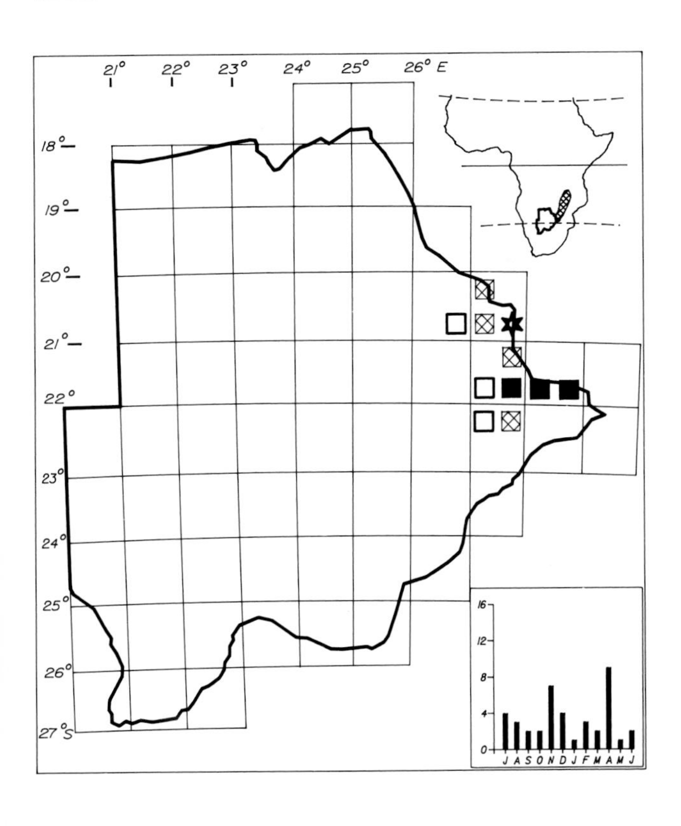

Status

Sparse to locally very common resident in the east recorded between Mosetse and Tutume in the north and Selebi Phikwe, Bobonong and Mogapi in the southern part of its range. It has not been recorded in the nearby Tswapong nor Shoshong hills. It occurs in Botswana at the southwestern tip of its restricted range in eastern and south-central Zimbabwe. Usually in small groups.

Habitat

Lightly-wooded rock outcrops and granite boulders where it occurs on rock under the canopy of trees. Occurs at isolated outcrops rather than on hills. Its specialised habitat is vulnerable to tree felling and it has disappeared from one known site for this reason. Typical habitat is found at Selebi Phikwe Golf Course.

Analysis

11 squares (5%) Total count 41 (0,02%)

WHITEBROWED SCRUB ROBIN

Erythropygia leucophrys

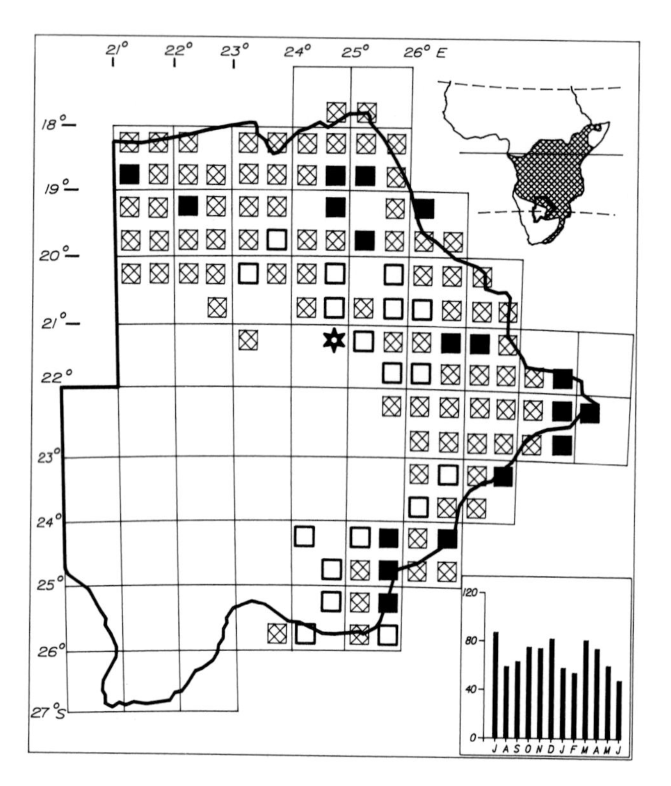

Status

Common to very common resident of the north and east. From the north it extends as far south as Tale Pan, Deception Valley and Makoba. From the east and southeast it extends as far west as the sources of easterly flowing rivers. It extends west along the Molopo River to as far as Medenham. Its range overlaps extensively with that of the Kalahari Scrub Robin in the north and east. They are not mutually exclusive but occupy different niches and sometimes occur alongside each other. Egglaying December. Solitary or in pairs.

Habitat

Bush and tree savanna in the high rainfall regions, usually in places with tall bushes, thickets and long grass. Also in all types of woodland with an understorey of dense vegetation and in riverine bush.

Analysis

115 squares (50%) Total count 829 (0,46%)

KALAHARI SCRUB ROBIN

Erythropygia paena

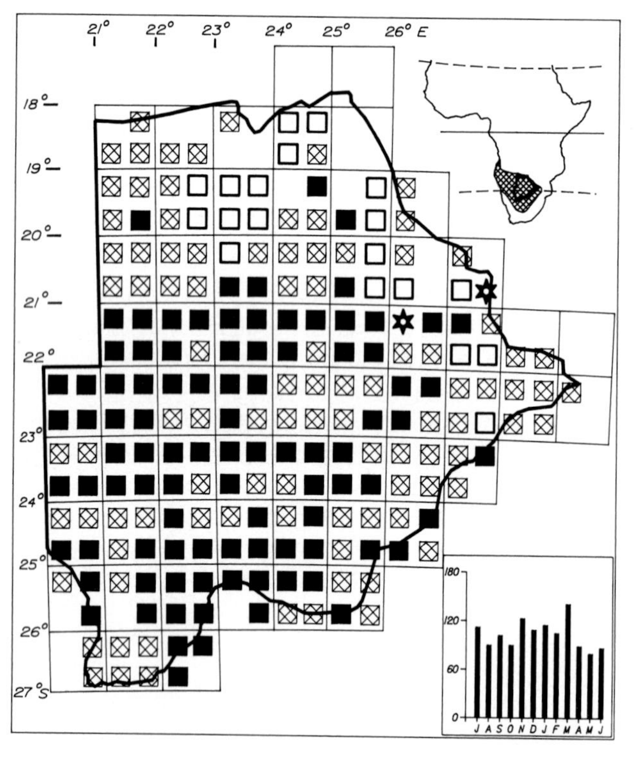

Status

Common to very common resident in all regions of the country. It is very common south of 21°S and west of 28°E, sparse to fairly common in the northeast, sparse in the Okavango Delta and fairly common in the northwest. It is a typical species of the Kalahari savannas and one of the commonest species in this habitat. It occupies a more open and arid niche than the Whitebrowed Scrub Robin. Egglaying July and all months from September to December. Solitary or in pairs.

Habitat

Open bush and tree savanna on Kalahari sand, edges of *Acacia* and mixed woodland, scrub savanna, scrub on the ecotone of Kalahari pans, dry riverine bush along seasonal drainage lines. In the high rainfall regions it often occurs in open areas with scattered bushes and trees and sparse ground cover. Often perches conspicuously on top of bushes and small trees.

Analysis

211 squares (92%) Total count 1294 (0,72%)

BEARDED SCRUB ROBIN

617

Erythropygia quadrivirgata

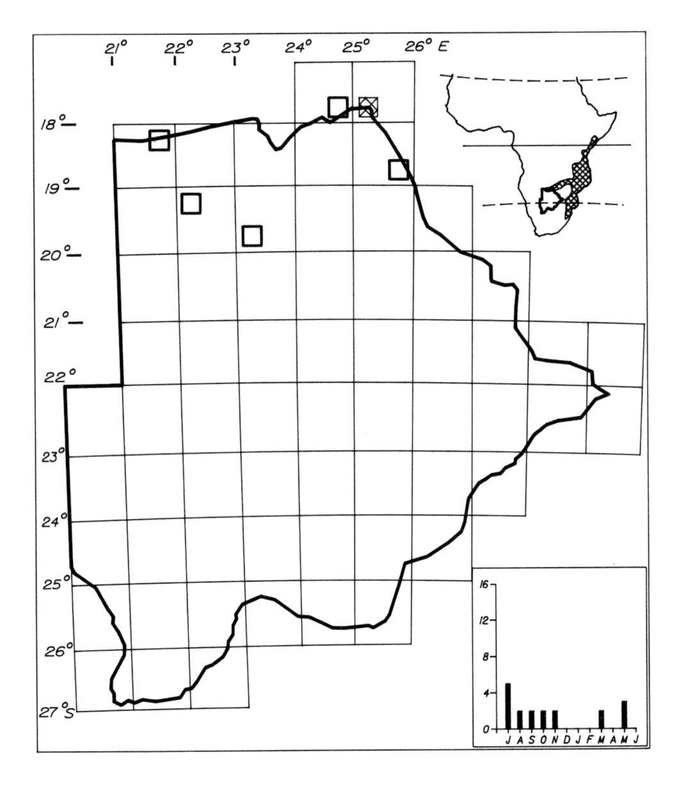

Status

Very sparse to uncommon resident north of 20°S. Most records are from Kasane and Shakawe. Poorly known and easily overlooked. In the northeast it has been recorded once as far south as Mpandamatenga. It is rare in the Okavango Delta as far south as southern Moremi from where there is also a past record. Usually solitary or in pairs.

Habitat

Thickets and tangled growth on the edges of forest and along the margins of rivers and streams.

Analysis

6 squares (3%) Total count 20 (0,011%)

GARDEN WARBLER

619

Sylvia borin

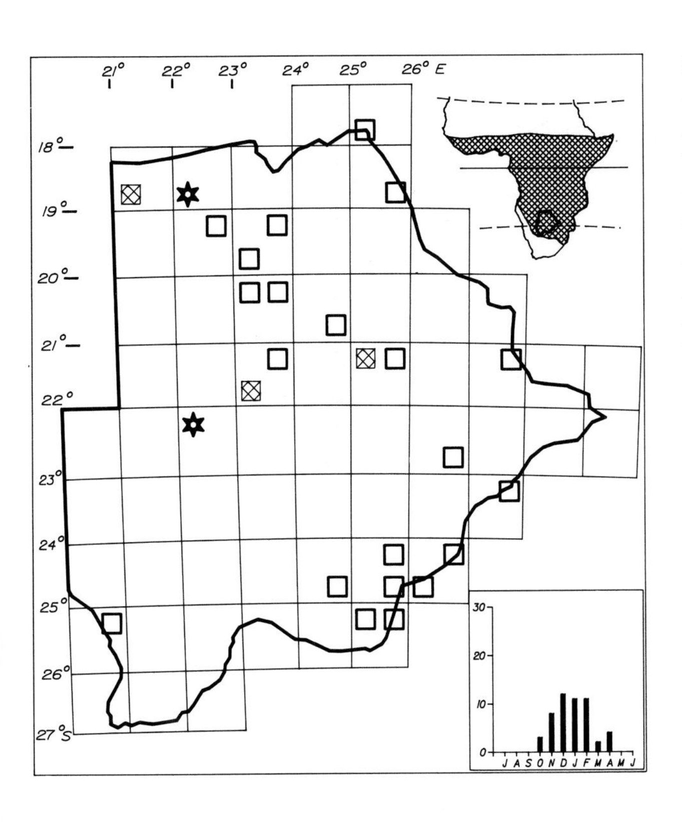

Status

Sparse to uncommon Palaearctic migrant mainly to the north and east but may occur in suitable habitat throughout the country. Most records are from Moremi, Orapa and Gaborone. Present from October to April (25th) and commonest between December and February. In the western half of the country it occurs at the southern limit of its migration at these longitudes. Usually solitary.

Habitat

Broadleafed woodland, riverine vegetation, tall thickets, riparian forest. Also in well-grown secondary vegetation near dams and in gardens as at Orapa, Kanye, Lobatse and Gaborone.

Analysis

26 squares (11%) Total count 57 (0,03%)

WHITETHROAT

620

Sylvia communis

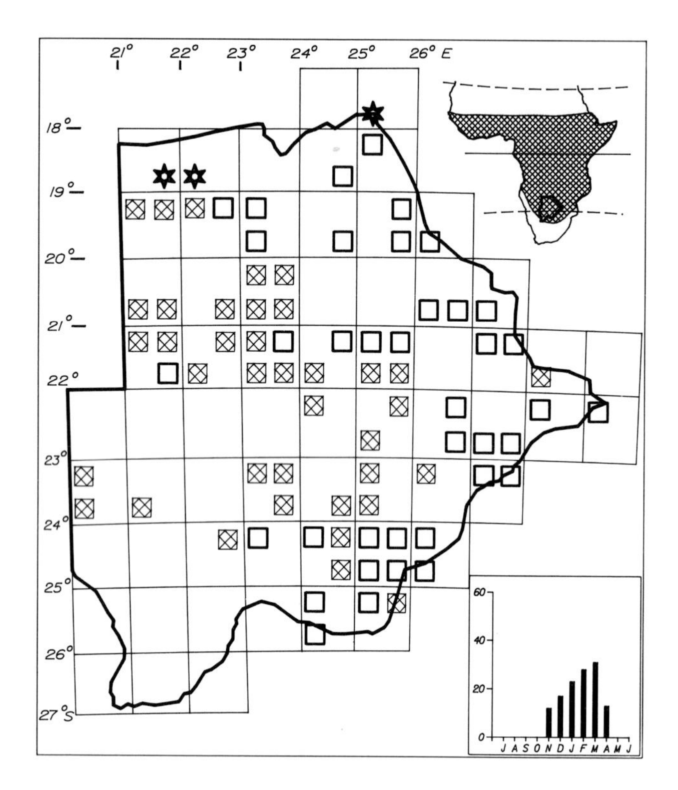

Status

Sparse to fairly common Palaearctic migrant to all areas except the southwest and occurring near the southern limit of its continental migration. Present from November to April (25th) but mainly from January to March. Widely and thinly dispersed except on northward passage when it is more common. There were 30 birds over 3 km² at Groot Laagte (2021C) in March and smaller congregations in several other areas at that time of year. Usually solitary.

Habitat

Deciduous thicket particularly *Acacia* and riverine, scrub and low bushes in semidesert, bush and tree savanna with dense vegetation. Commonest in the lower rainfall areas.

Analysis

79 squares (34%) Total count 134 (0,075%)

TITBABBLER

621

Parisoma subcaeruleum

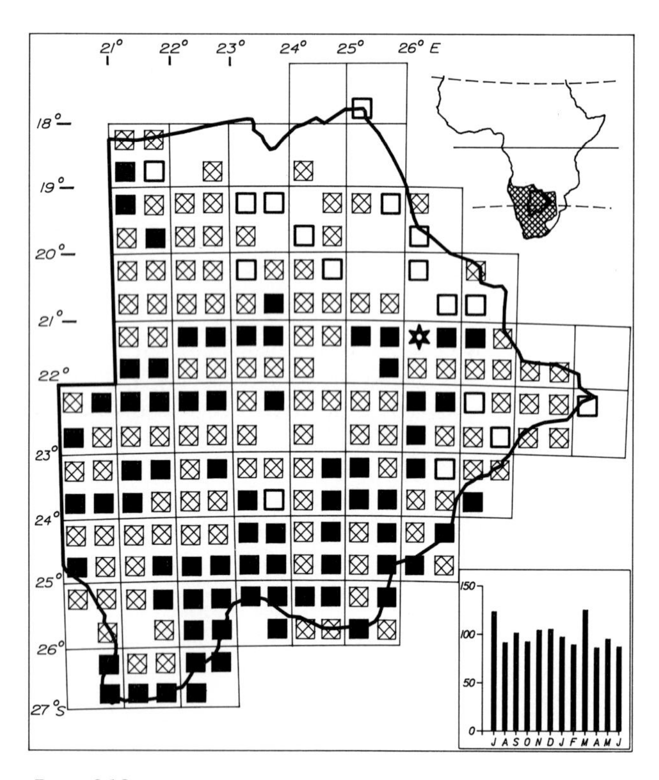

Status

Very common resident south of 21°S and west of 28°E. Sparse and uncommon in the northeast and fairly common to locally very common in the northwest. Recorded as very common in 31% of squares throughout the country. One of the most predictable species of the Kalahari. Egglaying September to February. Solitary or in pairs.

Habitat

Thorn tree and bush savanna, edges of deciduous woodland, semidesert with scattered bushes but rare where there is only low scrub. Also riverine *Acacia* and bush.

Analysis

199 squares (87%) Total count 1251 (0,7%)

ICTERINE WARBLER

625

Hippolais icterina

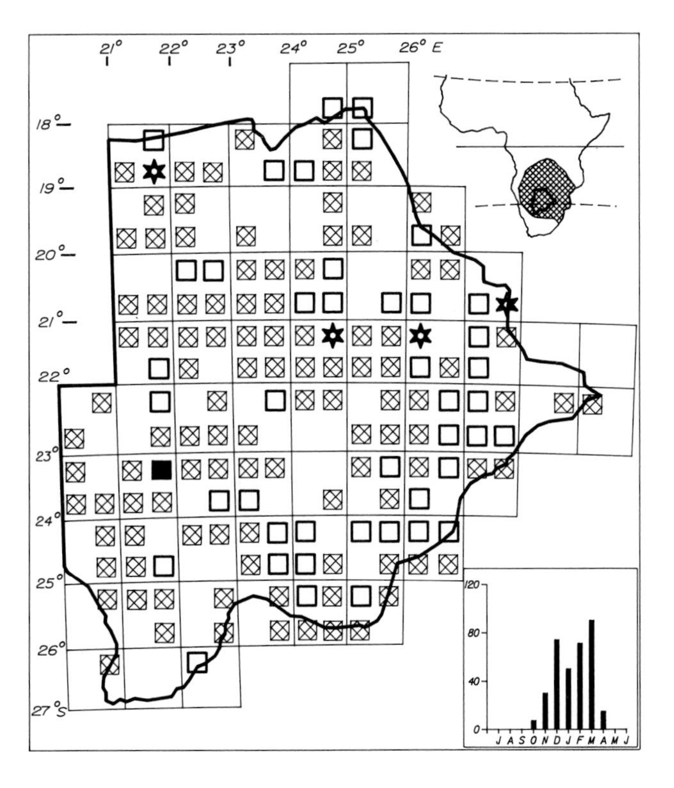

Status

Fairly common to common Palaearctic migrant throughout. Few birds arrive in late October but most do so in November. Common and widespread from December to March. Departure in April. Numbers fluctuate from year to year. Usually solitary, but loose groups in adjacent trees and bushes occur and several birds in one tree or bush is not uncommon.

Habitat

Mostly in *Acacia* trees and bushes in which it is more predictable than the Willow Warbler. Less commonly in mixed tree and bush savanna, but regular in riparian *Acacia* and semidesert bush. Often in or near woodland patches or tall trees (8 m or more).

Analysis

154 squares (67%) Total count 353 (0,19%)

OLIVETREE WARBLER

626

Hippolais olivetorum

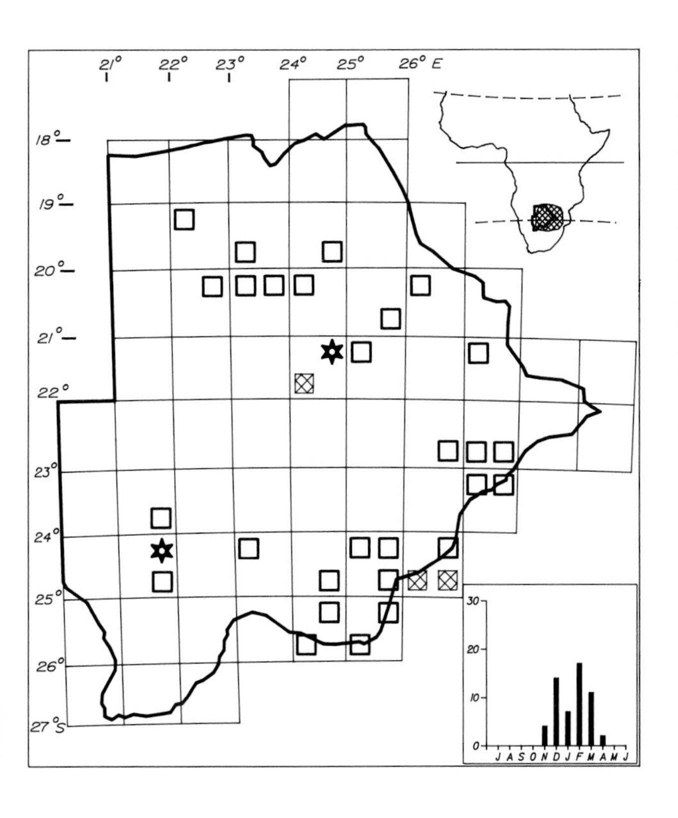

Status

Sparse to uncommon Palaearctic migrant mainly from December to March. Earliest date 21 November and latest 2 April. Easily overlooked as it is very skulking and difficult to see even when calling in a bush. The number of records suggest that Botswana is an important wintering area for this species which has a small world population. Usually solitary, but four birds were seen in adjacent bushes at Nxai Pan in November.

Habitat

Acacia thicket especially *A. mellifera*. Also other thorn and deciduous thicket in Kalahari savanna and around lake shores as at Lake Ngami or along dry river beds as at near Kokong. Usually where thicket or dense vegetation occurs over an extensive local area.

Analysis

33 squares (14%) Total count 58 (0,03%)

GREAT REED WARBLER

Acrocephalus arundinaceus

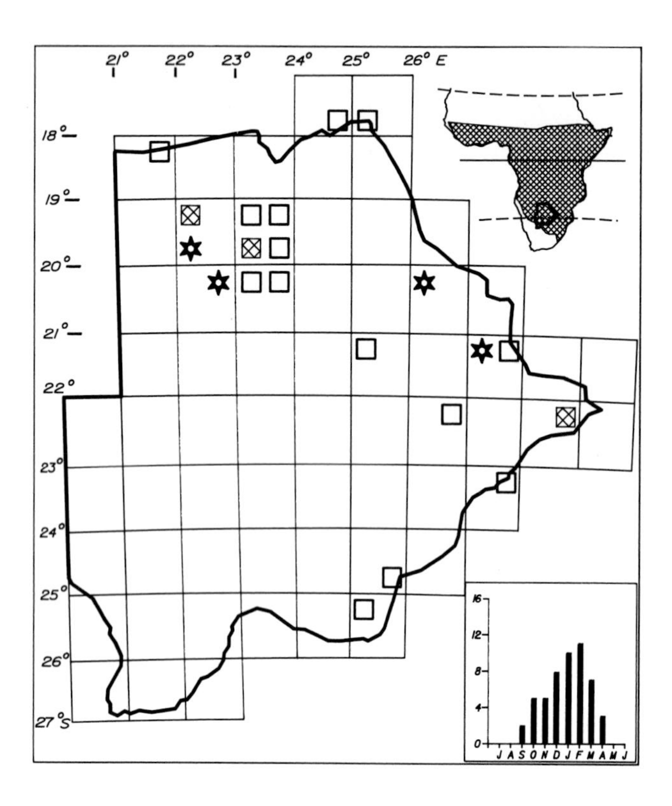

Status

Sparse Palaearctic migrant recorded only in the north and east during the drought period of this survey and may well only occur in these regions but more commonly in years of good rainfall. Usually arrives mid-October but is commonest from December to March. Earliest date 26 September, latest date 4 April. Numbers probably fluctuate from year to year. Usually solitary.

Habitat

Reedbeds and rank vegetation near water, in river valleys and rarely in gardens. There is less suitable habitat in Botswana for this species than for the Olivetree Warbler and it is likely to be less common except in the northern wetlands. The habitats of the two are totally different although their calls can be confused.

Analysis

21 squares (9%) Total count 54 (0,03%)

AFRICAN MARSH WARBLER

Acrocephalus baeticatus

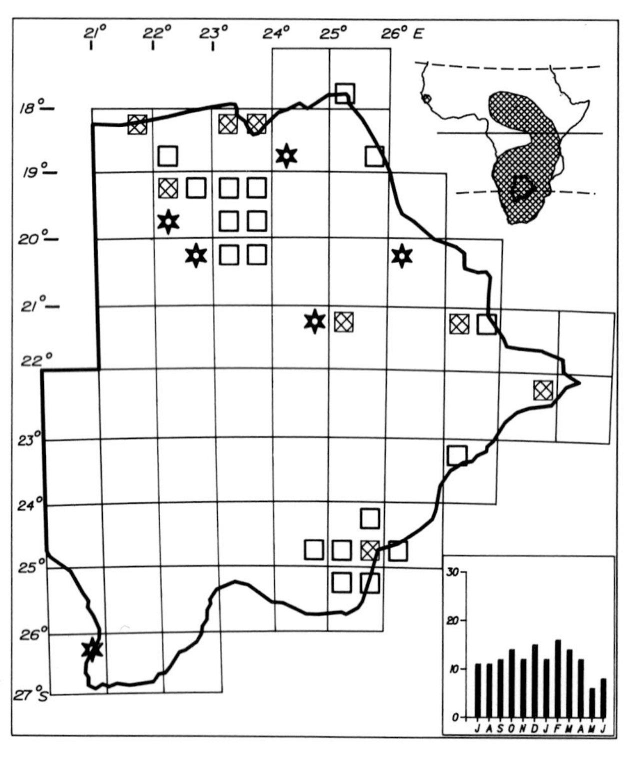

Status

Sparse to locally fairly common resident of wet habitats in the north and east. Probably more common than records suggest. Occurs throughout the year but records are fewer in May and June when there is little vocalisation. There is an old record from the Nossob Valley where it may occur in high rainfall years. Egglaying December and January. Solitary or in pairs.

Habitat

Reedbeds and tall rank vegetation on marshes, on the edges of rivers, lakes, lagoons and dams. Manmade water impoundments have increased the available habitat in the southeast.

Analysis

32 squares (14%) Total count 154 (0,09%)

EUROPEAN MARSH WARBLER

633

Acrocephalus palustris

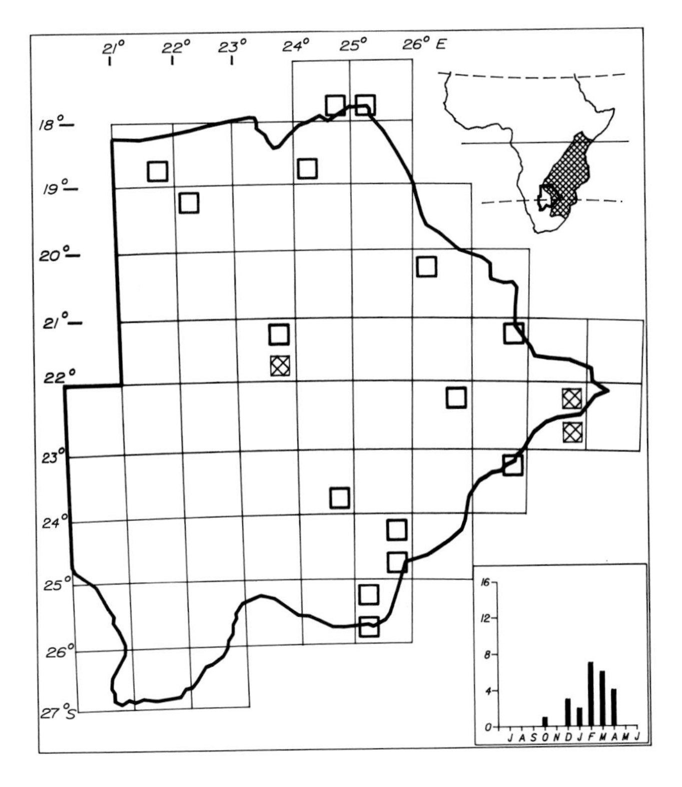

Status

Sparse to uncommon Palaearctic migrant not recorded in Botswana prior to this survey. Present from December to early April (4th) but with one record from late October. Most records fall in February and March either because of a northward passage or because it is more vocal at this time. Many of the records have been accepted by confirmation with taperecording. Usually solitary.

Habitat

Dense vegetation such as bushes and thickets. Appears to be confined to river valleys but does not require wet habitat. On the contrary most records are from vegetation away from water but it does occur in dry riparian thickets and in gardens. Two records are from Deception Valley—a fossil valley in the central Kalahari.

Analysis

18 squares (8%)   Total count 23 (0,013%)

EUROPEAN SEDGE WARBLER

634

Acrocephalus schoenobaenus

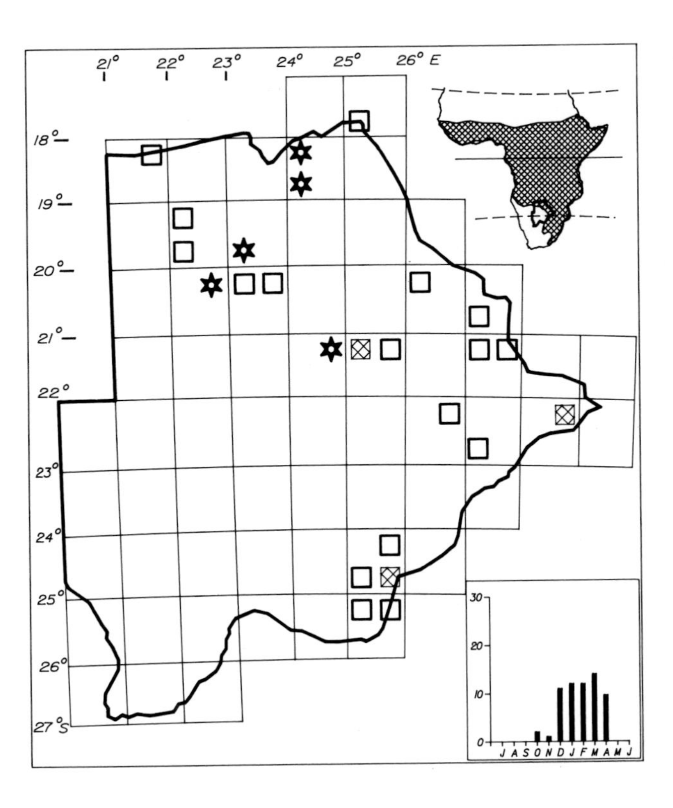

Status

Sparse to uncommon Palaearctic migrant to the north and east. Present mainly from December to April—earliest date 19 October, last date 8 April. Very localised in wet habitat. Numbers are small and there is no obvious passage movement. Not well known in Botswana. Often solitary, but 2–5 birds may be present at one locality.

Habitat

Reedbeds and rank aquatic growth on the edges of rivers, lakes, lagoons, dams and sewage ponds. Also in tall vegetation in marshes and swamps. Mostly skulks at the base of vegetation. Nearly always in moist conditions but will remain at a site which has recently dried out.

Analysis

25 squares (11%)   Total count 75 (0,04%)

LESSER SWAMP WARBLER

635

Acrocephalus gracilirostris

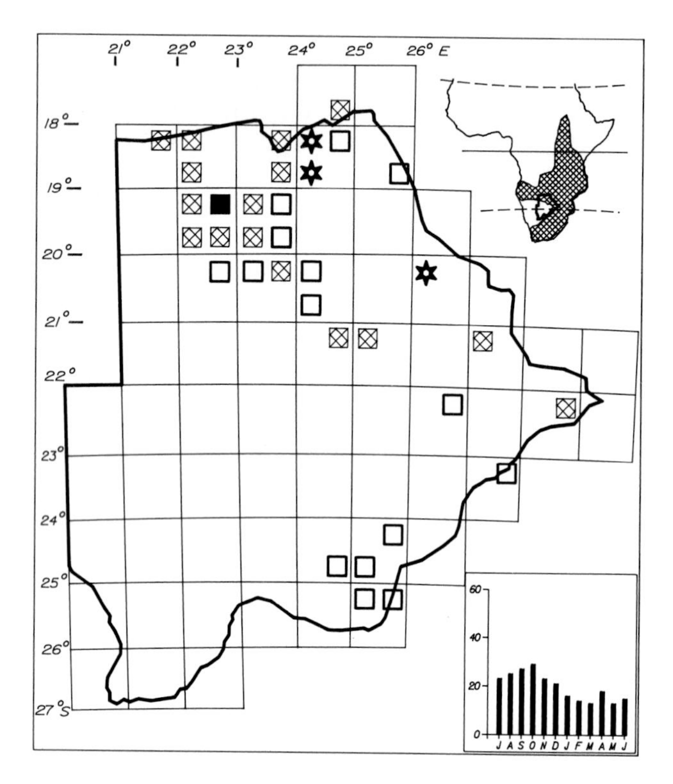

Alternative name

Cape Reed Warbler

Status

Common resident of the Okavango Delta and at Linyanti and Serondella. Sparse to fairly common on the periphery of these areas, along the Boteti River and in the east and southeast. Some local movements occur but it is mainly sedentary with pairs utilising the same territory for breeding in consecutive years if conditions remain suitable. Nestbuilding November. Solitary or in pairs.

Habitat

Tall reeds, rushes and sedges along the margins of rivers, lakes, marshes, lagoons and dams. In similar vegetation in swamps where it is also found in papyrus to a lesser extent than the Greater Swamp Warbler.

Analysis

35 squares (15%) Total count 247 (0,14%)

GREATER SWAMP WARBLER

636

Acrocephalus rufescens

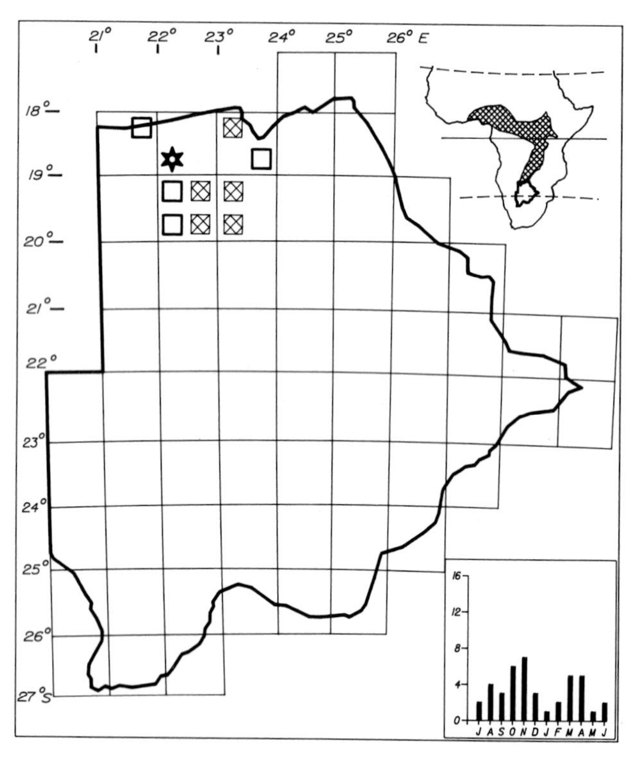

Status

Sparse to locally fairly common resident of the Okavango Delta and on the Kwando River and Savuti Channel swamps. These are the only known localities in southern Africa where the species occurs. It has been poorly studied and may be more common than present records indicate. Egglaying November. Solitary birds usually seen or heard.

Habitat

Papyrus and other tall aquatic vegetation in permanent swamps. Extensive areas of this habitat seem to be necessary to support a viable population and in the absence of such an environment anywhere near this region it is unlikely to extend its range. Its status is therefore vulnerable to destruction or abuse of this habitat.

Analysis

10 squares (4%) Total count 43 (0,02%)

LITTLE RUSH WARBLER

638

Bradypterus baboecala

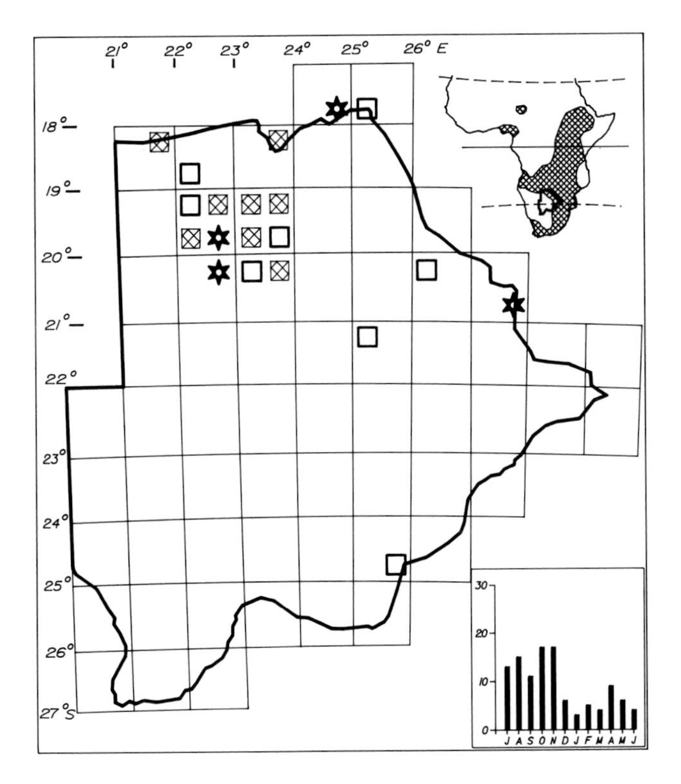

Alternative name

African Sedge Warbler

Status

Sparse to locally fairly common resident of the Okavango, Linyanti and Chobe river systems. Recorded in the east at Nata and Orapa with an old record from near Francistown. Regularly reported from Gaborone and may in future be found at other localities in the east where suitable habitat exists. Local seasonal movements may occur. It has an unique call but can be easily overlooked. Usually solitary.

Habitat

Marshes, swamps and tall aquatic vegetation on the margins of rivers, streams, lakes and dams. Usually skulks at the base of the vegetation.

Analysis

20 squares (9%) Total count 113 (0,06%)

WILLOW WARBLER

643

Phylloscopus trochilus

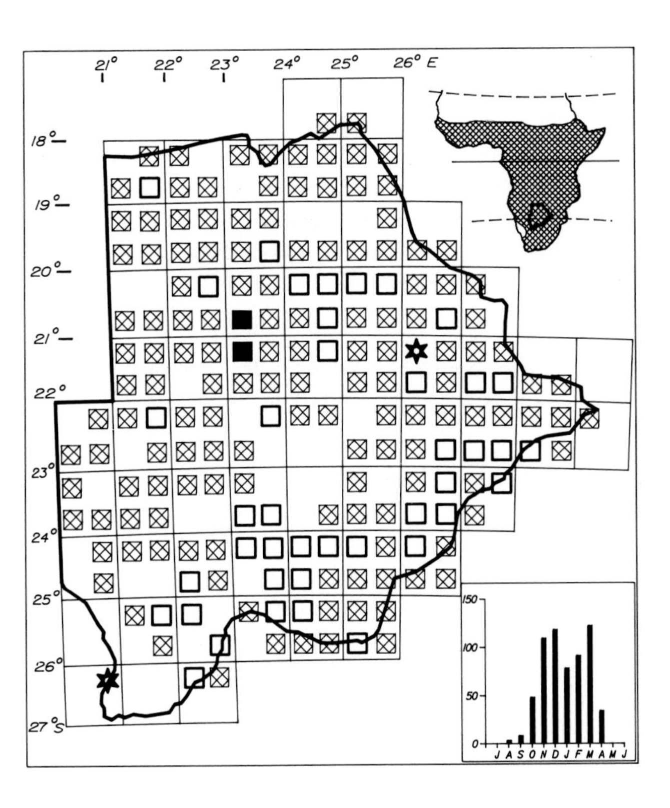

Status

Common Palaearctic migrant throughout the country. Occurs in good numbers from November to March but earliest arrivals usually appear in October and even late August in some years. The main northward passage is in March—latest date 29 April. Occasionally solitary but more often in loose groups up to 10 in one tree or bush. At times it is very common at a locality and can be numerically the most abundant warbler in an area.

Habitat

All types of woodland, tree and bush savanna, semidesert bush and scrub savanna. Edges of riparian forest, gardens and farmland. Occurs in a wide diversity of habitats even on woody shrubs on flat open grass plains and the interior of riparian forest.

Analysis

185 squares (80%) Total count 638 (0,36%)

BARTHROATED APALIS

645

Apalis thoracica

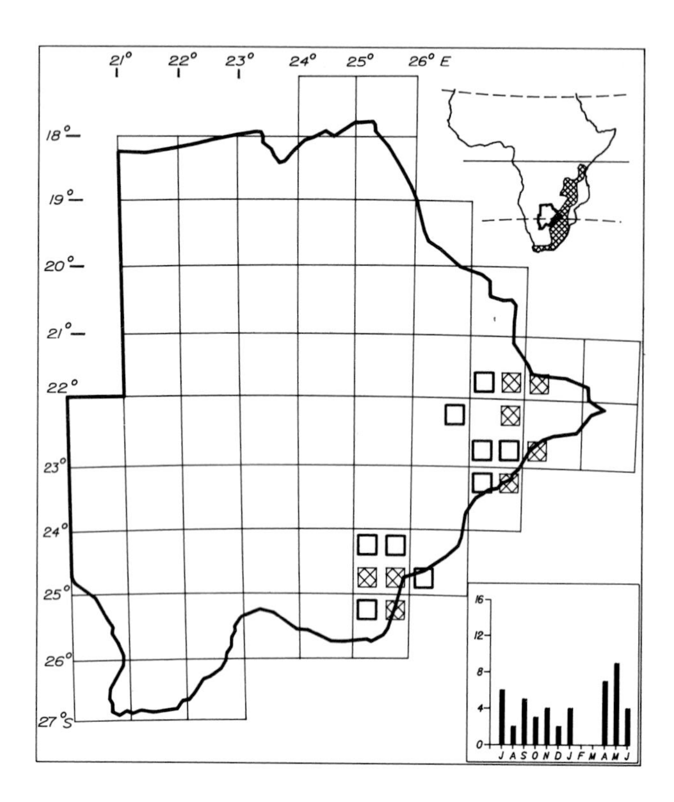

Status

Sparse to locally common resident of hilly areas of the east and southeast. Some seasonal movements may occur but it appears to be mainly sedentary in this region. Not well known nor studied in Botswana. Usually in pairs.

Habitat

Undergrowth at the base of cliffs or ravines, grassy hillsides with tangled vegetation and bush and tree cover, rank grass and scattered bushes along river banks in hilly terrain.

Analysis

17 squares (7%) Total count 60 (0,03%)

YELLOWBREASTED APALIS

648

Apalis flavida

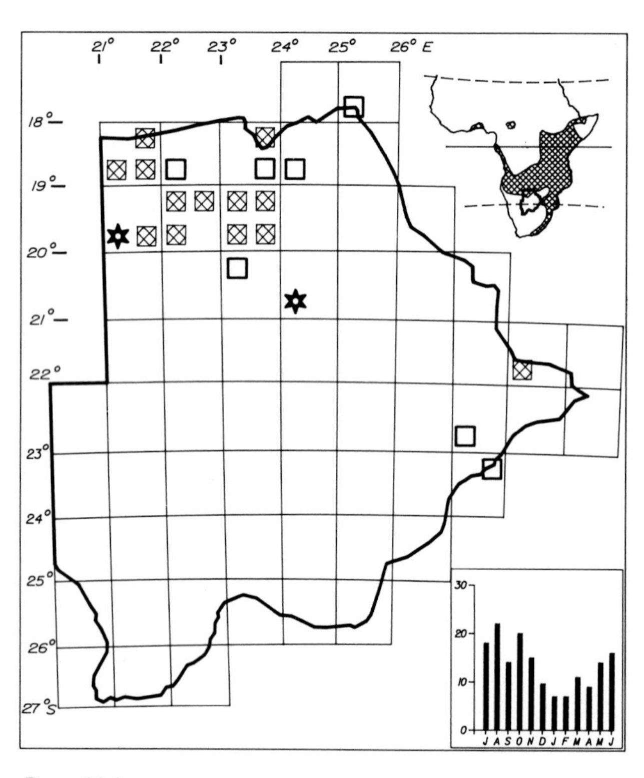

Status

Fairly common to locally common resident of the Okavango Delta, the northwestern woodlands, Linyanti, Savuti and Kasane. Reappears sparsely in the Tuli region where it occurs at the western limit of its range in the middle Limpopo Valley. It may occur in the northeastern woodlands from whence there are no current records, at least along drainage lines. Solitary or in pairs.

Habitat

Riparian forest usually near the edge, deciduous woodland in high rainfall areas and particularly in river valleys.

Analysis

22 squares (10%) Total count 174 (0,097%)

LONGBILLED CROMBEC

651

Sylvietta rufescens

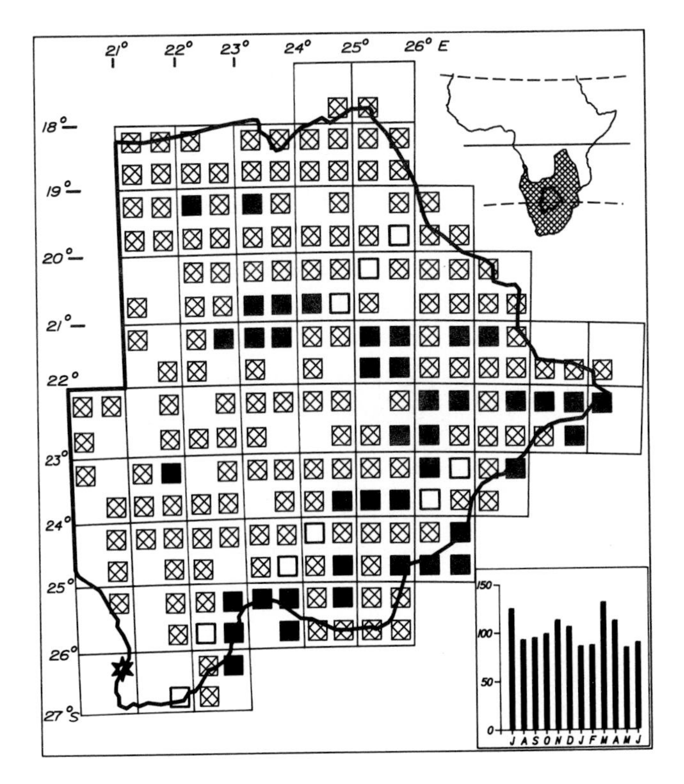

Status

Common to very common resident throughout the country. Reported as very common in 18% of squares. It is a typical species of the drier Kalahari savannas as well as woodland in higher rainfall areas. Egglaying from October to February. Solitary or in pairs. Often joins mixed bird parties.

Habitat

Any woodland but more usually open tree and bush savanna and semidesert bush; secondary growth on edges of woodland, riparian forest and floodplains; ecotone of plains and pans. Usually in lower or mid stratum where there is good cover from foliage, stems and branches.

Analysis

195 squares (85%) Total count 1249 (0,7%)

YELLOWBELLIED EREMOMELA

653

Eremomela icteropygialis

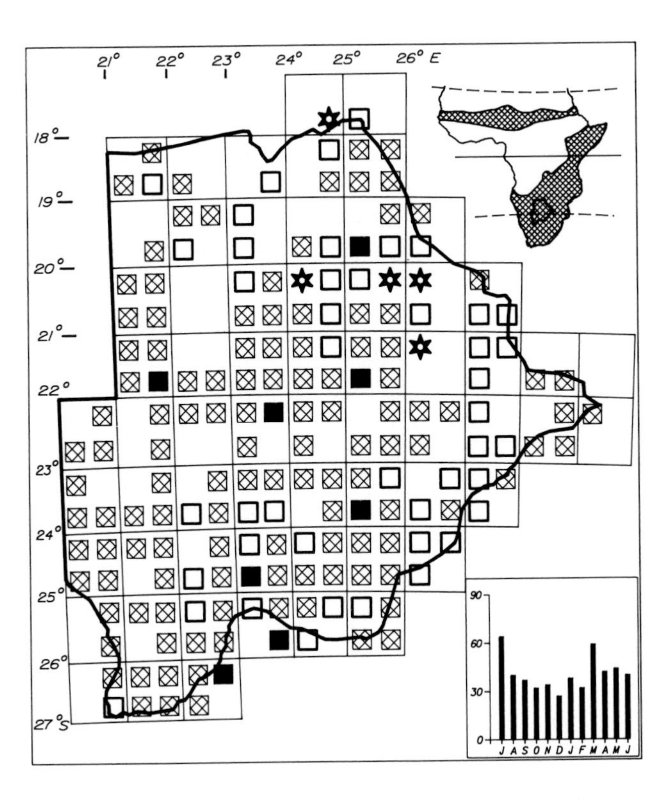

Status

Fairly common to common resident throughout the country except in wetlands and the higher rainfall areas of the east where it is sparse and unpredictable. Some local movements may occur. Egglaying October and November. Usually solitary or in pairs.

Habitat

Bush and scrub savanna in semidesert; *Acacia* and mixed woodland in the drier regions. Commonly in low shrubs and bushes and particularly along dry watercourses and on the ecotone of pans. In the higher rainfall areas it occurs in secondary growth on the edges of woodland and in open bush savanna.

Analysis

176 squares (76%) Total count 507 (0,28%)

GREENCAPPED EREMOMELA

655

Eremomela scotops

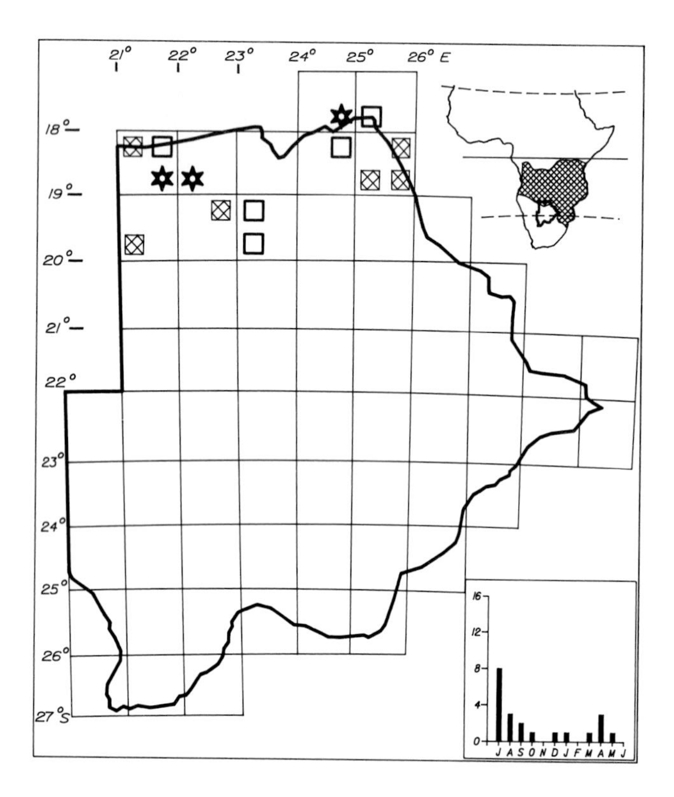

Status

Sparse to uncommon resident of the northern woodlands. Recorded as far south as the Aha Hills in the west and the Sibuyu Forest Reserve in the east. Commonest in the miombo woodland west of Mpandamatenga and in the western part of the Moremi Wildlife Reserve. Restrictions on getting out of one's vehicle in a national park is a possible reason for it not being recorded more often in the north. Solitary or in small loose groups.

Habitat

Miombo and *Baikiaea* woodland. Mainly in the canopy but often on leafy saplings or bushes in the understorey. Usually in areas of woodland with light ground cover of grass and shrubs.

Analysis

14 squares (6%) Total count 28 (0,016%)

BURNTNECKED EREMOMELA

656

Eremomela usticollis

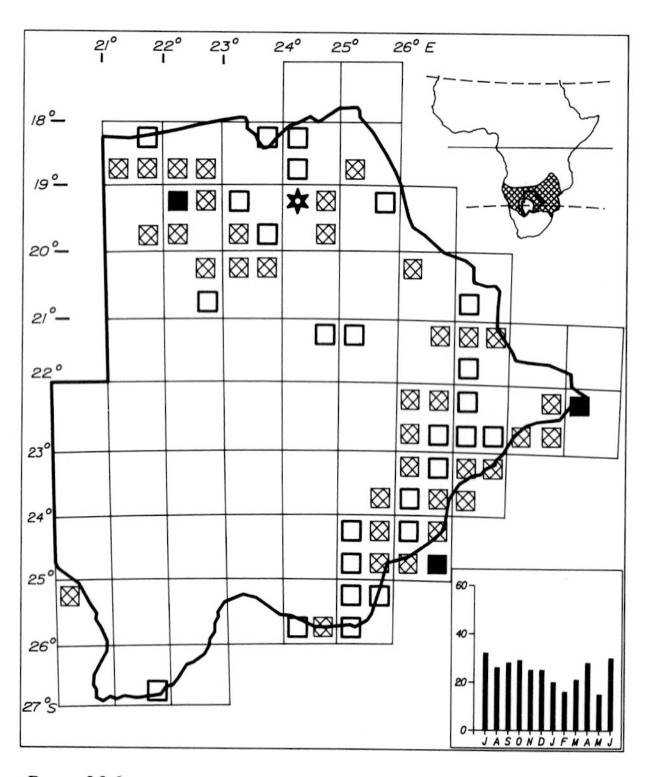

Status

Sparse to locally common resident of the north, east and southeast. Extends along the Molopo River to Boshoek in the east. Recorded once at Khuis on the western Molopo and once in the Nossob Valley. Very common in some riparian areas as at Gomare, Pont's Drift and the Marico River east of Gaborone. Egglaying November and December. Usually in pairs or small loose groups.

Habitat

Acacia tree and bush savanna in the higher rainfall areas, riverine trees and bush. Less commonly in open mixed woodland with scattered bushes.

Analysis

66 squares (29%) Total count 307 (0,17%)

BLEATING BUSH WARBLER

Camaroptera brachyura

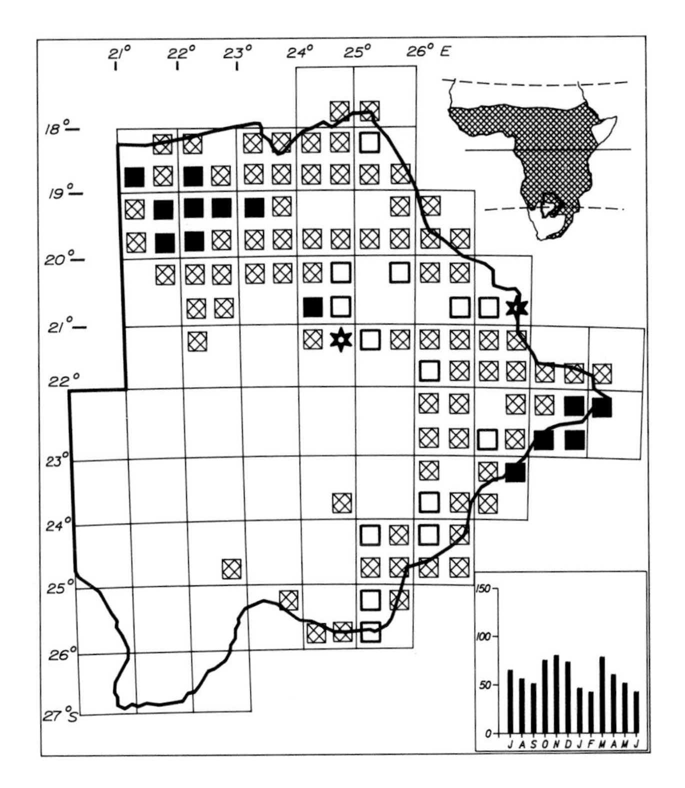

Status

Common to very common resident of the north and east. Sparse to common in the southeast extending along the Molopo River to Bray and to west of Khakhea. The southern birds are at the southern limit of the species range at these longitudes. Usually in pairs.

Habitat

Bushes, thickets and tangled vegetation in woodland and tree savanna, dense riparian vegetation. Commonest in the high rainfall regions.

Analysis

106 squares (46%) Total count 735 (0,41%)

BARRED WARBLER

Calamonastes fasciolatus

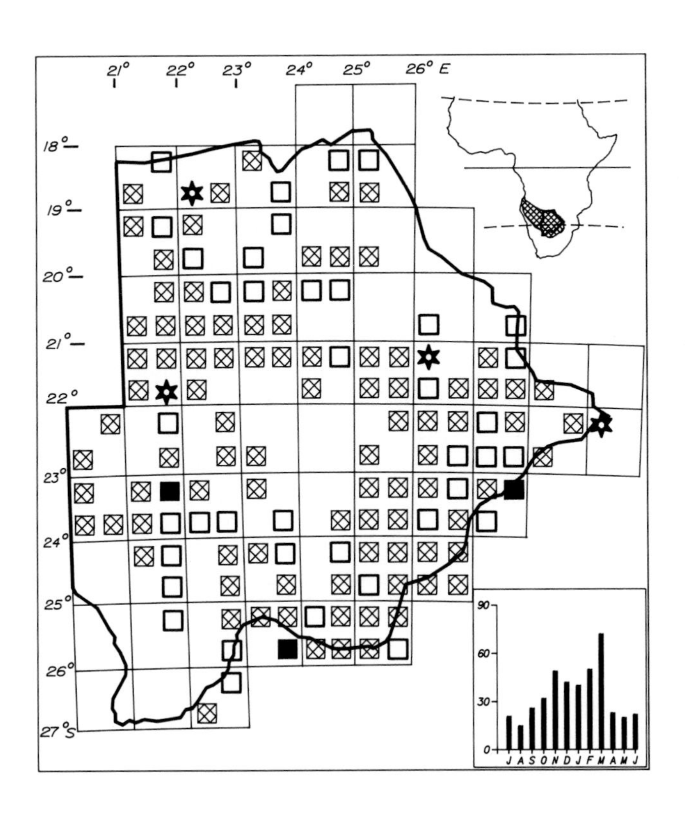

Status

Sparse to common resident in all regions except the southwest where it is apparently mainly absent in the duneland. The gaps in records in the central Kalahari also need further investigation. In the northeast its range overlaps that of the Stierling's Barred Warbler whose habitat dominates. Nest-building March. Solitary or in pairs.

Habitat

Acacia and mixed deciduous woodland but not miombo, *Baikiaea* nor mopane. Tree and bush savanna in areas of good rainfall and in semidesert but not in the most arid conditions. Often in areas with bushes and dense cover.

Analysis

137 squares (60%) Total count 428 (0,24%)

STIERLING'S BARRED WARBLER

Calamonastes stierlingi

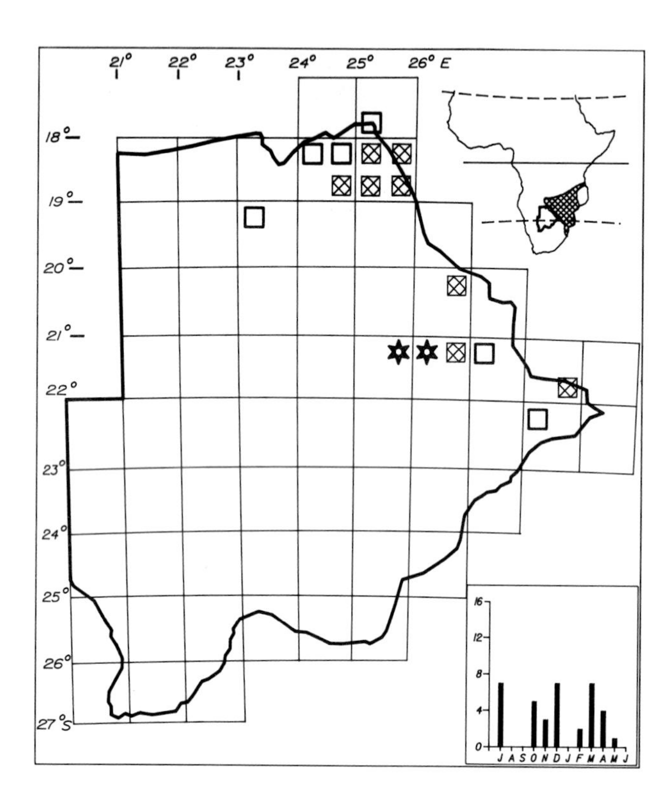

Status

Common resident of the northeastern woodland and sparse on its periphery to as far west as the eastern part of the Moremi Wildlife Reserve. Reappears patchily in the east near Nata and extends southward to the northern Tuli region. Although its range overlaps that of the Barred Warbler, they occur in different habitats and are not considered to be in direct competition. Solitary or in pairs.

Habitat

Miombo and *Baikiaea* woodland where it is found in bushes and shrubs in the understorey as well as in the canopy of the trees. Also on the edge of woodland in bushes, thickets and dense vegetation.

Analysis

16 squares (7%) Total count 40 (0,02%)

FANTAILED CISTICOLA

Cisticola juncidis

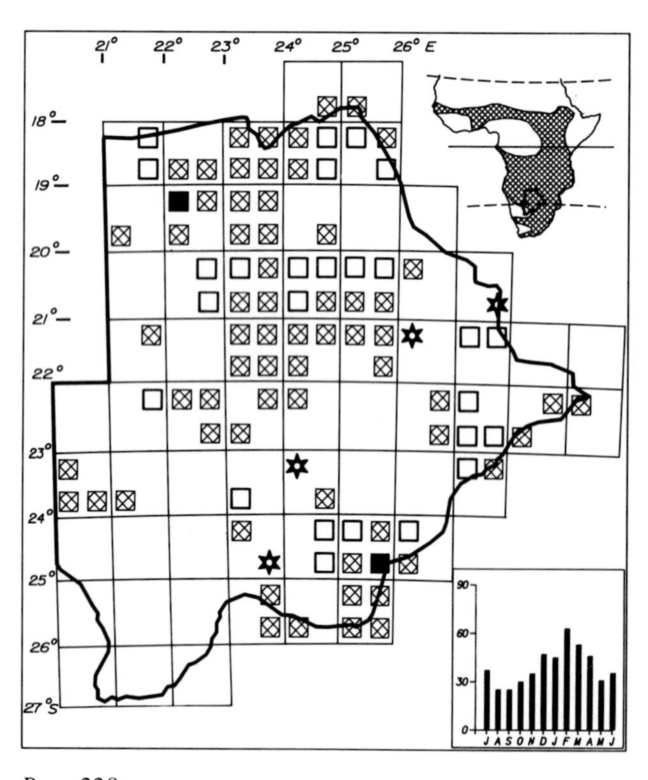

Status

Fairly common to common resident of the higher rainfall areas of the north and southeast. There are significant movements into the drier regions including the central Kalahari in years of good rainfall and a contraction of range during drought. Its apparent scarcity in the east needs further study. Egglaying January. Solitary or in pairs.

Habitat

Open grassland which is moist or seasonally wet, remaining there even in years of poor rainfall but retreating to more reliably moist situations in drought. Floodplains, fossil valleys, edges of marshes, dambos and vegetated Kalahari pans where it often remains in the moist grass in the centre after the pan has dried.

Analysis

97 squares (42%) Total count 497 (0,28%)

DESERT CISTICOLA

665

Cisticola aridula

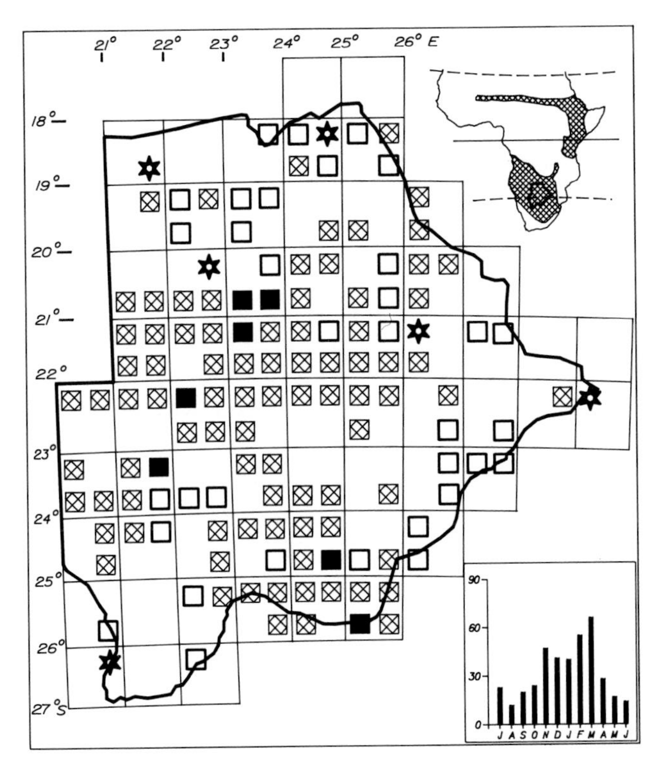

Status

Fairly common to common resident south of 20°S and west of 26°E. It is the typical cisticola of Kalahari grassland. Extends northeastwards on plains north of Nata and around Nxai Pan. Otherwise it is sparse or uncommon in the north and east. In the southeast it is sparse east of Gaborone but very common on the plains around Jwaneng and Phitsane Molopo. May occur alongside the Fantailed Cisticola in the higher rainfall areas and on some western pans. Egglaying November to March. Solitary or in pairs.

Habitat

Open short grassland, plains, dry grassed pans and open well-grassed bush and scrub savanna in semidesert. Sometimes utilises small shrubs for attaching the nest in grassland.

Analysis

132 squares (57%) Total count 405 (0,23%)

TINKLING CISTICOLA

671

Cisticola rufilata

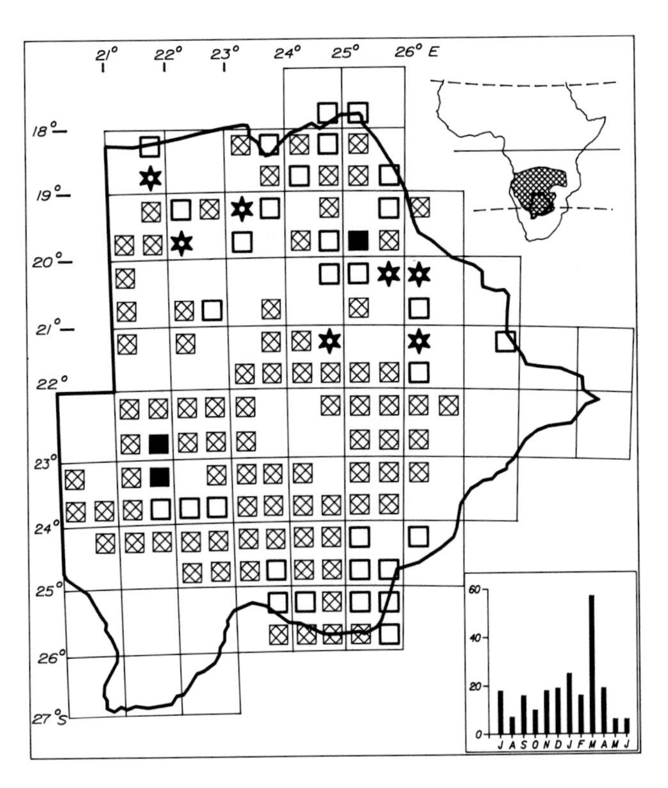

Status

Fairly common to common resident of the central and western Kalahari. Sparse or uncommon north of 21°S and in the southeast. Its range reciprocates that of the Rattling Cisticola which appears to replace it almost entirely in the east. Absent in the southwest. Egglaying January. Solitary or in pairs.

Habitat

Semidesert bush and scrub savanna, in open areas of stunted bushes or secondary growth in grassland, edges of pans on the ecotone with woodland. In the north it occurs on the edge of woodland and in tree savanna in dry sandy areas, in mopane bush and in disturbed areas with secondary growth on cultivation or at roadsides.

Analysis

124 squares (54%) Total count 231 (0,13%)

RATTLING CISTICOLA

Cisticola chiniana

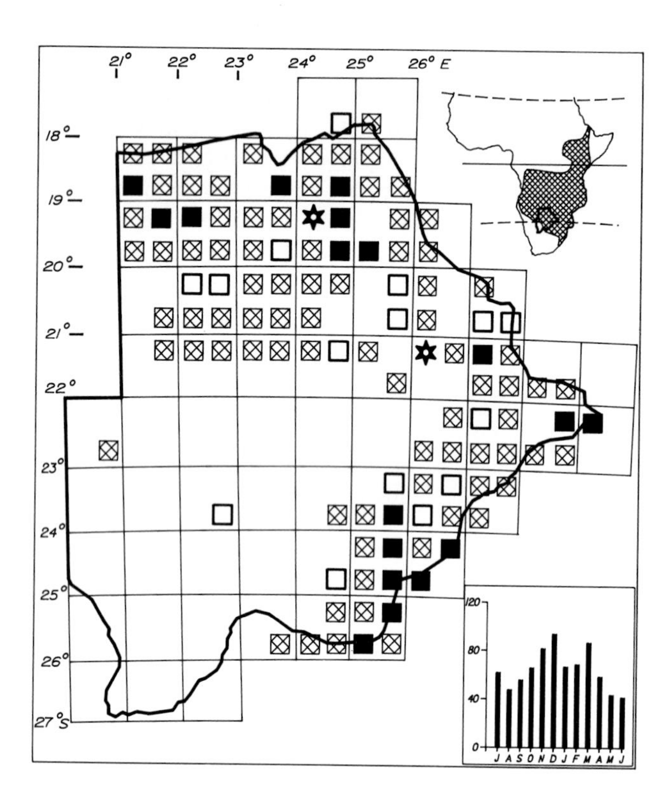

Status

Fairly common to very common resident of the north and east to as far south as Ramatlabama. Extends westwards along the Molopo River valley almost to Bray. In the west there are isolated records from Kang and from south of Kalkfontein (Tsootsha). Its status in the western woodland needs more detailed investigation. Egglaying December to March. Solitary, in pairs or in family groups.

Habitat

Tree and bush savanna, edges of woodland, secondary vegetation where there are well-established bushes, bushy growth or isolated thicket in open areas of tall grass.

Analysis

115 squares (50%) Total count 793 (0,44%)

REDFACED CISTICOLA

Cisticola erythrops

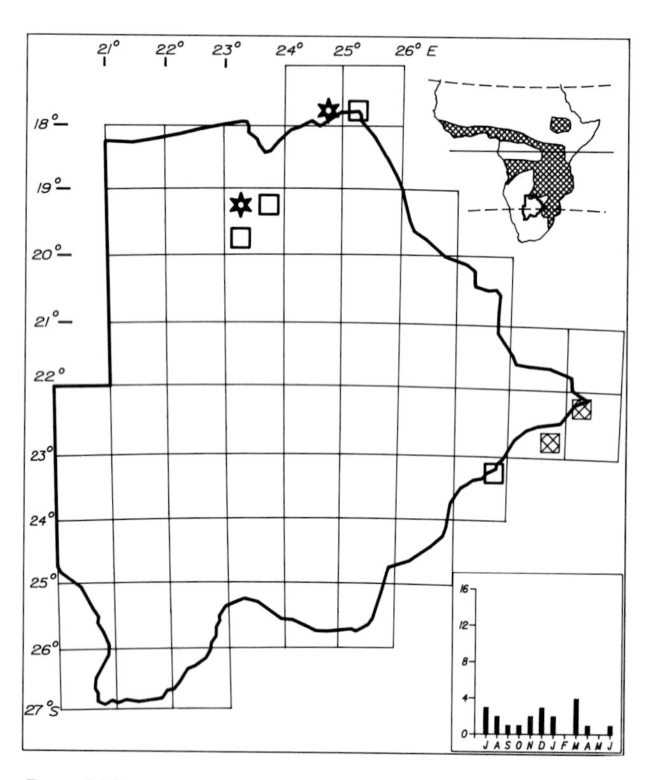

Status

Sparse and very localised resident. Most commonly recorded near Pont's Drift from where it extends south along the Limpopo River to Martin's Drift. Next most commonly recorded at Kasane. Occurs in the Okavango Delta in the Moremi region and may well be distributed more widely in this area. At present it is very poorly known in Botswana. Solitary or in pairs.

Habitat

Tall grass and rank vegetation along rivers and waterways, on the edges of marshes and swamps. Never far from permanent water. It extends into adjacent rank growth on floodplains and in river valleys.

Analysis

7 squares (3%) Total count 24 (0,01%)

BLACKBACKED CISTICOLA

675

Cisticola galactotes

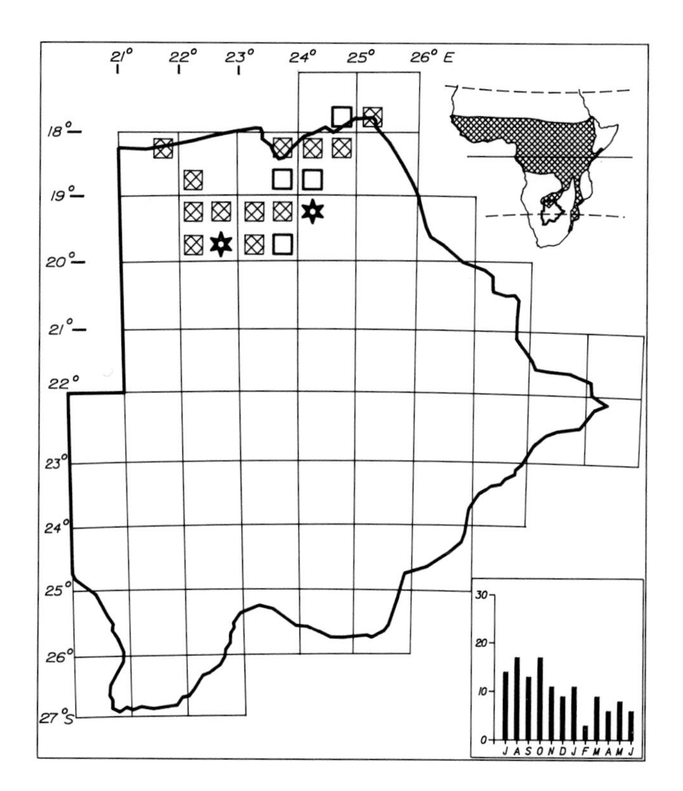

Status

Sparse to fairly common resident of the Okavango, Linyanti and Chobe river systems. Its range is very similar to that of the Chirping Cisticola and confusion between the two does occur. Their habitats and vocalisations are quite different. Local seasonal movements occur. Usually in pairs or solitary.

Habitat

Tall grasses, sedges and reeds on floodplains, marshes, edges of swamps and lagoons—not necessarily at or over water. It will remain in completely dried out marshes when stranded by drought as occurred at Savuti and Khwai rivers during this survey. It tolerates drier conditions than the Chirping Cisticola.

Analysis

18 squares (8%) Total count 133 (0,07%)

CHIRPING CISTICOLA

676

Cisticola pipiens

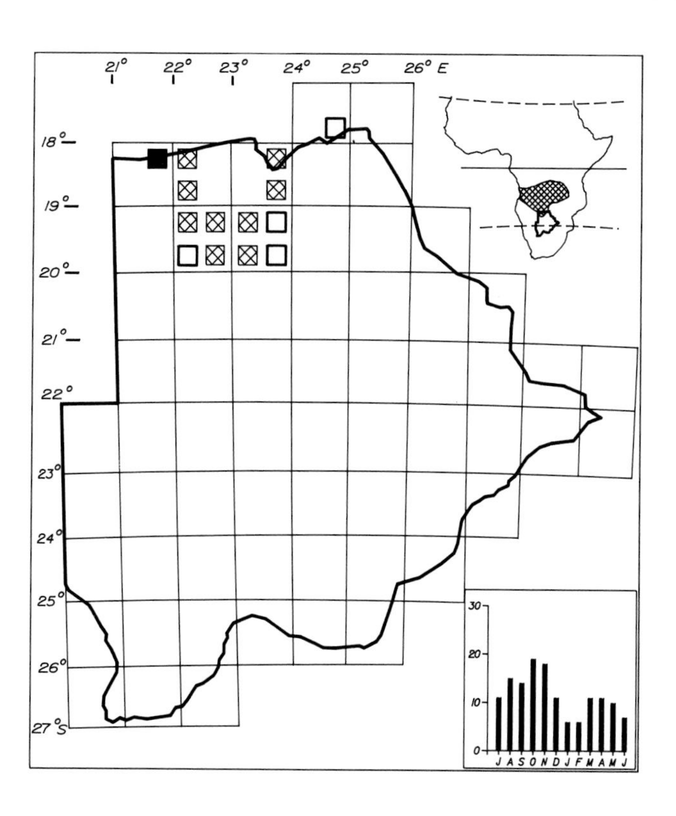

Status

Fairly common to locally very common resident of the Okavango and Linyanti river systems. Sparse on the periphery of these wetlands and on the Chobe River at Serondella. Common at Shakawe. It occurs in Botswana at the southern limit of its continental range. Generally more numerous and more frequently recorded than the Blackbacked Cisticola but it is less widespread. Solitary or in pairs.

Habitat

Reeds, papyrus and tall aquatic vegetation. It is almost always found in vegetation standing in water and is distributed over areas of permanent water with little migration to seasonal wetlands—in contrast to the Blackbacked Cisticola.

Analysis

14 squares (6%) Total count 142 (0,08%)

LAZY CISTICOLA

679

Cisticola aberrans

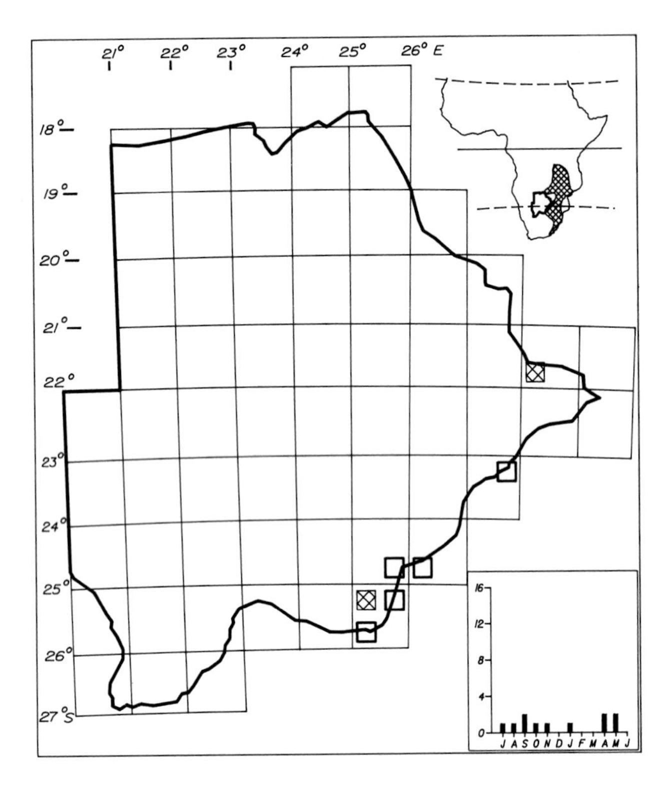

Status

Sparse to uncommon resident of the hilly regions of the southeast. Recorded once from near Bobonong where it may be more common in nearby hills than indicated. Poorly known in Botswana where it occurs at the western limit of its range in southern Africa. Solitary or in pairs.

Habitat

Well-grassed steep hillsides or slopes with scattered trees and protruding rocks, bases of gullies and valleys with tall rank vegetation, wooded mountainsides with rocks under tree canopy.

Analysis

7 squares (3%)   Total count 16 (0,009%)

NEDDICKY

681

Cisticola fulvicapilla

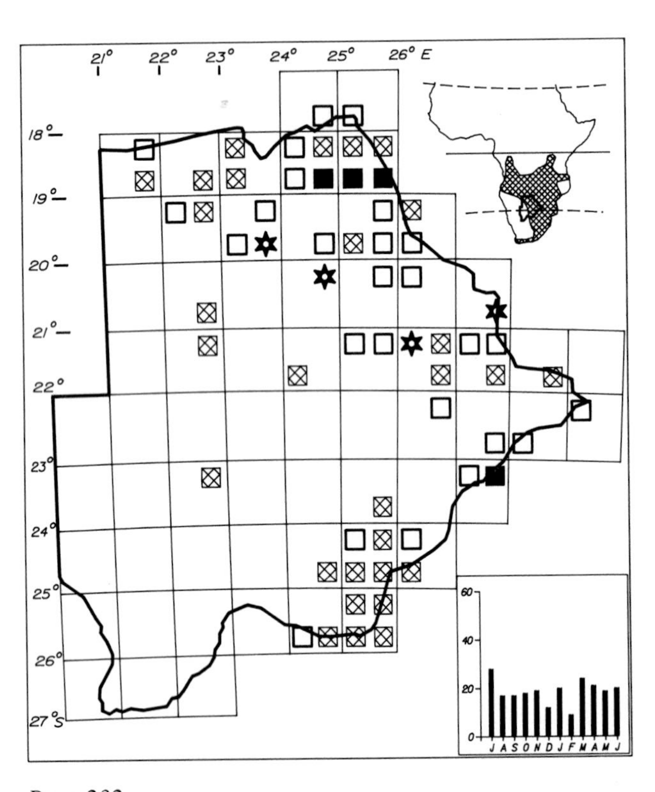

Status

Common to very common resident of the northern woodlands especially in the northeast becoming sparse and patchily distributed further south. In the southeast it is sparse to common and extends westward to the edge of the semidesert at Jwaneng. Recorded once north of Kang in the midwest where more detailed study of its status is required. Usually in pairs or solitary.

Habitat

Tall grass and scattered bushes in the understorey and on the edges of miombo and *Baikiaea* woodland in the north. In mixed *Acacia* and broadleafed woodland and tree and bush savanna in the east and southeast. It also occurs in secondary growth on the edges of woodland and dense vegetation in open areas. Also in gardens, plantations and edges of cultivation.

Analysis

63 squares (27%)   Total count 237 (0,13%)

TAWNYFLANKED PRINIA

Prinia subflava

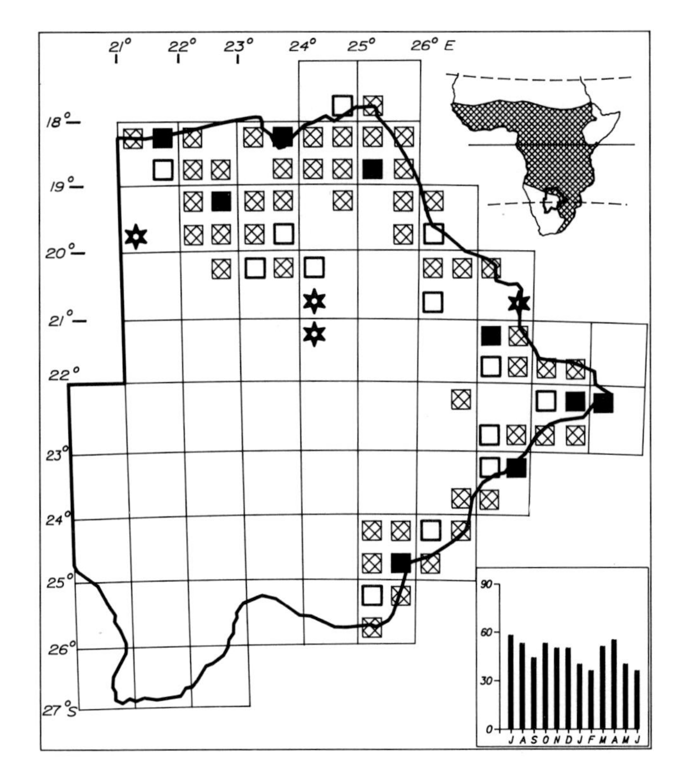

Status

Fairly common to very common resident north of 21°S and in the Limpopo drainage in the east and southeast. Its distribution is closely associated with the river systems and is limited to the higher rainfall areas. It has not been recorded from the Makgadikgadi Pans except at the Nata River mouth. Usually in pairs or small groups.

Habitat

Long grass, sedges and reeds along the banks of rivers and extending into adjacent floodplains and marshes. Rank grass on the edges of woodland, cultivation and disturbed ground such as roadsides and bridges. Luxuriant vegetation around dams and water impoundments and occasionally in gardens.

Analysis

72 squares (31%) Total count 586 (0,33%)

BLACKCHESTED PRINIA

Prinia flavicans

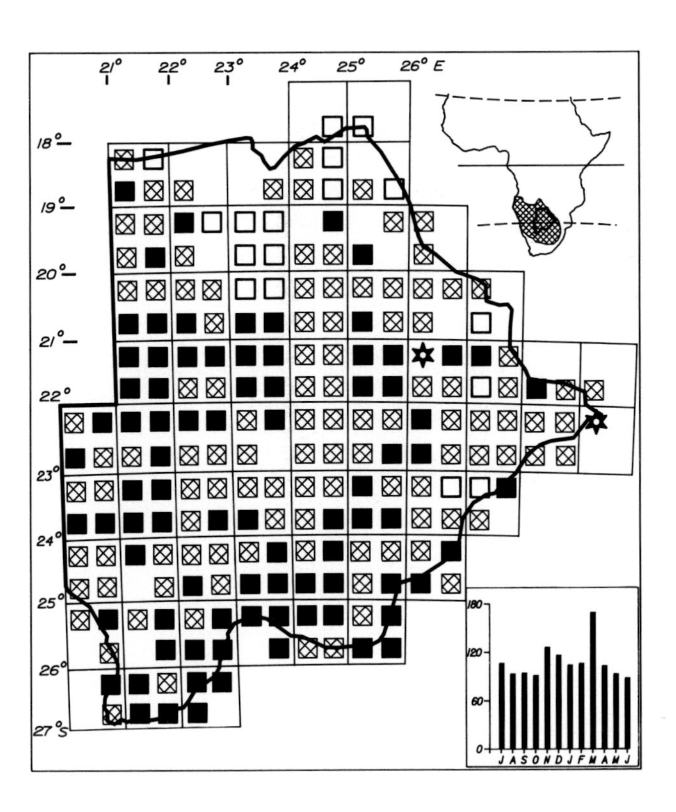

Status

Common to very common resident throughout the country except where it is replaced in the wetlands by the Tawnyflanked Prinia. Generally very common and ubiquitous south of 20°S where it is a characteristic species of the Kalahari. Usually in pairs or family parties.

Habitat

Any type of savanna from arid scrub to bush and tree savanna in the higher rainfall areas. Found in rank grass, low bushes, scrub vegetation and secondary growth, open grassland with scattered trees or bushes and in woody shrubs. Occurs mainly in the undergrowth but perches readily in small trees and on tall bushes.

Analysis

212 squares (92%) Total count 1336 (0,75%)

RUFOUSEARED WARBLER

688

Malcorus pectoralis

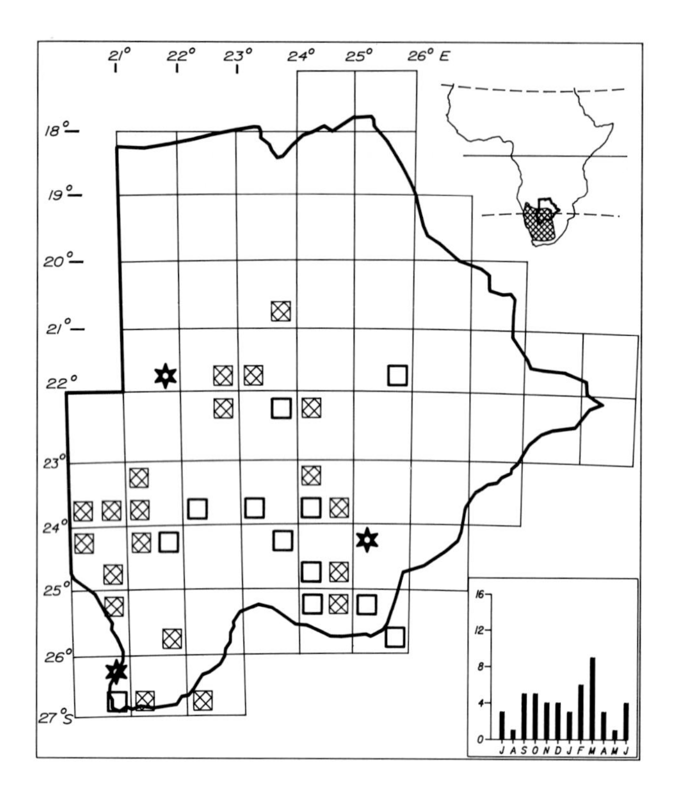

Status

Sparse to uncommon resident of the south and west extending northwards into the central Kalahari Game Reserve to as far as Deception Valley and east to Makoba. It is patchily distributed, easily overlooked and not well known. Occurs in pairs or solitary, but several pairs can be present at one locality.

Habitat

Low scrub on the edges of pans and plains, open short grass with scattered stunted bushes, e.g. *Grewia* and *Catophractes*. It occurs in continuous semidesert scrub in the central Kalahari where it has also been found in stunted bushes on the edge of woodland.

Analysis

35 squares (15%) Total count 53 (0,03%)

SPOTTED FLYCATCHER

689

Muscicapa striata

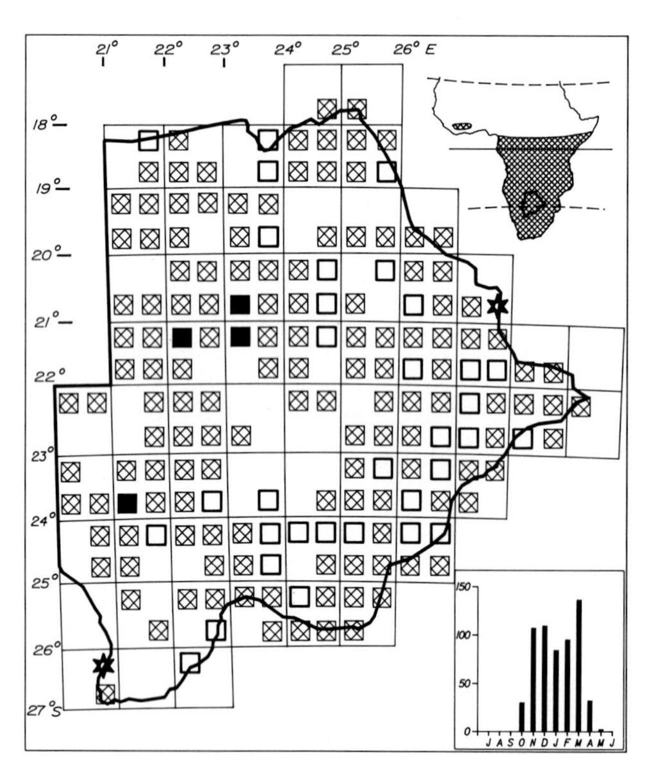

Status

Fairly common to common Palaearctic migrant to all parts of the country. Sparse in the central Kalahari and southwest. In many areas it is the commonest flycatcher during some of the summer months. Arrives in late October and departs in the first half of April (earliest date 4 October, last date 5 May). Numbers are augmented by birds still on passage in December and March. Usually solitary.

Habitat

Edges of all types of woodland, tree and bush savanna, riparian trees, gardens and plantations.

Analysis

175 squares (76%) Total count 616 (0,34%)

BLUEGREY FLYCATCHER 691

Muscicapa caerulescens

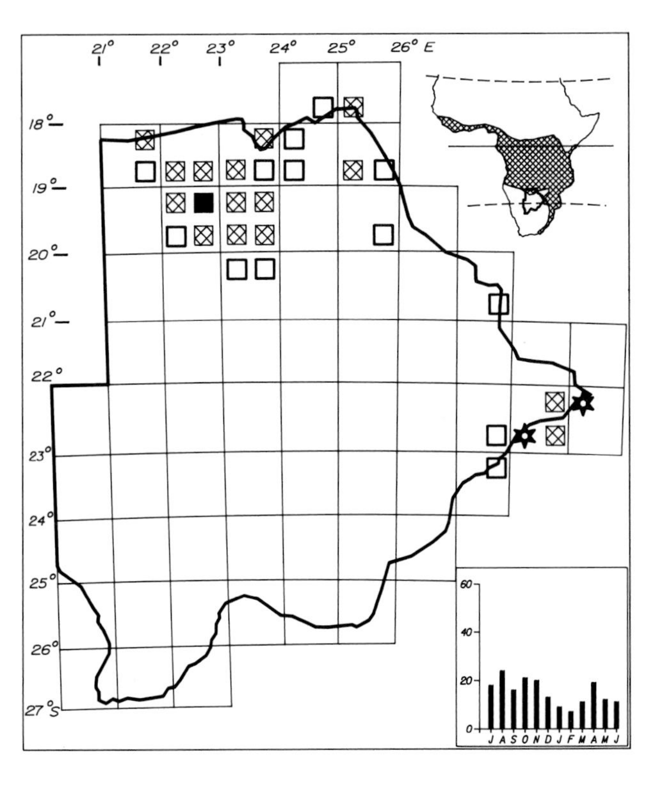

Status

Fairly common to very common resident of the Okavango Delta. Sparse to fairly common in the Chobe, Linyanti, Savuti and eastern Tuli areas. Patchily recorded in the northeast. Occurs in Botswana at the western limits of its range in southern Africa. Egglaying October and December. Usually solitary or in pairs.

Habitat

Riparian fringing forest and adjacent tall woodland. Also in well-grown secondary vegetation, thickets and bushes on the edge of woodland.

Analysis

31 squares (13%) Total count 185 (0,10%)

FANTAILED FLYCATCHER 693

Myioparus plumbeus

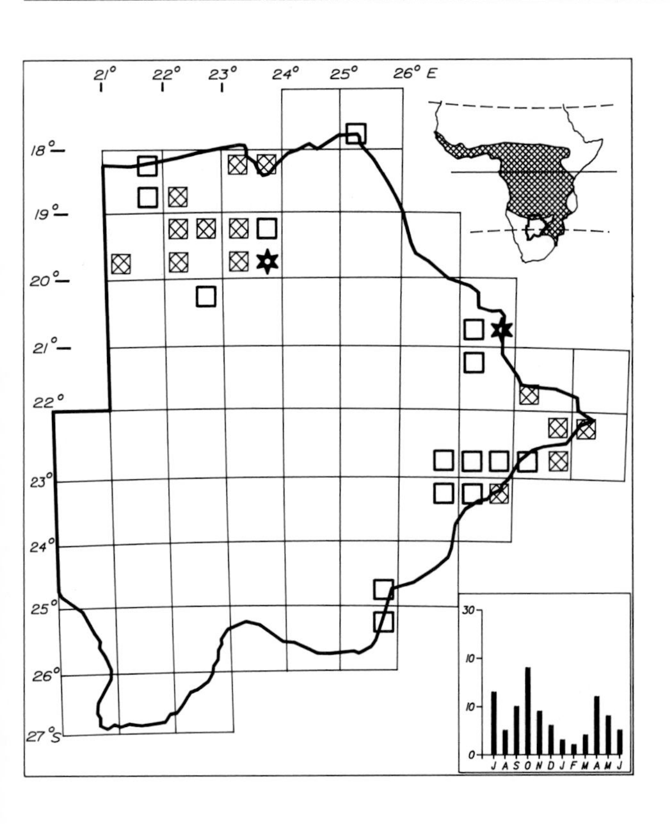

Alternative name

Leadcoloured Flycatcher

Status

Sparse to uncommon resident of the Okavango, Linyanti, Chobe, Shashe and Limpopo river systems. Extends westwards along the Limpopo drainage to as far as Serowe. Reappears in the southeast at Gaborone and Lobatse. Easily overlooked. Usually solitary or in pairs.

Habitat

Dense cover in woodland on alluvial soils, edges of riverine forest, tree savanna with thicket, bush savanna in river valleys.

Analysis

31 squares (13%) Total count 106 (0,06%)

BLACK FLYCATCHER 694

Melaenornis pammelaina

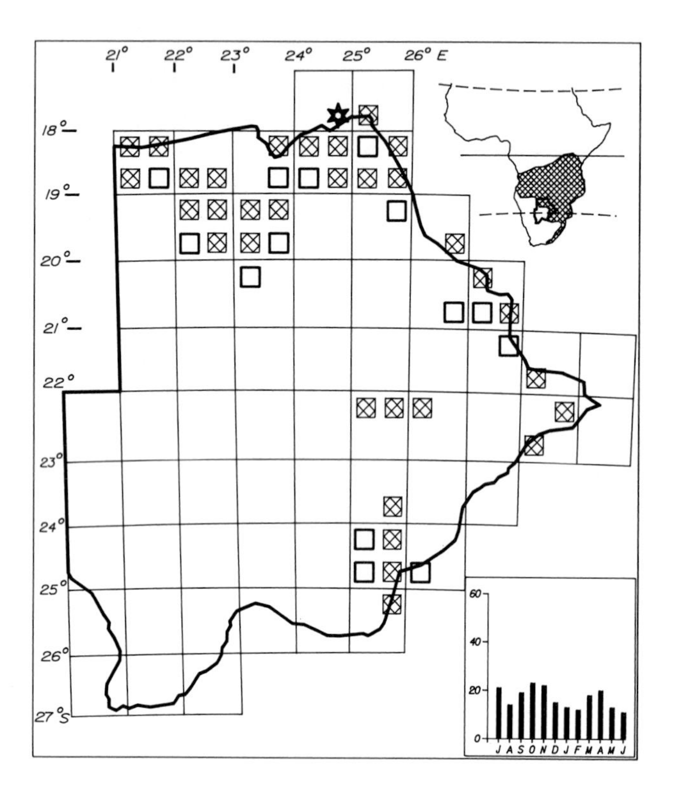

Status

Uncommon to fairly common resident of the northern woodlands. Extends south along the eastern border to the Limpopo River. Occurs in eastern Kalahari woodlands west of Serowe. Reappears in the southeast and extends west to as far as Moshaneng. Occurs in Botswana at the western limit of its southern African range. Egglaying January. Solitary or in pairs.

Habitat

Miombo and *Baikiaea* woodland, riparian *Acacia*, well-developed mopane woodland and mixed broadleafed woodland. Usually in open areas under tree canopy with little understorey. Also in mature tree savanna and on edges of woodland.

Analysis

47 squares (20%) Total count 210 (0,12%)

MARICO FLYCATCHER 695

Melaenornis mariquensis

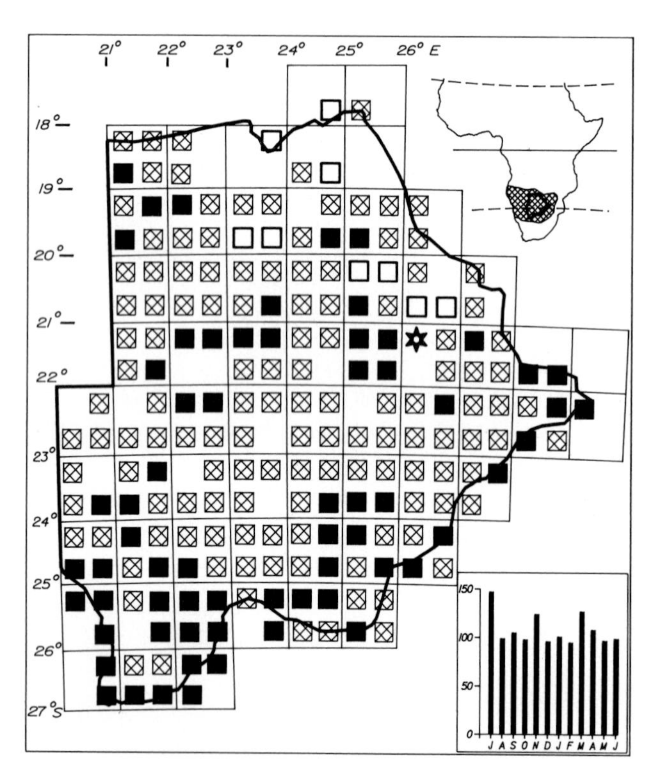

Status

Common to very common resident throughout the country. Sparse or absent in heavily wooded areas of the northeast and north of the Okavango Delta. It is the most common flycatcher in the central and southern Kalahari. Recorded as very common in 28% of all squares. Botswana lies at the centre of its total range in the arid savannas of southern Africa. Egglaying September to April, mainly November and December. Solitary or in pairs.

Habitat

Thorn tree and bush savanna. Almost totally absent from miombo, *Baikiaea* and mopane except where mixed with *Acacia*. Occurs in open areas with scattered trees, tall bushes or secondary growth and with bare or lightly-grassed groundcover. Often feeds on the ground. Rarely in wide open grassland with low bushes which is occupied by the Chat Flycatcher. Often perches on telegraph wires and fence posts like a shrike.

Analysis

202 squares (88%) Total count 1351 (0,76%)

PALLID FLYCATCHER

696

Melaenornis pallidus

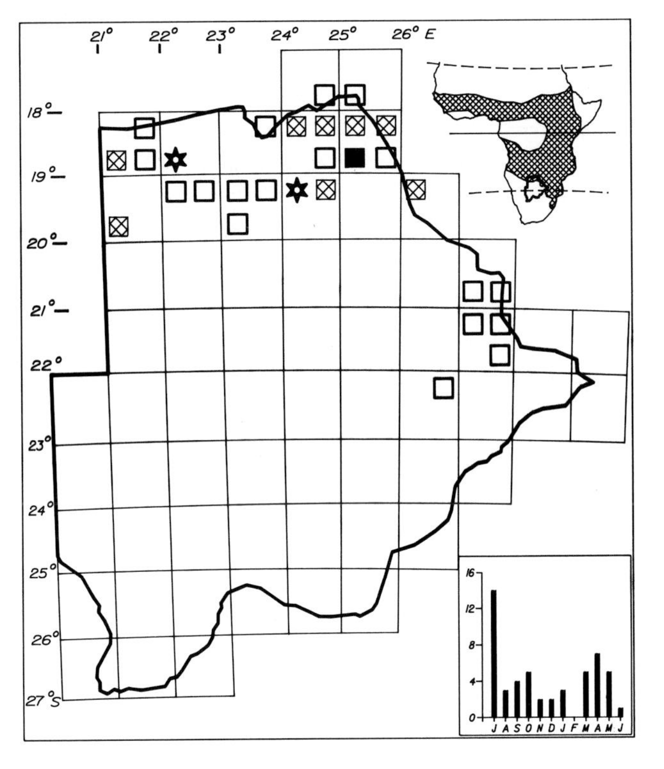

Alternative name

Mousecoloured Flycatcher

Status

Fairly common resident of the northeastern broadleafed woodlands. Sparse to uncommon in the north and northwest and in the east around Francistown. Recorded once at Serowe. It may subsequently be found on the upper reaches of the Nata, Nkange and Tutume rivers but is probably restricted there by mopane intrusion and human habitation. Solitary or in pairs.

Habitat

Miombo, *Baikiaea* and other well-established broadleafed woodland particularly where there is bush and shrub undergrowth. Usually found in midstratum. Uncommonly on the edge of woodland and in open areas with tall bushes.

Analysis

29 squares (13%) Total count 57 (0,03%)

CHAT FLYCATCHER

697

Melaenornis infuscatus

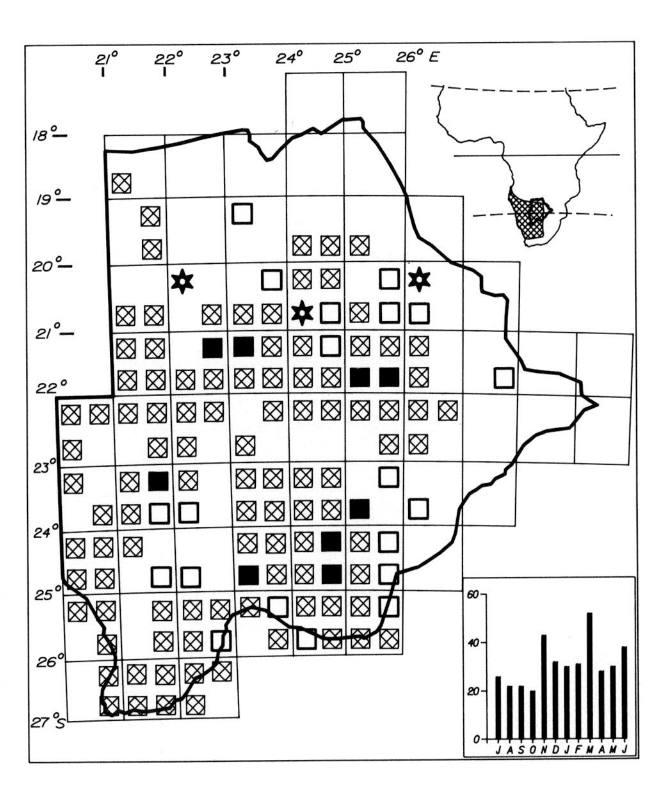

Status

Fairly common to locally very common resident of the central and southern Kalahari south of 21°S and west of 27°E. Extends northwards through the Makgadikgadi Depression to Nxai Pan where it can be locally common. It is uncommon in the northwest along the Xaudum Valley. Egglaying November. Usually in pairs or family groups.

Habitat

Open grassland areas in semiarid bush savanna or *Acacia* savanna, grass plains with scattered low bushes or trees. Perches conspicuously on bushes or woody shrubs, often in pairs.

Analysis

131 squares (58%) Total count 402 (0,22%)

FISCAL FLYCATCHER — 698

Sigelus silens

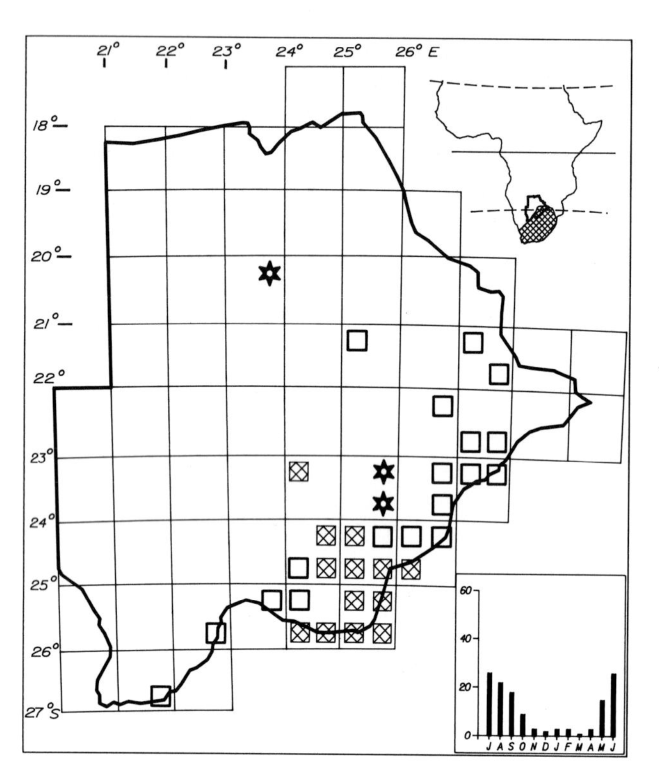

Status

Mainly a nonbreeding winter visitor from May to September from breeding grounds in South Africa. There is also a small resident population in the southeast which occurs at the northern extremity of the breeding range as far north as Kanye and Molepolole. In winter it occurs more widely in the southeast and in some years reaches as far north as Orapa and Francistown in June to August. Usually solitary or in pairs.

Habitat

Tree and bush savanna, mainly in *Acacia* but also in mixed deciduous trees, in small patches of woodland or clumps of trees adjacent to open grassland. Occasionally in gardens.

Analysis

34 squares (15%) Total count 153 (0,08%)

CHINSPOT BATIS — 701

Batis molitor

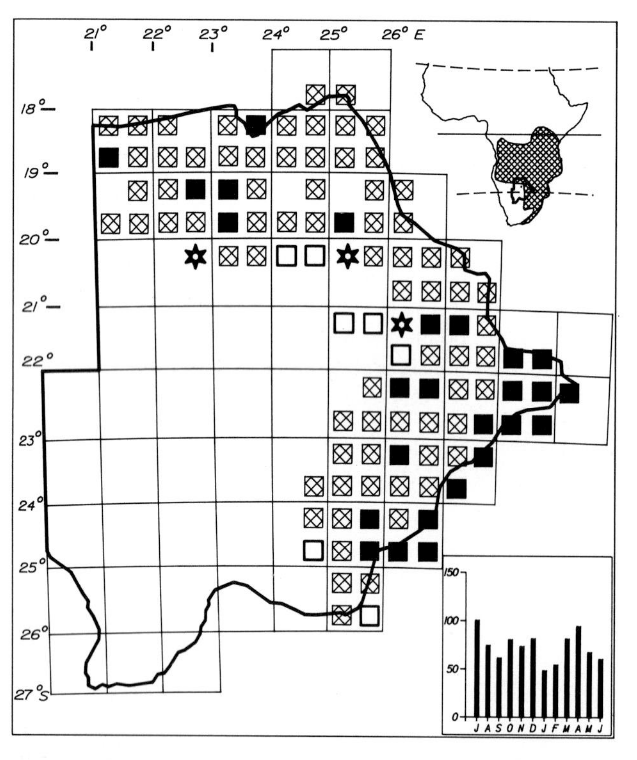

Status

Common to very common resident mainly north of 20°S and east of 25°E. Around Orapa and the western Makgadikgadi the Pririt Batis predominates and on the whole the two species are allopatric with few areas of overlap. Hybrids have not been proven. Egglaying October to January, mainly December. Usually in pairs.

Habitat

Any woodland and tree and bush savanna in the higher rainfall areas. Also riverine trees and bushes but not in evergreen forest. Habitat varies from upper stratum of miombo and *Baikiaea* woodland to bushes in river valleys.

Analysis

108 squares (47%) Total count 908 (0,51%)

PRIRIT BATIS — 703

Batis pririt

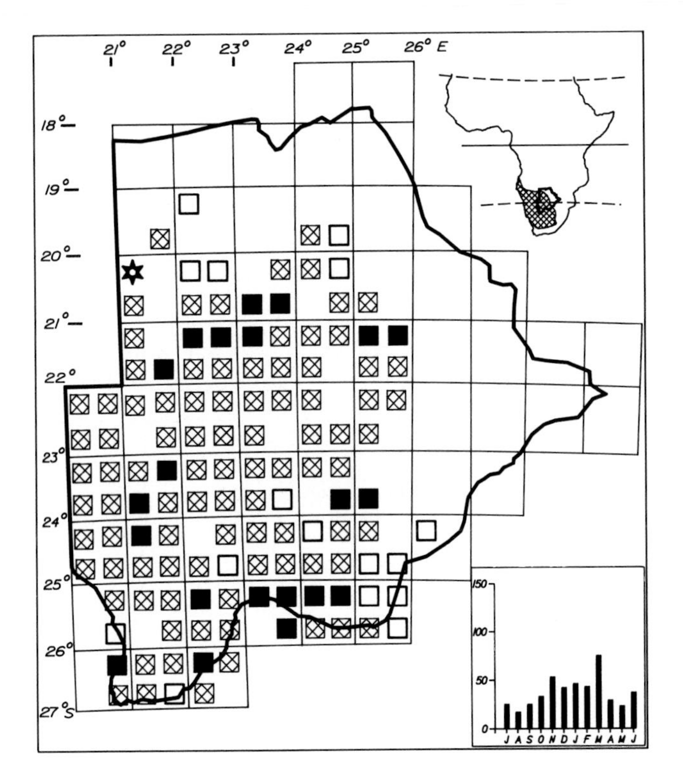

Status

Common to very common resident of the Kalahari south of 20°S and west of 25°E and extending eastwards to 26°E at Letlhakane and Gaborone. Recorded in the northwest as far as Gomare. Overlaps with the Chinspot Batis in the southeast, east and at Nxai Pan. Egglaying December. Usually in pairs.

Habitat

Tree and bush savanna on Kalahari sand in semidesert. Common in well-wooded areas of the west where it occurs in the upper stratum as well as in adjacent bushes. Its habitat utilisation is similar to and as diverse as the Chinspot Batis but in drier environs.

Analysis

128 squares (56%) Total count 475 (0,27%)

FAIRY FLYCATCHER — 706

Stenostira scita

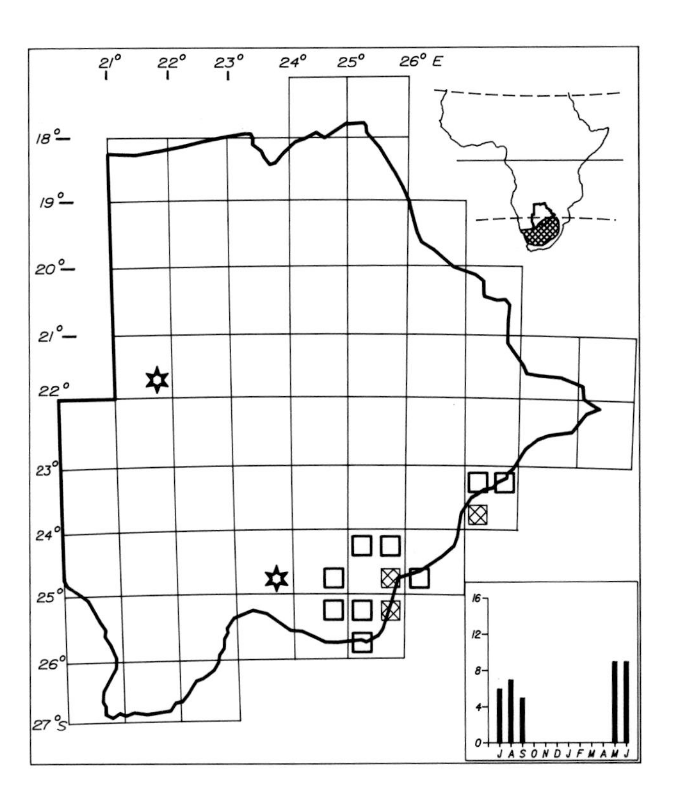

Status

Uncommon nonbreeding winter visitor to the southeast between May and September from its breeding grounds in South Africa. Recorded as far north as Martin's Drift and may occur further north in some years (as occurred to Zimbabwe in 1966). Easily overlooked and not well known. A more detailed study of its status in the east and along the Molopo River is required. Usually solitary.

Habitat

Bush and tree savanna especially along river valleys. Usually in patches of trees or clusters of tall bushes. Also occurs in gardens and dense cover on the edge of woodland.

Analysis

14 squares (6%) Total count 43 (0,02%)

PARADISE FLYCATCHER

710

Terpsiphone viridis

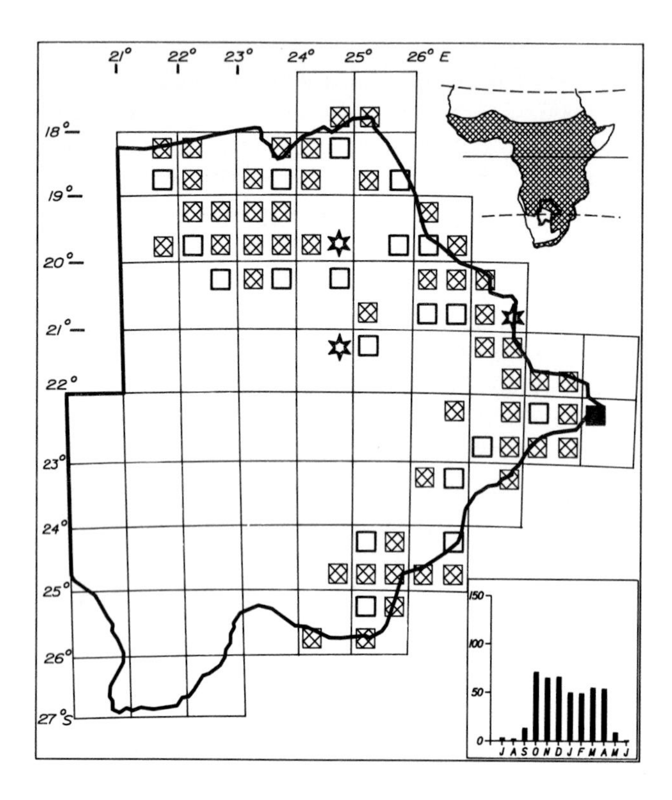

Status

Fairly common breeding intra-African migrant to the north and east. Sparse at the periphery of its range. Present mainly from October to April with some arrivals in September and departures in May. A few birds occur between June and August possibly overwintering or birds of the *T. v. granti* subspecies of South Africa. Egglaying October to January, mainly November and December. Usually in pairs.

Habitat

Riparian forest and well-developed woodland where there are large shady trees or thicket. Also in riverine *Acacia*, gardens, parks and plantations.

Analysis

72 squares (31%) Total count 460 (0,26%)

AFRICAN PIED WAGTAIL

711

Motacilla aguimp

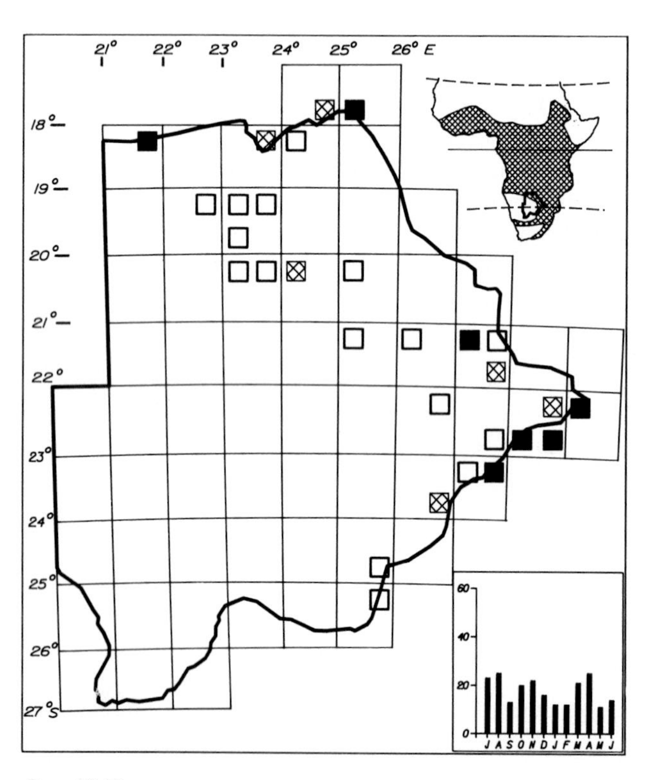

Status

Highly localised resident of the north and east. In the Okavango region it is very common at Shakawe but sparse elswhere. Very common at Kasane but sparse or uncommon along the Chobe and Linyanti rivers. Very common at Shashe Dam and along the Limpopo River but sparse elswhere in the east and southeast. Local movements occur. Solitary or in pairs.

Habitat

Shores of rivers and dams. Confined to areas of permanent water but wanders to adjacent wet areas at times. Usually on sand, shingle, pebbles or rock and less frequently than the Cape Wagtail on wet soil and mud.

Analysis

27 squares (12%) Total count 220 (0,12%)

CAPE WAGTAIL

713

Motacilla capensis

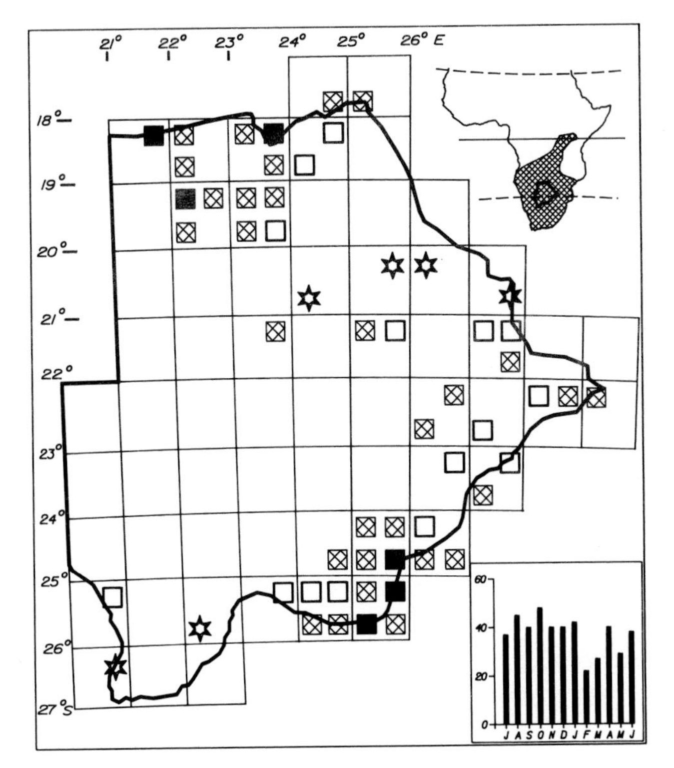

Status

Fairly common to very common resident of the Okavango, Linyanti and Chobe river systems and in the southeast. Extends along the Molopo River to as far as Bray. Recorded once from Nossob Camp in the southwest but there was no other sighting from this region during this survey. Sparse to fairly common in the east in scattered localities. Local movements occur. Usually in pairs, solitary or small loose groups.

Habitat

Shores of rivers, lakes, lagoons and dams. Swamps, marshes, sewage ponds, gardens and parks. Associated with moist ground and is thus more widespread than the African Pied Wagtail.

Analysis

56 squares (24%) Total count 467 (0,26%)

YELLOW WAGTAIL

714

Motacilla flava

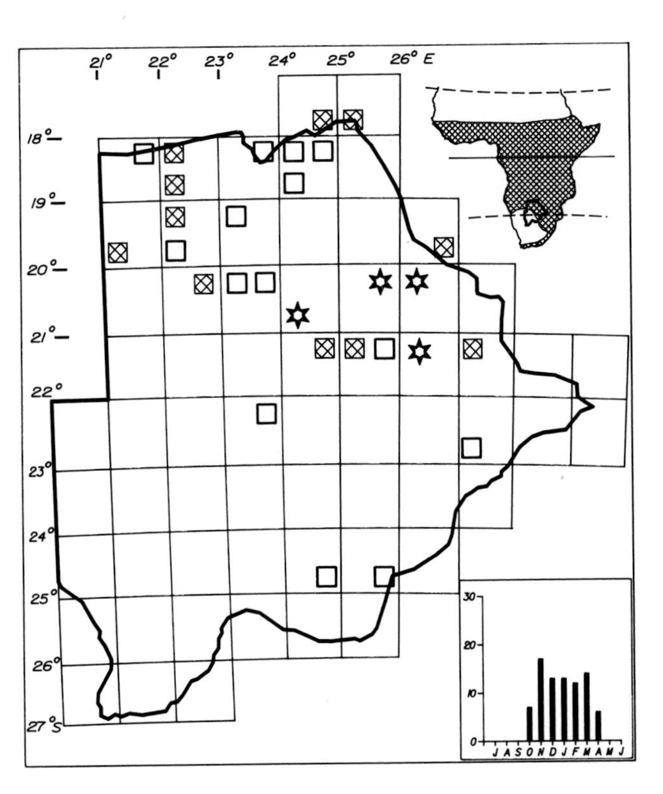

Status

Sparse to locally common Palaearctic migrant from October to April. Erratically recorded. Numbers fluctuate from year to year. Locally common at Shakawe, Kasane, Gomare, Shashe, Mopipi and Orapa. Rare south of 22°S. Numbers are often higher in February and March when flocking occurs prior to northward migration. Several subspecies occur but subspecific variations have not been studied in detail for this survey. Of the races reported *M.f. flava* and *M.f. thunbergi* are the most frequent. Solitary or in small loose groups.

Habitat

Short grass near water including village greens (Shakawe), sewage ponds, floodplains, edges of marshes and lawns. Also at puddles on untarred roads.

Analysis

30 squares (13%) Total count 92 (0,05%)

RICHARD'S PIPIT

716

Anthus novaeseelandiae

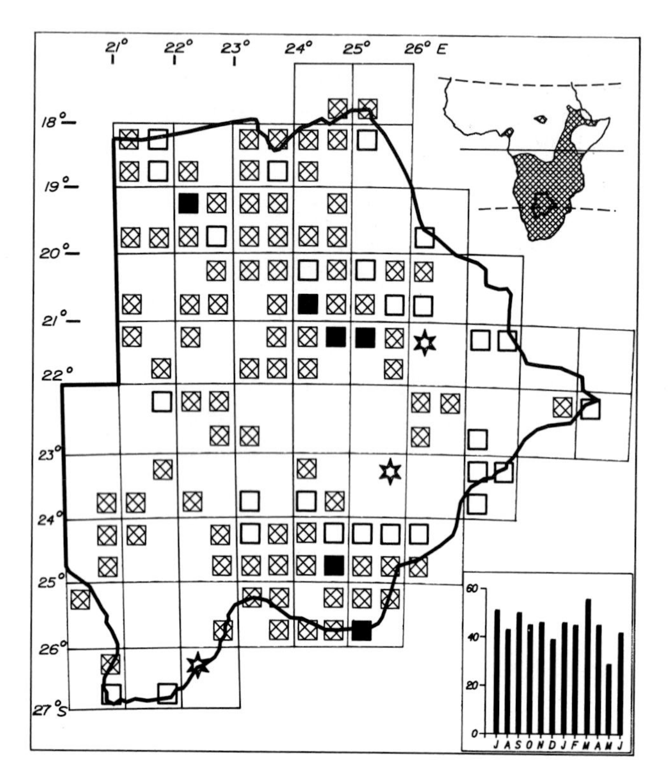

Alternative Name

Grassveld Pipit

Status

Sparse to locally very common throughout the country. Several races occur whose movements are complex. It appears that birds north of 23°S are present mainly from July to January. During March and April there are increased numbers throughout. South of 23°S there is an influx from May to September. However birds are present in many areas in all months. At times abundant on the floodplain of Lake Ngami or the Makgadikgadi grassland. Solitary, pairs or loose gatherings.

Habitat

Open short grassland. Different races may show different preferences when breeding or on passage. Mostly on moist ground—floodplains, marshes, shores of pans, lakes, dams and rivers. Also cleared ground, roadsides, airfields, dry grass plains and pans.

Analysis

118 squares (51%) Total count 559 (0,31%)

LONGBILLED PIPIT

717

Anthus similis

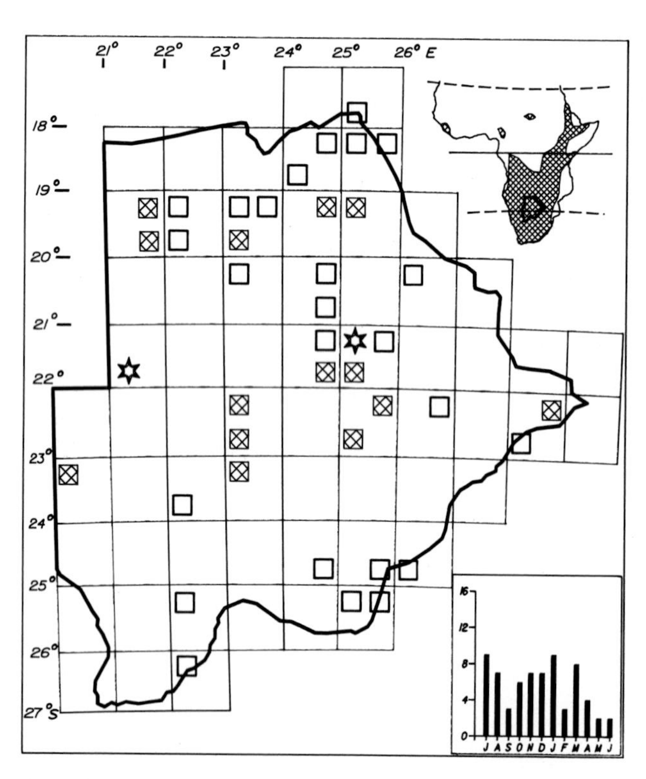

Status

Sparse to fairly common resident. Widely but patchily recorded throughout the country. It appears to be a good indicator of woodland. It is not well known and warrants detailed study. Probably sedentary. Usually in pairs.

Habitat

Broadleafed woodland in the north especially near the edges and cleared areas e.g. borrow pits. In the central Kalahari it occurs at patches of woodland on rock ridges or in bush savanna with stony ground. Also on the ecotone of dry Kalahari pans where there are scattered bushes and trees on the rising ground of the apron. In the southeast it is scarce on slopes of wooded hills and in tree and bush savanna.

Analysis

40 squares (17%) Total count 72 (0,04%)

PLAINBACKED PIPIT

718

Anthus leucophrys

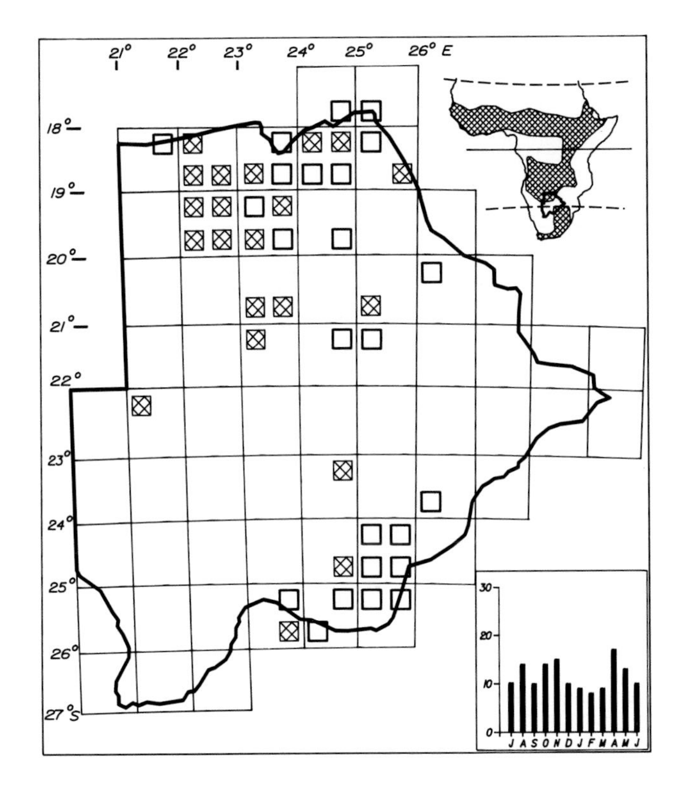

Status

Sparse to fairly common resident mainly north of 21°S and in the southeast. These appear to represent two populations which occur throughout the year in each region. Its apparent absence in the central Kalahari may be due to the absence of suitable habitat. Usually in pairs or small loose groups.

Habitat

In the north it occurs mainly on or near floodplains. Elsewhere on flat sparsely-grassed ground in tree and bush savanna, on poor shallow soils with stunted vegetation, open short grassland with termite mounds, old cultivation or overgrazed ground with secondary growth.

Analysis

44 squares (19%) Total count 145 (0,08%)

BUFFY PIPIT

719

Anthus vaalensis

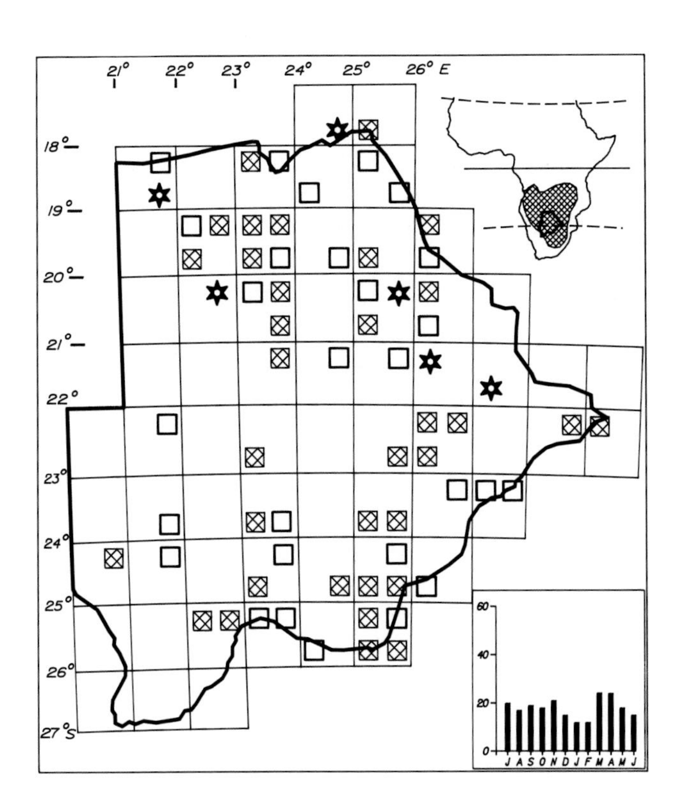

Status

Sparse to fairly common resident throughout the country. It occurs in all months of the year in all areas but it is highly nomadic and unpredictable in occurrence. There is no evidence of regular movement in Botswana whereas it is a dry season visitor to Zambia. Usually in pairs or small loose groups.

Habitat

Flat open areas of short grass usually on richer and deeper soils than the Plainbacked Pipit, such as alluvial soils. In open spaces in tree and bush savanna particularly recently burnt grassland. Short grass in wide fossil valleys, flood-plains, dry grassed pans and airfields.

Analysis

67 squares (29%) Total count 233 (0,13%)

STRIPED PIPIT

Anthus lineiventris

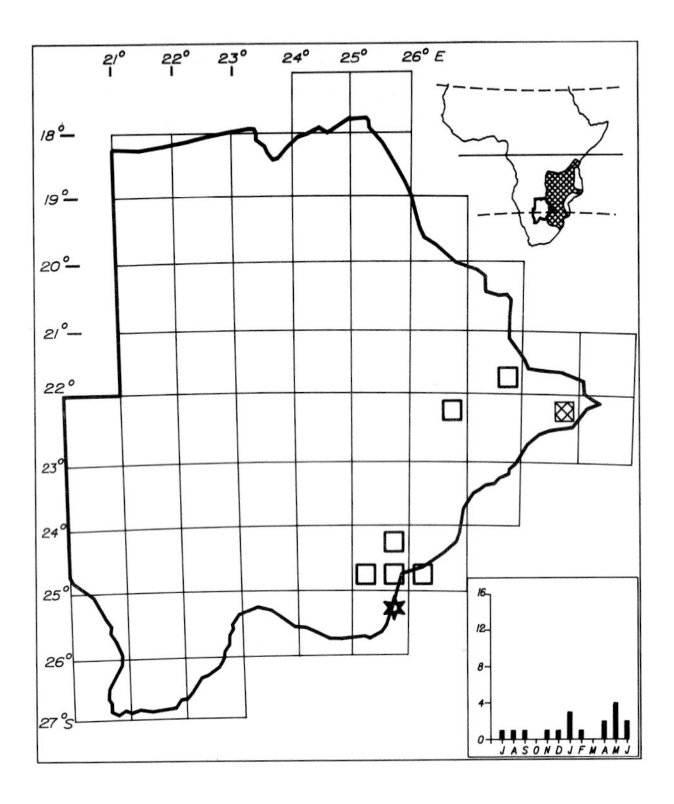

Status

Sparse and very localised resident of hilly regions of the southeast and east. Occurs in Botswana at the western limits of its range in Zimbabwe and Transvaal. Poorly known. Solitary or in pairs.

Habitat

Wooded hillsides usually on steep slopes where rocks are exposed under the trees. Forages on the ground under the canopy and in rock strewn gullies with scattered trees and bushes.

Analysis

8 squares (4%) Total count 19 (0,01%)

TREE PIPIT

Anthus trivialis

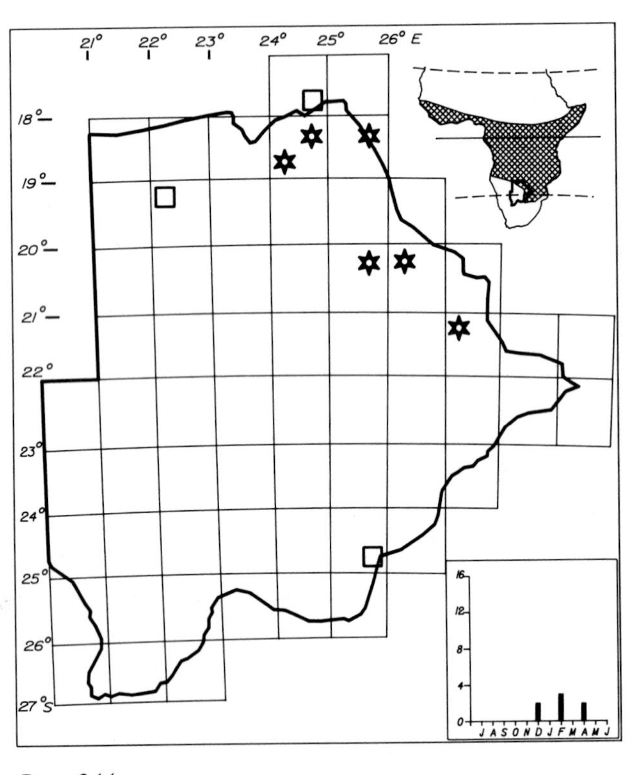

Status

A very scarce Palaearctic migrant which has been recorded twice in the north and once in the southeast during this survey. Dates of sightings fall between 16 December and 18 April. Past records, confined to the northeast, are shown on the map. Probably occurs annually in small numbers. Solitary.

Habitat

Broadleafed woodland such as miombo and *Baikiaea* in the north. Also in tree savanna and mixed woodland.

Analysis

9 squares (4%) Total count 9 (0,004%)

BUSHVELD PIPIT

723

Anthus caffer

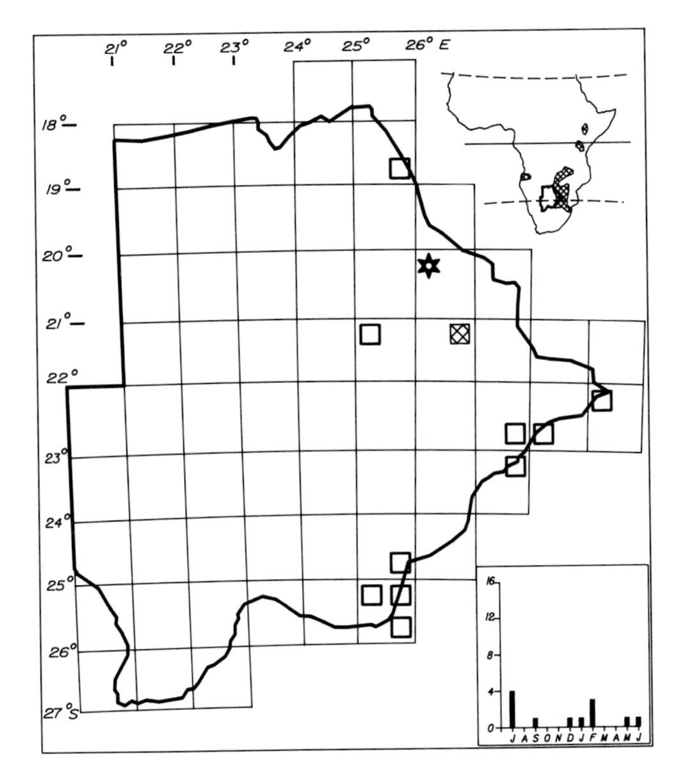

Status

Sparse resident of the southeast, east and northeast. Occurs in Botswana at the western limit of its southern African range. At Orapa it has been recorded in December, January and May suggesting it may be resident there. Solitary, in pairs or small loose groups.

Habitat

Tree and bush savanna in areas of grass with intervening bare ground or loose stones and rocks. Also on edges of woodland and on ground under woodland canopy where there is poor groundcover.

Analysis

11 squares (5%) Total count 16 (0,009%)

ORANGETHROATED LONGCLAW

727

Macronyx capensis

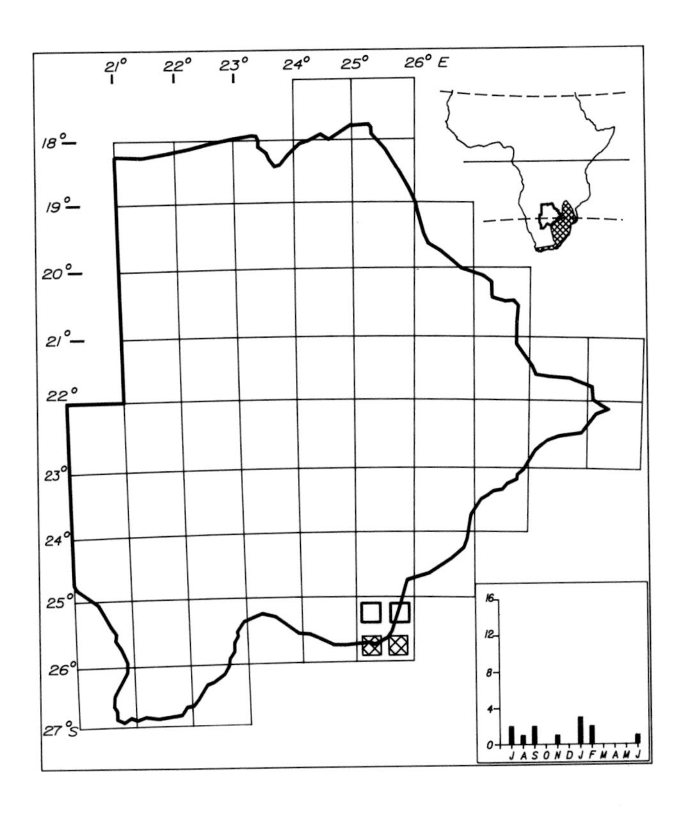

Status

Fairly common resident of the extreme southeast where it occurs at the northwestern limit of its South African range. Most records are from Phitsane Molopo, Good Hope and Ramatlabama. Scarce in adjacent squares immediately north and unlikely to occur more extensively. Usually in pairs.

Habitat

Moist grassland in river valleys, marshy ground, seasonally inundated grassland near dams, pans and rivers; usually in short grass with tussocks under which it places its nest.

Analysis

4 squares (2%) Total count 15 (0,008%)

PINKTHROATED LONGCLAW

730

Macronyx ameliae

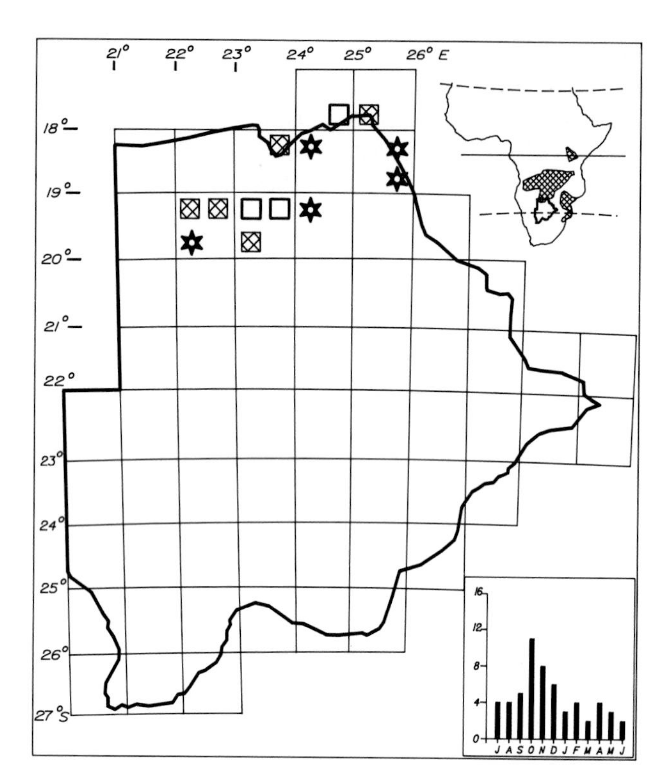

Status

Sparse to fairly common resident of the Okavango, Linyanti and Chobe river systems. In the past it extended as far south as Mpandamatenga in the northeast and may occur there again in high rainfall years. It occurs at the southern limit of its continental range at these longitudes. Present in all months but local movements occur. Usually in pairs.

Habitat

Moist grassland, floodplains, marshes; usually where the grass is fairly tall and tussocky. Moves to adjacent areas when the grass has been burnt. Generally in more permanently wet conditions than the Orangethroated Longclaw.

Analysis

13 squares (6%) Total count 63 (0,04%)

LESSER GREY SHRIKE

731

Lanius minor

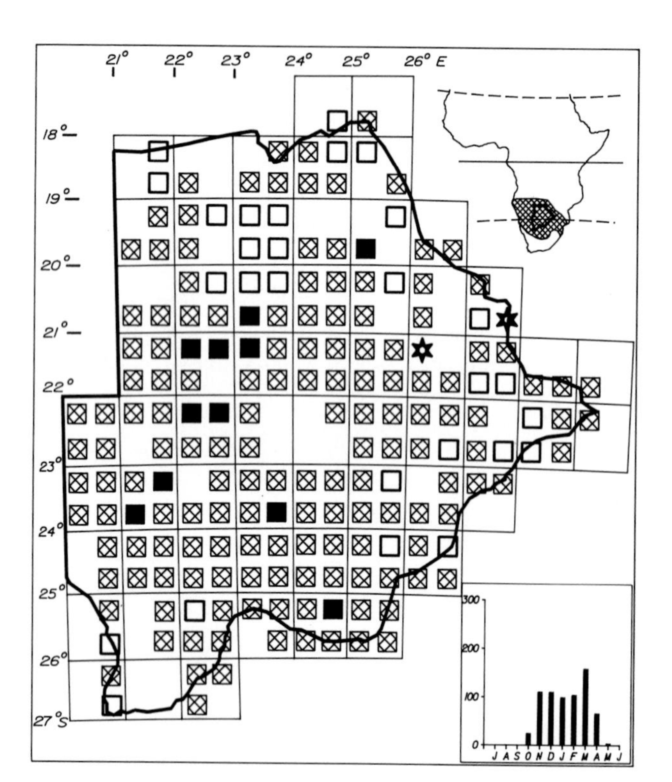

Status

Common Palaearctic migrant to all regions of the country. Generally more common in semiarid areas south of 20°S than in the more lush savannas of the north. Present between October (15th) and early May with greatest numbers between November and April. Botswana is a major wintering area for this species. Usually solitary.

Habitat

Open tree and bush savanna, specifically in more open areas than the Redbacked Shrike. Open semidesert scrub savanna with widely scattered low bushes or stunted trees, ecotone of Kalahari pans. Perches conspicuously on isolated bushes in open savanna and commonly on telephone wires and poles along roads.

Analysis

189 squares (82%) Total count 706 (0,39%)

FISCAL SHRIKE

732

Lanius collaris

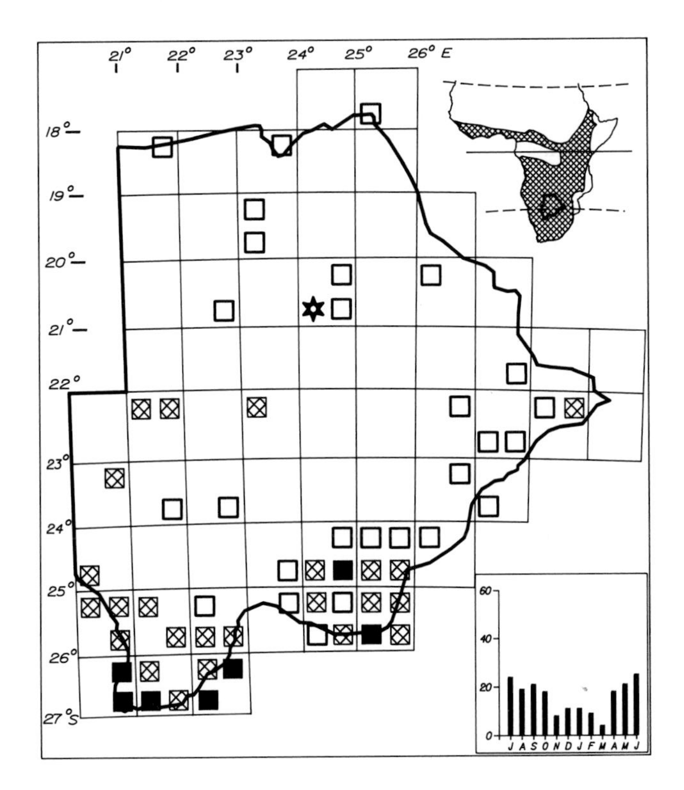

Status

Sparse to common resident of the southeast and southwest. It is very common at Jwaneng, Phitsane Molopo and along the western Molopo Valley. Its numbers are increased in winter by birds moving in from the south. Very sparse and unpredictable in the east, west and north where it occurs mainly in winter but the origin of these birds is uncertain. Egglaying August and September. Solitary or in pairs.

Habitat

Open tree and bush savanna, riverine bush, cultivation and occasionally gardens. Favours open spaces such as grass plains, wide river valleys and *Acacia* bushes along drainage lines in semidesert.

Analysis

59 squares (26%) Total count 218 (0,12%)

REDBACKED SHRIKE

733

Lanius collurio

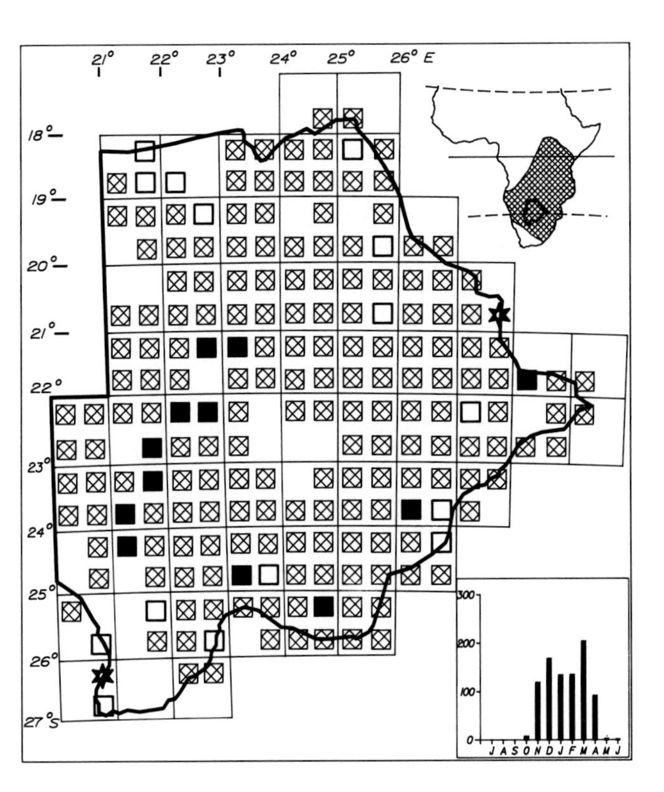

Status

Common to very common Palaearctic migrant occurring throughout the territory. Its main arrival is in early November (earliest date 25 October) and departure mid-April (latest date 3 May). Peak passages are in December and March. Fully moulted birds are seen from February onwards. Males predominate and approximately 30% of all birds on arrival are in juvenile plumage. Usually solitary, but occasionally in groups of 5–20 birds.

Habitat

Tree and bush savanna, edges of woodland, cultivation, gardens, telephone wires along roads. More common than Lesser Grey Shrike in heavily bushed areas. Rare within woodland.

Analysis

200 squares (87%) Total count 906 (0,51%)

LONGTAILED SHRIKE

735

Corvinella melanoleuca

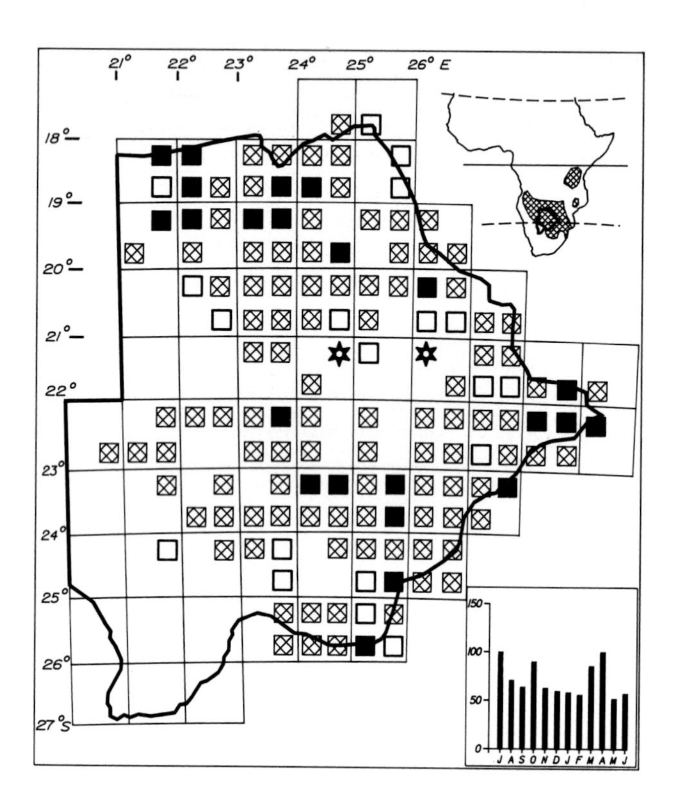

Status

Fairly common to very common resident in all regions except the southwest. Its status in the southern part of the northwest needs further study—its apparent absence there may be due to poor field coverage. Occurs in Botswana at the western limit of its range in southern Africa. Egglaying all months from November to February. Usually in small groups of 3–10 birds.

Habitat

Tree and bush savanna especially *Acacia* savanna. Commonest in moist areas and in river valleys but also occurs in semidesert conditions well away from water.

Analysis

142 squares (62%) Total count 906 (0,51%)

TROPICAL BOUBOU

737

Laniarius aethiopicus

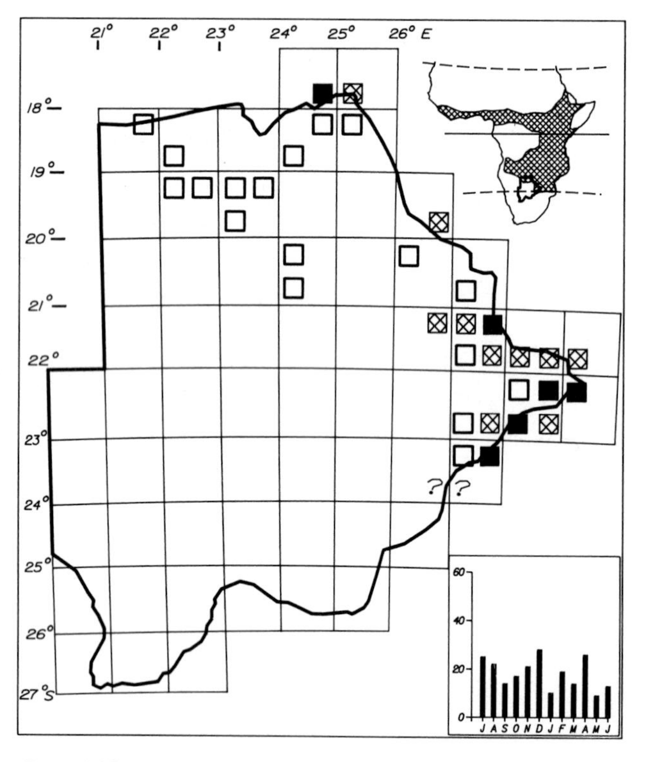

Status

Fairly common to very common resident along the Limpopo and Shashe river drainage and at Kasane and Serondella. Sparse in the Okavango Delta and along the Boteti River. Extends along the eastern border north of Francistown to as far as Tsebanana. Its range may overlap that of the Southern Boubou in the upper Limpopo basin south of the Tropic of Capricorn where further research is needed to clarify each species' status. Egglaying November. Usually in pairs or solitary.

Habitat

Dense cover along water courses, thickets, riparian vegetation, gardens.

Analysis

36 squares (16%) Total count 230 (0,13%)

SWAMP BOUBOU

738

Laniarius bicolor

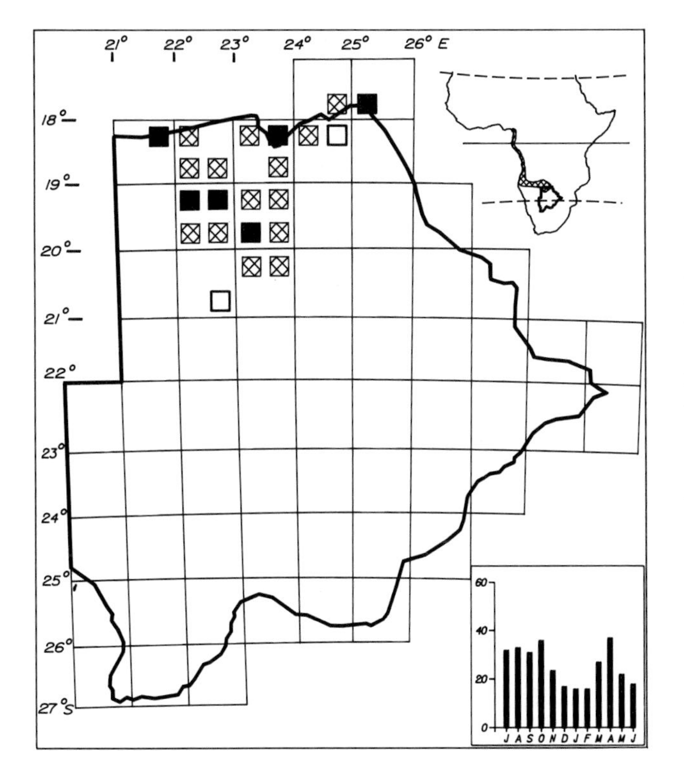

Status

Fairly common to very common resident of the Okavango, Linyanti and Chobe river systems to which it is confined. Sparse at Lake Ngami during this survey but it may occur there more commonly in wetter rainfall cycles. Occurs in Botswana at the southern limit of its continental range. Egglaying July, September, February and March. Usually in pairs or solitary.

Habitat

Reed and papyrus beds, dense riverine vegetation, thicket, riverside gardens. Near or over water in swamps, marshes, on floating islands, at lagoons and along rivers and other waterways.

Analysis

22 squares (10%) Total count 319 (0,18%)

CRIMSONBREASTED SHRIKE

739

Laniarius atrococcineus

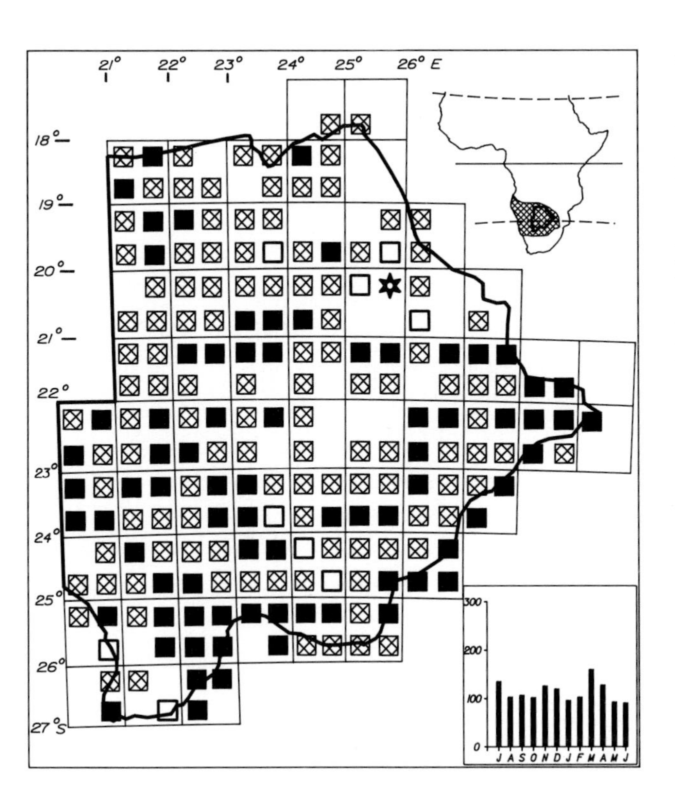

Status

Common to very common resident throughout the country. Commonest in the southern part of its range. It is the most commonly recorded shrike in Botswana. It is absent from the broadleafed woodland of the northeast although its range extends east into the adjacent Hwange National Park in Zimbabwe. Egglaying October to February, mainly November. Host to the Black Cuckoo. Usually in pairs.

Habitat

Tree and bush savanna. Usually in the cover of bush, thicket or tangled growth. Commonest in dry savanna, but also occurs in lush conditions along river valleys but not in evergreen forest or riparian trees.

Analysis

199 squares (86%) Total count 1419 (0,79%)

PUFFBACK 740

Dryoscopus cubla

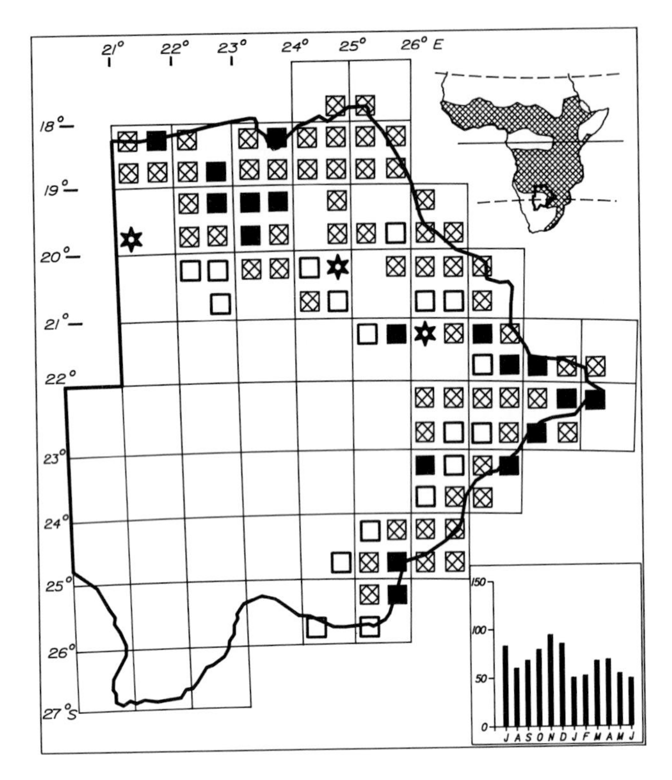

Status

Fairly common to very common resident of the north and east. It occurs in Botswana at the southern and western limits of its continental range. Its distribution in Botswana illustrates very well the limiting effect of the arid Kalahari savanna on birds of tropical woodland affinities. Egglaying November and January. Solitary or in pairs. Often a member of woodland bird parties.

Habitat

Canopy of deciduous woodland and riverine forest. Bush savanna where there are some mature trees. Occurs in the high rainfall areas and does not extend into dry Kalahari savanna.

Analysis

98 squares (43%) Total count 829 (0,46%)

BRUBRU 741

Nilaus afer

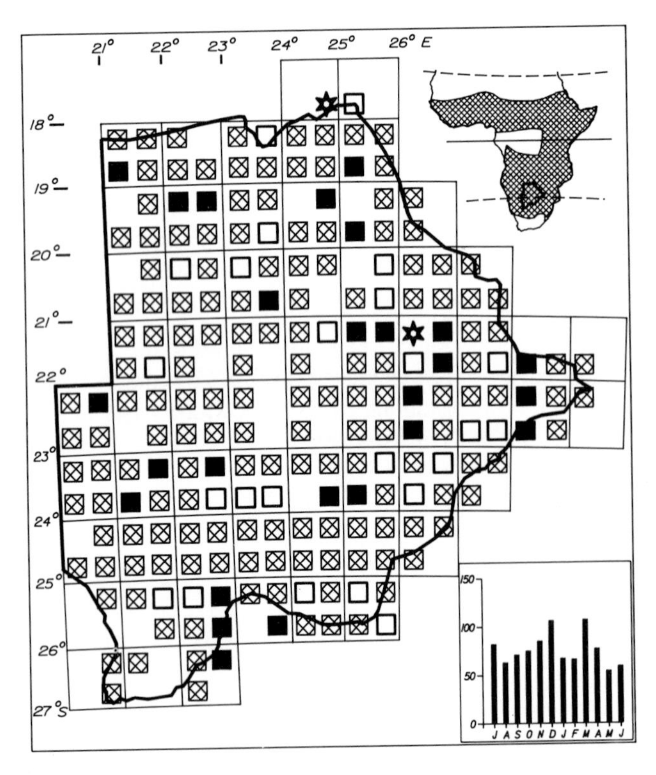

Status

Fairly common to very common resident throughout the country. Recorded as very common in 11% of squares and is equally as common in semidesert as in rich woodland. Several races occur. Egglaying October, November, January and February. Usually in pairs.

Habitat

Acacia and broadleafed woodland, tree and bush savanna in high and low rainfall areas. It is found only where there are trees and does not occur in semidesert scrub or in areas with only low bushes.

Analysis

206 squares (89%) Total count 938 (0,53%)

THREESTREAKED TCHAGRA 743

Tchagra australis

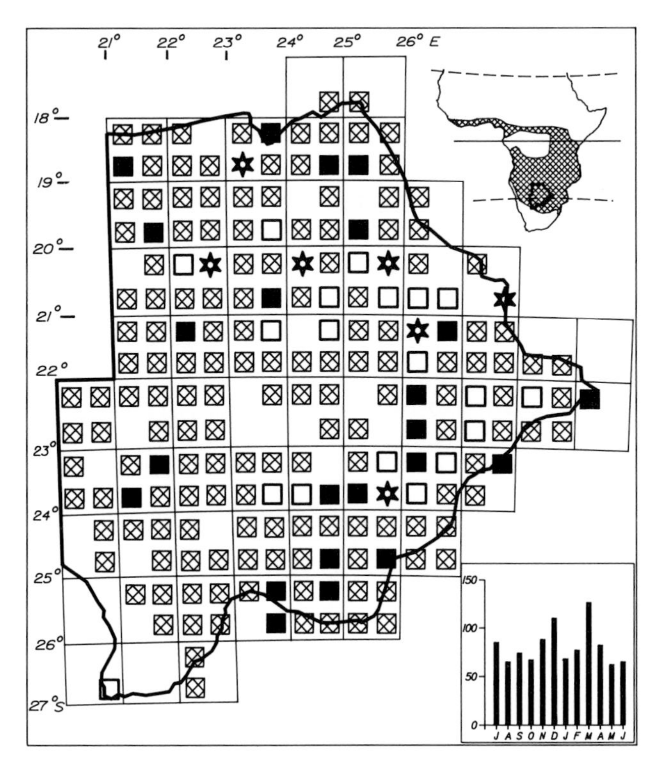

Alternative name

Brownheaded Tchagra

Status

Uncommon to very common resident throughout the country. Sparse and mainly absent in the southwest corner. Recorded as very common in 10% of squares in a wide variety of habitats. Occurs in Botswana near the southern limit of its continental range. Egglaying in November. Solitary or in pairs.

Habitat

Tree and bush savanna in high and low rainfall areas. Also in broadleafed and *Acacia* woodlands near the edge and in isolated patches. Usually in leafy trees or in thick bushes where it remains well hidden. Skulks low down in vegetation and forages on the ground.

Analysis

198 squares (86%) Total count 1014 (0,57%)

BLACKCROWNED TCHAGRA 744

Tchagra senegala

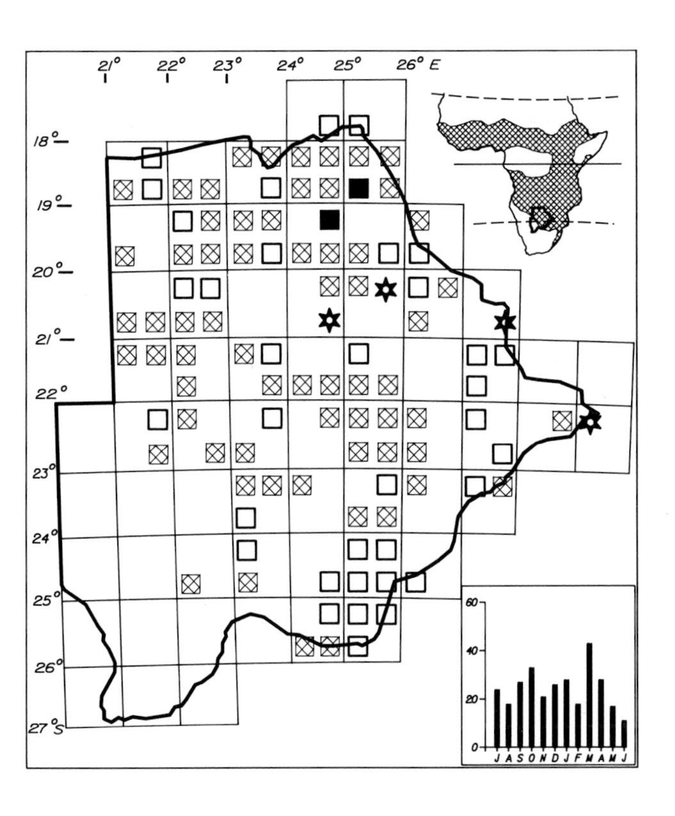

Status

Sparse to common resident in the north, east, central and southeastern regions. Mainly absent in the west and southwest and in the Makgadikgadi region. Commonest in the northern woodlands. Previously regarded as a species of woodland or rich savanna—this survey has revealed a widespread presence in the central Kalahari. Often occurs alongside the Threestreaked Tchagra. Egglaying December. Solitary or in pairs.

Habitat

Broadleafed woodland and tree savanna particularly in areas with good ground cover. In miombo woodland more common than the Threestreaked Tchagra. Also occurs in stunted *Terminalia* and scrub savanna in the central Kalahari.

Analysis

105 squares (46%) Total count 307 (0,17%)

BOKMAKIERIE 746

Telephorus zeylonus

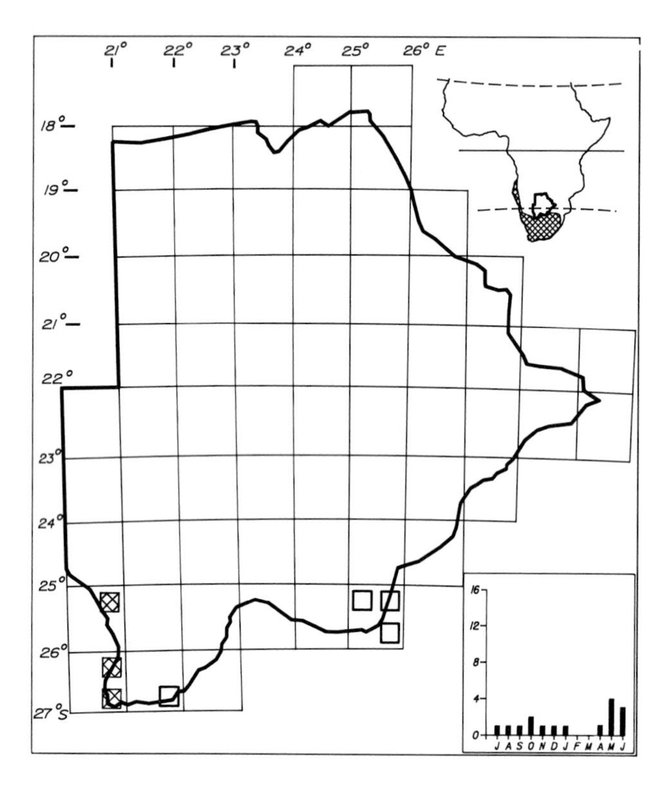

Status

Sparse resident of the southeast. Uncommon in the southwest from Nossob Camp south to Bokspits and west to Khuis. Occurs at the northern and eastern limit of its range in South Africa and Namibia. May be more common in winter. The current records are new to the avifauna of Botswana. They may represent a cyclic extension of range but it seems more likely that its limited presence had not been previously detected. Solitary or in pairs.

Habitat

Bushes in grassland, open savanna and along river valleys. Also recorded in gardens and on fence posts on farmland and at roadsides.

Analysis

7 squares (3%) Total count 18 (0,01%)

ORANGEBREASTED BUSH SHRIKE 748

Telephorus sulfureopectus

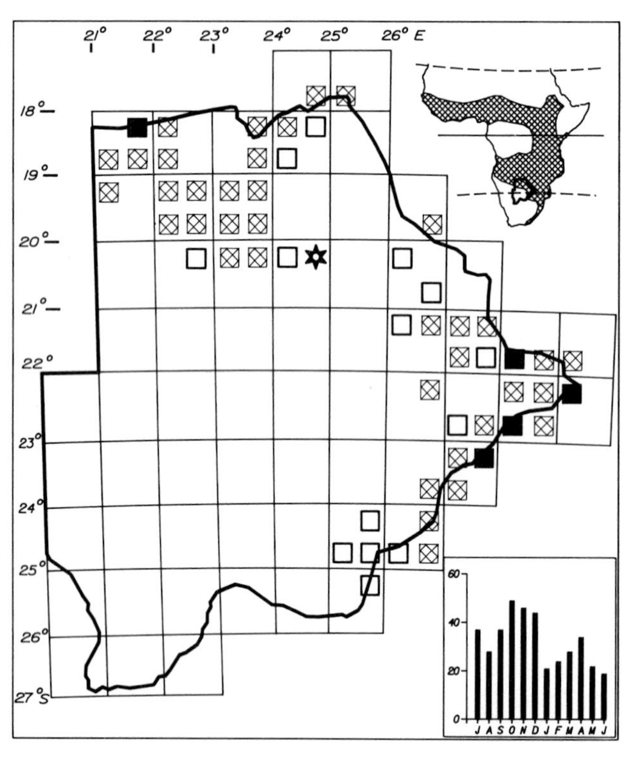

Status

Uncommon to very common resident in the north and east. Commonest in the valleys of the Limpopo, Shashe, Okavango, Chobe and Linyanti rivers. Extends south to Lobatse but is scarce in the southeast. Egglaying December and January. Solitary or in pairs.

Habitat

Riparian forest and riverine vegetation. Extends into adjacent *Acacia* and broadleafed woodland and tree savanna in river valleys. Usually in the cover of dense growth, bushes and thicket.

Analysis

57 squares (25%) Total count 397 (0,22%)

GREYHEADED BUSH SHRIKE

Malaconotus blanchoti

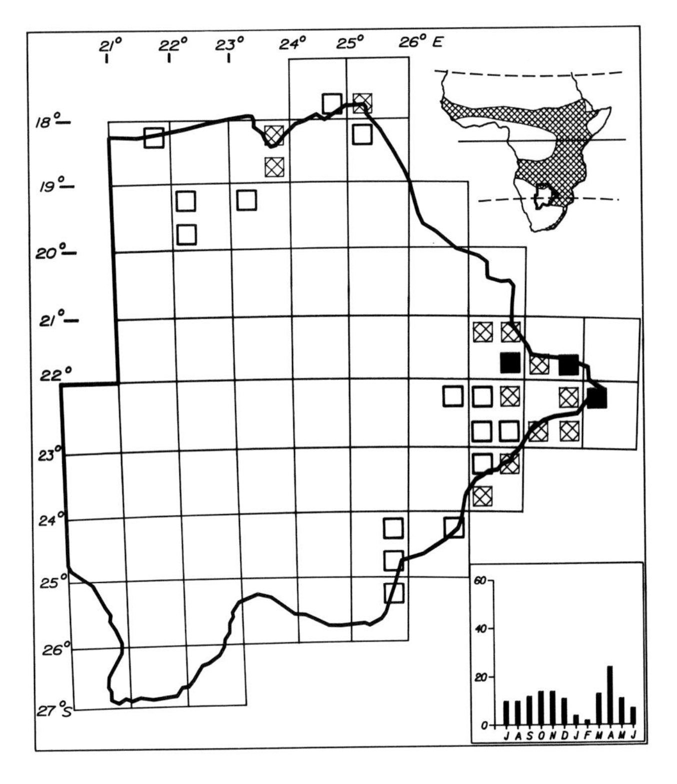

Status

Sparse to locally very common in the east from Francistown to Martin's Drift. Very common along the Shashe River to its confluence. Sparse in the north where it has been recorded as common only at Linyanti and Kasane. Extends south along the Limpopo drainage to Gaborone and Lobatse. It is not well known in Botswana and more detailed study is merited. Solitary or in pairs.

Habitat

Riverine forest and thicket, woodland in river valleys where it is found in tangled growth and creepers under the canopy and in bushes on the edges. It appears to be less common than the Orangebreasted Bush Shrike in this habitat but it is also more easily overlooked.

Analysis

30 squares (13%) Total count 139 (0,08%)

WHITE HELMETSHRIKE

Prionops plumatus

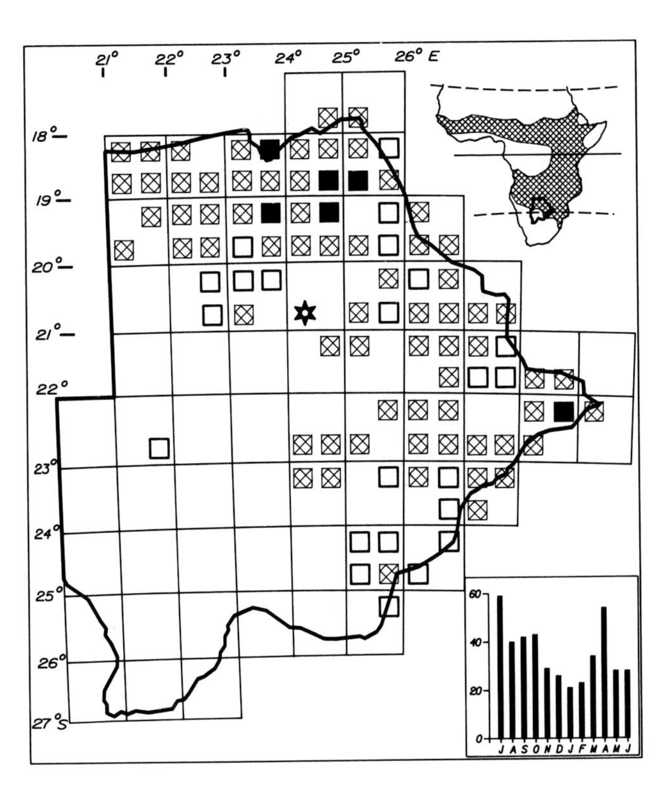

Status

Fairly common to very common resident of the north and east. Extends from the east into the central Kalahari woodlands. The isolated record from Takatshwane in the west suggests a more extensive western range in the past and warrants further investigation. It occurs in Botswana at the southern and western limit of its range in Africa. Egglaying October to February. Occurs in flocks of 5–15 birds.

Habitat

All types of woodland but mainly broadleafed including mopane. Also tree and bush savanna even where the vegetation is stunted by low precipitation.

Analysis

98 squares (43%) Total count 439 (0,25%)

REDBILLED HELMETSHRIKE

754

Prionops retzii

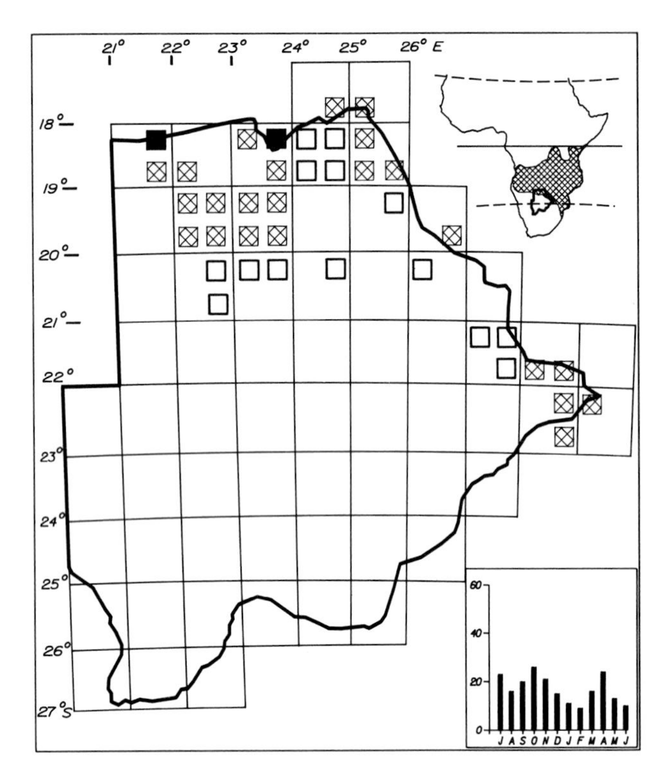

Status

Fairly common to locally very common resident of the Okavango Delta and the Linyanti, Chobe and Nata river systems. Fairly common in the east along the lower Shashe and Limpopo drainage. Occurs at the southern and western limit of its African range. Less widespread than the White Helmetshrike. Egglaying November. Parasitism by the Thickbilled Cuckoo, a rare species, has not yet been demonstrated in Botswana. Occurs in flocks of 5–15 birds.

Habitat

Broadleafed woodland including mopane, usually where the trees are well established and occur as continuous mature stands. Extends into adjacent tree savanna including *Acacia* and into mixed woodland.

Analysis

39 squares (17%) Total count 212 (0,12%)

WHITECROWNED SHRIKE

756

Eurocephalus anguitimens

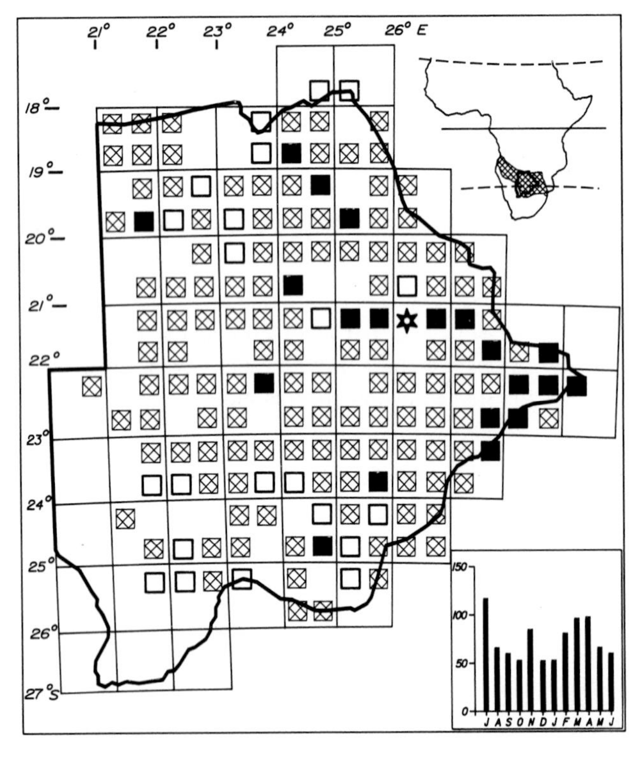

Status

Common to very common resident throughout the country except in the southwest and far west where its absence represents the western limit of its southern African range. Recorded as very common in 9% of squares mainly in the east and north. Egglaying November, December and January. Usually in pairs, often several pairs occur together in the same vicinity.

Habitat

Tree and bush savanna particularly semidesert thorn savannas on Kalahari sands. Also on the edges of woodland in high and low rainfall areas. Never occurs within woodland with touching canopies. Uses open areas for perching and hawking.

Analysis

164 squares (71%) Total count 873 (0,49%)

WATTLED STARLING

760

Creatophora cinerea

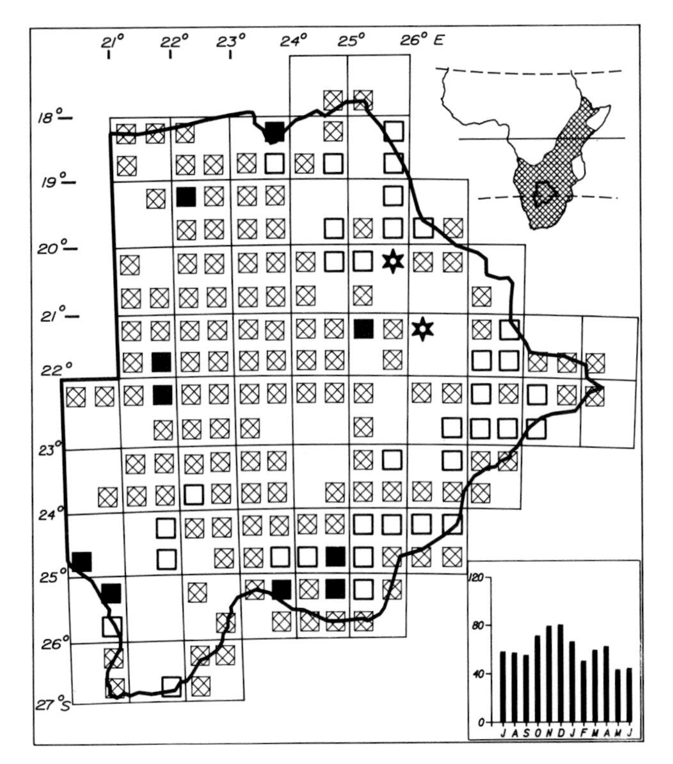

Status

Uncommon to locally very common breeding African migrant. Present throughout the year in many areas. North of 20°S it is mainly a nonbreeding visitor. Breeds from November to March in 'cities' in the central Kalahari and other dry woodland. Moves to other areas of Botswana and to neighbouring countries in the nonbreeding season. Several hundred birds may occupy a 'city' consisting of many nests in adjacent trees over an area up to one square kilometre. Usually in small (5–15 birds) or large (50–200 birds) flocks.

Habitat

Breeds in semiarid woodland and tree savanna including remnant riverine *Acacia* along fossil valleys e.g. Okwa. Occurs in a wide variety of habitats in the nonbreeding season including villages, towns, parks, gardens, sewage ponds and farmland. A large flock was once found on the ground in the shade of stunted mopane in scorching midday heat in December. Breeding may be linked to the breeding of locusts as a food source.

Analysis

170 squares (74%)   Total count 744 (0,42%)

PLUMCOLOURED STARLING

761

Cinnyricinclus leucogaster

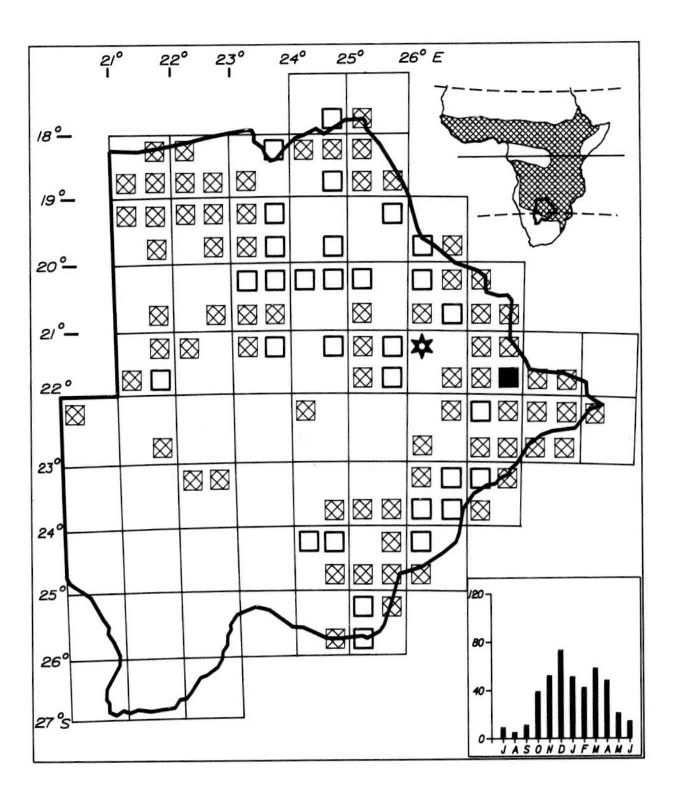

Status

Uncommon to common breeding intra-African migrant mainly to northern and eastern regions. Thinly distributed in central, western and southwestern regions. Recorded in all months but occurs mainly between October and April. Its movements are complex and not well understood in southern Africa and merit special study. Egglaying November to February. Occurs in pairs, small groups or flocks of 15–50 birds.

Habitat

Broadleafed woodland, riparian forest, mixed woodland—usually where there are trees with fruit or berries. Also in gardens in towns where it will breed if undisturbed (as at Jwaneng).

Analysis

105 squares (46%)   Total count 442 (0,25%)

BURCHELL'S STARLING 762

Lamprotornis australis

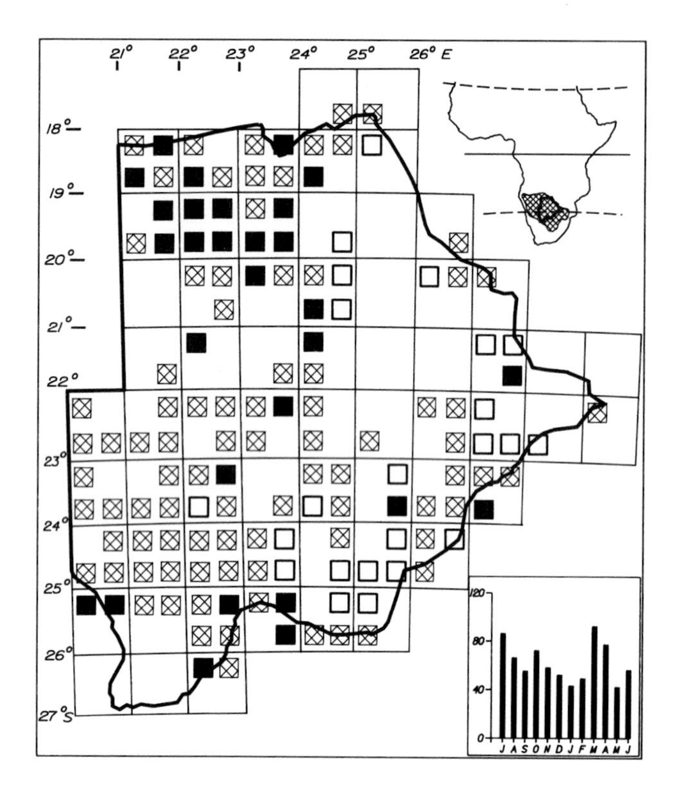

Status

Very common resident of the Okavango Delta and northwestern woodlands. Common and widespread south of 22°S but sparser in the east. It is locally common or very common in river valleys in all regions of the country such as along the Nossob, Molopo, Limpopo, Shashe, Nata, Boteti, Okavango, Chobe and Linyanti rivers. Absent from the Makgadikgadi Pans and northeastern mopane woodlands except in the Tsebanana area near the Nata River confluence. Some seasonal movements occur. Egglaying January. Usually in groups of 2–10 birds.

Habitat

Acacia woodland and tree savanna especially in areas where there are good stands of *Acacia erioloba* such as along river valleys and parklike areas which occur at Good Hope and in the western woodlands. Also in mixed woodlands with mature trees and open areas with bare or lightly-vegetated ground cover. Often near human habitation in towns and villages particularly those with *Acacia erioloba*.

Analysis

137 squares (60%) Total count 773 (0,43%)

LONGTAILED STARLING 763

Lamprotornis mevesii

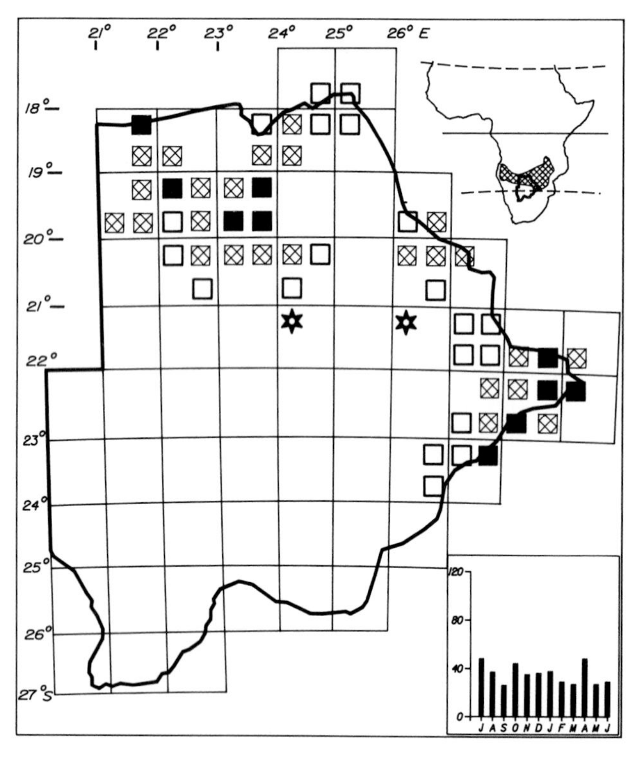

Status

Fairly common to very common resident of the Okavango, Linyanti and Boteti river systems. Sparse in the northeast. Very common along the lower Shashe and Limpopo rivers to as far south as Martin's Drift. Sparse to fairly common in adjacent areas of the east. Sparse around Francistown but common around Nata and Tsebanana. Local movements may occur but not as extensively as in the Burchell's Starling. Egglaying February. Usually in pairs or small groups, but roosts communally in large flocks at times.

Habitat

Tall trees on alluvial soils, mopane and riparian *Acacia*. Also in other tall mature stands such as *Combretum* and *Adansonia*. Usually where there is bare ground or poor understorey. Forages by walking on the ground. It is absent from stunted mopane.

Analysis

57 squares (25%) Total count 415 (0,23%)

GLOSSY STARLING

764

Lamprotornis nitens

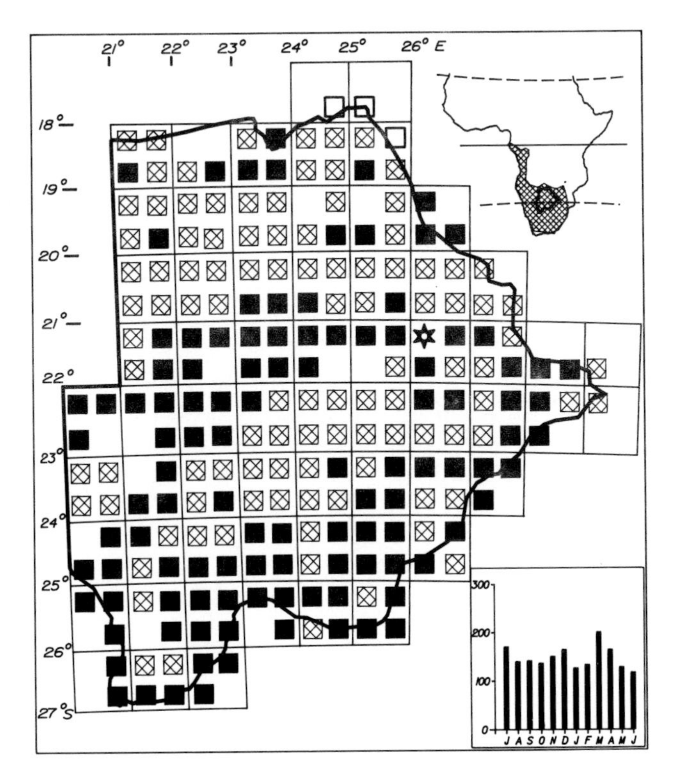

Status

Common to very common resident throughout. It is the commonest starling and one of the most common and widespread species in the country. Generally less common north of 20°S than south of this parallel. Its range overlaps extensively in the north and east with the Greater Blue-eared Starling but the two are separated ecologically. Egglaying in all months from October to April. Solitary, in pairs, sometimes in large flocks in the nonbreeding period.

Habitat

Tree and bush savanna of all types including semidesert with only a few isolated trees. Absent from treeless areas such as plains and scrub savanna but appears at Kalahari pans to drink and forage. Also occurs on the edge of any woodland and in a variety of habitats in towns and villages such as gardens, parks, plantations, cultivation, cattle pens, boreholes and farmland.

Analysis

217 squares (94%) Total count 1834 (1,03%)

GREATER BLUE-EARED STARLING

765

Lamprotornis chalybaeus

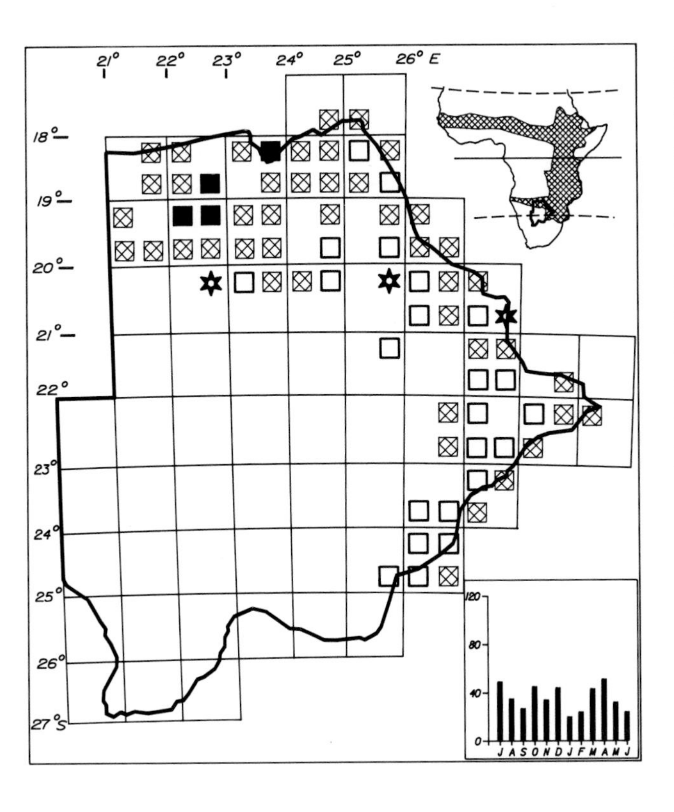

Status

Fairly common to very common resident of the Okavango, Linyanti and Chobe river systems. Uncommon in adjacent areas of the north. In the east it is sparse to fairly common along the Nata, Shashe and Limpopo drainage south to Mochudi, Gaborone, Tlokweng and the Marico River but it is very sparse in the southern squares. In the northwest it is usually the commonest of the short-tailed glossy starlings in summer but elswhere in their overlapping range it is less common than the Glossy Starling. Seasonal movements occur. Egglaying March. Usually in pairs, less often in small flocks.

Habitat

Broadleafed woodland and tree savanna where there are mature trees. Also in riparian forest and riverine *Acacia*.

Analysis

74 squares (32%) Total count 435 (0,24%)

SHARPTAILED STARLING 767

Lamprotornis acuticaudus

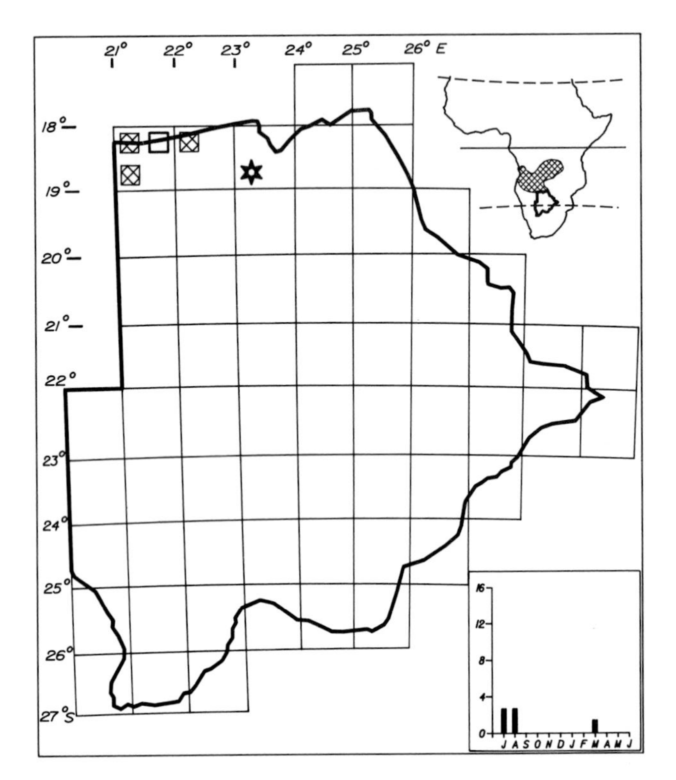

Status

Poorly known and of uncertain status. It has been recorded in the extreme northwest on five occasions since 1980. One previous record from 1823C in 1977. Four of the records are for July and August and one in March. Occurs at the southern limit of its range in Angola, northern Namibia and Zambia. Probably a regular seasonal visitor as it is known to wander like other members of the genus. A detailed study of this species in the northwest is needed. Usually in small flocks of 5–30 birds.

Habitat

Deciduous woodland and tree savanna. Usually on the edge of woodland with adjacent open areas such as plains, floodplains and marshes.

Analysis

5 squares (2%) Total count 6 (0,003%)

REDWINGED STARLING 769

Onychognathus morio

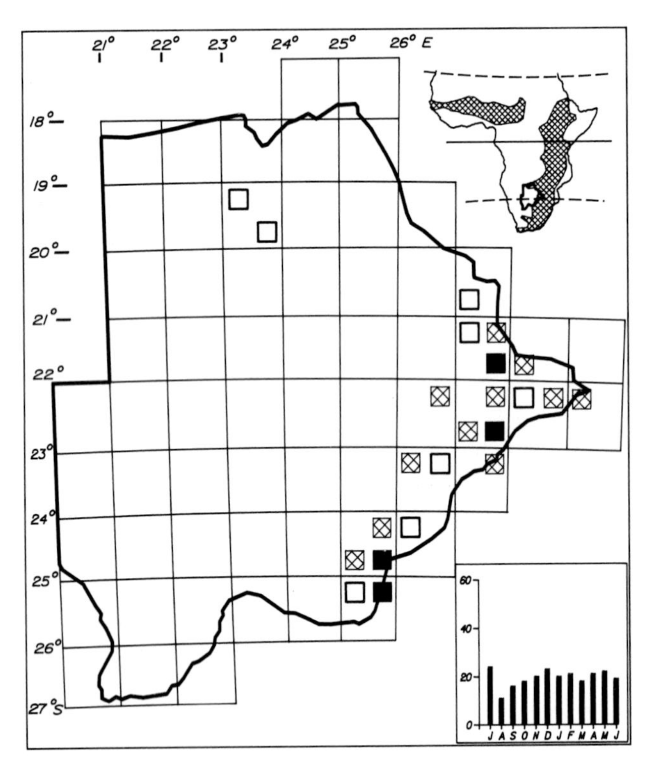

Status

Uncommon to locally very common resident of the east south of Plumtree. Very common in rocky areas at Selebi Phikwe, Tswapong Hills, Gaborone and Ootse. Rare in the southern part of the Okavango Delta where it may occur only as a wanderer. Egglaying October and January to April. Usually in pairs or small flocks.

Habitat

Rocky hills and wooded gorges in mountainous regions. Breeds on cliff ledges. Also in areas of granite boulder outcrops in the northern part of its range. Has adapted successfully to urban environments where it breeds regularly on tall buildings and feeds in gardens and parks.

Analysis

24 squares (10%) Total count 248 (0,14%)

YELLOWBILLED OXPECKER 771

Buphagus africanus

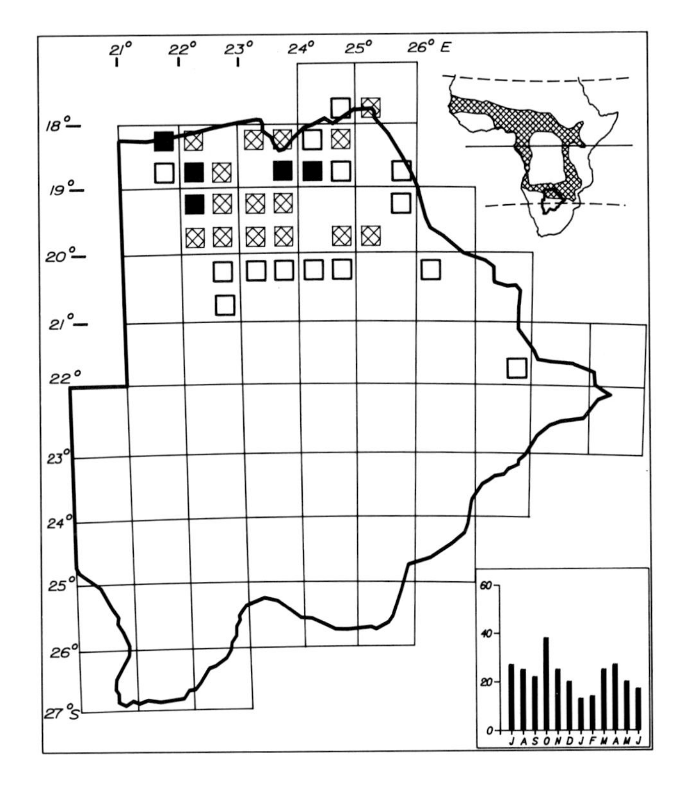

Status

Fairly common to very common resident in the Okavango, Linyanti and Chobe regions south to Nxai Pan and Maun. Sparse on the periphery of the northern game herds as at Lake Ngami, Makalamabedi, Nata and Mpandamatenga. Recorded once near Selebi Phikwe. Occurs at the southern limit of its continental range at these longitudes. Egglaying October. Usually in small groups of 4–8 birds.

Habitat

Deciduous woodland and tree savanna in the tropical rainfall areas of the north. Nests in tree holes and forages on game animals such as buffalo, giraffe, rhinoceros, hippopotamus and the larger antelopes, also on cattle and donkeys in the northwest.

Analysis

34 squares (15%) Total count 276 (0,15%)

REDBILLED OXPECKER 772

Buphagus erythrorhynchus

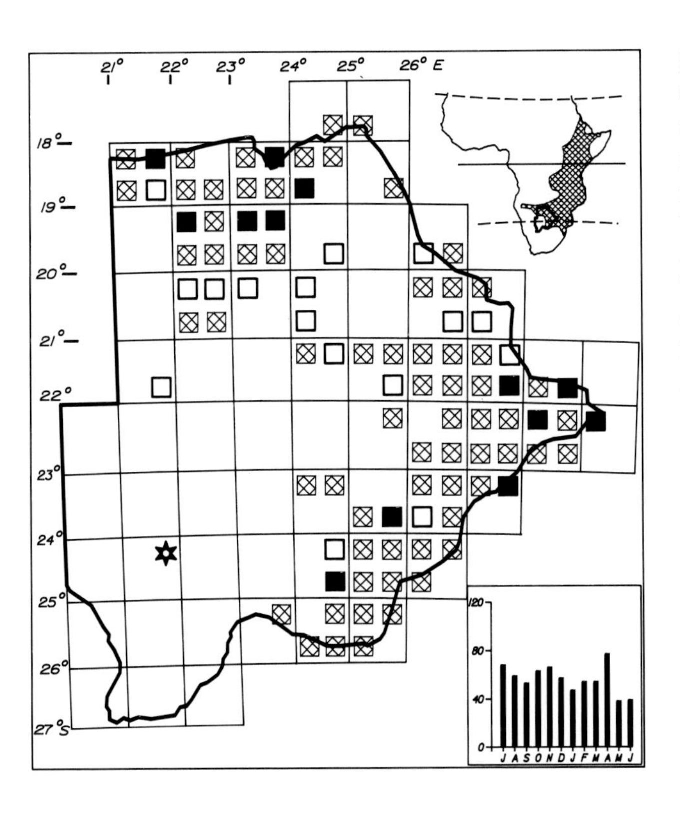

Status

Common to very common resident of the north and east south to Ramatlabama and westwards along the Molopo River to as far as Bray. It is sparse and unpredictable in the northeast and in the Makgadikgadi region. Overlaps extensively with the Yellowbilled Oxpecker in the north and occasionally the two are found together on the same animal. The past record from Tshane in the west may indicate a recent contraction of range. Occurs at the western limit of its range in southern Africa. Egglaying November. Usually in groups of 4–15 birds.

Habitat

Broadleafed and *Acacia* woodland and tree savanna. Extends into semidesert where there are trees for nesting. Forages on game animals in the north. South and west of Francistown it is found almost exclusively on cattle and donkeys. Often near towns and villages and at cattle posts and kraals if there are trees nearby.

Analysis

96 squares (42%) Total count 702 (0,39%)

MARICO SUNBIRD

779

Nectarinia mariquensis

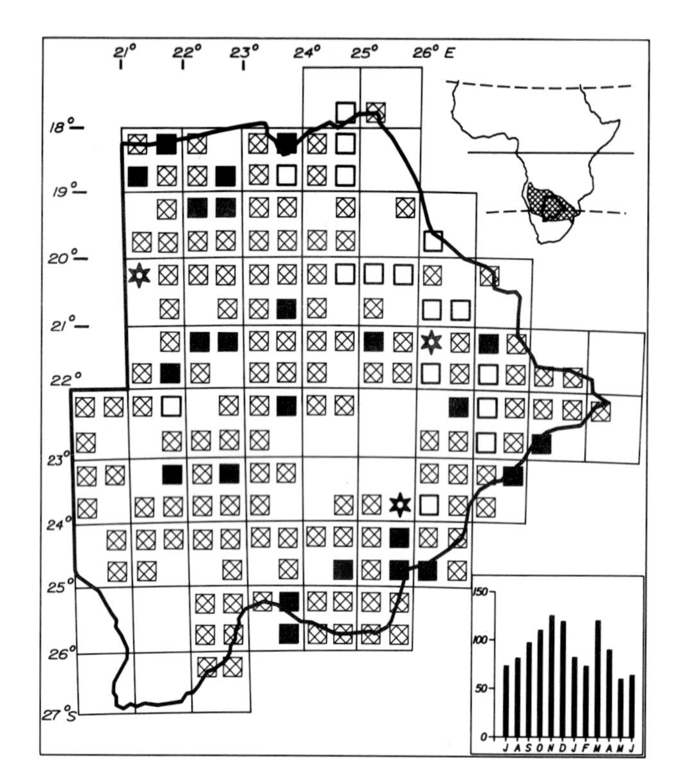

Status

Fairly common to very common resident throughout the country except in the southwest where it is mainly absent. Sparse and uncommon in the northeastern woodlands. It is the characteristic sunbird of the Kalahari savannas. Very common in 10% of squares. Egglaying September to January. Solitary or in pairs.

Habitat

Acacia and mixed deciduous woodlands, tree and bush savanna on Kalahari sand especially where there are flowering plants such as mistletoes, succulents or creepers. Also in gardens, parks, villages and on farms.

Analysis

167 squares (73%) Total count 1121 (0,63%)

WHITEBELLIED SUNBIRD

787

Nectarinia talatala

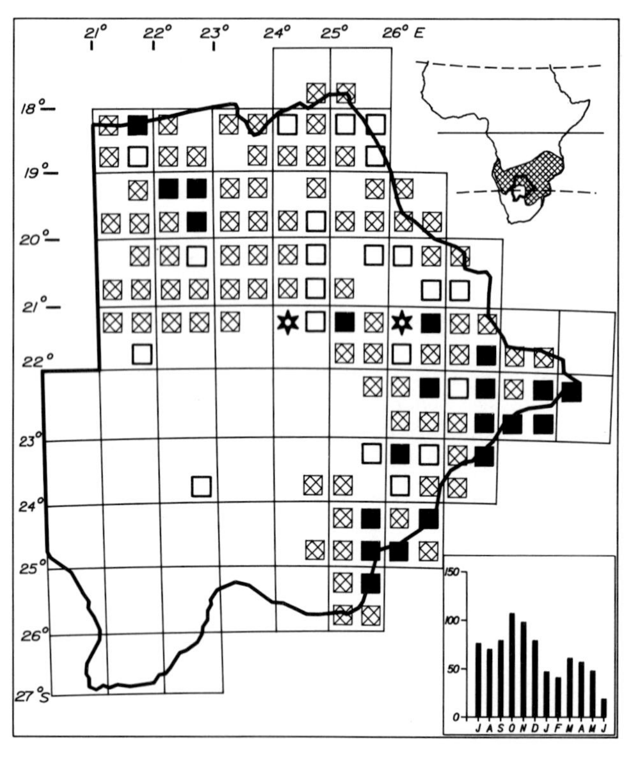

Status

Fairly common to very common resident of the north and east. Occurs in Botswana at the western limits of its southern African range. Recorded once at Kang and its status in the western woodlands needs further investigation. Very common generally in the east and southeast possibly as a result of an increase in gardens with flowering plants in these well-populated regions. Egglaying September to January, mainly September and November. Solitary or in pairs.

Habitat

Any woodland and tree and bush savanna in the higher rainfall areas. Regular in gardens and parks. Also in riverine vegetation including *Acacia* along watercourses.

Analysis

122 squares (53%) Total count 830 (0,47%)

DUSKY SUNBIRD

788

Nectarinia fusca

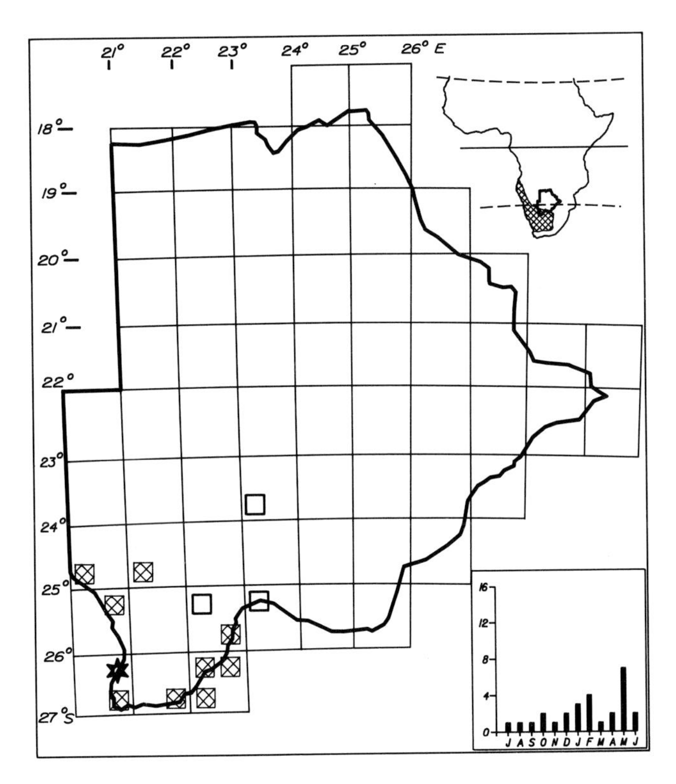

Status

Uncommon to locally fairly common resident of the southwest where it occurs at the eastern limit of its continental range. The species is known to wander in the nonbreeding season but confusion readily occurs with immature sunbirds of other species. Thus all records outside its normal range must at present be treated with circumspection. Solitary or in pairs.

Habitat

Rock outcrops with stunted trees and succulent vegetation, bush and tree savanna near watercourses or small escarpments and less frequently on duneland. Mainly in arid or semidesert regions with bare, stony or sandy soil and poor ground cover.

Analysis

13 squares (6%) Total count 30 (0,017%)

SCARLETCHESTED SUNBIRD

791

Nectarinia senegalensis

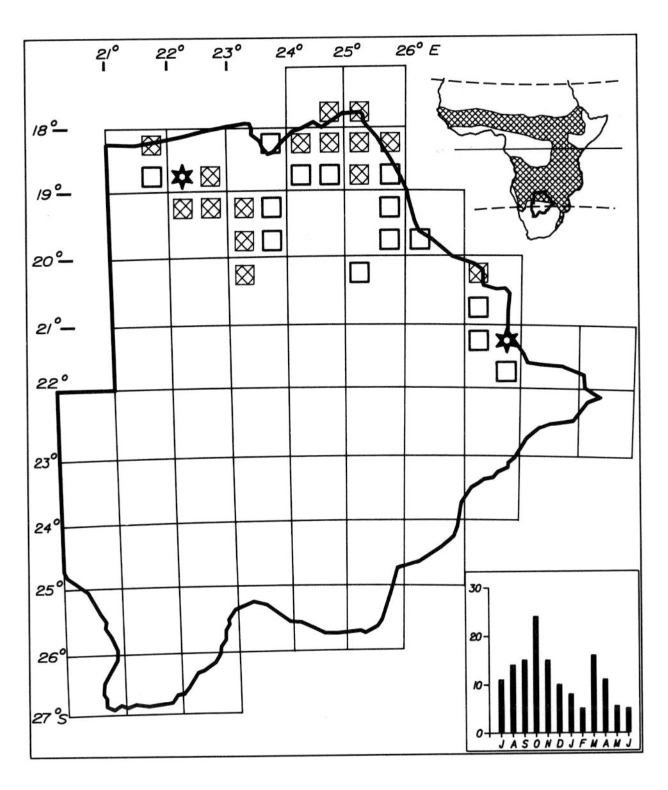

Status

Sparse to fairly common resident of the north and east to as far south as Selebi Phikwe. Its distribution appears to follow the drainage of the major rivers in the highest rainfall areas. Some irregular seasonal movements are suspected. Occurs in Botswana at the southern limit of its continental range at these longitudes. Egglaying in October and December. Solitary or in pairs.

Habitat

Deciduous tree and bush savanna and *Acacia* on alluvial soils in richly vegetated areas with tropical flowering plants such as *Erythrina*. Edges of woodland, well-established secondary growth in previously disturbed areas, riverine bush and gardens.

Analysis

31 squares (13%) Total count 150 (0,08%)

BLACK SUNBIRD

792

Nectarinia amethystina

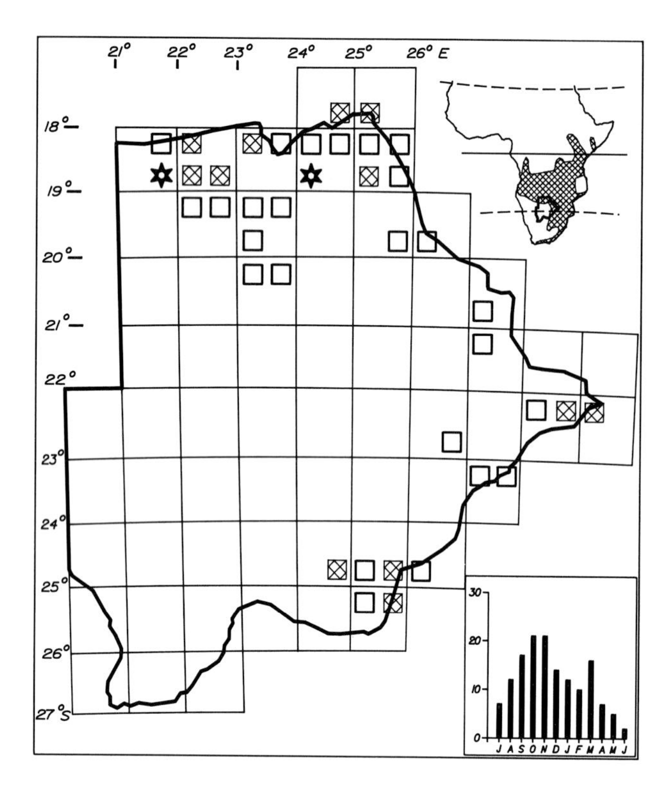

Status

Sparse to fairly common resident of the Okavango, Linyanti and Chobe regions. Sparse and unpredictable in the east and southeast to as far south as Lobatse where it is common. Seasonal movements occur and numbers are highest from August to March. Egglaying September to November. Solitary or in pairs.

Habitat

In the north it occurs in miombo and *Baikiaea* woodland usually in clearings with flowering bushes and creepers, on the edges of riverine forest, tree savanna with rich undergrowth and in gardens. In the east and southeast it occurs in mature woodland, riverine bush and richly vegetated savanna along watercourses, gardens and farms.

Analysis

39 squares (17%) Total count 154 (0,08%)

COLLARED SUNBIRD

793

Anthreptes collaris

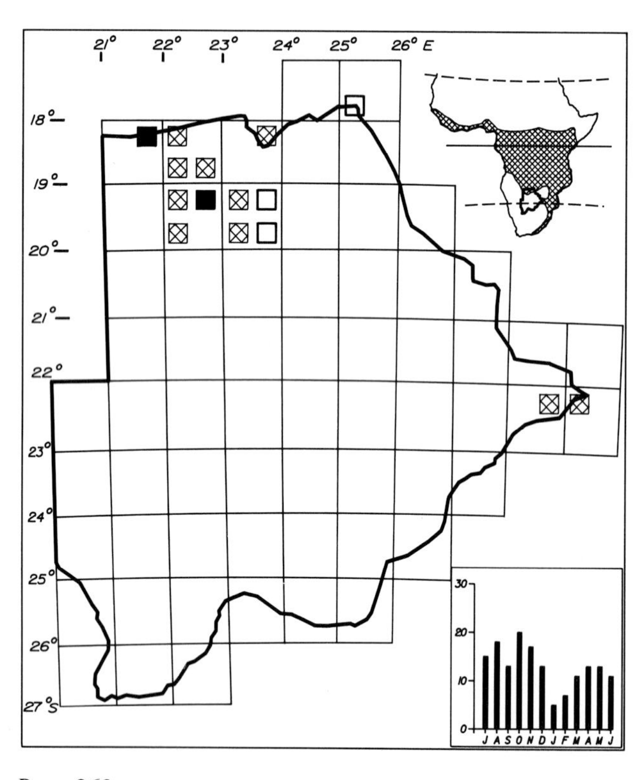

Status

Fairly common to locally very common resident of the Okavango and Linyanti river systems. Sparse on the periphery of this region and on the Chobe River at Kasane. Reappears near the Shashe/Limpopo confluence where it is fairly common and occurs as a western extension of its range in the middle Limpopo Valley. Solitary or in pairs.

Habitat

Riparian forest and lush riverine bush. Rare away from riverine vegetation.

Analysis

14 squares (6%) Total count 162 (0,09%)

CAPE WHITE-EYE

796

Zosterops pallidus

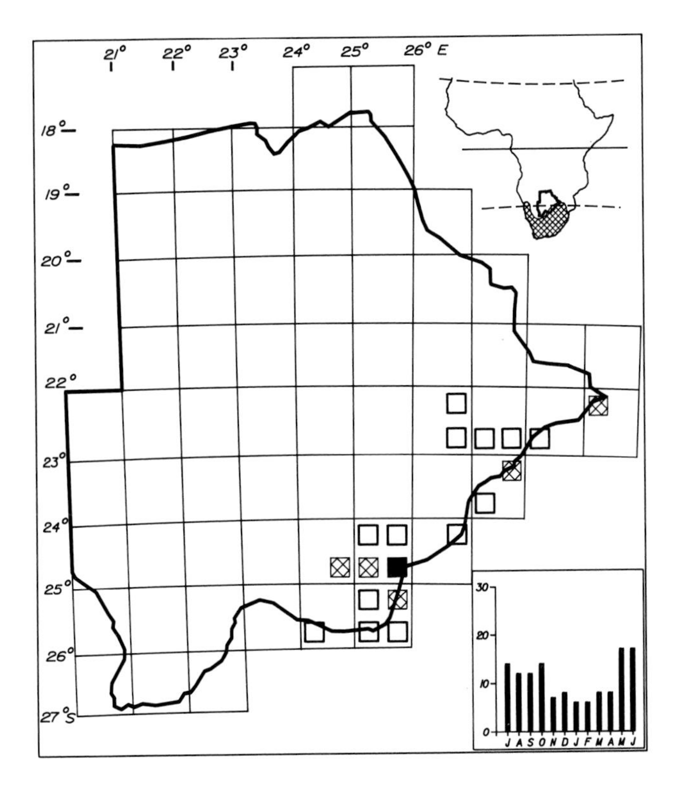

Status

Sparse to uncommon resident of the southeast except at Kanye and Lobatse where it is common and Gaborone where it is very common. It occurs in Botswana at the northern limit of its southern African range at these longitudes. Sparse and irregular in the east to as far north as Serowe and Pont's Drift. Egglaying October, November and March. Usually in small groups of 3–8 birds or in pairs when breeding.

Habitat

Dense vegetation and shady trees along river banks, in woodland and in tree and bush savanna. Also riparian *Acacia* in leaf and in gardens.

Analysis

19 squares (8%) Total count 144 (0,08%)

YELLOW WHITE-EYE

797

Zosterops senegalensis

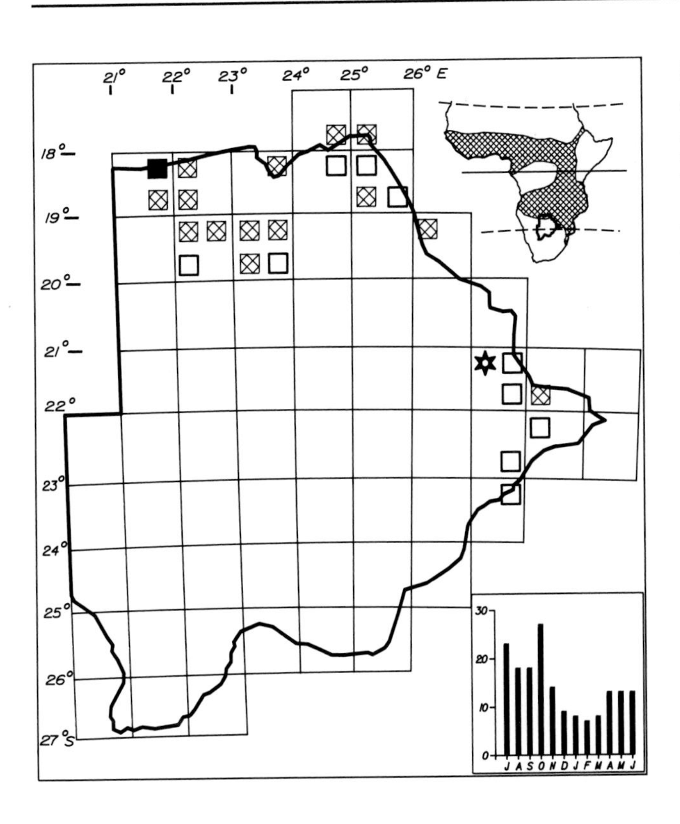

Status

Sparse to common resident of northern woodlands. Very common at Shakawe and Mohembo. Reappears in the east at Francistown and extends south to as far as Martin's Drift (just north of the Tropic of Capricorn). Its range overlaps that of the Cape White-eye near the Tropic of Capricorn and around Palapye but more detailed study of this overlap is required. Usually in small groups or in pairs when breeding.

Habitat

Miombo and *Baikiaea* woodlands, riparian forest and riverine woodland. Forages mainly in the canopy and occurs at or near the ground less commonly than the Cape White-eye.

Analysis

26 squares (11%) Total count 178 (0,10%)

REDBILLED BUFFALO WEAVER

798

Bubalornis niger

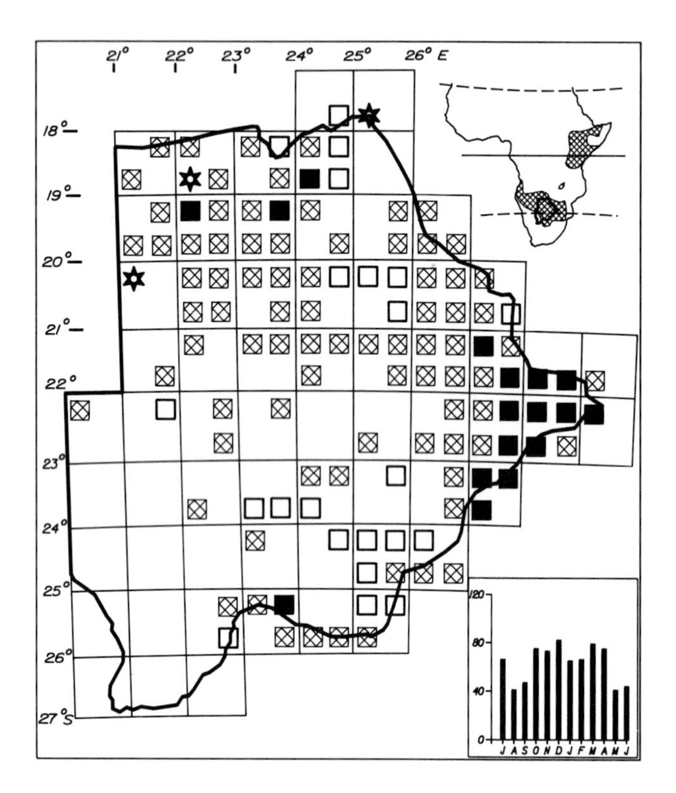

Status

Common to very common resident of the north and east. Sparse to uncommon in the southeast and central and western areas below 22°S. Extends along the Molopo River to as far as Tshabong and is absent in the southwest and west. Local seasonal movements occur. Egglaying January and February. Usually in flocks of 5–50 birds.

Habitat

Acacia savanna in areas with tall trees and open ground, stands of *Acacia* in river valleys, open park-like woodland. Also in open mixed woodland. Nests colonially in large conspicuous stick nests in tall trees including Baobabs in the north and east.

Analysis

123 squares (53%) Total count 776 (0,44%)

WHITEBROWED SPARROWWEAVER

799

Plocepasser mahali

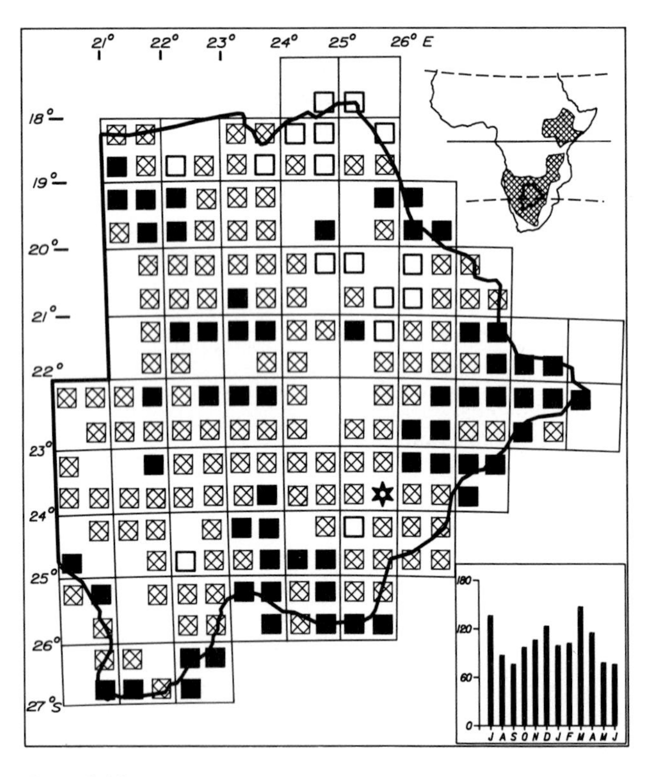

Status

Fairly common to very common resident throughout the country. Reported as very common in 27% of squares spread in all regions. Nests colonially in untidy conspicuous grass nests. Some local movements occur away from breeding sites. Egglaying December and January. Usually in loose groups of 6–20 birds or in pairs.

Habitat

Tree and bush savanna particularly *Acacia* in western and central regions and mopane in the east in open grassy areas. Also occurs in overgrazed areas with poor grass cover. Sparse in woodland such as the miombo and *Baikiaea* of the northeast.

Analysis

196 squares (85%) Total count 1292 (0,73%)

SOCIABLE WEAVER 800

Philetairus socius

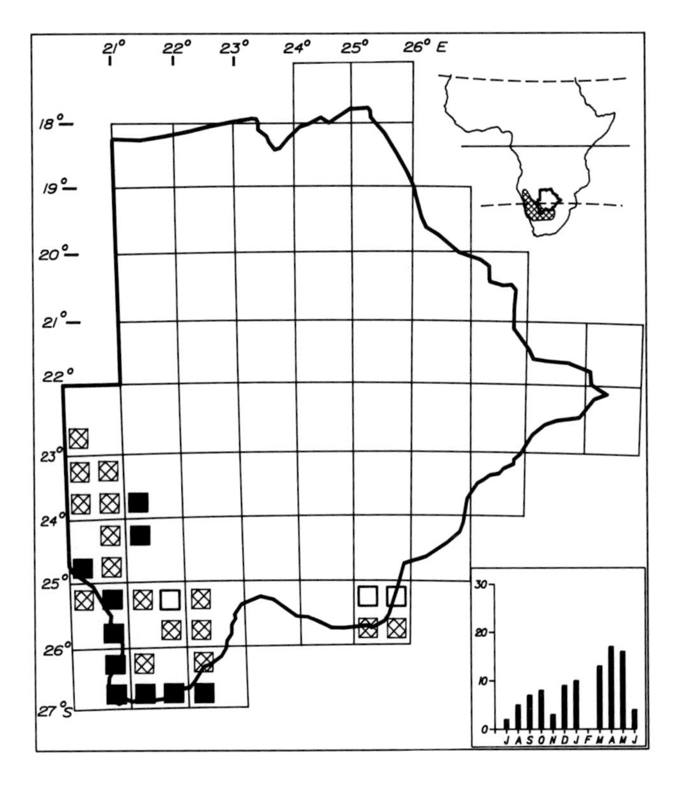

Status

Uncommon to locally very common resident of the southwest. Sparse to uncommon in the southeast. Very common in the Nossob and western Molopo valleys. Occurs in Botswana at the eastern limit of its southern African range. Breeds in communal grass nests which are often huge and nearly always conspicuous. Usually in flocks of 10–40 birds, occasionally in hundreds.

Habitat

Open areas of *Acacia* woodland especially *Acacia erioloba* where there is patchy or good grass cover in duneland and river valleys in low rainfall regions. Also in thorn tree savanna. Feeds on the ground in semidesert and savanna. Nests are usually placed along branches of trees but fence posts and telegraph poles are also used regularly.

Analysis

29 squares (13%)   Total count 102 (0,06%)

HOUSE SPARROW 801

Passer domesticus

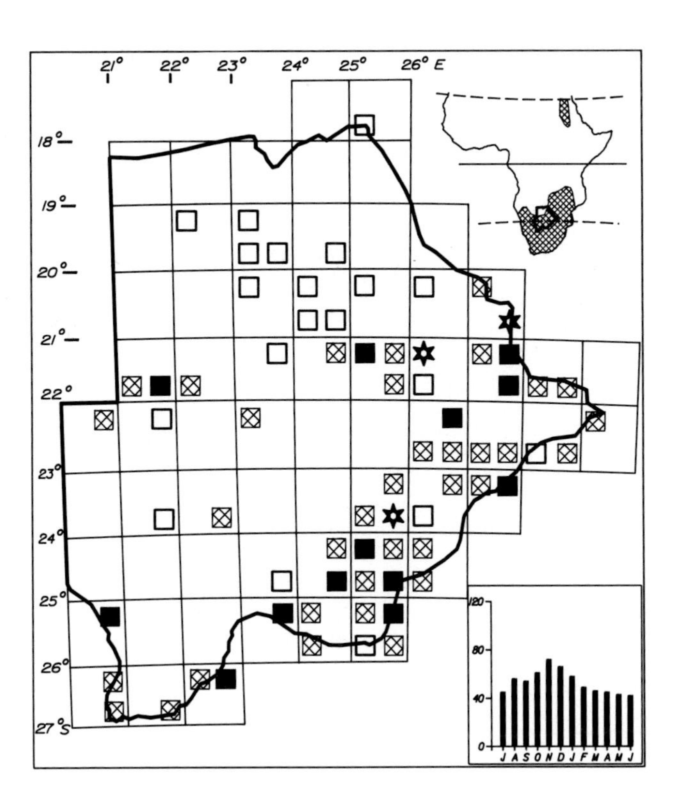

Status

A locally very common resident of the larger towns and residential areas. Commonest in the inhabited areas of the east. Sparse north of Orapa. This species was introduced into South Africa and subsequently spread northwards reaching Zimbabwe and Botswana in the late fifties and Zambia in the mid-sixties. Expected gradually to become more widespread with increasing human habitation. Egglaying September to November but probably in all months in some years. Usually in pairs or small groups.

Habitat

Towns, villages and residential areas such as hotels and game camps where there are brick or stone buildings. Less common in settlements with mud, wattle, wood and thatch construction only. Nests under eaves, in roof spaces, in hollow pipes of fences, on garage walls and a variety of other solid construction.

Analysis

71 squares (31%)   Total count 654 (0,36%)

GREAT SPARROW 802

Passer motitensis

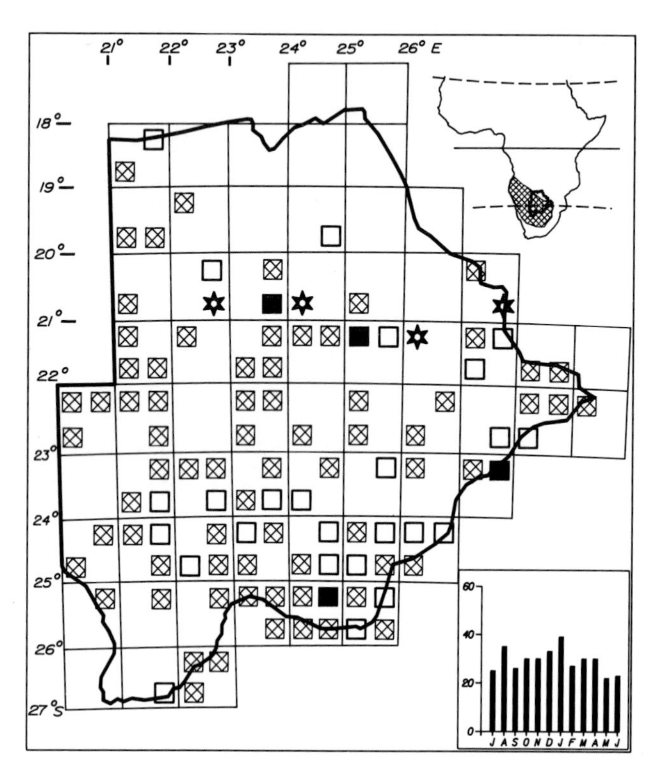

Status

Sparse to common resident south of 20°S. Extends northwards in the west to as far as Shakawe and is common at Gomare. Also reaches Nxai Pan. Reported as very common near Makalamabedi, Orapa, Martin's Drift and Sita Pan. There is no evidence of movements. Egglaying October to February. Usually in pairs or solitary.

Habitat

Tree and bush savanna, mainly *Acacia* or thorn savanna in semidesert areas. Often at cattle posts in the Kalahari, in *Acacia* shading cattle kraals along the Molopo River, and at remote villages. Also on the edges of towns but unusually near buildings or in urban gardens.

Analysis

105 squares (46%) Total count 364 (0,20%)

CAPE SPARROW 803

Passer melanurus

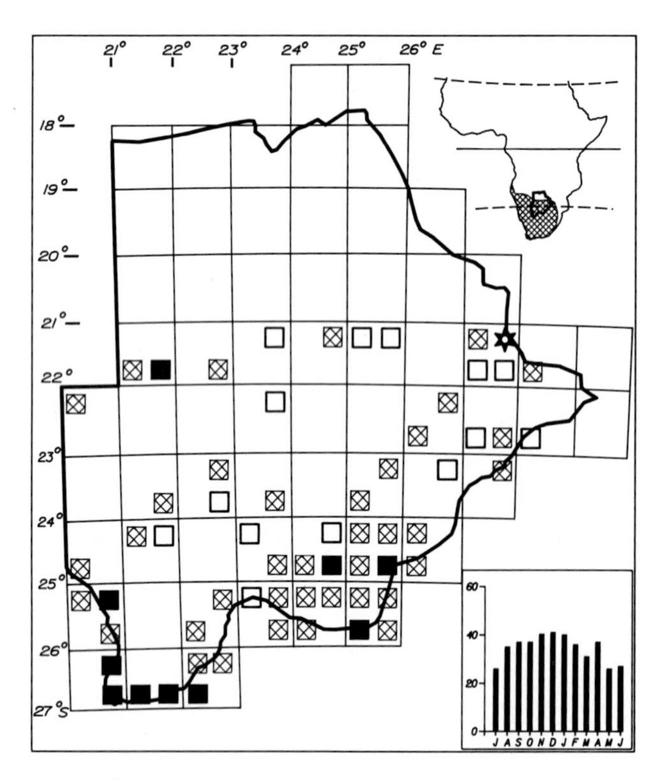

Status

Common to very common resident of the southeast and the river valleys of the southwest. Generally uncommon north of 24°S but it is common at some towns and villages such as Ghanzi, Serowe, Francistown and Mopipi but apparently not at Orapa, Palapye, Kang or Selebi Phikwe. Further investigation is required as it has adapted to human habitation successfully in many areas. Egglaying all months September to February. Common host of the Diederik Cuckoo. In pairs or flocks of up to 50 birds.

Habitat

Tree and bush savanna in high and low rainfall regions. Also in riverine bush, *Acacia* along dry watercourses, fossil valleys, farms, parks, towns and villages. Nests in gardens and under the eaves of buildings as well as in trees and bushes in savanna.

Analysis

63 squares (27%) Total count 453 (0,25%)

GREYHEADED SPARROW

804

Passer diffusus

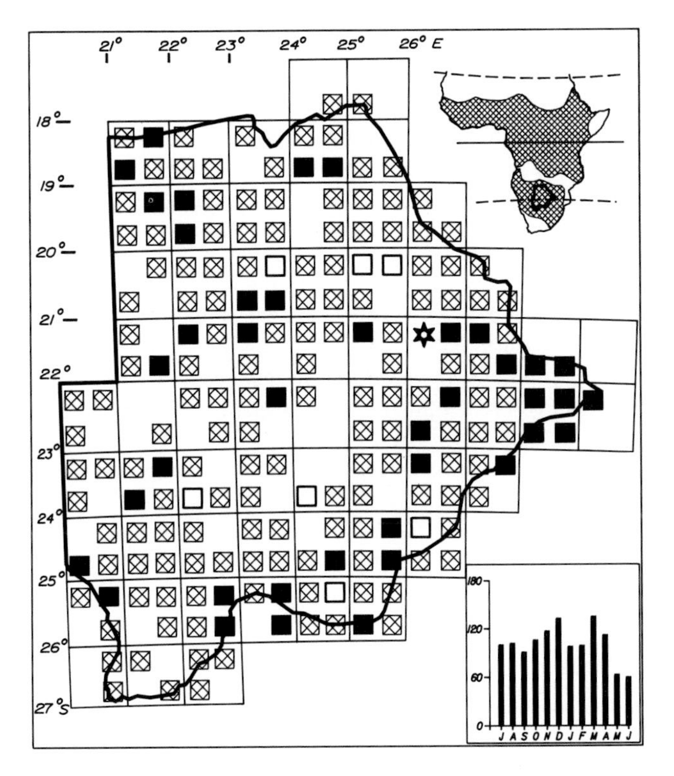

Status

Fairly common to very common resident throughout the country. Most common in the east but may be very common in any well-wooded area. Absent in treeless habitats. Very common in 17% of squares. Irregular movements occur. Egglaying January to March. Usually in pairs or small flocks, but nonbreeding flocks may be 50–100 birds.

Habitat

All types of woodland, tree and bush savanna. Regularly in gardens in urban and rural habitation. Occasionally nests under eaves but also in hollow posts and holes in trees or walls.

Analysis

193 squares (84%) Total count 1254 (0,7%)

YELLOWTHROATED SPARROW

805

Petronia superciliaris

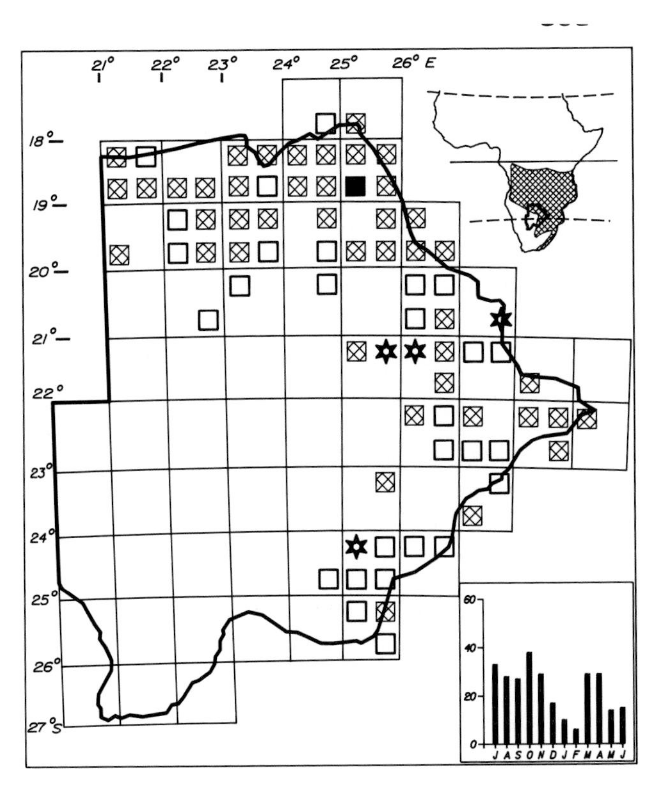

Status

Sparse to common resident of the north and east. It is most common in the well-wooded areas of the northeast and is sparse or uncommon south of 23°S. Mainly sedentary but short distance local movements may occur. Usually in pairs or solitary.

Habitat

Broadleafed woodland particularly miombo and *Baikiaea* in the northeast. Also in tree and bush savanna in the higher rainfall areas and in river valleys. Forages on the ground. Usually in wooded areas with bare, leaf-littered or lightly-grassed ground under the trees.

Analysis

76 squares (33%) Total count 288 (0,16%)

SCALYFEATHERED FINCH **806**

Sporopipes squamifrons

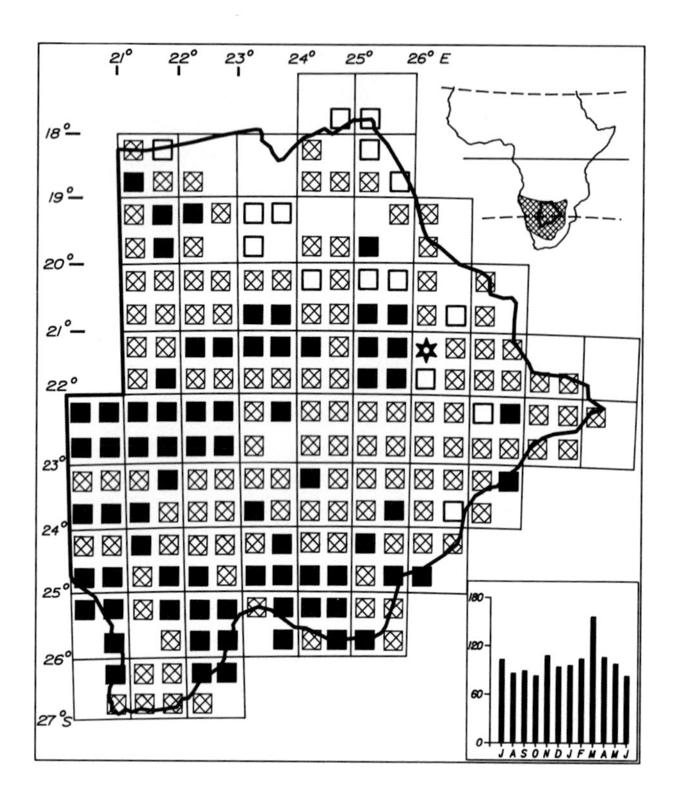

Status

Common to very common resident throughout the country except in the northeast where it is uncommon and unpredictable. It is one of the typical birds of the Kalahari. Commonest in the drier regions of the south, west and north. Egglaying December to April, mainly April, also suspected to breed throughout the year. Usually in small flocks of 4–10 birds.

Habitat

Acacia bush and scrub savanna and a variety of thorn savanna. Also in gardens, old cultivation, grass plains with few scattered bushes or woody shrubs. Nests in thorn trees and bushes, often utilising *Acacia mellifera* when this is available.

Analysis

208 squares (90%) Total count 1260 (0,71%)

THICKBILLED WEAVER **807**

Amblyospiza albifrons

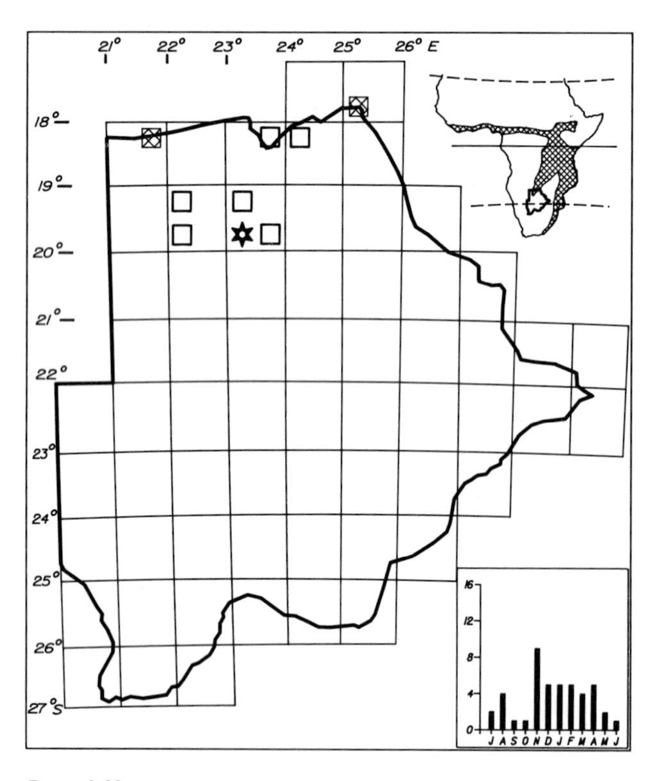

Status

Sparse resident of the Okavango, Linyanti and Chobe river systems. Recorded as fairly common at Shakawe and Kasane. Local seasonal movements and postbreeding dispersal occur. Not well known in Botswana and merits special study. Egglaying April although it is likely to breed in earlier summer months as well. Usually in small groups, several pairs breed together at one site.

Habitat

Tall reedbeds in lagoons, marshes, floodplains or ponds, usually near riparian forest or thicket on the fringes of woodland. Less commonly in extensive reedbeds or papyrus. May remain in these chosen breeding sites even when dried out. Moves to forest edge or dense marshland vegetation in the nonbreeding season.

Analysis

9 squares (4%) Total count 46 (0,03%)

SPECTACLED WEAVER

810

Ploceus ocularis

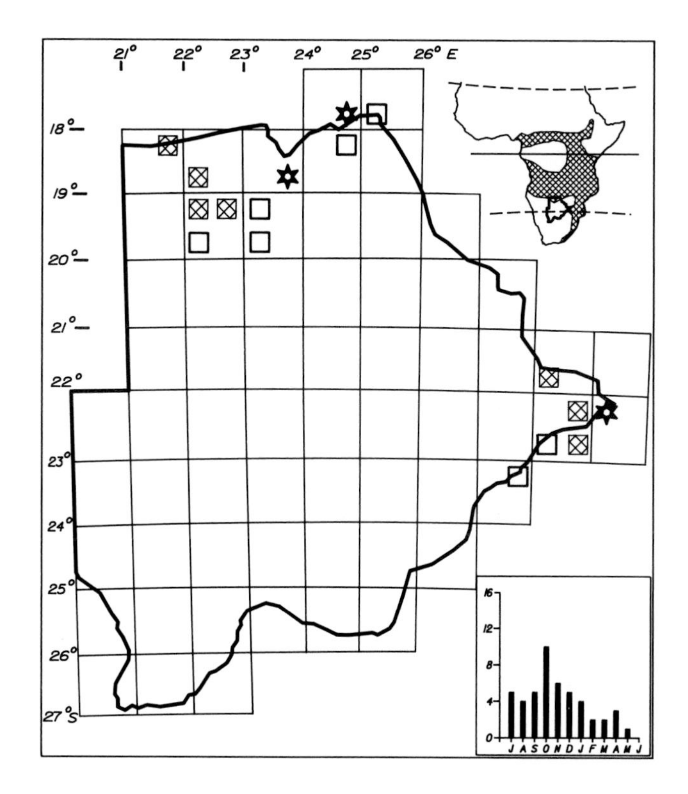

Status

Sparse to uncommon resident in the Okavango, Chobe, Shashe and Limpopo river systems. There are no recent records from Linyanti. In the north it occurs at the southern limit of its continental range at these longitudes and in the east at the western limit. Not well known. Egglaying December. Usually in pairs.

Habitat

Edges of riparian forest, riverine bush, open woodland near watercourses in areas with tangled vegetation and creepers. Often nests on branches overhanging water or dry river beds.

Analysis

17 squares (7%)   Total count 50 (0,03%)

SPOTTEDBACKED WEAVER

811

Ploceus cucullatus

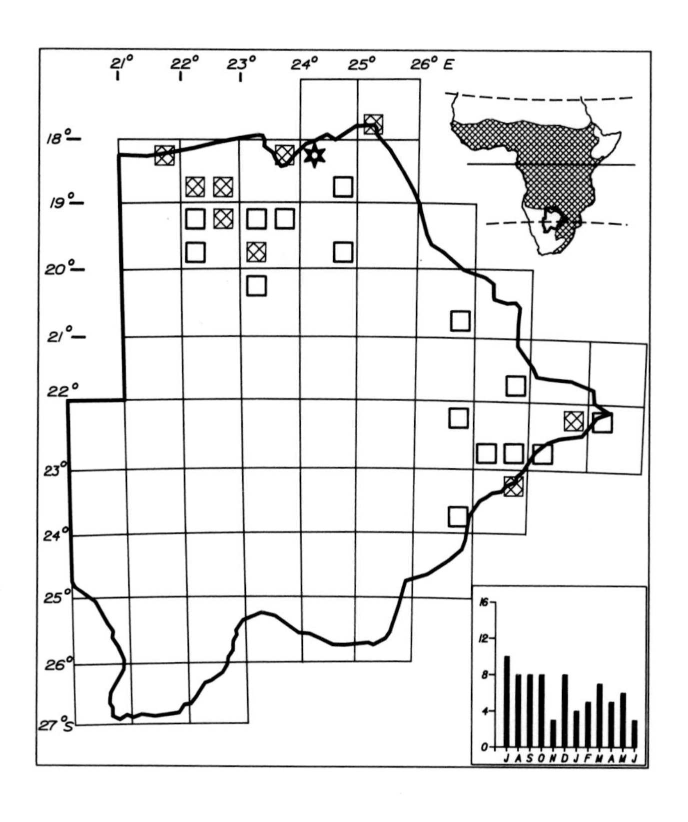

Status

Sparse to locally common resident in the Okavango, Linyanti and Chobe river systems. Sparse at Nxai Pan and east of Savuti Marsh. Reappears in the east at Selebi Phikwe and is sparse to locally uncommon along the Limpopo River to as far south as Buffels Drift. Recorded once at Mosetse (2026D). Usually in flocks of 10–20 birds.

Habitat

Reedbeds on the margins of rivers, lagoons and pools with adjacent riparian forest or woodland. Forages in nearby tree and bush savanna. Never far from water but may remain in wetland when it is seasonally dry.

Analysis

25 squares (11%)   Total count 81 (0,05%)

MASKED WEAVER

814

Ploceus velatus

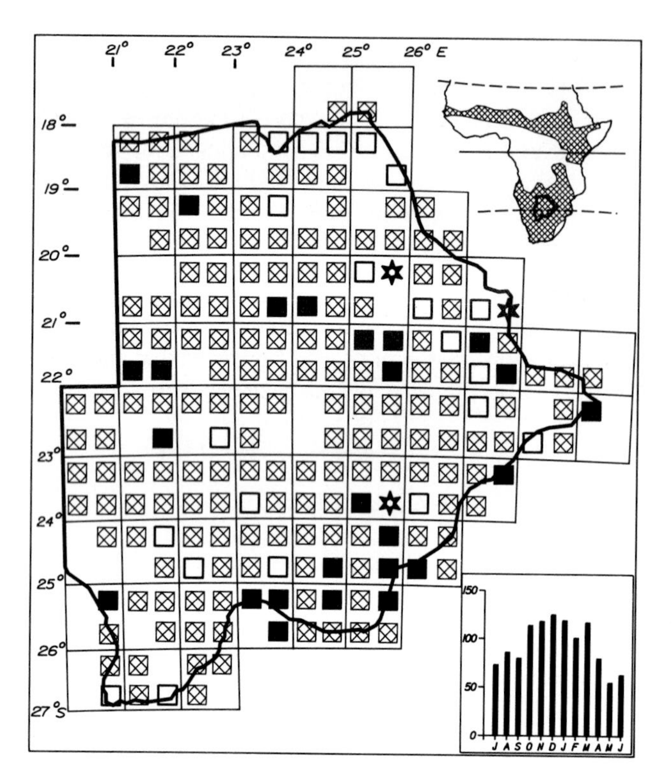

Status

Common to very common resident throughout the country. The commonest weaver in Botswana. Significant local movements take place which may lead to absences at some localities in winter. Very large nonbreeding flocks of 100–500 birds are sometimes seen in the central Kalahari. Egglaying all months from September to March, mostly in November. Usually in flocks of 10–50 birds, nests in colonies with 3–30 nests in one tree.

Habitat

Breeds mainly in *Acacia* tree and bush savanna, also on the edges of broadleafed woodland, on farmland, in riverine bush and in urban or rural gardens. Isolated nesting colonies occur in semidesert. Large nonbreeding flocks are found in semidesert tree, bush and scrub savanna. Smaller nonbreeding flocks are found in the breeding habitat.

Analysis

205 squares (89%) Total count 1166 (0,65%)

LESSER MASKED WEAVER

815

Ploceus intermedius

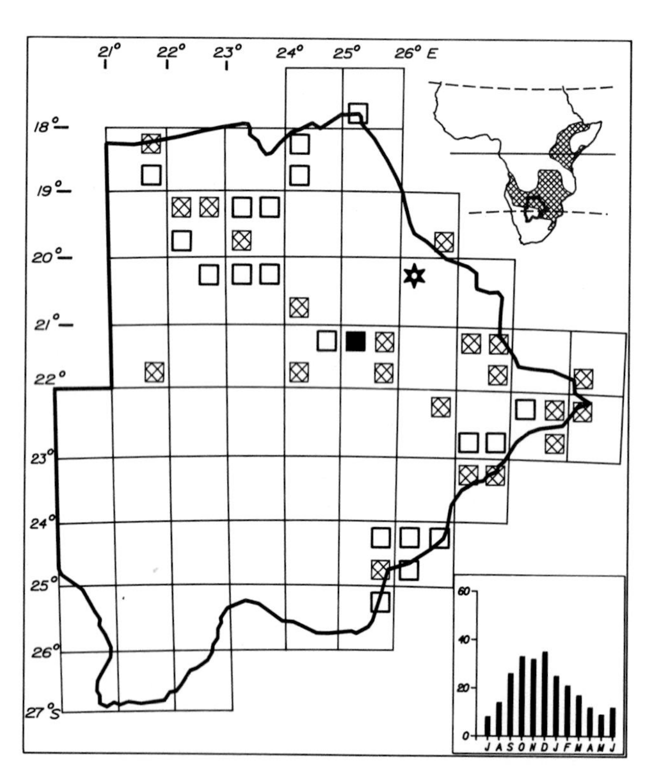

Status

Sparse to fairly common resident of the north and east. Sparse to uncommon in the southeast. Usually localised and unpredictable except in river valleys but there are some well-known breeding colonies in towns such as at an hotel in Gaborone and Ghanzi, and in Orapa town. Egglaying October. Usually in small flocks of 5–10 birds, but breeding colonies may have up to 30 nests in large trees.

Habitat

Tree and bush savanna, edges of deciduous woodland, riparian *Acacia*. Usually in river valleys but also in areas near water of ponds, dams, reservoirs and pools. Nests in colonies in trees and bushes, sometimes over water.

Analysis

43 squares (19%) Total count 250 (0,14%)

GOLDEN WEAVER

816

Ploceus xanthops

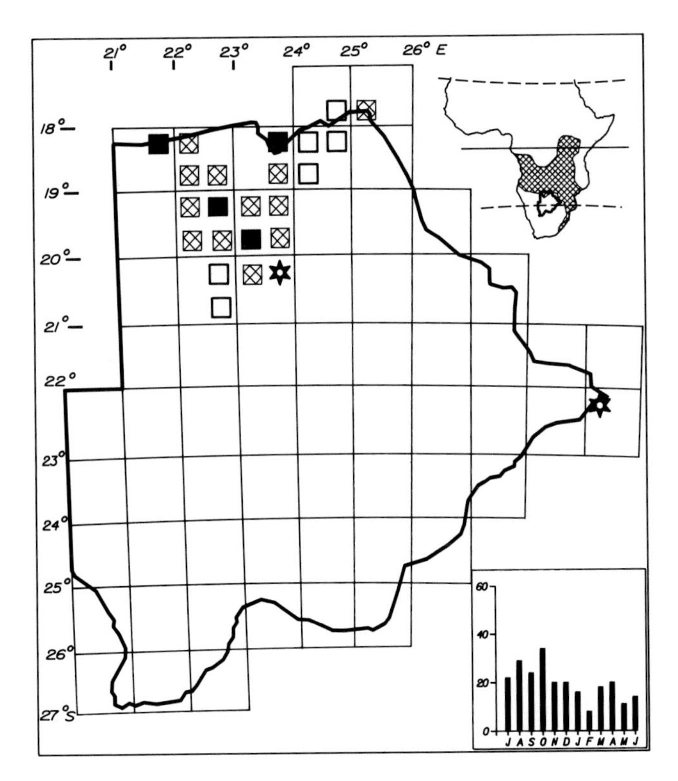

Status

Common to locally very common resident of the Okavango, Linyanti and Chobe river systems to as far south as Lake Ngami and Tale Pan. Further investigation is required to ascertain its status in the east from where there is an old record from Pont's Drift and an uncorroborated sighting at Francistown. Egglaying March. Usually solitary or in pairs.

Habitat

Reedbeds and rank vegetation near water. Forages in adjacent bush and tree savanna, woodland, riparian forest and gardens. Nests in reeds or on branches overhanging water, sometimes in bushes or trees near water.

Analysis

23 squares (10%) Total count 242 (0,14%)

BROWNTHROATED WEAVER

818

Ploceus xanthopterus

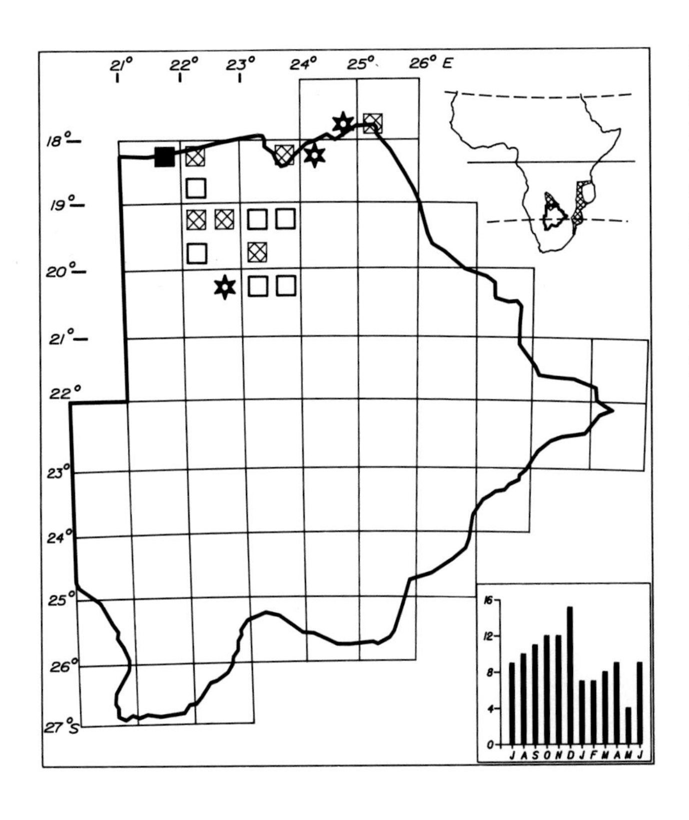

Status

Sparse to fairly common resident of the Okavango, Linyanti and Chobe river systems to as far south as Makalamabedi. Very common at Shakawe. Occurs at the southern limit of its central African range at these longitudes. Not well known and is a good candidate for detailed study. Active nests in November. Usually in small flocks of 5–10 birds.

Habitat

Reedbeds and tall aquatic vegetation in swamps, marshes, rivers and lagoons. Forages in adjacent riparian forest, thicket and dense growth in riverside gardens but not apparently in woodland and savanna as does the Golden Weaver.

Analysis

16 squares (7%) Total count 120 (0,07%)

REDHEADED WEAVER

Anaplectes rubriceps

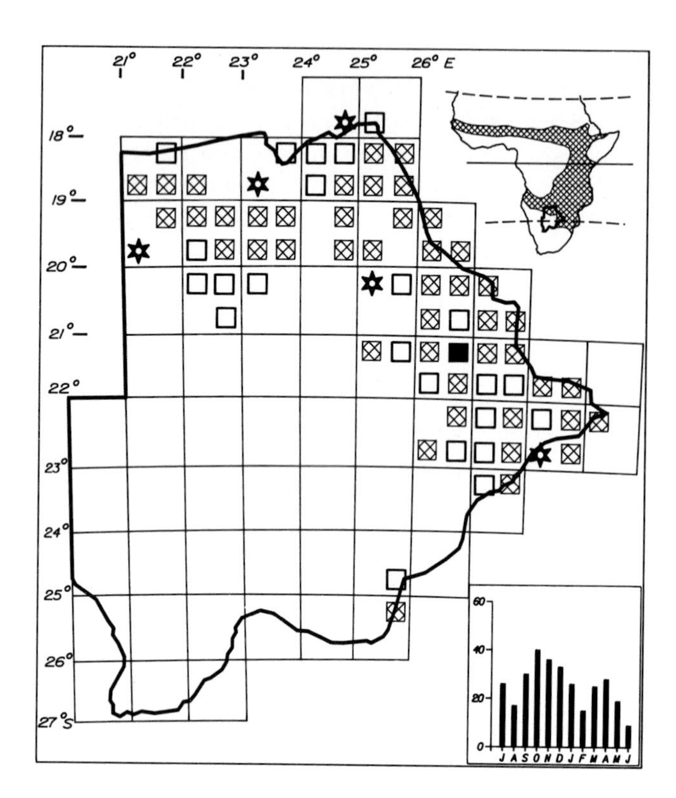

Status

Sparse to common resident of the north and east south to Martin's Drift. There is an isolated population at Lobatse and Gaborone at the western extremity of its range in western Transvaal. Its status along the Marico and Limpopo river drainage needs more detailed study. Egglaying November and December. Solitary or in pairs. Sometimes in mixed bird parties.

Habitat

Broadleafed and mixed woodlands in the high rainfall regions and on alluvial soils. Does not extend into the central Kalahari but occurs up to the Limpopo watershed in the east and to the southern limits of the wetland areas of the north.

Analysis

74 squares (32%) Total count 316 (0,18%)

REDBILLED QUELEA

Quelea quelea

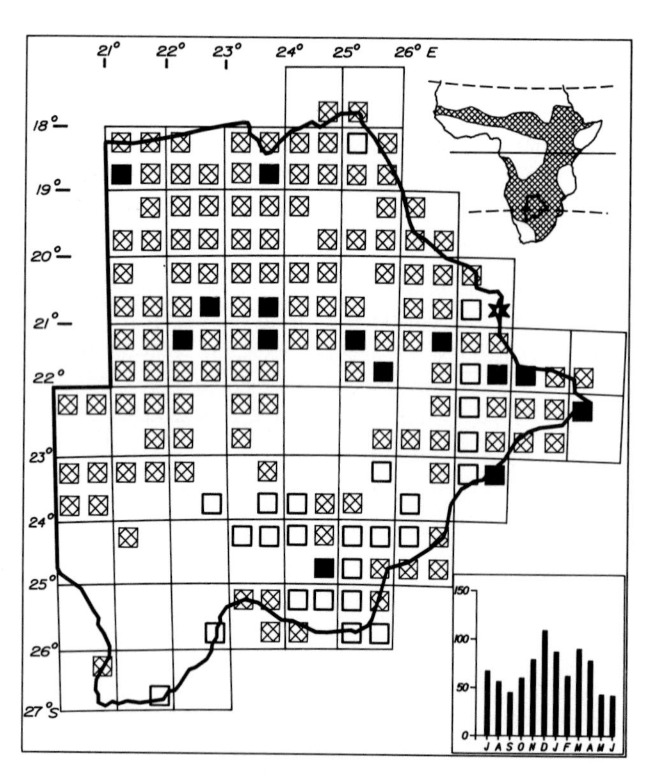

Status

Common to very common resident north of 24°S. Sparse to common in the southeast and mainly absent in the southwest and the central Kalahari. Highly nomadic and is seasonally abundant at some localities. Likely to occur in all areas but during drought its range appears to have been restricted from the low rainfall regions. No recent breeding records but males in breeding plumage are often seen in summer months. Usually in flocks of 10–100 birds, but flocks of hundreds, thousands and millions have been reported.

Habitat

Open grassland, tree and bush savanna, edges of woodland, floodplains, farmland, cultivation, gardens. Occurs in almost any habitat except forest and swamps. Notorious as a pest of grain crops.

Analysis

160 squares (70%) Total count 833 (0,47%)

RED BISHOP 824

Euplectes orix

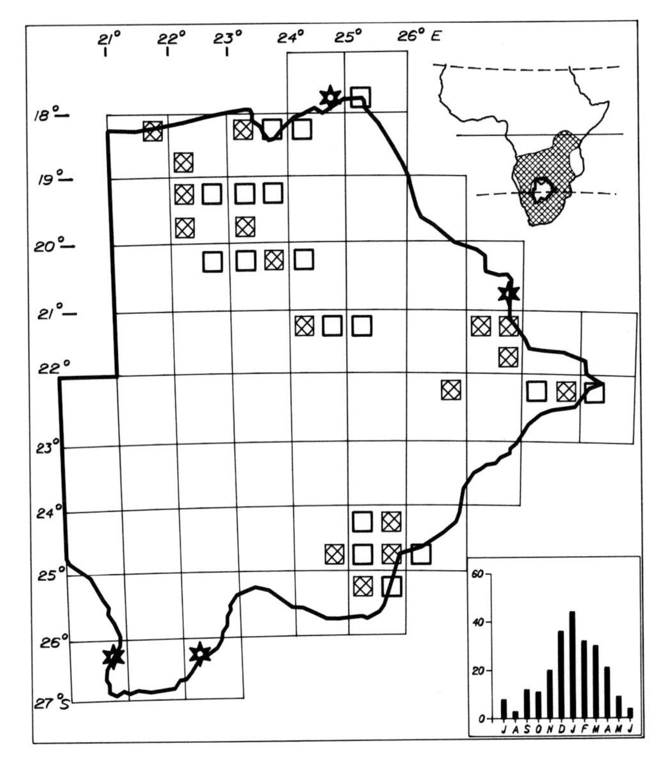

Status

Sparse to locally common resident of the Okavango, Linyanti and Chobe river systems to as far south as Lake Ngami and Makalamabedi. Reappears at Mopipi and Orapa and again in the east along the Shashe and Limpopo drainage to as far west as Serowe. In the southeast it occurs at the western limit of its range in the Transvaal. Local movements occur in the non-breeding season. Egglaying February and March. Usually in flocks of 5–50 birds. All male flocks sometimes occur in the early breeding season.

Habitat

Reedbeds and tall rank grasses on the edges of rivers, lakes and lagoons, and in tall vegetation in marshes, near sewage pond outlets and around farm dams. Even small areas, e.g. 50 m^2 of reeds are sufficient to attract it to breed at some sites.

Analysis

38 squares (16%) Total count 241 (0,13%)

GOLDEN BISHOP 826

Euplectes afer

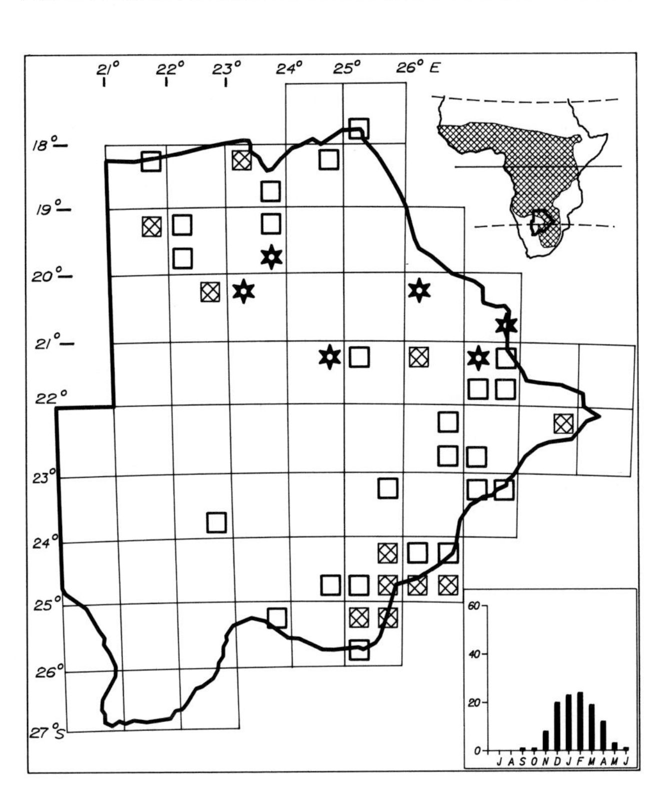

Status

Sparse to uncommon resident of the river systems of the north, east and southeast. Its distribution is similar to that of the Red Bishop but it is known to appear in unlikely spots in arid regions when suitable wet habitat becomes available. This suggests that it is nomadic and an opportunistic breeder. Mainly recorded when males are in breeding dress. Egg-laying December, January and February. Mainly in flocks of 5–20 birds, males are usually seen solitarily in breeding plumage.

Habitat

Tree and bush savanna, open grassland in river valleys, floodplains, rank vegetation around ponds, farm dams and water catchments near human habitation or at cattle posts. Breeds at wet depressions in savanna, marshes, tall vegetation along seasonal streams or near a variety of small artificial or natural water habitats.

Analysis

42 squares (18%) Total count 133 (0,07%)

REDSHOULDERED WIDOW

828

Euplectes axillaris

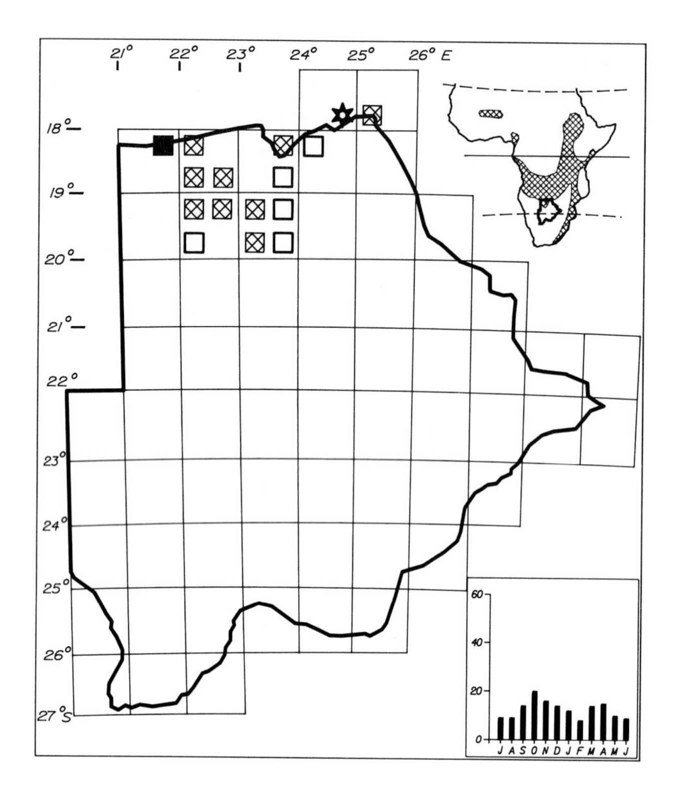

Status

Sparse to locally common resident of the Okavango, Linyanti and Chobe river systems. Occurs in Botswana at the southern limit of its continental range at these longitudes. Present throughout the year but it is most frequently reported in summer months when it is in breeding dress. Usually in flocks of 5–30 birds.

Habitat

Floodplains, marshes, seasonally inundated grassland and along drainage lines in open tree and bush savanna.

Analysis

16 squares (7%) Total count 156 (0,08%)

WHITEWINGED WIDOW

829

Euplectes albonotatus

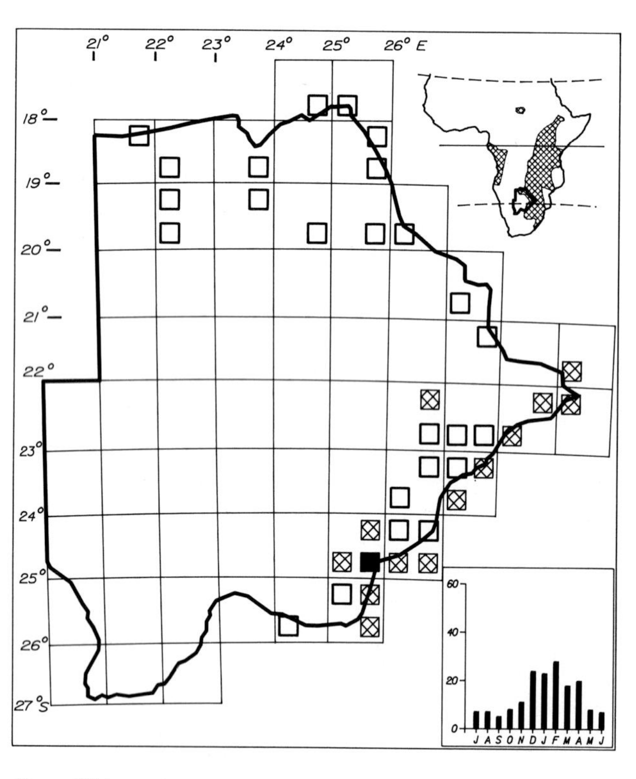

Status

A sparse and patchily distributed resident in the north. In the east and southeast it is sparse to locally fairly common. Extends west along the Molopo River to as far as Boshoek. Egglaying December, January and February. Usually in flocks of 5–30 birds.

Habitat

Open areas in tree and bush savanna, especially on fallow lands and rank vegetation surrounding cultivation and roadsides. Usually on grassland which is seasonally wet. Also in rank growth and tall grass in dry river beds, along drainage lines and around seasonal pools.

Analysis

39 squares (17%) Total count 179 (0,10%)

LONGTAILED WIDOW 832

Euplectes progne

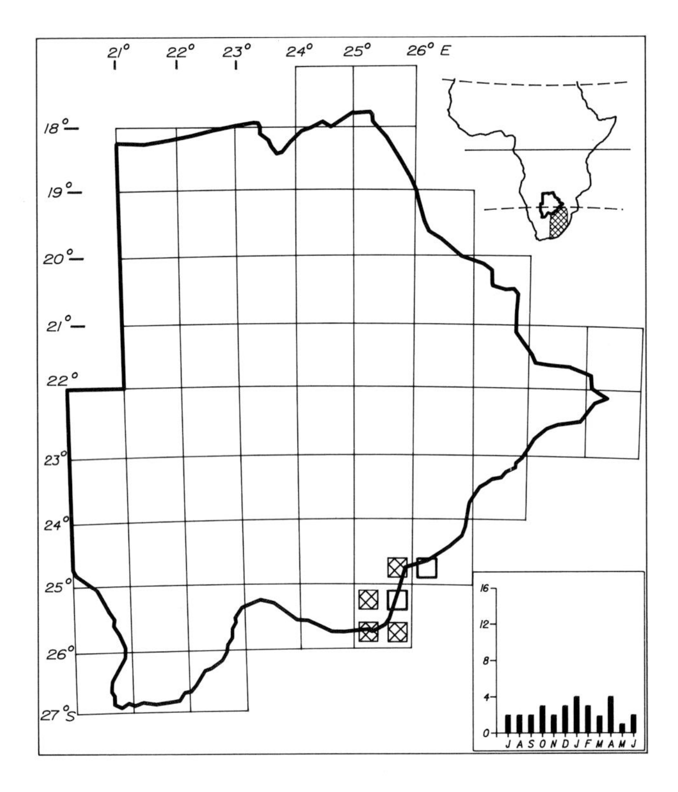

Status

Sparse to fairly common resident of the southeast corner of the country where it occurs at the western limit of its range in South Africa. There may have been a recent extension of its range into Botswana as it was not recorded in Gaborone in the early 1970s. Mainly sedentary but local movements occur in the nonbreeding season. In breeding condition in summer months. Usually in flocks of 5–30 birds.

Habitat

Open grassland in river valleys, along drainage lines and around dams. Also on watershed plains, on marshy ground and at seasonally wet depressions in savanna.

Analysis

6 squares (3%) Total count 35 (0,019%)

GOLDENBACKED PYTILIA 833

Pytilia afra

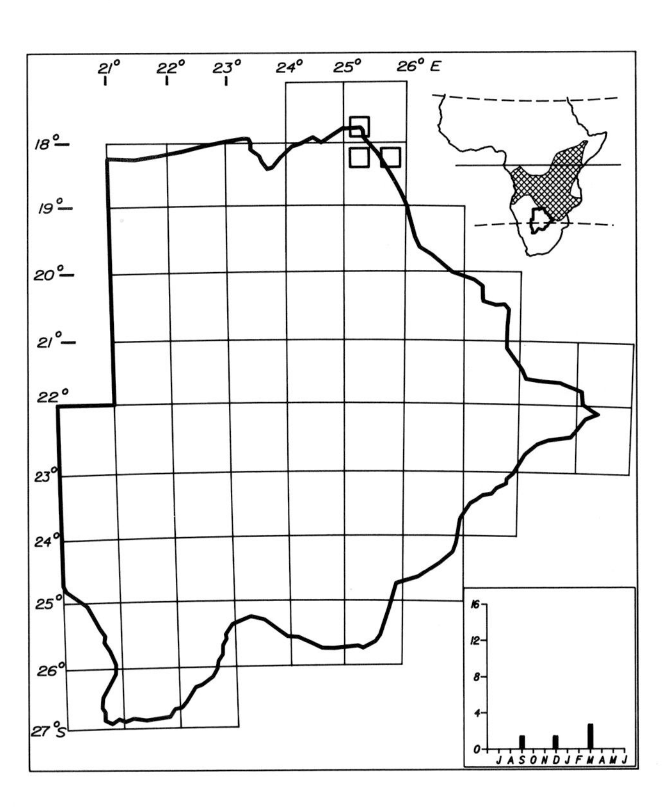

Status

Very sparse resident of the northeast where it occurs at the western limit of its range in Zimbabwe. Its status is poorly known as its occurrence in Botswana only came to light during this survey. Likely to be regularly present in the northeast where its parasite, the Broadtailed Paradise Whydah, is found. Movements may occur—with unconfirmed records from Serowe and near Maun. Usually solitary or in pairs.

Habitat

Bushes and dense vegetation on the edges of woodland in high rainfall regions. Also in thick cover on the edges of clearings, cultivation or open spaces.

Analysis

3 squares (1%) Total count 4 (0,002%)

MELBA FINCH

Pytilia melba

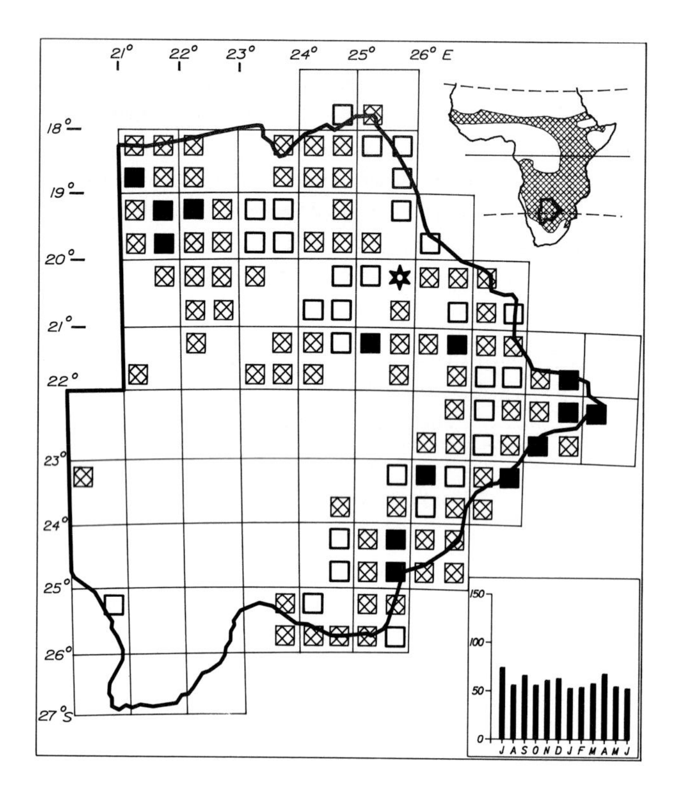

Status

Fairly common to locally very common resident throughout the north and east. It is absent from most of the central Kalahari, the southwest and west. Recorded once near Morwa Pan (2320A) and once near Nossob Camp (2520B) in the extreme west where it may occur at the eastern limit of its range in Namibia. Extends along the Molopo River to as far as Bray. Egglaying February and April. Usually in pairs.

Habitat

Thicket and dense vegetation in tree and bush savanna, riverine bush, bushes on the edge of woodland, rank growth along drainage lines and on the edge of disturbed ground, occasionally in gardens.

Analysis

116 squares (50%) Total count 743 (0,41%)

JAMESON'S FIREFINCH

Laganosticta rhodopareia

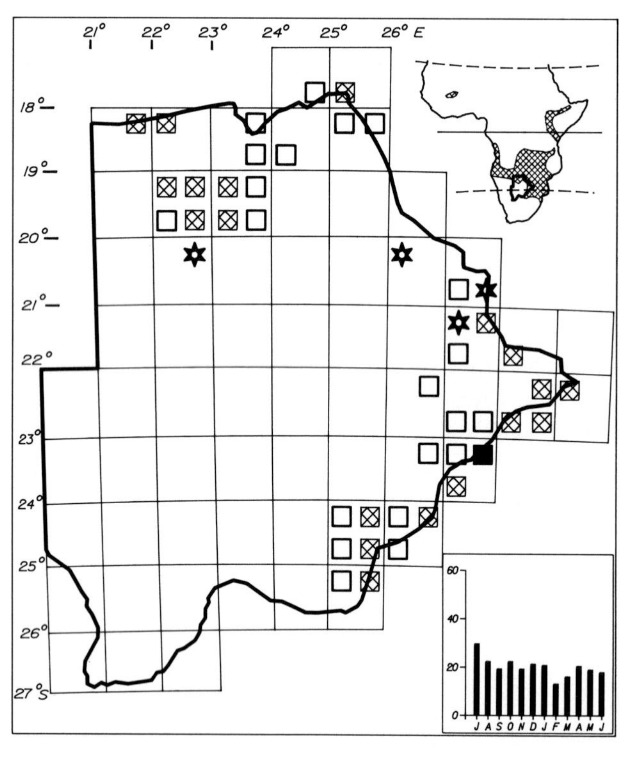

Alternative name

Pinkbacked Firefinch

Status

Sparse to fairly common resident of the north and east. Probably more common than recorded but it occurs in Botswana at the western limits of its range and movements to and from this area may occur in some years. Very common at Martin's Drift and common at Gaborone. Egglaying December. Usually in pairs or small groups.

Habitat

Tree and bush savanna in high rainfall areas where it occurs in rank grass, bushes and thickets especially along streams and gullies and on the edge of riparian forest. Also on the edges of deciduous woodland, edges of cultivation and in gardens.

Analysis

45 squares (20%) Total count 282 (0,16%)

REDBILLED FIREFINCH 842

Lagonosticta senegala

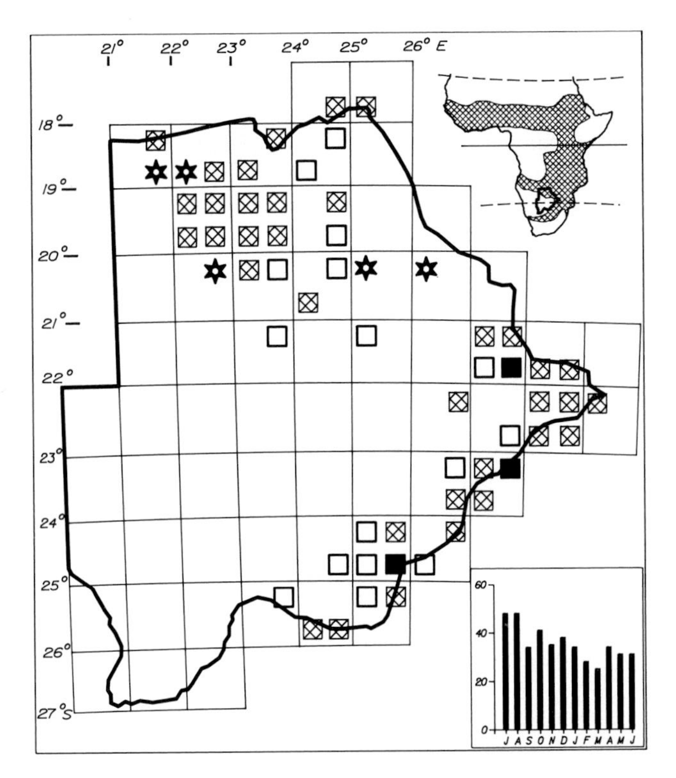

Status

Sparse to common resident of the north and east. Its distribution is similar to that of the Jameson's Firefinch but it is more widespread and common. It may tolerate more arid habitat than the Jameson's Firefinch and extend its range into semidesert in years of higher rainfall. The solitary egglaying record in April was parasitised by the Steelblue Widowfinch. Usually in pairs or small groups.

Habitat

Tree and bush savanna. Forages on the ground under cover of bushes and thicket, on adjacent bare ground and in rank grass. Occurs in similar habitat to the Jameson's Firefinch alongside which it is sometimes found. In semidesert it occurs in and under clumps of *Acacia* bushes in dry river valleys or along drainage lines where there is tangled growth and rank grass.

Analysis

59 squares (26%) Total count 453 (0,25%)

BROWN FIREFINCH 843

Lagonosticta nitidula

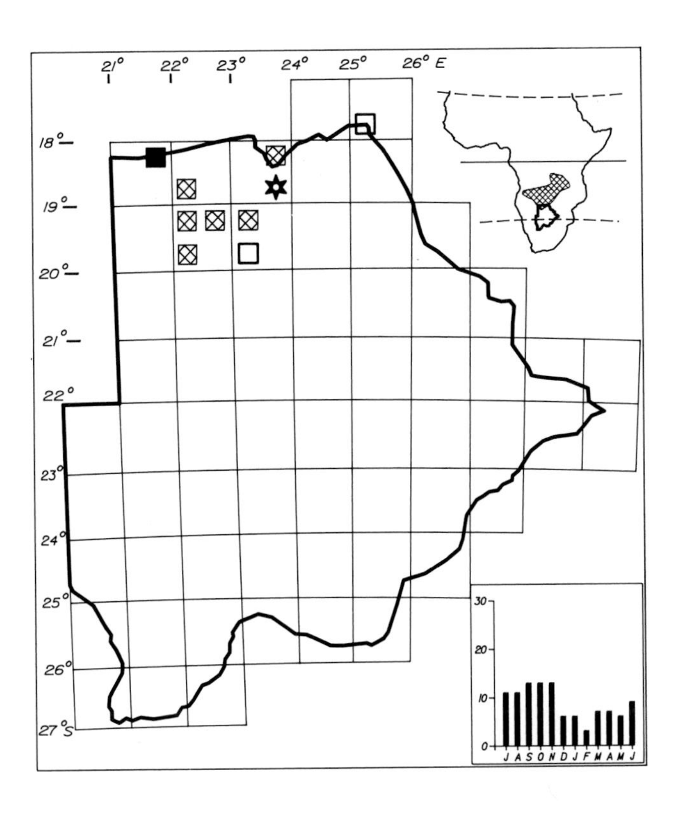

Status

Sparse to fairly common resident confined to the Okavango, Linyanti and Chobe river systems where it occurs at the southern limit of its continental range. Very common at Shakawe. It is present throughout the year and is considered to be sedentary. Usually in pairs or small groups.

Habitat

Edges of riparian forest and dense riverine vegetation where it occurs usually on leaf litter and in the groundstratum. Forages on the ground under thickets and bushes and infrequently in open areas under trees and on the edge of thicket. Occurs mostly in shade and is often on decomposing moist vegetation.

Analysis

10 squares (4%) Total count 108 (0,06%)

BLUE WAXBILL

844

Uraeginthus angolensis

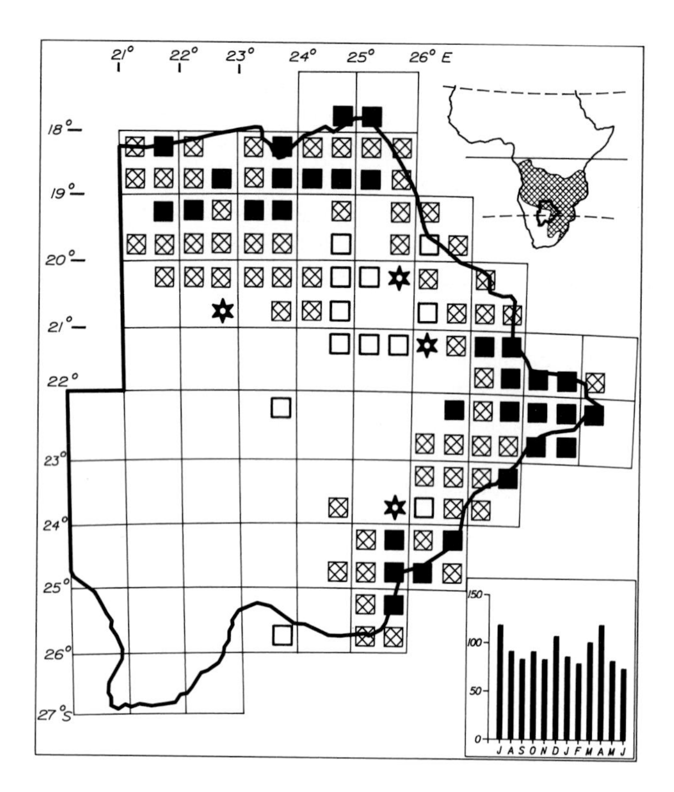

Status

Common to very common resident of the north and east. Sparse at Nxai Pan and in the Makgadikgadi and Orapa areas. Extends marginally into Kalahari savannas along drainage lines. Recorded west along the Molopo River as far as Boshoek. May occur in the western woodlands near Lehututu where special search needs to be made to establish its status. Egglaying January and April. Usually in small parties of 5–15 birds.

Habitat

Tree and bush savanna, particularly *Acacia* savanna, in open areas with scattered trees and bushes and open grassland. Clearings in woodland, edges of cultivation, riverine bush, gardens. In the more arid areas it occurs in thorn bushes along gullies and dry river beds.

Analysis

107 squares (47%) Total count 1146 (0,64%)

VIOLETEARED WAXBILL

845

Uraeginthus granatinus

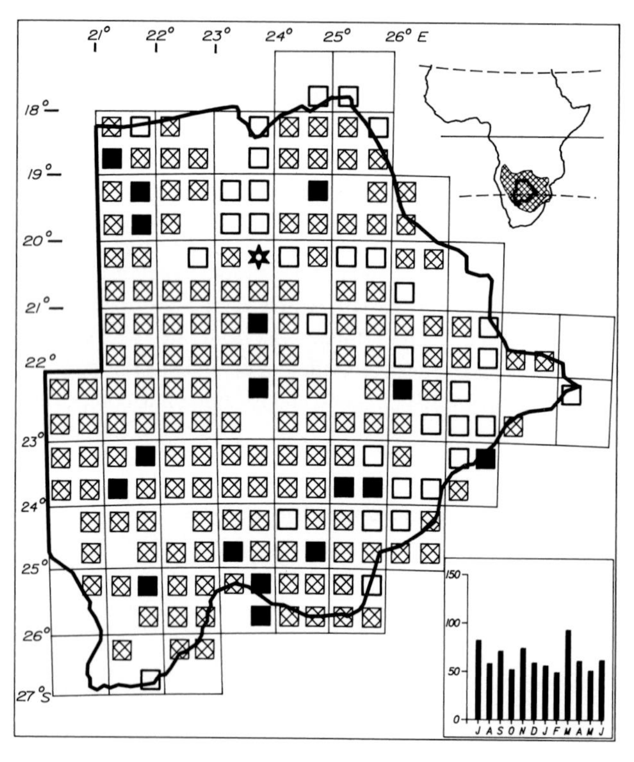

Status

Common to very common resident throughout the country except in the northern wetlands where it is sparse. The most widespread waxbill in Botswana and the typical family member in the Kalahari. Overlaps with the Blue Waxbill in the north and east but it is the less common species throughout the latter's range. Egglaying December. Host to the Shafttailed Whydah. Usually in parties of 4–10 birds, in pairs when breeding.

Habitat

Acacia tree and bush savanna typically in dry areas with bushy thorn trees such as *Acacia mellifera* and *Ziziphus mucronata*. On broken ground, in gullies with clumps of bushes and long grass, in open woodland with scattered bushes and in riverine bush.

Analysis

195 squares (85%) Total count 800 (0,45%)

COMMON WAXBILL 846

Estrilda astrild

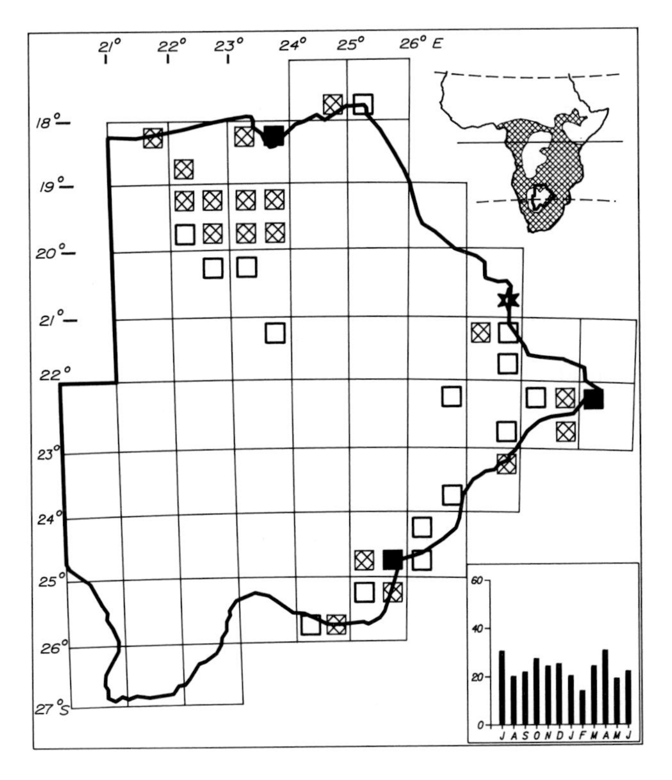

Status

Sparse to common resident of the Okavango, Linyanti and Chobe river systems. Recorded once near Deception Valley. Reappears in the east at Francistown and is sparse to fairly common along the Shashe, Motloutse, Limpopo and Ngotwane rivers to Lobatse and west along the Molopo River to Boshoek. Recorded as very common at Linyanti, Pont's Drift and Gaborone. Usually in small flocks of 5–15 birds, sometimes in flocks of up to 50 birds.

Habitat

Tall grasses and sedges in swamps, marshes, along margins of rivers, lagoons, lakes, dams, seasonal pans and ponds. Also in rank grass along drainage lines, ditches and irrigation. Usually in damp or wet habitat but wanders into gardens, cultivation, farmland and towns where it occurs in secondary growth and rank grass even when dry.

Analysis

37 squares (16%) Total count 307 (0,17%)

BLACKCHEEKED WAXBILL 847

Estrilda erythronotos

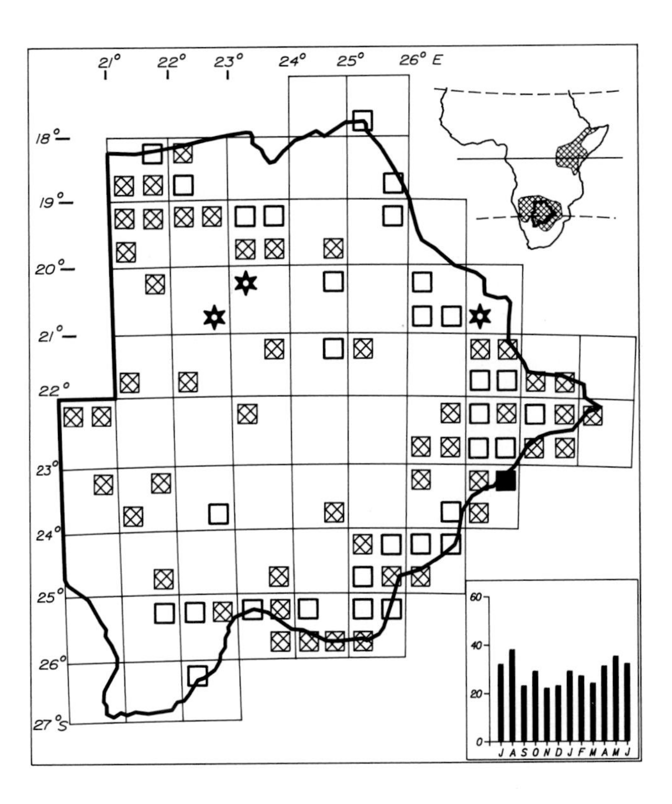

Status

Sparse to common resident in the northwest, east and southeast. Fairly common along the Molopo River to as far west as Tshabong. Elsewhere throughout the country it is patchily distributed and may be genuinely absent in the central Kalahari and the arid southwest. In pairs or small groups of 4–10 birds.

Habitat

Tree and bush savanna, particularly *Acacia*, in dry areas with low ground cover or on bare sand. Riverine thicket and riparian *Acacia*, edges of pans with stunted *Acacia* bushes, bushes on the edge of deciduous woodland and occasionally in gardens.

Analysis

84 squares (37%) Total count 373 (0,21%)

QUAIL FINCH

852

Ortygospiza atricollis

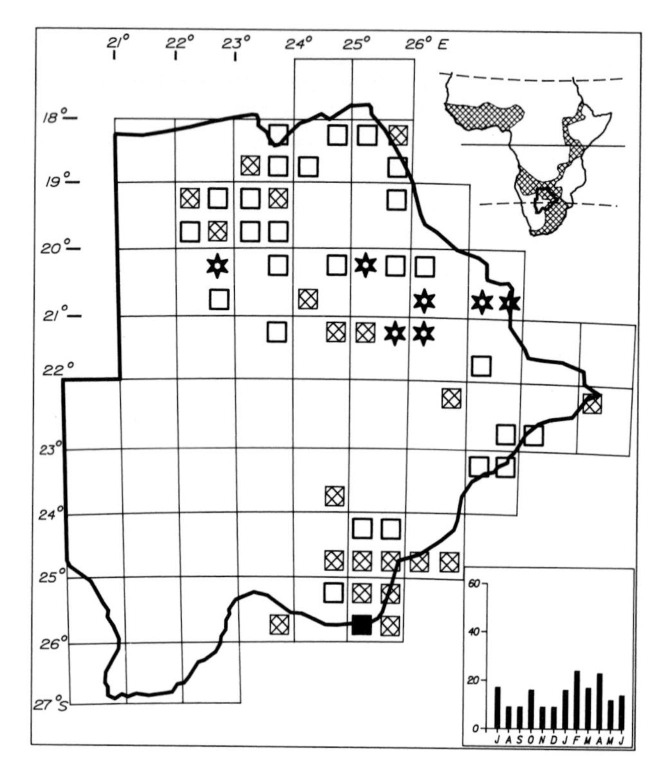

Status

Fairly common to common resident of the southeast to as far north as Salajwe. Sparse to locally common in the north and east. Common at Orapa and Jwaneng and very common at Phitsane Molopo. Its unexpected absence from most of the Makgadikgadi region during this survey is attributed to drought. It is expected to be more widespread in high rainfall years but is unlikely to spread into the arid savannas of the central and western Kalahari. Usually in groups of 4–30 birds.

Habitat

Short grassland in moist conditions. Floodplains, watershed plains, edges of lakes and dams, in river valleys, at seasonally inundated depressions in open savanna. It often remains in these habitats even when the ground has dried out during summer months but moves to more reliably moist habitats in winter.

Analysis

54 squares (23%) Total count 198 (0,11%)

ORANGEBREASTED WAXBILL

854

Sporaeginthus subflavus

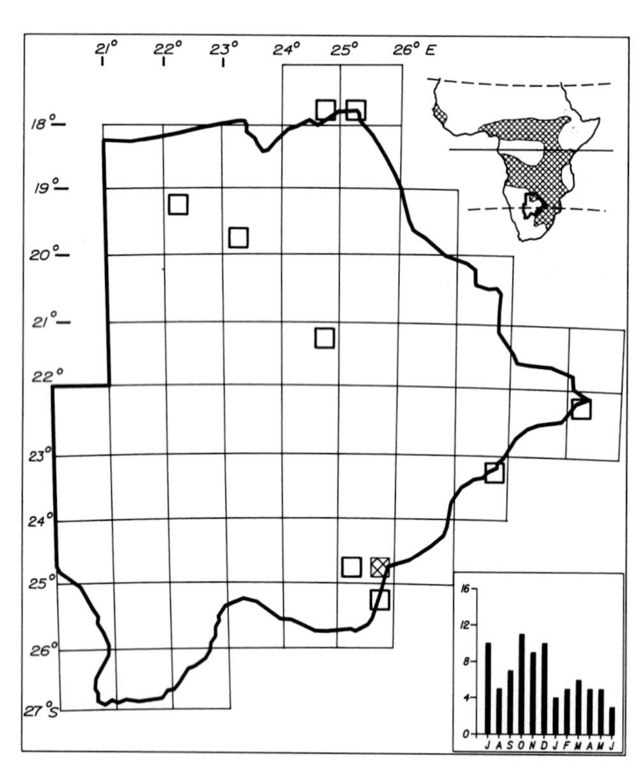

Status

Sparse and very localised resident in isolated localities in the north and east. Throughout its range in Botswana it occurs at the limits of its southern African distribution and the paucity of records may be attributable to the effects of drought causing a contraction in range. Sufficient suitable habitat exists for it to have been recorded more often particularly in the Okavango region. Usually in flocks of 4–20 birds.

Habitat

Reeds and tall vegetation in marshes and on the margins of rivers, dams and lagoons. Occurs at dams in Lobatse, Gaborone and Kanye.

Analysis

10 squares (4%) Total count 86 (0,05%)

CUTTHROAT FINCH

855

Amadina fasciata

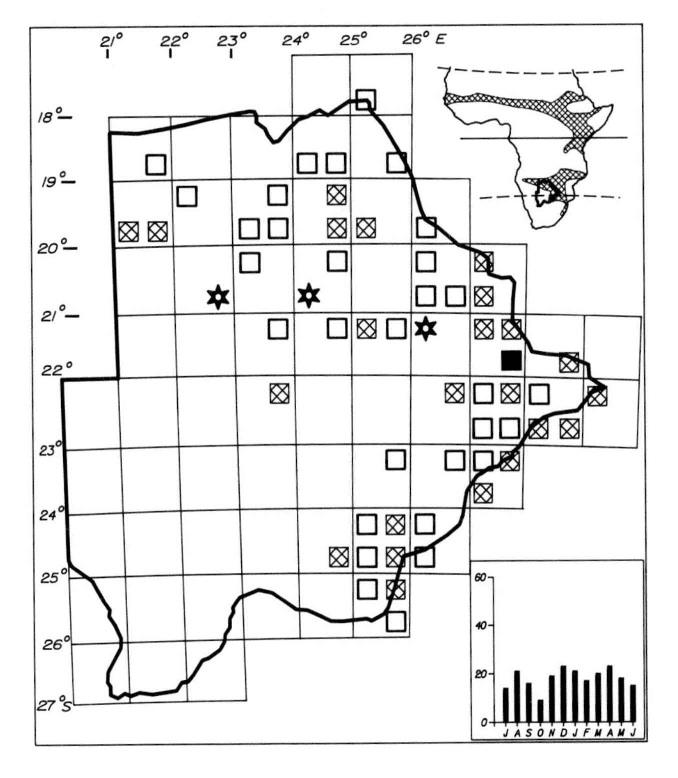

Status

Sparse to common resident of the east and southeast. Sparse to locally fairly common in the north to as far south as Deception and Okwa valleys. Common at Orapa and Francistown and very common at Selebi Phikwe. Occurs mainly in the high rainfall areas but extends into the margins of the central Kalahari in valleys. Egglaying February, March and April. Usually in groups of 5–15 birds.

Habitat

Open sparsely covered ground in tree and bush savanna and on the edge of deciduous woodland. Also in gardens, parks and football fields in towns and villages. Utilises old weaver nests for roosting and breeding.

Analysis

58 squares (25%) Total count 277 (0,15%)

REDHEADED FINCH

856

Amadina erythrocephala

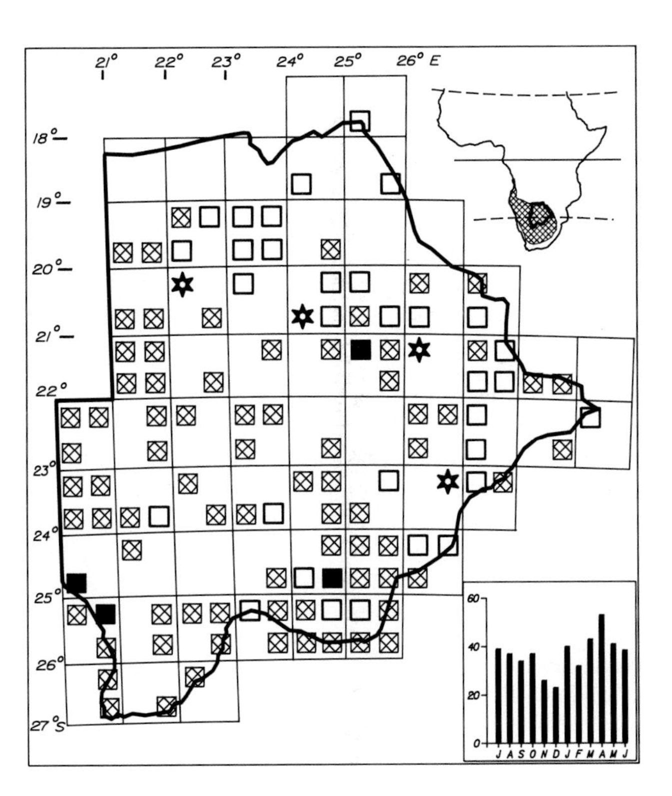

Status

Fairly common to common resident in all regions. Sparse or absent in extreme northern areas and sparser than the Cutthroat Finch in the east. Patchy and unpredictable in many areas. It is found most commonly along the Molopo and Nossob valleys. Nomadic. Local and seasonal movements are irregular. Egglaying April, June and July. Usually in flocks of 4–20 birds, occasionally occurs in flocks of 50–200 birds.

Habitat

Semidesert grassland where it forages mainly on bare patches between grass tufts. Open areas in tree and bush savanna especially thorn and *Terminalia* savanna. Also on disturbed ground such as at borrow pits, boreholes, dam walls, shores of pans, football fields. Occasionally in gardens when adjacent to dry grassland.

Analysis

116 squares (50%) Total count 472 (0,26%)

BRONZE MANNIKIN 857

Spermestes cucullatus

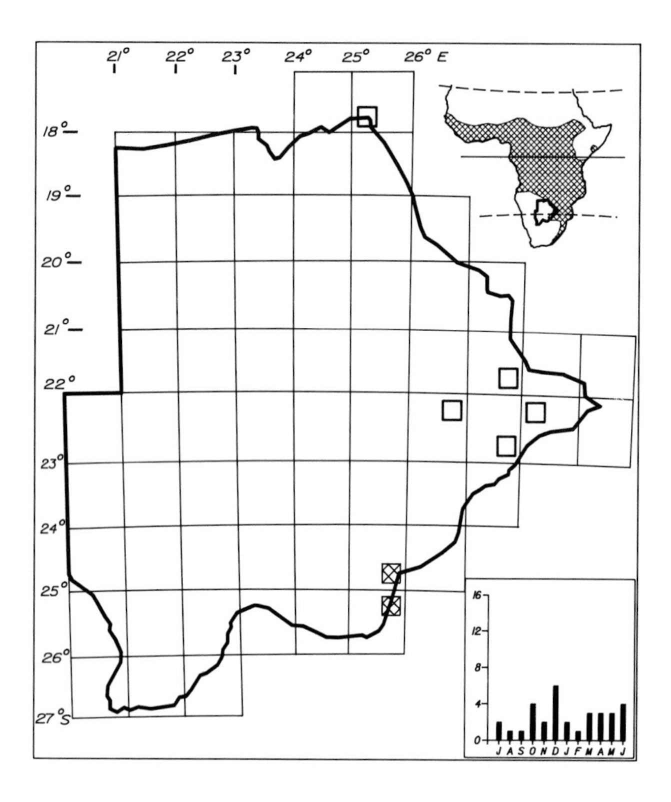

Status

Uncommon in the Lobatse and Gaborone areas. Sparse at Kasane. Very sparse in the east where it has been recorded only once at each of the localities shown. Its status is uncertain. Not recorded in Botswana prior to this survey. Occurs at the western limit of its range in southern Africa. Likely to be sedentary once established in an area. Usually in pairs or family groups.

Habitat

Secondary growth and rank vegetation in well-wooded areas. Thickets, bushes and dense growth in river valleys. Also in rank grass in gardens and around neglected cultivation. Usually near water.

Analysis

7 squares (3%) Total count 38 (0,02%)

PINTAILED WHYDAH 860

Vidua macroura

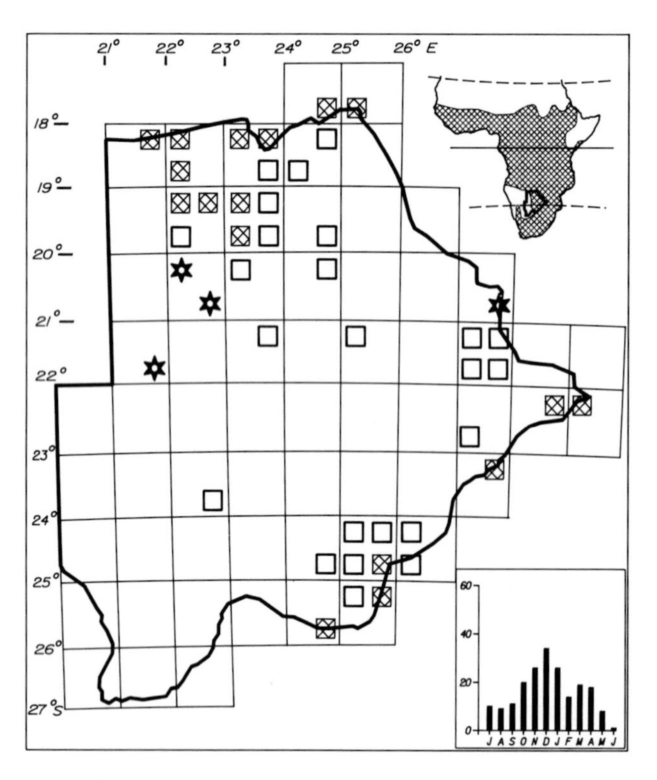

Status

Fairly common to common resident of the Okavango, Linyanti and Chobe river systems. Sparse elswhere in the north to as far south as Orapa and Deception Valley. In the east and southeast it is sparse to fairly common. Recorded twice at Kang. Nomadic. It may be more widespread in high rainfall years. It is polygamous and a parasitic breeder whose main host is the Common Waxbill. Usually in flocks of 10–30 birds with a ratio of 5–6 females to one male when breeding.

Habitat

Tree and bush savanna in high rainfall areas usually in moist open conditions such as floodplains, inundated grassland, river valleys. In the early breeding season it selects areas with green grass and remains in these areas when the grass is desiccated. Also in gardens and a variety of disturbed ground with secondary growth, e.g. roadsides, edges of cultivation and near sewage ponds.

Analysis

44 squares (19%) Total count 215 (0,12%)

SHAFTTAILED WHYDAH

861

Vidua regia

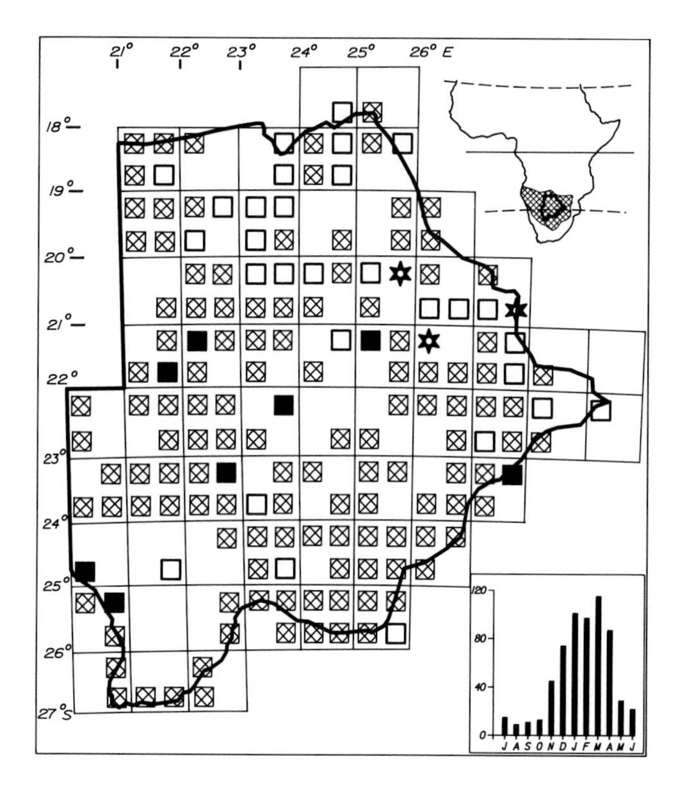

Status

Common resident in all regions. Uncommon in the Okavango Delta and adjacent wetlands of the north. A typical Kalahari savanna species whose range and abundance closely matches that of its breeding host, the Violeteared Waxbill. Nomadic; mainly in winter like the Pintailed Whydah. Usually in flocks of 10–50 birds when breeding. All male flocks in breeding plumage occur in March and April.

Habitat

In low rainfall areas it occurs mainly in *Acacia* and *Terminalia* tree and bush savanna, semidesert scrub savanna and the edges of dry woodland. In high rainfall regions it occurs in the drier types of tree and bush savanna, open areas on the edge of woodland and *Acacia* along drainage lines. Feeds on the ground in sparsely vegetated or bare areas.

Analysis

162 squares (70%) Total count 669 (0,37%)

PARADISE WHYDAH

862

Vidua paradisaea

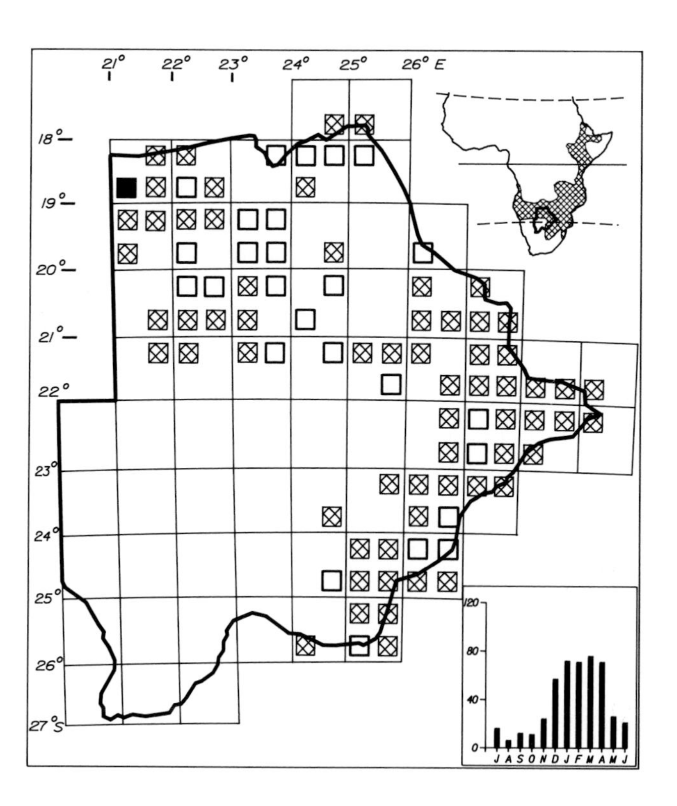

Status

Fairly common to common resident of the north and east to as far south as Ramatlabama and Boshoek. Extends into the margins of the central Kalahari along the Kuke fence and at Lephepe, Salajwe and Jwaneng. Recorded mostly between November and May when males are in breeding dress. As with other viduines its winter movements are poorly known. Parasitises the Melba Finch. Usually in small flocks of 5–30 birds.

Habitat

Tree and bush savanna in high rainfall regions. Open areas in woodland and in the large open spaces frequented by the Pintailed Whydah and the Shafttailed Whydah. Males often perch prominently on top of bushes, trees, fence posts and telephone wires along roads and tracks. Also in gardens and on farmland.

Analysis

90 squares (39%) Total count 480 (0,27%)

BROADTAILED PARADISE WHYDAH

863

Vidua obtusa

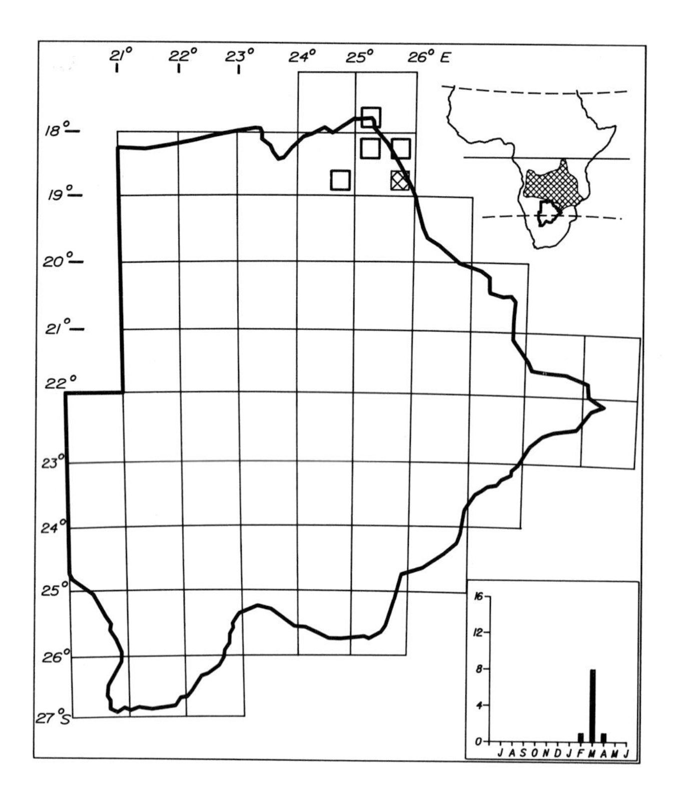

Status

Sparse to uncommon resident of the northeast corner of the country. An anticipated (Smithers 1964) recent addition to the avifauna of Botswana. Its breeding host, the Goldenbacked Pytilia, is confined to this area, both being at the extreme southwestern limit of their tropical distribution. As with the Paradise Whydah most records are attributable to males in breeding dress. Poorly known and warrants further study in Botswana. Usually in small flocks.

Habitat

Broadleafed woodland where it occurs mainly in open patches and on the edge. Deciduous tree savanna. Feeds on the ground. Males perch conspicuously in the breeding season like the Paradise Whydah.

Analysis

5 squares (2%) Total count 9 (0,005%)

BLACK WIDOWFINCH

864

Vidua funerea

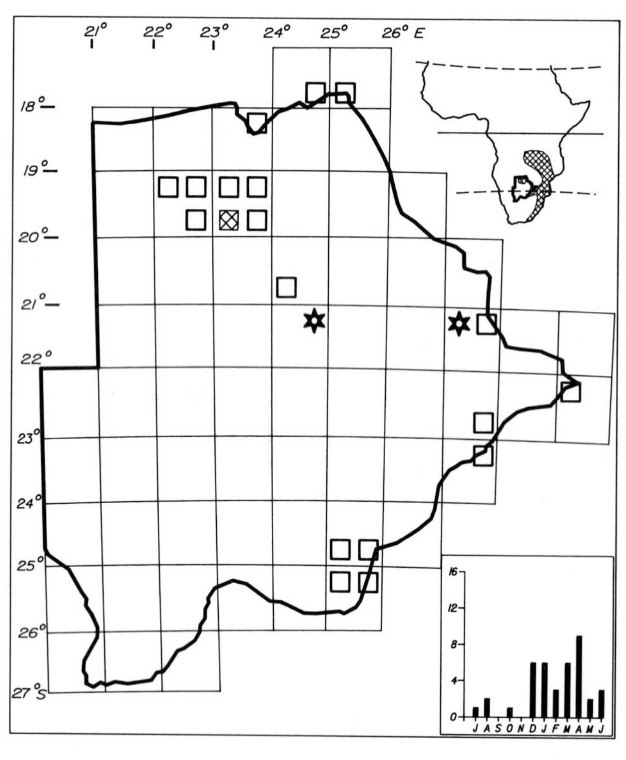

Status

Sparse resident of the north, east and southeast. The species is included in the avifauna of Botswana controversially and is mentioned here to encourage research into the matter. The distribution of birds identified on bill and leg colour as breeding males is shown on the map but the subspecies *V. c. okavangoensis* of the Steelblue Widowfinch has the same bill and leg colour. Furthermore, the breeding host of *V. funerea* does not occur in Botswana; *V. funerea* occurs on the east side of Zimbabwe; Payne (1985) never recorded *V. funerea* in Botswana during his specialised study of this family; *V. wilsoni* may occur in northern Botswana. Nominate *V. chalybeata* has been seen alongside purported *V. funerea* in the Okavango and Chobe areas. The range of *V. funerea* in the western Transvaal makes it possible for it to occur in the east and southeast of Botswana.

Habitat

Tall grass and secondary vegetation on the edges of woodland and in tree and bush savanna. Usually near water and in high rainfall areas.

Analysis

21 squares (9%) Total count 53 (0,03%)

PURPLE WIDOWFINCH

865

Vidua purpurascens

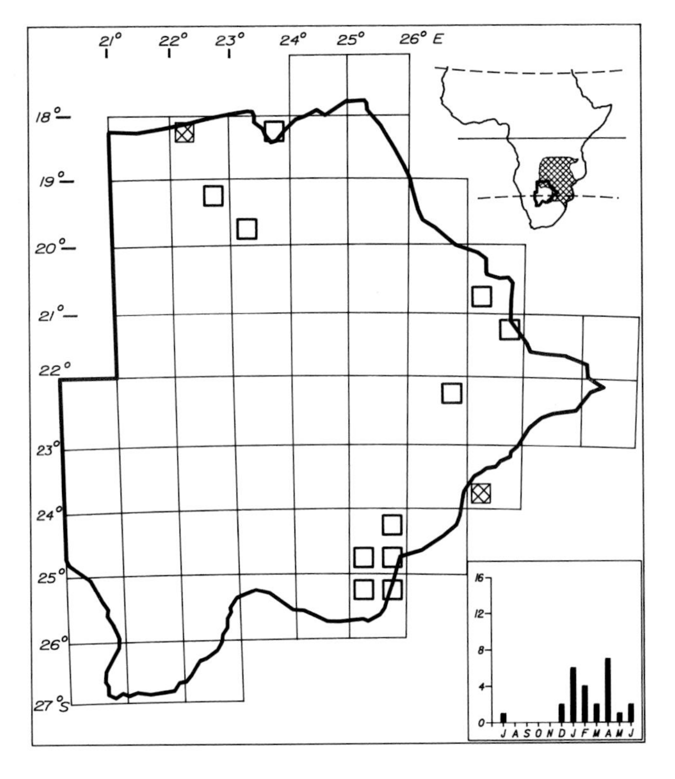

Status

Sparse to uncommon resident in the Okavango Delta and at Linyanti. Very sparse to uncommon in the east and southeast to as far south as Kanye and Lobatse. Its range overlaps that of its host, Jameson's Firefinch. Thinly distributed, unpredictable and poorly known. All records are of males in breeding plumage from December to June.

Habitat

Tree and bush savanna in high rainfall areas, usually where the vegetation is lush and often near water. Males sing from the tops of trees. Forages on bare ground or between grass tufts.

Analysis

13 squares (6%)   Total count 29 (0,02%)

STEELBLUE WIDOWFINCH

867

Vidua chalybeata

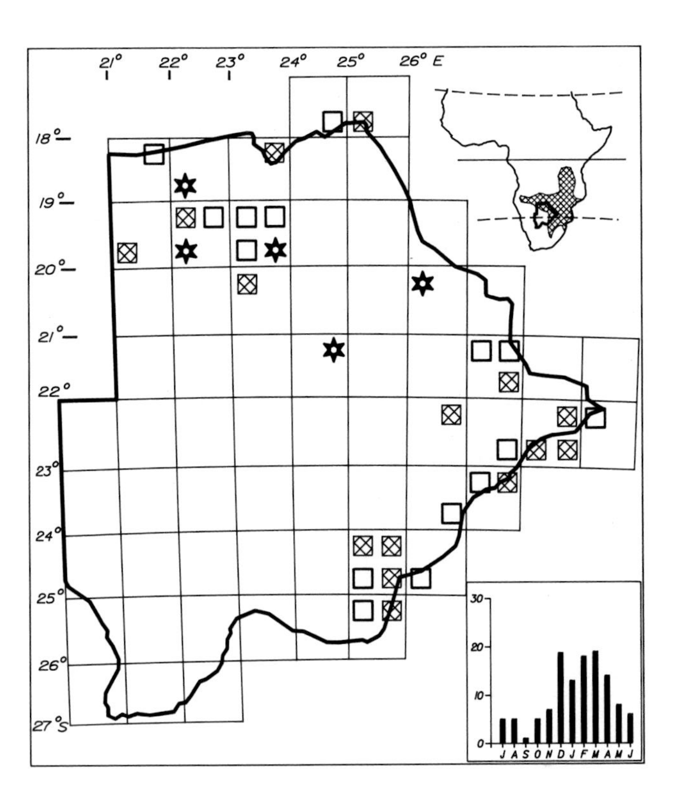

Status

Sparse to fairly common resident in the Okavango, Linyanti and Chobe regions. Reappears at Francistown and extends southwards in the east to Lobatse and Kanye. Generally more common and widespread than the Purple Widowfinch. Its range closely matches that of its host, the Redbilled Firefinch. Most frequently recorded between December and April when males are in breeding plumage. Nomadic in winter. Confusion exists between the subspecies *V. c. okavangoensis* of this species and the Black Widowfinch. Usually in flocks of 5–20 birds with several territorial males.

Habitat

Riverine *Acacia*, open deciduous woodland, tree and bush savanna in river valleys, floodplains with scattered bushes. Usually near water. Forages on sparsely-covered or bare ground in open areas amongst bushes and secondary growth.

Analysis

35 squares (15%)   Total count 136 (0,07%)

YELLOWEYED CANARY 869

Serinus mozambicus

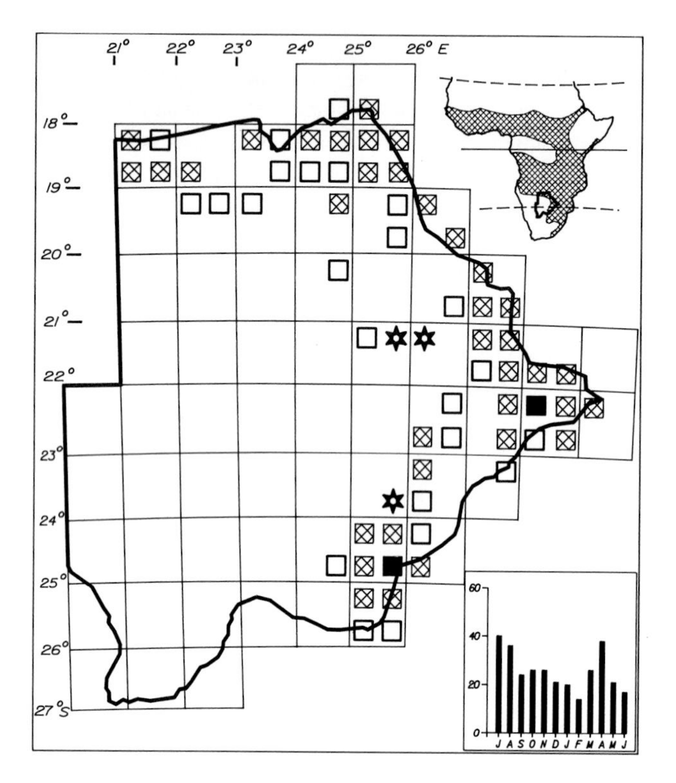

Status

Sparse to locally common resident of the north, east and southeast. It occurs in Botswana at the western periphery of its range in southern Africa and is confined to the high rainfall and well-wooded areas. It is very sparse in the Okavango Delta. It is common to very common at some localities in the east and southeast. Seasonal movements occur with flocks gathering at times in winter months. Egglaying December. Usually in pairs or small groups of 4–10 birds, winter flocks of 20–40 birds.

Habitat

Mature broadleafed or mixed woodland, tree and bush savanna mainly in river valleys, riverine bush, usually near water. Also in gardens, on farms and in secondary growth on edges of cultivation or around dams.

Analysis

65 squares (28%) Total count 331 (0,18%)

BLACKTHROATED CANARY 870

Serinus atrogularis

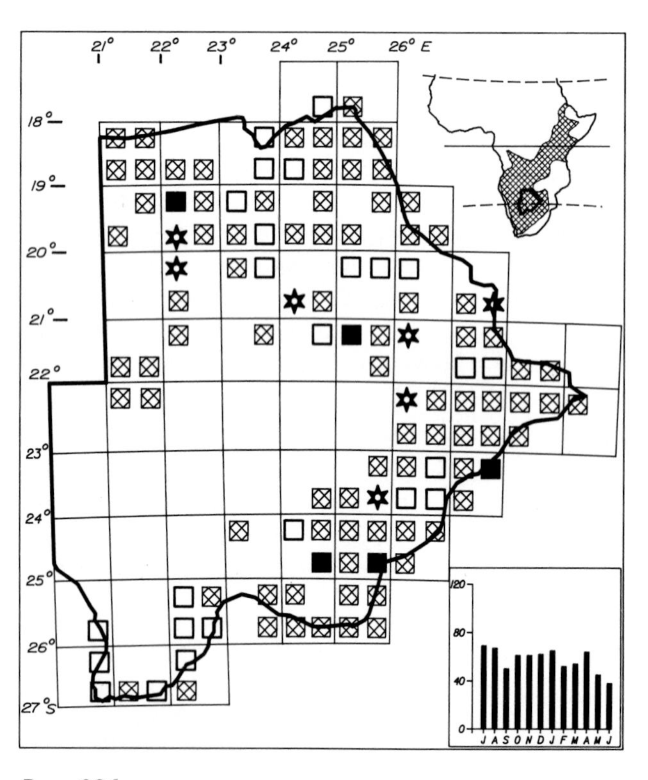

Status

Fairly common to common resident of the north, east and southeast. Extends into semidesert around habitation as at Ghanzi, Lephepe, Jwaneng, Salajwe, Morwamosu, Khokong. Extends west along the Molopo River to the Nossob Valley but is generally sparse in the southwest. Absent from most of the central and western regions. Local movements occur. Egglaying January. Usually in small groups of 5–15 birds.

Habitat

Tree and bush savanna, open areas on edges of woodland, riverine bush, edges of cultivation, gardens, farms, secondary growth around villages, cattle posts, boreholes and dams. Feeds on the ground on seeds and seeding plants and weeds such as occur around leaking water tanks, pipes and pumps. Never far from water.

Analysis

119 squares (52%) Total count 723 (0,41%)

YELLOW CANARY 878

Serinus flaviventris

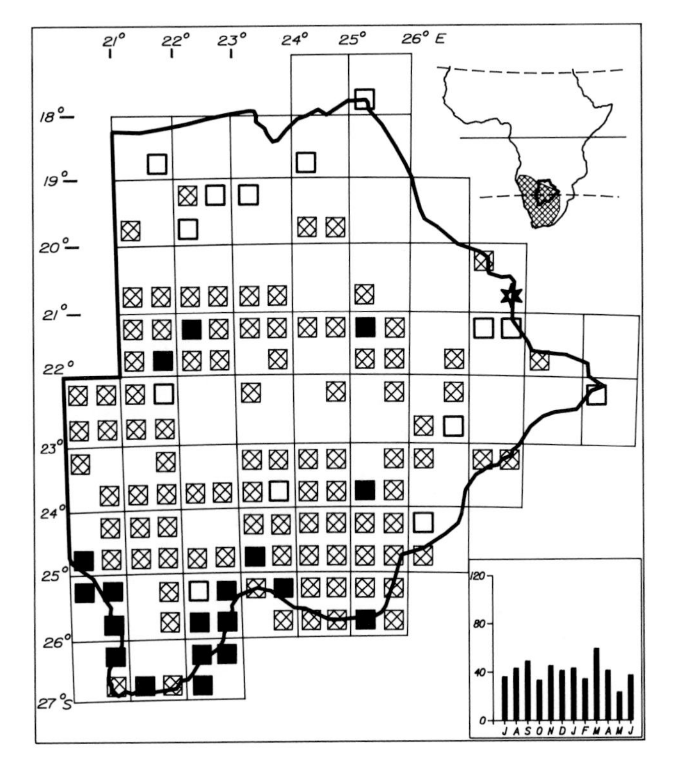

Status

Fairly common to very common resident south of 21°S and west of 26°E. Very common in the southwest. In the north and east it is sparse to uncommon. It is the common canary of the Kalahari savannas. Its range overlaps very little with that of the Yelloweyed Canary. Seasonal movements occur and large flocks of 20–100 birds on migration are seen in March, April and May and sometimes in winter. No clear pattern of movement is shown as it is present in all months of the year and such movements may be mainly local and related to breeding. Egglaying February. Usually in pairs or small groups of 4–10 birds. All male flocks may occur on migration.

Habitat

Kalahari tree and bush savanna, mainly in *Acacia* and *Terminalia*. Patches of woodland, plains with scattered trees, scrub savanna, low shrubs on the edges of pans, dunes with sparse grass and bush cover and on well-vegetated dunes. Also in gardens, cultivation and fallow lands.

Analysis

126 squares (55%) Total count 520 (0,29%)

STREAKYHEADED CANARY 881

Serinus gularis

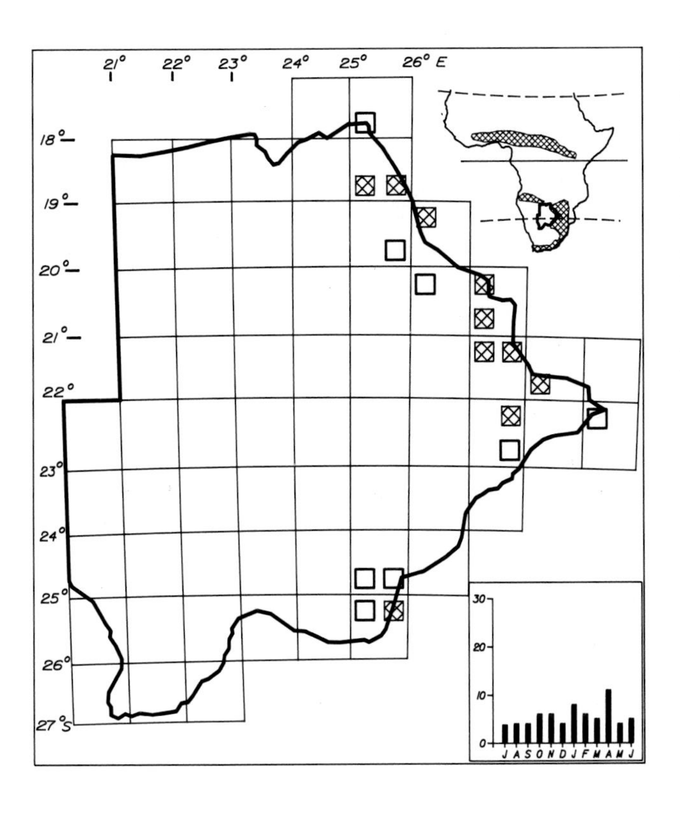

Status

Sparse to locally fairly common resident on the eastern periphery of the country where it occurs at the western limit of its range in southern Africa. Most frequently recorded around Francistown. Its absence along the upper reaches of the Limpopo River requires further investigation. It is easily overlooked and is not well-known. Seasonal movements are thought to occur. Usually in small groups of 4–8 birds, but occurs in pairs when breeding.

Habitat

Open broadleafed and mixed *Acacia* woodland, edges of woodland, riverine bush. Usually in river valleys and in areas with tall trees interspersed with bushes.

Analysis

18 squares (8%) Total count 69 (0,038%)

BLACKEARED CANARY

Serinus mennelli

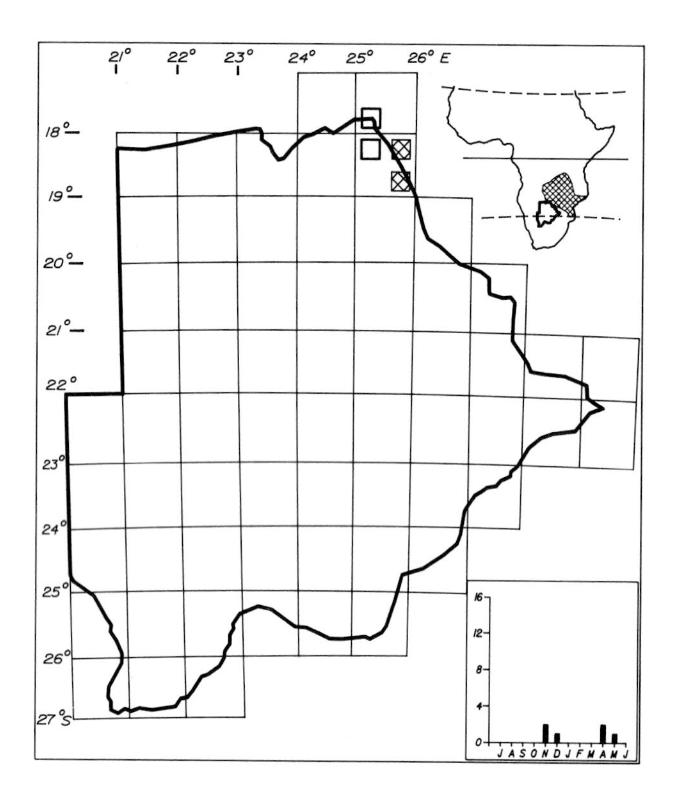

Status

Very sparse resident of the northeastern woodlands. Occurs at the extreme western tip of its Zimbabwean distribution and apparently disjunct from its Zambian range. Poorly known and warrants special study in Botswana. It is likely to be sedentary. It is not clear why its range does not extend westward into the available and suitable woodland. Usually in small groups. Sometimes as members of mixed bird parties.

Habitat

Baikiaea and miombo woodland where it occurs under the canopy. Feeds on the groundstratum. Seeks refuge in bush or saplings or the base of the canopy.

Analysis

4 squares (2%) Total count 7 (0,004%)

GOLDENBREASTED BUNTING

Emberiza flaviventris

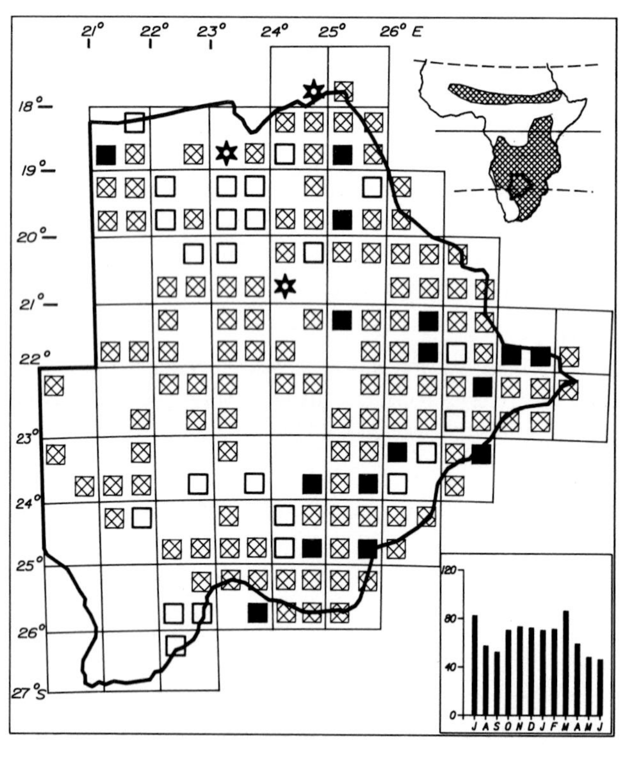

Status

Fairly common to common resident throughout the country except in the southwest corner. Commonest in the northeast, east and southeast where it is locally very common at some times of the year. Considerable movements occur with flocks moving through localities. The extent of these movements have not been studied and may be mainly postbreeding dispersal in March and April. However, part of the population is suspected to have movements to and from Zambia and Zimbabwe. Egglaying in all months from October to January. Usually solitary or in pairs, sometimes in flocks of 10–30 birds.

Habitat

Any woodland, tree and bush savanna. Usually under the canopy of trees and not in open areas without shade, such as scrub savanna.

Analysis

153 squares (67%) Total count 805 (0,45%)

CAPE BUNTING

885

Emberiza capensis

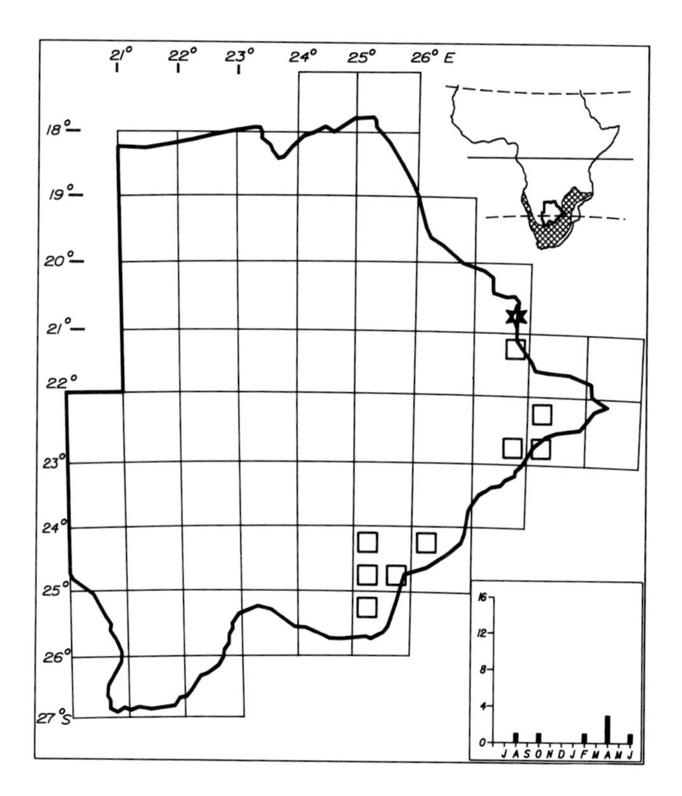

Status

Very sparse resident of rocky escarpments in the southeast, the Tswapong Hills and adjacent hilly areas to Francistown. Poorly known and warrants special study. Occurs at the western limit of its range in Zimbabwe and the Transvaal. Usually solitary or in pairs.

Habitat

Escarpments and hillsides with exposed rock or stony ground. Also rocky gorges and clefts made by rivers exposing rock.

Analysis

10 squares (4%) Total count 12 (0,006%)

ROCK BUNTING

886

Emberiza tahapisi

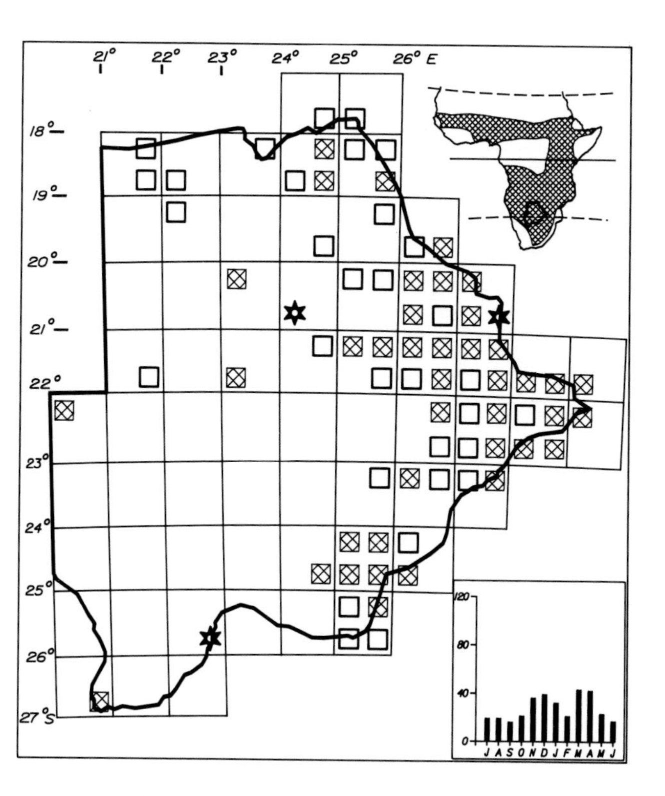

Alternative name

Cinnamonbreasted Rock Bunting

Status

Fairly common to common resident of the east and southeast. Sparse and unpredictable in the north. Occurs as far south as Ghanzi and Mamuno in the west and recorded twice at Bokspits in the southwest. Present throughout the year but most records fall between October and April. Local movements occur and intra-African migration is suspected. Usually in small groups of 3–10 birds, but in pairs when breeding.

Habitat

Rock outcrops, stony hillsides, gorges and escarpments when breeding. At other times may be found in a wide variety of habitat not associated with rocks—edges of woodland, tree and bush savanna, mopane scrub, edges of plains, dry river beds, dams, sewage ponds, cultivation and gardens. Usually on bare or sparsely-covered ground.

Analysis

75 squares (33%) Total count 348 (0,19%)

LARKLIKE BUNTING

Emberiza impetuani

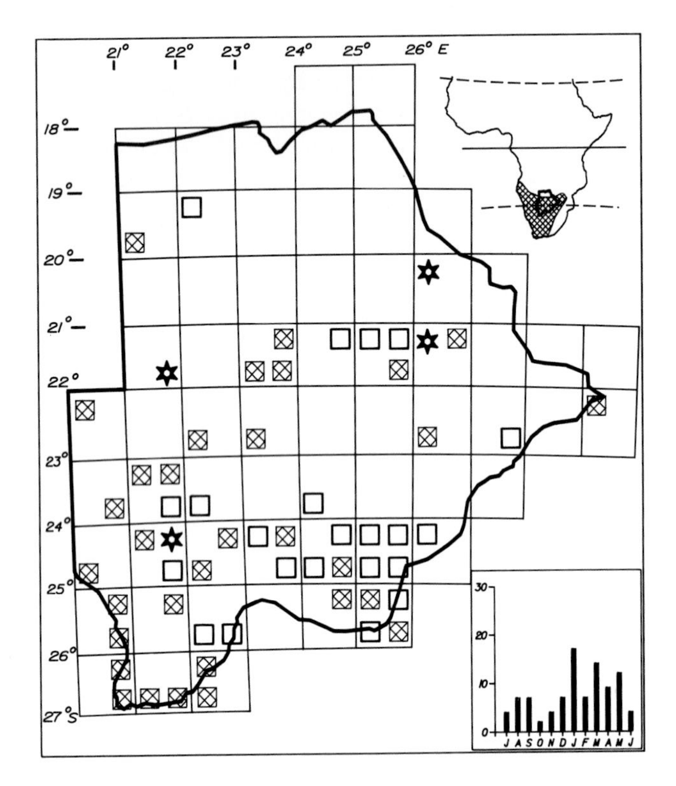

Status

Sparse to fairly common nomad south of 21°S. Common in the Nossob Valley. Thinly distibuted and unpredictable. More common in some years (e.g. 1981, 1987 and 1988 in this survey) than others. Recorded once near the Aha Hills and once at Gomare. Usually in small flocks of 5–15 birds, occasionally in flocks of 50–100 birds.

Habitat

Open tree and bush savanna in the dry regions. Usually on bare or sparsely-vegetated ground—bare dunes, escarpments in dry river valleys, rock or stone outcrops, stony hillsides. Often near water at cattle posts, boreholes and human habitation in arid areas.

Analysis

57 squares (25%) Total count 120 (0,06%)

Appendix A
Rare species

Recorded and confirmed as occurring in Botswana on fewer than 10 occasions and in no more than four squares.

GARGANEY 110
Anas querquedula 4 records

The records are: an undated record from Lobatse (Smithers 1964), 9 February 1975 from Gaborone, 29 November 1981 from Jwaneng, February 1985 from Gaborone. Palaearctic migrant. It may occur sporadically as far south as Botswana when there is drought in its usual nonbreeding areas further north. The Botswana records are the most southerly in Africa. Regular in Zambia in the 1970s.

EGYPTIAN VULTURE 120
Neophron percnopterus 2 records

A record from 2327B in January 1990 has been accepted on photographic evidence. This would be on the western edge of its previous range in the Transvaal although it is considered to be extinct there now. A sighting on the Boteti River (2024B) in December 1976 is considered to be the only other authentic record for Botswana.

HONEY BUZZARD 130
Pernis apivorus 5 records

Recorded from squares 1923C, 2022B, 2226B and 2426C (two records). Palaearctic migrant. It occurs in Botswana at the western limit of its previously described nonbreeding range in southern Africa. All records fall between 10 October and 16 April.

CROWNED EAGLE 141
Stephanoaetus coronatus 1 record

Recorded once on 24 July 1989 from square 1725C which is at the western tip of its range in northern Zimbabwe. It may be a sparse but regular resident in riverine forest along the Zambezi River and extend irregularly into dense woodland in the northeast.

PALMNUT VULTURE 147
Gypohierax angolensis 4 records

There are old records from Tsotsoroga Pan (1824C), Shorobe (1923D) and Toten (2022B). Claimed from 1923A in April 1972 and from 2125B in 1988. No recent confirmed record. It possibly still occurs.

EUROPEAN MARSH HARRIER 164
Circus aeruginosus 3 records

From 1725C on 2 December 1988, Lake Ngami on 19 November 1989 and 2525C on 24 January 1988. A Palaearctic migrant which migrates as far south as Botswana in small numbers in some years. It may be more common than reported in the north in years of good rainfall.

BLACK HARRIER 168
Circus maurus 6 records

An old visual record from Lake Ngami is now considered dubious. Two old records from the southwest (Gemsbok National Park and '50 miles' west of Tshane) are accepted in the light of the other available records which are all from the southwest. There are two records from the Nossob Valley (2520D and 2620B) and one from Mabuasehube (2521B). Only 2 of the records are from the past 10 years—both were in May, 1985 and 1989. Occurs in the southwest at the northern limit of its range in South Africa.

SOOTY FALCON 175
Falco concolor 1 record

Recorded from near Njuga Hills (2024B) on 26 December 1988. Another record in 1989 is under consideration. Possibly an irregular Palaearctic migrant on passage from breeding grounds in the Red Sea area to winter mainly in Madagascar. The number of inland records in subsaharan Africa have increased during the past 20 years.

CORNCRAKE 211
Crex crex 6 records

Four records from Smithers (1964) are from the Boteti River, Lake Ngami, 'Tati' (March 1874) and Palatsiwe Pan. During this survey two further records have been added—Gaborone, 14 December 1980 and Stevensford (2227D), 17 February 1982. Palaearctic migrant which, on the above evidence, may occur sparsely in the north and east between December and March in some years.

LUDWIG'S BUSTARD 232
Neotis ludwigii 4 records

Three recent records are from the Nossob Camp (2520B) area

of the southwest between October and December in 1983 and 1988. One past record from the Nossob Valley (Maclean 1970). Occurs at the eastern limit of its range in South Africa and Namibia and has only been recorded within sight of the boundary of the country.

WHITEBELLIED KORHAAN 233

Eupodotis cafra 2 records

To the record in Smithers (1964) from '5 miles' south of Kanye is added a 1983 record from near Mmathethe in the same square (2525A). The sight record reported in Smithers (1964) from Kaotwe Pan (2223A) is in abeyance until more is known of this species' distribution. Occurs in 2525A at the northern limit of its range in South Africa.

TEREK SANDPIPER 263

Xenus cinereus 4 records

Recorded on 10 October 1970 and November 1971 from Lake Ngami. The two recent records are from 1821B on 11 October 1987 and from 1725C in September 1988. Two purported records from Gaborone have not been substantiated. Normally a coastal Palaearctic migrant in southern Africa which occurs as a vagrant inland.

SPOTTED REDSHANK 267

Tringa erythropus 1 record

One record from 1724D on 1 November 1987 was accepted. A rare Palaearctic vagrant.

KNOT 271

Calidris canutus 2 records

Two birds at Lake Ngami in 1975. Two sightings at Mogobane Dam (2425D) in November 1978. There have been no records in the past 10 years but it is a species easily overlooked. Usually a coastal Palaearctic migrant in southern Africa.

LONGTOED STINT 275

Calidris subminuta 1 record

One bird at Lobatse sewage ponds on 11 November 1984. A very rare Palaearctic migrant to southern Africa.

PECTORAL SANDPIPER 279

Calidris melanotos 4 records

First recorded in Botswana from Mwaku Pan (2022D) on 3 January 1971. Also collected on the Boteti River at Toromoja (2124B) in April 1971. One bird at Mogobane Dam (2425D) on 26 November and one on 17 December 1978. Palaearctic migrant. No records in the past 10 years. Also rarely recorded in Zimbabwe and South Africa.

TEMMINCK'S STINT 280

Calidris temminckii 1 record

Recorded once in March 1986 at Savuti (1824C). A very rare Palaearctic migrant in southern Africa.

GREAT SNIPE 285

Gallinago media 5 records

From 1824B on 20 March 1967, 1923C on 19 October 1985 and December 1989, 1823A on 28 March 1985 and 1725C on 2 December 1988. A Palaearctic migrant which is easily overlooked and possibly a regular migrant to the northern wetlands annually. Scarce but widespread migrant to Zimbabwe, not recorded in South Africa.

BARTAILED GODWIT 288

Limosa lapponica 8 records

Two records prior to this survey from Lake Ngami in January (Hall) and on 23 October 1969 (Fraser). Also one from Mogobane Dam (2425D) on 9 October 1978. Five recent records—1923D in October 1987, 1725C on 26 December 1988, 2425D in October and November 1988 and 2125A in February 1989. Palaearctic migrant. Usually migrates to coastal areas of southern Africa. Not as common inland as the Blacktailed Godwit which is known to invade southern African areas in some years.

WHIMBREL 290

Numenius phaeopus 9 records

Recorded only from Gaborone and Orapa. Palaearctic migrant not previously recorded in Botswana though known as a scarce passage migrant inland in Zimbabwe and South Africa. Present in Gaborone in November and December 1985, September, October and November 1986 and September and October 1988. Present near Orapa October and November 1989. The above records suggest it is a regular passage migrant in very small numbers through north and east Botswana.

GREY PHALAROPE 291

Phalaropus fulicaria 1 record

Recorded from Tshono Pan (2424D) on 3 March 1961. Palaearctic migrant which is very rare and unpredictable inland and for which all future records should be substantiated with certainty.

ROCK PRATINCOLE 306

Glareola nuchalis 4 records

Only recorded from 1725C—Kasane and the adjacent Zambezi River. Undated record from the Chobe rapids (Smithers 1964), October 1972, 25 August 1988 and 29 September 1988. Occurs on its known breeding distribution along the Zambezi River. May occur elsewhere in northern Botswana sporadically as it is known to appear rarely away from large rivers in Zimbabwe.

LESSER BLACKBACKED GULL 313

Larus fuscus 9 records

From Lake Ngami on 27 September 1970, between 24 and 31 May 1971, between 24 February and 3 March 1981, 16 April 1985 and 19 November 1989. From Nata 12 October 1973 and 14 March 1986. From Jwaneng 30 October 1981. Previously recorded once from the Liambezi Lake (1724C) in the Caprivi in December 1961. Palaearctic migrant which may be expected to occur sparsely in some years at its extreme southern limit inland. Occurs as far south as Durban on the east coast.

BLACKHEADED GULL 319

Larus ridibundus 1 record

Recorded from 1724D on 10 April 1988. A very rare Palaearctic migrant barely known in southern Africa. Easily confused with Greyheaded Gull.

GULLBILLED TERN 321

Gelochelidon nilotica 4 records

Two specimens collected at Lake Dow (2124B) in January 1959. Sight records from Lake Ngami on 21 June 1970 and from 1923A in April 1972 and 17 June 1975. No recent records. A vagrant Palaearctic migrant normally to as far south as East Africa but which occurs further south in some years.

GREEN LOURIE 370

Tauraco persa 2 records

From 1823D in March 1987 and December 1988. Occurs at the western tip of its restricted range along the Zambezi Valley in northern Zimbabwe. The paucity of records suggest it is localised and possibly relict in northern Botswana. Status uncertain.

ROSS'S LOURIE 372

Musophaga rossae 1 record

Specimen collected 16 km east south east of Ikoga (1822C) on 7 October 1974. Several recent purported records (even sight records) have been unable to substantiate its presence in the Okavango Delta conclusively. It is unmistakable if well seen. Its occurrence would be at the extreme south of its range in western Zambia and possibly southern Angola but is at least 200 km from the nearest substantiated record outside Botswana.

THICKBILLED CUCKOO 383

Pachycoccyx audeberti 3 records

Specimen collected from the Khwai River (1923B) on 16 July 1969. Two recent records from 2229A on 23 May 1986 and 15 March 1987. Possibly occurs as a local migrant in some years to the north and east but could be resident and overlooked. Parasitises the Redbilled Helmetshrike and is likely to occur only within the range of this species.

EMERALD CUCKOO 384

Chrysococcyx cupreus 5 records

All records are from 1725C—one in December, two in January, one in February and one in April. An intra-African migrant which occurs in the northeast at the western tip of its range in Zimbabwe. It is found in riverine forest along the Zambezi River and could occur in this habitat along the Chobe River.

GRASS OWL 393

Tyto capensis 2 records

Specimen collected in 2425D in 1971 is in the National Museum, Gaborone. A sight record in 1978 from the same locality—Ngotwane grassland near Broadhurst sewage ponds—was accepted because of the previous specimen. No recent record. May occur in localised areas of the southeast or east.

BÖHM'S SPINETAIL 423

Neafrapus boehmi 5 records

Alternative name Batlike Spinetail

From Shakawe (1821B) on 24 March 1985 and December 1989, from Linyanti Camp (1823B) in May 1986 and 30 July 1987 and from 1922B in 1986. A rare resident occurring at its southern limit in the extreme north of the country probably only near riverine forest but possibly also in rich woodland. Easily overlooked.

OLIVE BEE-EATER 439

Merops superciliosus 2 records

Recorded once from 1725C in March 1974 and once on 18 January 1984. These records may refer to wanderers from a population which breeds in the Zambezi Valley or to migrants from Madagascar or to birds which breed in Namibia and Angola. Status uncertain but likely to remain rare in Botswana.

SLENDERBILLED HONEYGUIDE 479

Prodotiscus zambesiae 3 records

From 1821C on 22 June 1986, from 1821D on 25 October 1987 and from 1825A on 2 November 1986. A rare resident of the northern woodlands occurring at the southern limit of its range in Africa at these longitudes. Elusive and easily overlooked.

AFRICAN BROADBILL 490

Smithornis capensis 1 record

In 1923A on 13 October 1988 a bird was seen in breeding display. Possibly an overlooked sparse resident in riverine forest in the Okavango Delta and along northern perennial rivers.

STARK'S LARK 511

Spizocorys starki 10 records

Eight records are shown on a map in Smithers (1964). Specified localities are Khakhea (2423D), Kokong (2423A), Tshane (2421B), Lephepe (2325B) and Mabeleapudi (2522C). Occurs on short grass surrounding pans in the dry season but absent in February and March. One recent record from 2622A in December 1987. Also recorded by Maclean (1970) in the Nossob Valley. Easily overlooked.

BLACK SAWWING SWALLOW 536

Psalidoprocne holomelas 1 record

From 1922B on 30 August 1987. Breeding intra-African migrant to southern Zimbabwe and eastern South Africa between September and April. Status uncertain. The Eastern Sawwing Swallow has been reported as an isolated record from Sesheke in Zambia in August about 150 km from this current Botswana record.

RUFOUSBELLIED TIT 556

Parus rufiventris 1 record

From 1821B in July 1986. Possibly occurs irregularly in the extreme northwest at the western and southern limit of its range in northern Namibia and southern Zambia.

CAPE ROCK THRUSH 581

Monticola rupestris 1 record

A specimen reported in Smithers (1964) collected at Kanye. This species might occur in the southeast at the western limit of its range in the western Transvaal, possibly only as part of a seasonal movement. The resident rock thrush in this area is the Shorttoed.

MOUNTAIN CHAT 586

Oenanthe monticola 5 records

From 2520A in April 1986, 2521D in September 1988, 2620D in October 1985, 2622A in August 1988 and a record from Nossob Valley by Maclean (1970). Possibly a sparse resident in the southwest at the eastern limit of its range in Namibia and northern limit of its range in South Africa. It could also occur irregularly in the southeast at the northern limit of its range.

SICKLEWINGED CHAT 591

Cercomela sinuata 6 records

From 2321D on 24 April 1986, 2424B on 17 April 1981, 2424D on 29 March 1981, 2525A on 15 February and 28 June 1981 and an undated record from 2525A in 1983. It may be a rare resident in the southeast or appear in some years as a seasonal visitor. Its occurrence in this region is at the northern limit of its range in South Africa.

WHINCHAT 597

Saxicola rubetra 1 record

One record from Xaxaba (1923C) on 27 March 1987 is accepted. All other claims from the Okavango Delta have not been substantiated. Only occurs as a rare Palaearctic vagrant.

RIVER WARBLER 627

Locustella fluviatilis 3 records

Recorded from the edge of the dry Tati River in eastern Francistown on 29 March to 2 April 1985. Reported from the same locality on 16 March 1989. Palaearctic migrant which may occur in eastern Botswana in some years but is elusive and easily overlooked. Smithers (1964) suggests that it might also occur in the northern wetlands.

PALECROWNED CISTICOLA 668

Cisticola brunnescens 3 records

Specimen collected in 1923A in 1975. Two sight and sound records from 1823C and 1823D on 25 and 26 April 1989. There may be a discrete resident population in the northern wetlands at the southern limit of its range in Zambia. Known from eastern Zimbabwe and eastern South Africa.

LEVAILLANT'S CISTICOLA 677

Cisticola tinniens 1 record

From 2525B on 17 July 1981. Occurs at the western limit of its range in South Africa. Unlikely to occur except near the border of the country in the southeast.

CROAKING CISTICOLA 678

Cisticola natalensis 5 records

Recorded by Smithers (1964) at Mpandamatenga and vicinity (1825D). The type locality is 'Bamangwato District' i.e northeastern Botswana in 1882. Four recent records from 1824B in all months from July to October 1988. Occurs in this area at the western limit of its range in Zimbabwe. Possibly an overlooked sparse resident or occurs in some years as a range expansion from the east.

CAPE BATIS 700

Batis capensis 1 record

From 2327B in August 1988. This sighting was corroborated. Known from some localities in the adjacent western Transvaal but rare there. There may be a relict population in the Limpopo Valley from whence the current record was obtained.

GREY WAGTAIL 715

Motacilla cinerea 1 record

From Puku Flats in 1725D on 13 December 1984. An occasional rare Palaearctic vagrant to southern Africa. This is the first Botswana record.

MOUNTAIN PIPIT 901

Anthus hoeschi 1 record

Specimens were collected of four birds at Francistown between 14 and 22 October 1965—the specimens are now in Museum Alexander Koenig, Bonn. A migration of this species is postulated through northeastern Botswana from wintering grounds along the Zaire/Zambezi watershed to the Drakensberg massif in southeastern Africa. Not recorded since. Difficult to distinguish from Richard's Pipit in the field.

SOUZA'S SHRIKE 734

Lanius souzae 1 record

A female collected at Serondella (1724D) on 16 September 1967. Occurs at the southern limit of its Zambian range and eastern limit of its range in Angola and Namibia. No recent records. Found in *Baikiaea* woodland.

SOUTHERN BOUBOU 736

Laniarius ferrugineus 7 records

From Lobatse (2525B) in October to December 1984, near Kopong (2425B) on 20 April 1985, two sightings in January 1986 and one undated subsequent record from the Ngotwane River in 2425D. During this survey the distinction between this species and the Tropical Boubou which were previously treated as conspecific has been investigated by Hunter (1985, 1988). Birds in the Ngotwane and Marico river catchment are probably all Southern Boubou while those on the Limpopo River may all be Tropical Boubou. Possibly only occurs in the southeast—at the western limit of its range. Care is required with all records in the east below the Tropic of Capricorn.

LESSER BLUE-EARED STARLING 766

Lamprotornis chloropterus 5 records

Specimen collected at Serondella (1724D) on 14 September 1967. Subsequent records are all from 1725C—undated record 1975, April and July 1982 and 18 January 1984. There are also two records by Fletcher (in a letter to Smithers) from Francistown in 1962 but these are not mentioned by Smithers (1964). Occurs at the western limit of its range in Zimbabwe and is not likely to occur further west.

PALEWINGED STARLING 770

Onychognathus nabouroup 2 records

From 2420C on 10 August 1985 and 2521B on 30 September 1983. Occurs at the eastern edge of its range in Namibia. These are the first records for Botswana. A rare resident or a seasonal visitor in some years to the southwest of the country.

COPPERY SUNBIRD 778

Nectarinia cuprea 3 records

All records are from Serondella (1724D)—3 November 1987, 8 March 1988 and 26 April 1989. An intra-African migrant occurring in northeastern Botswana at the western limit of its range in Zimbabwe during the breeding season (November to March in Zimbabwe).

PURPLEBANDED SUNBIRD 780

Nectarinia bifasciata 5 records

From 1725C on 3 November 1987, 6 March and 6 December 1988, 1923A on 6 May 1973 and 1923B on 21 July 1981. Unsubstantiated record from 1821B. Possibly an overlooked sparse resident in the northeast. Status uncertain in the Okavango Delta. Occurs at the western limit of its range in Zimbabwe and southern limit of its range in Zambia.

CHESTNUT WEAVER 812

Ploceus rubiginosus 6 records

From 1821B on 22 April 1990, Gomare (1922B) in June, July and August 1988, Nokaneng (1922D) in 1988 and on the Thamalakane River 10 km south of Maun (2023A) on 4 January 1988. These records are the first for Botswana. Occurs in the northwest at the eastern limit of its range in Namibia but these records extend its previously known range considerably eastwards. Status uncertain. Poorly known throughout its range. In the canopy of *Acacia tortilis* and *Acacia erioloba* and feeding on the ground and on unharvested millet and sorghum.

CUCKOO FINCH 820

Anomalospiza imberbis 2 records

Alternative name Cuckoo Weaver

Undated record from Gazuma Pan (1825B) by Smithers. The only record in the past 10 years is from 1926A on 10 May 1986. Occurs at the western edge of its range in Zimbabwe and may occur rarely along the eastern border of Botswana seasonally or in high rainfall years.

REDHEADED QUELEA 822

Quelea erythrops 1 record

Recorded once in 1825A on 6 April 1985. An unexpected vagrant in Botswana. An intra-African migrant mainly to coastal areas of Mozambique between September and April from equatorial Africa.

REDCOLLARED WIDOW 831

Euplectes ardens 1 record

Adult male in breeding plumage in 2425D on 5–7 February 1986. Occurs at the northwestern limit of its range in the Transvaal. Possibly some range extension in some years.

LOCUST FINCH 853

Ortygospiza locustella 1 record

A specimen collected in 1923A in 1975 is in the American Museum of Natural History, New York, U.S.A. No recent claims. Rare south of 15°S in Zambia and absent from the middle Zambezi Valley. Possibly still exists as a relict population in the northern wetlands of Botswana.

Appendix B

Species which may occur

This is a list of birds which may occur in the country, (a) for which the current evidence is not conclusive, or (b) whose current known range in neighbouring countries suggests that they might occur in Botswana as an extension of this range. All future records of these species should be submitted with supporting evidence, e.g. photographs and good field notes.

MADAGASCAR SQUACCO HERON 73

Ardeola idae

Claimed once from 1725C but without adequate supporting evidence. Possible as a rare vagrant from Madagascar.

SHOEBILL 82

Balaeniceps rex

Two birds claimed to have been seen near South Gate, Moremi Wildlife Reserve (1923B) in 1979. One bird claimed to have been seen near Khwai River Lodge in 1988. There is no conclusive evidence to support either record. An elusive species of large swamps in southcentral and east Africa. Improbable in Botswana.

LONGLEGGED BUZZARD 151

Buteo rufinus

One claim from 1725C during the past 10 years not accepted. Would occur only as a rare Palaearctic vagrant. Sometimes confused with Steppe Buzzard in Africa south of the equator.

AUGUR BUZZARD 153

Buteo augur

Claimed once from the Tsodilo Hills but unsubstantiated. Possible as a wanderer in the north. The Tsodilo Hills provide likely habitat for such wanderers.

GREY KESTREL 184

Falco ardosiaceus

All old records have been rejected or withdrawn. One record under consideration for 1989. Unlikely to occur.

SHELLEY'S FRANCOLIN 191

Francolinus shelleyi

No claim—range in Zimbabwe abuts northeastern boundary of Botswana.

REDNECKED FRANCOLIN 198

Francolinus afer

Several claims from the north have been rejected. There is no reason to change the rationale expressed by Smithers (1964) that this species will not occur.

BLACKRUMPED BUTTONQUAIL 206

Turnix hottentotta

Alternative name Hottentot Buttonquail

No recent claims. Possible in the east north of Francistown.

STRIPED CRAKE 216

Aenigmatolimnas marginalis

Two records—April 1988 in 1922B and May 1989 in 1923C have not been accepted for lack of adequate evidence. Very likely to occur as an intra-African migrant to the northern wetlands, but it is secretive in habits.

BUFFSPOTTED FLUFFTAIL 218

Sarothrura elegans

No claims. An intra-African migrant which reaches southern parts of South Africa from the tropics. There are suitable areas of Botswana for this species to utilise on passage or in which it might even remain to breed, e.g. riparian forest in the Okavango Delta and along the northern rivers, the Limpopo Valley and dry thickets in low-lying areas such as the Nokaneng thickets.

BLACK TERN 337

Chlidonias niger

Several claims from Lake Ngami and Makgadikgadi Pans but none substantiated. Easily confused with Whitewinged Tern in nonbreeding dress. Palaearctic migrant normally occurring along the Namibian coast—rare inland. Only one record accepted in the Transvaal.

ROSYFACED LOVEBIRD 367

Agapornis roseicollis

Recently recorded in the Nossob Camp square (2520B) on the South African side of the border. May occur occasionally within a restricted range in Botswana in the far west at the eastern limit of its range in Namibia.

CAPE EAGLE OWL 400

Bubo capensis

Several claims from different areas have been rejected. May possibly occur in the east as an extension of range from the Matobo Hills in Zimbabwe.

BRADFIELD'S SWIFT 413

Apus bradfieldi

A sighting in 2620B on 11 April 1986 admitted the species as likely to occur in Botswana. Further evidence is required to substantiate the species on the Rarities List. Most likely to occur in the southwest at the eastern limit of its range in Namibia.

CROWNED HORNBILL 460

Tockus alboterminatus

No claims. Occurs in the middle Zambezi Valley and in part of the Hwange National Park. May possibly therefore occur in miombo or on the edge of riparian forest in northern Botswana—probably not in areas occupied by Bradfield's Hornbill.

OLIVE WOODPECKER 488

Mesopicos griseocephalus

No claims. Known from Katima Mulilo and Victoria Falls in Zambia thus could occur in riparian forest in the eastern sector of the Chobe River.

REDTHROATED WRYNECK 489

Jynx ruficollis

One claim rejected. From its main range in southcentral and eastern Transvaal a few records extend towards the eastern border of Botswana. Could occur as a vagrant in the east.

ANGOLA PITTA 491

Pitta angolensis

No claims. There are early 20th-century vagrant records in the western Transvaal. Normally an intra-African migrant to southern Mozambique and the middle Zambezi Valley. Possible in poorly explored areas of the Chobe, Okavango and Limpopo Valleys at altitudes below 1000 m.

MELODIOUS LARK 492

Mirafra cheniana

All previous records have been disproven or are doubted. The Patlana Flats record of 1961 (Smithers 1964) and the specimen collected at Nata (Ginn 1976) have been found to be Monotonous Larks (K. Hustler in *litt.*). A record from 2122A in 1970 is rejected for lack of suitable habitat. A sight record from 2525B in January 1988 is the only possible authentic record as there is *Themeda* grassland in this region. Status uncertain and great care is required fully to substaniate future records.

BLACKEARED FINCHLARK 517

Eremopterix australis

Recorded by Maclean (1970) in Botswana along the Nossob Valley. No recently confirmed Botswana records. It could occur, if only sparsely, in the southwest at the northern limit of its range in South Africa.

ANGOLA SWALLOW 519

Hirundo angolensis

Recorded once from the Caprivi Strip and occurs along the Okavango and Kunene Rivers in southern Angola. May occur as a vagrant in the extreme north of Botswana in some years.

WHITENECKED RAVEN 550

Corvus albicollis

One record from 2227D in 1983 not substantiated. Occurs in the hilly regions of northwestern Transvaal and southwestern Zimbabwe. Thus might extend its range to hilly regions of eastern Botswana.

SOMBRE BULBUL 572

Andropadus importunus

No claims. Occurs in riverine forest along the middle Limpopo Valley and might reach eastern Botswana in part of the Limpopo drainage.

EUROPEAN WHEATEAR 585

Oenanthe oenanthe

Three claims of this Palaearctic migrant in November and December not accepted. It can be confused with immature Capped Wheatear which are common at that time of year. Recorded from Nossob Valley (Maclean 1970). Only one record accepted in the Transvaal. Would only occur in Botswana as a rare Palaearctic vagrant.

TRACTRAC CHAT 590

Cercomela tractrac

Recorded in the Nossob Valley (Maclean 1970). No Botswana record—a claim from near Mabuasehube not substantiated. Could occur in the extreme southwest at the eastern limit of its Namibian range.

KAROO CHAT 592

Cercomela schlegelii

Recorded from Nossob Valley (Maclean 1970) as an extension of the northern limit of its South African range. Unlikely to occur in Botswana.

CAPE ROBIN 601

Cossypha caffra

Occurs in southwestern Transvaal up to the southeast border of Botswana. Well adapted to human habitation and is considered likely to extend its range into southeastern Botswana. Subsequently recorded at Gaborone (Wall 1992).

COLLARED PALM THRUSH 603

Cichladusa arquata

One claim from Linyanti (1823B) not substantiated. Occurs in the middle Zambezi Valley at the Victoria Falls and for 30 km upstream. Could occur in palm areas of the extreme northern wetlands.

LAYARD'S TITBABBLER 622

Parisoma layardi

Appears on the first Botswana checklist but no record in Botswana has been traced or substantiated. Unlikely to occur.

EUROPEAN REED WARBLER 630

Acrocephalus scirpaceus

Recently accepted as occurring in northern Namibia and may thus also occur in northern Botswana. This Palaearctic migrant is indistinguishable in the field on song and morphology from the resident and intra-African migrant African Marsh Warbler *A. baeticatus*. Mist-netting would be required to establish its status in Botswana.

REDFACED CROMBEC 650

Sylvietta whytii

Claims from 1725C, 1821B and 1823B have not been substantiated. Occurs in eastern Zimbabwe in the canopy of *Baikiaea* and miombo woodland. Possible in northeastern Botswana.

REDCAPPED CROMBEC 652

Sylvietta ruficapilla

Recorded once on Nampini Ranch west of the Victoria Falls in Zimbabwe in the canopy of miombo woodland—the only record for southern Africa. Normally occurs as a northern miombo endemic not occurring south of 15°S. The single Zimbabwean record is contrary and is mentioned only because it occurred near the Botswana border.

PIED STARLING 759

Spreo bicolor

No claims. Occurs in southwestern Transvaal up to the southeastern border of Botswana and can be expected to occur in adjacent areas of Botswana if only rarely. Well adapted to open habitation such as farmland in the Transvaal.

MIOMBO DOUBLECOLLARED SUNBIRD 784

Nectarinia manoensis

Two old claims from the Francistown region have not been substantiated. Could occur in this area at the southwestern limit of its range in Zimbabwe.

BLUEBILLED FIREFINCH 840

Lagonosticta rubricata

None of the several claims from southeast Botswana could be substantiated after a review of photographic evidence and a specimen. Occurs in the Transvaal up to the Botswana border in the region of the Tropic and may subsequently be found to occur in the Tuli Block region

VIOLET WIDOWFINCH 866

Vidua wilsoni

A rare and very poorly known species of restricted distribution in northeastern Namibia, southern Angola and the Caprivi Strip. Its range might include adjacent areas of extreme northern Botswana.

BLACKHEADED CANARY 876

Serinus alario

Recorded in the Nossob Valley (Maclean 1970). One claim from 2622A is in abeyance. Possibly occurs in southwestern areas of Botswana in some years at the northern limit of its South African range.

Appendix C

Old records (pre-1980)

This section contains a list of old records from publications or from individuals. The numbers in brackets enumerate the number of records obtained from the named source in the specified year. The purpose of the list is to identify the sources used in this work so that persons who know of other sources can add their contribution to a future update of the avifauna of Botswana. The research for this list concentrated on the period from 1964 to 1979 on the assumption that most records before that period are included in Smithers (1964).

1953 – Paterson, M. (1).
1957 – Paterson, M. (1).
1962 – Fletcher, A.B. (424); Traylor, M.A. (41); Jackson & Steyn (1).
1963 – Winterbottom, J.M. (28).
1964 – Smithers, R.H.N. (860); Vernon, C.J. (10); Vernon & Thorn (131).
1965 – Fabian, D.T. (63); Child, G. (1).
1966 – Ginn, P.J. (1180); Clancey, P.A. (3); Jackson & Steyn (28); Ginn, P.J. (50).
1967 – Child, G. (44); Cooke, P. (78); Irwin, Niven & Winterbottom (294).
1968 – Ginn, P.J. (120); Tree, A.J. (3).
1969 – Boulton, R. (101); Tree, A.J. (8); Fraser, W. (11); Falcon College (64).
1970 – Jones, P.J. (13); Fraser, W. (3); Tree, A.J. (74); Day, D. (2); Maclean, G.L. (40).
1971 – Jones, P.J. (139); Jacka, R.D. (4); Dawson & Jacka (3); Tree, A.J. (1); Wilson, G. (1); Day, D. (1); Archer & Turner (1).
1972 – Child, G. (7); Jacka, R.D. (4); Dawson & Jacka (1); Newman, K.B. (190); Jones, P.J. (1); Jones, C. (49); Mills, M.G.L. (1); Steyn, P. (1); Beesley, J. (2).
1973 – Payne, R.B. (31); Stanyard, D. (157); Beesley, J. (13); Jones, C. (35); Boulton, R. (107); Newman, K.B. & Fabian, D.T. (196); Ginn, P.J. (68); Trevor, S.M. (1); Milewski, A.V. (4); Tree, A.J. (1).
1974 – Ginn, P. (2); Stanyard, D. (272); Mills, M.G.L. (2); Beesley, J. (8); Bennett, G. (34); Tree, A.J. (4); Gerhardt, J. (4).
1975 – Dowsett & Shephard (149); specimens (2); Dawson & Jacka (1); Dowsett, R.J. (86); Beesley, J. (8); Mills, M.G.L. (2); Gerhardt, J. (8); Liversedge, R. (1); Wilson, J. (1); Stanyard, D. (58); Fabian, D.T. (36); Dowsett, Smith & Steen (221); Dowsett, Astle & Steen (58); Dowsett & Steyn (46); Beasley, A. (130); Newman, K.B. (94); Winifred Carter Expedition (198).
1976 – Beesley, J. (4); Start, J.M. (397); Gerhardt, J. (331); Day, D. (1); Stanyard, D. (25).
1977 – Start, J.M. (116); Stanyard, D. (49); Fraser, W. (566); Day, D. (1); Hodgson, M.C. (3); Emanuel, V. (1).
1978 – Fraser, W. (136); Hodgson, M.C. (1); Wilson, J. (18); Archer & Turner (2).
1979 – Wilson, J. (15); Hodgson, M.C. (1); Mathews, N. (1); Jones, C. (1); Barnes, J.E. (1); Knox, A.J. (1).

Old lists were received by the Coordinator from Dowsett, R.J.; Fabian, D.T.; Jones, P.J. and Newman, K.B. From the archives of the Ornithological Association of Zimbabwe D. Rockingham-Gill donated old Botswana field cards from Beasley, A.; Boulton, R.; Cooke, P.; Ginn, P.J.; Vernon & Thorn; Vernon, C.J. From the archives of the Botswana Bird Club old records were obtained from Beesley, J.; Fletcher, A.B. (copy of a letter to Smithers); Gerhardt, J.; Owens, M. & D. (1980); Stanyard, D.; Start, J.M.; and Yellen, J. (1980). A copy of *The Birds of Makgadikgadi – a preliminary report* was kindly donated by P.J. Ginn. The remaining sources in the above lists were extracted from published works. The contribution of all these ornithologists is gratefully acknowledged.

Appendix D

Gazetteer

The Gazetteer incorporates places mentioned in the text, similar sounding names, interesting birdwatching localities and some localities mentioned in articles published in *The Babbler*. Significant geographical localities, e.g. Fly's Kop and many pans are entered for areas which are otherwise barren of names due to their remoteness. There has been an attempt to include at least one name in every square. Using a name for each square as a means of orientating the reader has been avoided. The terms Francistown East, Francistown West, Kanye North and Kanye South have been used in the text to highlight that Francistown and Kanye are on gridlines and future records should specify from which side of the lines the records are reported. Places which appear in the Gazetteer of localities in Smithers (1964) have been included as a means of maintaining the historical thread for old records.

The names follow in the first instance the map of the Republic of Botswana 1:1 500 000 (Department of Survey and Lands, Gaborone) Edition 4, 1984. Localities and names in 1:250 000 regional maps (Department of Survey and Lands, Gaborone) have been used for finer detail but the spelling and exact location may have changed since the publication and should be used with this caveat in mind. It is partly for this latter reason that Square numbers and not grid references are used as the reference for the Gazetteer.

Locality	Square
Aha Hills	1921C
Artesia	2326C
Baines Baobabs	2024B
Baines Drift	2228B
Barolong Farms	2525C
Bathoen Dam (Kanye)	2425C
Bere	2221D
Betsaa (Betsha)	1822D
Boatlanane	2325D
Bobonong	2128C
Bodibeng	2022D
Bogogobo	2621D
Bokspits	2620D
Bonwalenong	2227C
Bonwapitse	2326B
Boritse Pan	2323C
Boshoek	2524C
Bosobogolo Pan	2522A
Bosoli Pan	2027D
Bray	2523B
Broadhurst (+ S.P.)	2425D
Bromfield Pan	1925B
Buffels Drift	2326D
Buitsivango (part)	2121C
Bushman Pits	2024A
Cacanika (Lagoon)	1923A
Chadibe	2327B
Chanoga	2023B
Chao (Tchau)	1922B
Charles Hill	2220A
Chief's Island (part)	1923A/C
Chief's Island (part)	1922B
Chilwero	1725C
Chinamba	1824D
Chinamba Hills	1824D
Chobe Lodge	1725C
Chubukwani	2221A
Chukutsa Pan	2124B
Cream of Tartar Pan	1825D
D'kar (Dekar)	2121D
Daly's Pan	2524C
Damara Pan	2222A
Danega	1922C
Deception Pan	2123B
Delta Camp	1923C
Dibajakwena Dam	2525A
Dibejama	1821A
Dibete	2326C
Dikbaadskolk	2520D
Dikbos	2622C
Dikgomo di kae	2424D
Dikogonye	2325C
Dinokwe	2326B
Dithope Game Ranch	2325B
Ditshegwane	2424B
Drotsky's Cave	1921C
Dukwe cordon fence	2026B
Dutlwe	2323D
Epukiro	2021C
Eretsa	1822D
Etsa Pan	2323A
Etsha	1922A
Fly's Kop	2621A
Foley's Siding	2127C
Francistown (west)	2127A
Francistown (east)	2127B
Gabba Pan	2021B
Gaborone	2425D
Gadikwe (Lagoon)	1923A
Gai Pan	2422A
Ganaleina	2124D
Gazuma Pan (Kazuma)	1825B
Gcoha Hills	1824A
Gcoverega (Joverega)	1924A
Gcwihaba Hills	2021A
Gemsbok Pan	2121C
Ghanzi	2121D
Gokoni Pan	1824B
Gomare (Gumare)	1922A
Gomodimo	2223D
Good Hope (+ Pan)	2525A
Gope	2224D
Groot Laagte (part)	2121A
Groot Laagte (part)	2021C
Gubatsaa Hills	1824C
Gubujango (Kobojango)	2128D
Guma Lagoon	1822C
Gutsha Pan	2025C
Gweta	2025A
Haina Hills	2023C
Hanahai (part)	2222A
Horseshoe Bend Camp	1823A
Hukuntsi	2321D
Ikgoga	2322B
Ikoga	1822C
Impalila Island	1725C
Inkonkwane Pan	2124B
Inkotsang	2525B
Janeng (Hills)	2525B
Jedibe Lagoon	1922B
Jiare	1926C
Jolley's Pan	1825D
Joverega	1924A
Jwage	2322B
Jwaneng	2424D
Jwaneng Hill (Jaaneng)	2523A
Kachekabwe	1824A
Kachikau	1824A
Kakulwane	1825A
Kalakamati	2027C
Kalamare	2226D

Place	Grid
Kalkfontein (Tsootsha)	2220B
Kameelsleep	2520D
Kang	2322D
Kangwa	2422B
Kangyane Pan	2322C
Kanye (north)	2425C
Kanye (south)	2525A
Kanyu	2024B
Kaokwe	1821B
Kaole Pan	2420A
Kaotwe Pan	2223A
Karakubis	2220B
Kareng	2022C
Kasane	1725C
Katima Mulilo (Zambia)	1724A
Kaucaca Pans	1925D
Kavimba	1824B
Kazungula	1725C
Kedia Hill	2124B
Kgale Hill	2425D
Kgokong (Kokong)	2423A
Kgoro Pan	2424A
Kgoro Pan	2525A
Khadu Pan	2021D
Khakhea (Kakia)	2423D
Khakhea Pan	2423C
Khama Khama Pan	1924D
Khiding Pan	2521B
Khisa	2522D
Khokowe Pan	2421C
Khudumelapye	2324D
Khuis	2621D
Khwai River (Lodge)	1923B
Khwebe Hills	2023C
Khwee Pan (Quee)	2223C
Khwekhwe Pan	2423A
Kilemon Pan	2320C
Koanaka Hills	2021A
Kobe Pan	2320A
Kodibeleng	2326A
Kokong	2423A
Kolonkwaneng	2621D
Komana	2023A
Kome Pan	2422B
Kooroomoorooi Pan	2425B
Kopong	2425B
Kubu Camp	1923D
Kubu Island	2025A
Kubugaswe Pan	2320A
Kuchwe	2324A
Kudiakam Pan	2024B
Kuke fence	West 21S
Kuke Quarantine Camp	2122A
Kule	2220C
Kumchuru	2223B
Kutse Game Rerserve (most)	2324A/C
Kutse Pan	2324B
Kwang Pan	2520B
Kwikamba	1824A
Lake Ngami	2022B
Lake Xau (Dow)	2124B
Lamont Pan	2122B
Langpan	2420B
Lechwe Flats	1724D
Lehututu	2321D
Lentsweletau	2425B
Lephepe	2325B
Lepokole Hills	2128C
Leporung Dam	2524D
Lerala	2227D
Lesholoago Pan	2422C
Letiahau	2122D
Letlhakane	2125B
Letlhakeng	2425A
Letsheng	2227C
Liambezi Lake	1724C
Linyanti Camp	1823B
Livingstone's Cave	2425B
Lloyd's Camp	1824C
Lobatse (Lobatsi)	2525B
Lokalane	2322A
Lokgwabe (Lochwabe)	2421B
Lokolong	2622C
Lone Tree	2322A
Lorolwane	2524A
Lovers Leap	2622C
Mababe	1924A
Mababe Depression	1824C
Mababe Scout Camp	1824C
Mabeleapodi	2522C
Mabete Pan (Pitsane)	2525B
Maboane	2424B
Mabuasehube Pan	2421D
Mabule	2524D
Machaneng	2327A
Mahalapye	2326B
Maikelelo F.R. (part)	1825A/C
Maitengwe	2027A
Maitloaphuduhudu	2321B
Majwaneng	2227D
Makalamabedi	2023B
Makaleng	2027C
Makhe Well	2226A
Makoba	2126C
Makobelo Pan	2126A
Makobeng	2227B
Makopong	2522B
Makwate	2327A
Malapo	2223B
Maleshe	2522C
Mamuno	2220A
Mannyelanong Hill	2425D
Manyane Pan	2321B
Maralaleng	2522D
Martin's Drift	2327B
Maruapula S.P	2425D
Masieding	2321B
Mashatu	2229A
Mashi a potsana	2422B
Masotswana Dam	2525A
Masotswana Valley	2525A
Masowe Pan	2326A
Matetleng Pan	2320D
Mathathane	2228B
Matsebi	1922D
Matshakana Pan	2424D
Matsitama	2126B
Maun	1923C
Maxwo cattle post	2225D
Mazeamanong	2224A
McCarthysrus	2622B
Medenham	2523D
Mengwe	2027A
Meratswe	2324B
Metlobo	2524B
Metseamanong	2224C
Metsematlhoko	1925D
Middlepits	2621D
Mmakgori	2524D
Mmakudukanye Hill	2129C
Mmashoro	2126C
Mmathethe	2525A
Mmatshumo	2125B
Mochabeng	2022D
Mochudi	2426A
Modipane	2426C
Moeding College	2525B
Moeng	2227D
Moeng College	2227D
Mogapi	2227B
Mogapinyana	2227B
Mogobane Dam	2425D
Mogogaphate	2128C
Mohembo	1821B
Moijabana	2226C
Mokatako	2525C
Mokgware Hills	2226D
Molalatau	2228B
Molapebelo Pan	2320C
Molapo	2123D
Molepolole	2425B
Molopo Farms (part)	2523A/B
Molopowabojang	2525B
Molotswae Gate	2125D
Monamodi Pan	2522A
Monong Pan	2321C
Mookane	2326D
Moorcroft's Pan	2622A
Mopipi	2124B
Moremaoto	2024A
Moremi Camp	1923A
Moremi South Gate	1923B
Moreswe Pan	2324C
Morupule	2227C
Morwa Pan	2320A
Morwamosu	2423A
Moselebe	2523B
Moselesele Park	2425D
Mosetse	2026D
Moshaneng	2425C
Moshatong Hill	2228D
Mosimane	2521A
Mosopa (Moshupa)	2425C
Mosopha	2227C
Mosu	2125B
Motlhatlogo	2022D
Motopi	2024A
Motshegaletau	2226C
Motsobonye Pan	2423A
Mpaathutlwa Pan	2521B
Mpandamatenga	1825D
Mporota	1923C
Mumpswe	2025B
Nanoga-onne	1924C
Nata	2026A
Nata Ranches	1925D
Ncojane	2320A
Ncojane Ranches (part)	2320B

Place	Grid
Ngoma (Bridge)	1724D
Ngotwane Dam	2425D
Ngotwane Siding	2425D
Ngwako Pan	2022D
Ngwezumba	1824B
Nhane	2424D
Njuga Hills	2024D
Nkange	2027A
Nkate	2026A
Nokaneng	1922C
Northern Wellfields	2424B
Nossob Camp (Nosop)	2520B
Ntwetwe Pan (main)	2025C
Nunga	1825C
Nunga Pan	1825D
Nunga Valley	1825D
Nxai Pan	1924D
Nxamaseri	1821D
Nxaunxau	1821C
Odi	2426C
Odiakwe	2025A
Ohe Pan	2321D
Okavango Camp	1923A
Okwa Pan	2221B
Oliphants Drift	2426B
Omawaneno	2522D/C
Orapa	2125A
Ootse (Ootsi)	2525B
Paje	2226B
Palapye	2227C
Parr's Halt	2327A
Passarge	2123A/B
Patlana Flats	2022B
Pelotshetlha	2525A
Peolwane Pan	2223A
Peter Pan	2123A
Phepheng	2522D
Phitsane Molopo	2525C
Phuduhudu	2024B
Pink Pan	2122D
Pioneer Gate	2525B
Piper Pans	2123C
Pitsane	2525B
Pitsikolo Pan	2423A
Plumtree (Zimbabwe)	2027D
Poha Pan	1824B
Polentswa Pan	2520A
Pont's Drift	2229A
Puku Flats	1724C
Qangwadum	1921D
Qhaaxwha Island	1922A
Quee Pan (Khwee)	2223C
Quoxo	2224C
Rakops	2124A
Rakuku	1923B
Ramatlabama	2525D
Ramotswa	2425D
Ramsden	2121C
Rappel's Pan	2620D
Rooibok	2122C
Rysana Pan	2125A
Salajwe	2324D
Sangoshe	1822A
Sankora	1921A
Santantadibe	1923C
Santawani (Lodge)	1923B
Sarakoma cattle post	2225B
Savuti Channel	1823D
Savuti Marsh	1824C
Sayo Pan	1924B
Sebele	2425D
Sebina	2027C
Sefare	2327B
Sefophe	2227B
Sehitwa	2022B
Sekandoko	1822A
Sekoma (Sekhuma)	2423D
Selebi Phikwe	2127D
Seleka Hill	2227D
Selenia Pan	2325B
Selinda Spillway	1823C
Semane	2424C
Semane Pan	2423B
Semolale	2128D
Sepako	1926D
Sepupa (Sepopa)	1822C
Serondella	1724D
Seronga	1822C
Serorome Valley	2326A/B
Serowe	2226B
Serule	2127C
Sesatswe	2420D
Setata	2125C
Shakawe	1821B
Shashe (+ Dam)	2127A
Sherwood Ranch	2227D
Shonso Shonso	1821A
Shorobe	1923D
Shoshong	2326B
Shoshong Hills (most)	2326B
Sibuyu Forest Reserve	1825D
Sita Pan	2524B
Sowa Pan (see Sua)	
Stevensford	2227C/D
Stoffel's Pan	1925B
Sua Pan (north) (Sowa)	2026A
Sua Pan (mid)	2025D
Sua Pan (south)	2126A
Sua Spit (most)	2026C
Sukwane	2024C
Takatokwane	2424A
Takatshwane Pan	2221D
Tale Pan	2022D
Tamafupa	1926A
Tamasane	2227A
Taroafupa Pan	1926A
Tati Concessions	2127B
Tati Siding	2127A
Tchau (Chao) Island	1922B
Thabatshukudu	2025D
Thakadu	2126B
Thamaga	2425D
Thamagu	2025D
Tlalambele Gate	2126A
Tlhabala	2226A
Tlokweng	2425D
Tlokweng Dam	2425D
Tlokweng Gate	2426C
Tokong Pan	2422D
Tonota	2127A
Topisi	2227A
Toromoja	2124B
Toteng	2022B
Tsamaya	2027D
Tsaro (Lodge)	1923B
Tsau (Tsao)	2022A
Tsau Hills	2122B
Tsaugara	1925C
Tsebanana (+ Pan)	1926D
Tsetsebjwe	2228A
Tsetseng	2323C
Tsgobe	2124C
Tshabong	2622A
Tshane (+ Pan)	2421B
Tshesebe (Tsessebe)	2027D
Tsholofelo S.P.	2425D
Tshono Pan	2424D
Tshotswa Pan	2421A
Tshukudu Pan	2220C
Tshweu Pan	2223D
Tsienyane	2124A
Tsodilo Hills	1821D
Tsoe (Tsoi)	2024C
Tsootsha (Kalkfontein)	2220B
Tsotsoroga Pan	1824C
Tswaane Pan	2323D
Tswaneng	2524D
Tswanyaneng Pan	2525C
Tswapong Hills	2227C/D
Tuli Circle	2129C
Tuli Circle	2229A
Tutume	2027A
Twee Rivieren	2620B
Txichira Lagoon	1922B
Ukwi Pan	2320D
Union's End	2420C
Werda	2523A
Xade	2223A
Xaixai (Caecae)	1921C
Xaka	2224B
Xakanaka (Camp)	1923A
Xalagena	2126D
Xamshiko	1822B
Xanagas	2220A
Xani	1925A
Xaudum (part)	1921B
Xaxaba (Camp)	1923C
Xhorodomo Pan	2124B
Xhumaga	2024B
Xhumu Island	1923A
Xobega	1923A
Xugana (Lodge)	1923A
Zanzibar	2228C
Zonye Pan	2321A
Zweizwe Pan	1824C

Appendix E

Rivers and fossil valleys

Rivers

Bonwapitse 140 km 2226C/D 2326B/D 2327C
Major tributary of Limpopo, rises in Shoshong Hills, flows SE.

Boro 60 km 1923C
One of Okavango outflows, major tributary of Thamalakane, forms near S of Chief's Island.

Boteti 360 km 2023A/B 2024A/C 2124A/B 2125A 2025C
Main Okavango outflow to Makgadikgadi (Ntwetwe Pan), old discharge to Lake Xau now channeled to Mopipi Dam.

Chobe 90 km 1824B 1724D 1725C
Major tributary of Zambezi, flows E, outflow of faulted Kwando/Linyanti system.

Dikolakolana 70 km 2325D 2425B 2426A
N tributary of Ngotwane, flows E.

Jao 100km 1822D 1922B
Central main channel of the Okavango Delta flowing SE.

Khwai 70 km 1923A/B
N outlet of Okavango Delta, unpredictable flow, flows E.

Kwando 70 km 1823A/B
Enters from Caprivi, forms SW border of Zambia as Mashi River, becomes the Linyanti by NE turn in 1823B.

Lake 20 km 2022B
Flows into Lake Ngami, formed by confluence of Nhabe with Kunyere River at Toteng.

Lepashe 80 km 2026D/B
Makgadikgadi feeder, flows W into SE Sua Pan.

Limpopo 350 km 2426B 2326D 2327A/B/C 2227D 2228B/C/D 2229A
E border from near Tropic of Capricorn to Shashe confluence, major river drainage of E, flows NE.

Linyanti 110 km 1823B 1824A 1724C
Flows NE from Linyanti Swamp to Liambezi Lake (Caprivi), filters through swamp to join Chobe River.

Lotsane 200 km 2226A/B/C/D 2227A/C/D 2228C
Major tributary of Limpopo, rises near Serowe and Mokgware Hills, flows E through Palapye to Limpopo.

Maitengwe 90 km 2027A 2026B 1926D/C
Tributary of Nata, rises in Zimbabwe and flows N forming Botswana border then W into the Nata River.

Marico 60 km 2426C/B
Forms Limpopo by joining Crocodile at Oliphants Drift, arises in Transvaal, forms part of SE Botswana border.

Metsemotlhaba 120 km 2425C/D/B 2426A
Major tributary of Ngotwane, rises in hills N of Kanye, flows NE via Thamaga, joins S of Mochudi.

Mhalatswe 110 km 2226D 2326B 2327A
Major tributary of Limpopo, rises on S of Mokgware Hills, flows SE through Mahalapye to Limpopo.

Molopo 550 km 2525C 2524C/D 2523D/B/A 2522B/D 2622B/A/C 2621C/D 2620D
Forms whole S border of Botswana, flows W, seasonal but has hardly flowed 1982 to 1990.

Mosetse 90 km 2027C 2026D/C
Makgadikgadi feeder, rises W of Shashe headwaters and flows W to Sua Pan entering S of Sua Spit.

Mosope 100 km 2027B 2026D 2126B 2026C
Makgadikgadi feeder, flows W into SE corner of Sua Pan.

Motloutse 250 km 2126B 2127A/C/D 2128C/D 2228B 2229A
Major tributary of Limpopo, rises SW of Francistown, flows E via Bobonong to join Limpopo in 2229A.

Nata 100 km 1926D/C 2026A
Arises in Zimbabwe, flows SW to Sua Pan, Makgadikgadi. Outlet is a major inland estuary—Nata Delta.

Ngotwane 250 km 2525A/B 2425D 2426C/A/B 2326D
Major tributary of Limpopo, rises in hills around Kanye, flows NE through Gaborone (Dam) to S of Buffels Drift.

Nhabe 50 km 2023A 2022B
Takes part of Okavango outflow to Lake Ngami, joins Kunyere River near Toteng to form Lake River.

Nkange 90 km 2027A 2026B/A
Tributary of Tutume, rises W of Shashe headwaters, flows NW to join Tutume just before confluence with Nata.

Nnywane 70 km 2525B 2425D
Major tributary of Ngotwane, rises near Lobatse, flows N via Ramotswa to join S of Gaborone.

Nossob 250 km 2420C 2520A/B/D 2620B/D
Forms SW border of Botswana, flows S, seasonal but mainly dry 1980 to 1990.

Nqoga 60 km 1822D 1922B
NE main channel of Okavango Delta.

Okavango 140 km 1821B 1822A/C
Enters from W Caprivi, Panhandle 140 km, Delta about 150 km radius, forms major wetland.

Ramatlabama 60 km 2525D/C
One of the headwaters of Molopo, flows W to join Molopo at Mokatako, forms first part of S country border.

Ramokgwebana 90 km 2027D 2127B/D
Major tributary of Shashe, flows S from Plumtree forming part of E border of Botswana E of Francistown.

Santantadibe 70 km 1923A/C
Tentative Okavango outflow to the SE, tributary of Thamalakane.

Savuti 80 km 1823D 1824C
Alternative outlet for Kwando/Linyanti system flowing SE to Savuti Marsh.

Sekhukhwane 90 km 2425B 2426A
Major tributary of Ngotwane, rises NE of Molepolole, flows E and joins N of Mochudi.

Semowane 80 km 2026D/B/A
Makgadikgadi feeder, rises near Tutume and flows W to Sua Pan entering S of Nata Delta.

Shashe 350 km 2027C 2127A/B/D 2128C/D 2129C 2229A
Major tributary of Limpopo, forms country border in part of E, major dam supply to Francistown.

Tati 130 km 2027C 2127A/B/D
Major tributary of Shashe near which it rises, flows SE through Francistown.

Thamalakane 100 km 1923B/D/C 2023A
Major outflow of Okavango Delta, flows SW through Maun, divides to form Boteti and Nhabe Rivers.

Thaoge 250 km 1822C 1922A/C 2022A
Main channel of the Okavango Delta to the SW, mainly clogged since 1850s, forms W edge of Delta.

Thune 150 km 2227B 2228A/B
Major tributary of Motloutse, flows E and joins Motloutse about 20 km W of Motloutse/Limpopo confluence.

Tsebanana 15 km 1926D
Arises in Zimbabwe, major tributary of Nata which it joins 15 km inside Botswana.

Tshokane 60 km 2227A/B/D
Major tributary of Lotsane, flows SE.

Tutume 100 km 2027C/A 2026B/A
Major tributary of Nata, rises W of Shashe headwaters, flows NW.

Fossil valleys

Buitsivango 50 km 2221B 2121D/C
Tributary of Deception, runs NW from Kang/Ghanzi road as continuation of Hanahai.

Deception 230 km 2124A 2123B/D/C 2122D/C
Major valley running from central Kalahari W to Ghanzi farms.

Gaotlhobogwe 40 km 2425A
Tributary of Moshaweng, runs NW to Letlhakeng.

Groot Laagte 140 km 2021C/D 2121A/B
Bifid valley crossing W part of Kuke fence N of Ghanzi farms.

Hanahai 50 km 2222A 2221B
Runs NW from near Okwa, crosses Kang/Ghanzi road and becomes the Buitsivango.

Khekhe 50 km 2422D 2522B 2523A
Tributary of Moselebe, runs S between Werda and Makopong.

Letlhakane 120 km 2125A/B/D
Runs SE from N of Orapa to S of Makoba, crosses Serowe/Orapa road near Letlhakane.

Malotswana 60 km 2423C 2523A
Tributary of Moselebe, runs SW.

Metatswe 140 km 2325C/A 2324B/A
Tributary of Quoxo, runs NW from just S of central Kalahari Game Reserve.

Moselebe 280 km 2525B/A 2524B/A 2523B/A
Major valley running W from S of Lobatse to W of Werda.

Moshaweng 150 km 2425C/A 2424B 2324D
Tributary of Metatswe, runs N through Khudumelapye.

Naledi 170 km 2324B/D/C 2424A/B/D
Runs S from N of Salajwe to SW of Jwaneng.

Okwa 400 km 2223B/A 2222B/A 2221B/A 2220B/A
From central Kalahari Game Reserve W to Namibia near Mamuno, major valley crossing Kang/Ghanzi road.

Quoxo 250 km 2124C 2123D 2223B 2224A/C 2324A/C
Major valley running N from Kutse Game Reserve into central Kalahari Game Reserve.

Sekhutlana 90 km 2524D/C/A
Tributary of Moselebe, runs W.

Selokolela 60 km 2425C 2525A 2524B
Tributary of Moselebe, runs SW.

Serorome 130 km 2325D/B 2326A/B/D
Runs from E edge of Kalahari to near Bonwapitse River, crosses main road just N of the Tropic of Capricorn.

Takatshwane 60 km 2221D 2222C/A
Tributary of Okwa, runs NE across Kang/Ghanzi road.

Ukhwi 100 km 2424C 2524A 2523B
Tributary of Moselebe, runs SW.

References

This is a list of selected references. It includes all notes and articles in the 19 editions of the *Babbler* which are relevant to distribution but excludes specific records in special sections of the journal such as migrants, breeding or unusual sightings. The last were extracted direct into the database if they were not already there. Also included are references relevant to Botswana from ornithological journals and publications from southern and southcentral Africa over the past 25 years—since the publication of Smithers (1964). As there is no bibliography for birdwatchers in Botswana, articles have been added from any source which may provide assistance in research. Some references have been included because they lend support to the assessments made in the text but are not quoted therein because they did not formulate the assessment.

Acocks, J.P.H. 1988. Veld types of South Africa. *Mem. Bot. Surv. Sth. Afr.* No.57.

Aldiss, D. & Hunter, N.D. 1985. A first record of Southern Boubou in Botswana. *Babbler* 10: 37–38.

Aldiss, Don 1986. 1986—The year of the Crane? A survey of Wattled Crane in Botswana. *Babbler* 11: 39–40.

Aldiss, Don 1986. The Okwa Valley: not quite paradise. *Babbler* 12: 23–24.

Aspinwall, Dylan R. 1982. Movement analysis of the Fierynecked Nightjar in Zambia. *Babbler* 4: 11–13.

Aspinwall, D.R. 1986. Some interesting bird records from northern Botswana. *Babbler* 11: 13–14.

Aspinwall, D.R. 1989. Crowned Eagle *Stephanoaetus coronatus* near Kasane. *Babbler* 18: 36–37.

Aspinwall, D.R. 1989. Spurwinged Plover *Vanellus spinosus* in northern Botswana. *Babbler* 18: 34–36.

Avery, G., Brooke, R.K. & Komen, J. 1988. Records of the African Crake *Crex egregia* in western southern Africa. *Ostrich* 59: 25–29.

Banfield, G.E.A. 1986. Another Bulawayo River Warbler. *Honeyguide* 32: 92.

Barnes, J.E. 1987. Black Sparrowhawk breeding in southeast Botswana. *Babbler* 14: 20–21.

Beesley, J.S.S. & Irving, N.S. 1976. The status of the birds of Gaborone and its surroundings. *Botswana Notes and Records* 8: 231–261.

Bell, Colin 1987. Interference at heronries in the Okavango Delta. *Babbler* 14: 2–3.

Benson, C.W. & Benson, F.M. 1977. *The birds of Malawi.* Montfort Press, Limbe.

Benson, C.W., Brooke, R.K., Dowsett, R.J. & Irwin, M.P.S. 1971. *The Birds of Zambia.* Collins, London.

Benson, C.W. & Irwin, M.P.S. 1966. The *Brachystegia* avifauna. *Ostrich Suppl.* 6: 297–321.

Benson, C.W. 1981. Migrants in the Afrotropical region south of the equator. *Bokmakierie* 33: 27–28.

Benson, C.W. 1981. Ecological differences between Grass Owl and Marsh Owl. *Bull. Brit. Orn. Club* 101: 372–376.

Blaker, D. 1966. Notes on the sandplovers *Charadrius* in Southern Africa. *Ostrich* 37: 95–102.

Borello, W.D. 1984. A note on sightings of the Black Harrier in the Kalahari Gemsbok National Park, South Africa. *Babbler* 8: 50–51.

Borello, W.D. 1985. First official Cape Vulture protected breeding site. *Babbler* 10: 43–44.

Borello, W.D. 1987. Vulture distribution in Botswana. *Babbler* 13: 11–23.

Borello, Wendy & Remi 1987. A new Cape Vulture breeding site in Botswana. *Bokmakierie* 39: 50.

Borello, W.D. & Borello, R.M. 1983. The Black Sparrowhawk *Accipiter melanoleucus*—an account of a first sighting in Botswana. *Babbler* 6: 21–22.

Borello, W.D. & Borello, R.M. 1988. Possible extension of (known) range of the Dickinson's Kestrel *Falco dickinsoni* in Botswana. *Babbler* 15: 18–20.

Botswana Society 1976. *Proceedings of the Symposium on the Okavango Delta and its future utilization.* National Museum, Gaborone.

Brewster, Chris 1986. A birding weekend at the Tsodilo Hills. *Babbler* 12: 17–18.

Brewster, Chris 1988. Birding along the Thamalakane and Boteti Rivers. *Babbler* 15: 26–28.

Brewster, C.A. 1989. Observations on widowfinches in Ngamiland. *Babbler* 17: 34–35.

Brewster, C.A. 1989. Chestnut Weaver *Ploceus rubiginosus*: a new species for Botswana. *Babbler* 17: 36–37.

Brooke, R.K. 1968. On the status of the Yellowthroated Sandgrouse south of the Zambezi. *Ostrich* 39: 33–34.

Brooke, R.K. 1968. On the distribution, movements and breeding of the Lesser Reedhen *Porphyrio alleni* in Southern Africa. *Ostrich* 39: 259–262.

Brooke, R.K. & Irwin, M.P. Stuart 1969. The status of the Honey Buzzard in Rhodesia. *Ostrich* 40: 135.

Brooke, R.K. 1984. South African Red Data Book—Birds. *South African National Scientific Programmes Report* No.97, CSIR, Pretoria.

Brown, L.H. & Seeley, M.K. 1973. Abundance of the Pygmy Goose *Nettapus auritus* in the Okavango Swamps, Botswana. *Ostrich* 44: 84.

Brown, Leslie H., Urban, Emil K. & Newman, Kenneth 1982. *The birds of Africa*, Volume 1. Academic Press, London.

Brown, R.C. 1974. Climate and climatic trends in the Ghanzi District. *Botswana Notes and Records* 6: 133–146.

Bruderer, Bruno & Bruderer, Heidi 1993. Distribution and habitat preference of Redbacked Shrikes *Lanius collurio* in southern Africa. *Ostrich* 64: 141–147.

Butchart, D. & Mundy, P. 1981. Cape Vultures in Botswana. *Babbler* 1: 6–7.

Butchart, Duncan 1988. Some notes on the Broadbilled Roller. *Babbler* 15: 29–30.

Cade, T.J. 1965. Survival of the Scalyfeathered Finch *Sporopipes squamifrons* without drinking water. *Ostrich* 36: 131–132.

Campbell, A. & Child, G. 1971. The impact of man on the environment of Botswana. *Botswana Notes and Records* 3: 91–110.

Campbell, A. 1978. *The Guide to Botswana.* Winchester Press, Johannesburg.

Child, G. 1968. *A preliminary checklist of the birds of the Chobe Game Reserve. Ecological Survey of Northern Botswana Appendix A.* FAO, Rome.

Child, Graham 1972. A survey of mixed 'heronries' in the Okavango Delta, Botswana. *Ostrich* 43: 60–62.

Child, Graham 1972. Water and its role in nature conservation and wildlife management in Botswana. *Botswana Notes and Records* 4: 253–255.

Clancey, P.A. 1964. The migratory status of the Pygmy Kingfisher in South Africa. *Ostrich* 35: 60.

Clancey, P.A. 1966. Subspeciation in southern African populations of the Sabota Lark *Mirafra sabota* Smith. *Ostrich* 37: 207–213.

Clancey, P.A. 1966. Racial variations in southern populations of *Caprimulgus rufigena* Smith. *Bull. Brit. Orn. Club* 86: 6–7.

Clancey, P.A. 1978. Some enigmatic pipits associated with *Anthus novaeseelandiae* from central and southern Africa. *Bonn Zool. Beitr.* 29: 148–164.

Clancey, P.A. 1980. *S.A.O.S. Checklist of southern African birds.* Southern African Ornithological Society, Pretoria.

Clancey, P.A. 1986. On the Mountain Pipit in Botswana. *Honeyguide* 32: 44.

Clancey, P.A. 1989. The status of *Cursorius temminckii damarensis*, Reichenow, 1901. *Bull. Brit Orn. Club* 109: 51–52.

Clancey, P.A. 1990. The Namibian subspecies of *Cisticola chiniana* (Smith), 1843. *Bull. Brit. Orn. Club* 110: 83–86.

Colebrook-Robjent, J.F.R. 1984. Nests and eggs of some African nightjars. *Ostrich* 55: 5–11.

Collar, N.J. & Stuart, S.N. 1985. *Threatened birds of Africa and related islands.* ICBP/IUCN, Cambridge.

Cooke, H.J. 1981. On the conservation of natural resources, with special reference to the Kalahari in Botswana. *Botswana Notes and Records* 13: 141–143.

Cooper, M.R. & Donnelly, B.G. 1983. Zoogeography and speciation in *Eremopterix*. *Honeyguide* 116: 20–25

Coppinger, M.P., Williams, G.D. & Maclean, G.L. 1988. Distribution and breeding biology of the African Skimmer on the upper and middle Zambezi River. *Ostrich* 59: 85–96.

Craig, A.J.F.K. 1982. The breeding season of the Red Bishop. *Ostrich* 53: 112–113.

Craig, A.J.F.K. 1988. The status of *Onychognathus nabouroup benguellensis*. *Bull. Brit. Orn. Club* 108: 144–147.

Cyrus, Digby & Robson, Nigel 1980. *Bird atlas of Natal*. University of Natal Press, Pietermaritzburg.

Da Camara-Smeets, Michelle 1987. Control of Quelea in Botswana. *Babbler* 13: 32–33.

Davidson, Ian 1982. Booted Eagle possibly breeding in the Richtersveld (Northwestern Cape) and further sight records from Namibia. *Ostrich* 53: 117.

Dawson, J.L. 1975. The birds of Kutse Game Reserve. *Botswana Notes and Records* 7: 141–150.

Dawson, J.L. & Jacka, R.D. 1975. *Some notes on the water birds of Lake Ngami*. Department of Wildlife, Gaborone.

Day, D.H. 1987. Birds of the Upper Limpopo River Valley. *Southern Birds* 14.

De Kock, A.C. & Watson, R.T. 1985. Organochlorine residue levels in Bateleur eggs from the Transvaal. *Ostrich* 56: 278–280.

De Villiers, J.S. 1972. The Yellowbilled Kite in South West Africa. *Ostrich* 43: 136.

Dean, W.R.J. 1976. Niche occupation of Rufouseared Warbler and Blackchested Prinia. *Ostrich* 47: 67.

Dean, W.R.J. 1978. Moult seasons of some Anatidae in the Western Transvaal. *Ostrich* 49: 76–84.

Dean, W.R.J. 1979. Population, diet and annual cycle of the Laughing Dove at Barberspan, Part 3: The annual cycle. *Ostrich* 50: 234–239.

Dean, W.R.J. 1988. The avifauna of Angolan miombo woodlands. *Tauraco* 1: 99–104.

Denman, E.B., Forrester, A.K. & Allen, H.E.K. 1976. Checklist of the birds found within the Orapa Security Area and the Mopipi Dam. *Botswana Notes and Records* 8: 263–268.

Douse, M.J. 1987. The National Conservation Strategy: what's in it for the birds? *Babbler* 14: 30–32.

Dowsett, R.J. 1985. Migration among southern African land birds. *Acta XIX Congressus Internationalis Ornithologicus* 1: 765–777.

Dowsett, R.J. 1985. Intra-African migrant birds in south-central Africa. *Acta XIX Congressus Internationalis Ornithologicus* 1: 778–790.

Dowsett, R.J., Backhurst, G.C. & Oatley, T.B. 1988. Afrotropical ringing recoveries of Palaearctic migrants 1. Passerines (Turdidae to Oriolidae). *Tauraco* 1: 29–63.

Dowsett-Lemaire, F. & Dowsett, R.J. 1987. European Reed and Marsh Warblers in Africa: migration patterns, moult and habitat. *Ostrich* 58: 65–85.

Dowsett-Lemaire, F. & Dowsett, R.J. 1988. Vocalisations of the green turacos (Tauraco species) and their systematic status. *Tauraco* 1: 64–71.

Earlé, Roy & Grobler, Nick 1987. *First atlas of bird distribution in the Orange Free State*. National Museum, Bloemfontein.

Earlé, R.A. 1981. Factors governing avian breeding in *Acacia* savanna, Pietermaritzburg, Part 1: extrinsic factors. *Ostrich* 52: 65–73.

Earlé, R.A. 1981. Factors governing avian breeding in *Acacia* savanna, Pietermaritzburg, Part 2: intrinsic factors. *Ostrich* 52: 74–83.

Earlé, R.A. 1987. Distribution, migration and timing of moult in the South African Cliff Swallow. *Ostrich* 58: 118–121.

Earlé, R.A. 1987. Moult and breeding seasons of the Greyrumped Swallow. *Ostrich* 58: 181–182.

Fothergill, Alastair 1983. A study of the mixed 'heronries' found at Cakanaca, Gcodikwe and Gcobega Lagoons. *Babbler* 5: 8–14.

Fraser, W. 1972. Birds at Lake Ngami, Botswana. *Ostrich* 42: 128–130.

Fraser, W. 1982. Observations on the birds of the Tuli Block, Botswana. *Bokmakierie* 34: 32–35.

Fry, C.H. 1980. An analysis of the avifauna of African northern tropical woodland. *Proc. 4th Pan-Afr. Orn. Congr.*: 113–124.

Fry, C.H., Hosken, J.H. & Skinner, D. 1986 Further observations on the breeding of Slaty Egrets *Egretta vinaceigula* and Rufousbellied Herons *Ardeola rufiventris*. *Ostrich* 57: 61–64.

Fry, C. Hilary, Keith, Stuart & Urban, Emil K. (Eds) 1986. *The birds of Africa*, Volume 3. Academic Press. London.

Gargett, V. 1977. A 13-year population study of the Black Eagles in the Matopos, Rhodesia, 1964–1976. *Ostrich* 48: 17–27.

Geldenhuys, J.N. 1976. Relative abundance of waterfowl in the Orange Free State. *Ostrich* 47: 27–54.

Geldenhuys, J.N. 1981. Moult and moult localities of the South African Shelduck. *Ostrich* 52: 129–134.

Ginn, P.J. 1976. Birds of Makgadikgadi: a preliminary report. *Wagtail* 15: 21–96.

Ginn, P.J., McIlleron, W.G. & Milstein, P. le S. (Eds) 1989. *The complete book of southern African birds*. Struik-Winchester, Cape Town.

Graham, Jan & Brian 1982. Some notes on Narina Trogon breeding at Linyanti. *Babbler* 4: 18–19.

Grobler, J.H. & Steyn, P. 1980. Breeding habits of the Boulder Chat and its parasitism by the Redchested Cuckoo. *Ostrich* 51: 253.

Hall, B.P. & Moreau, R.E. 1970. *An atlas of speciation in African passerine birds*. Trustees of the British Museum (Natural History), London.

Herholdt, J.J. 1988. The distribution of Stanley's and Ludwig's Bustards in southern Africa: a review. *Ostrich* 59: 8–13.

Hobbs, J.C.A. 1982. Some notes on garden birds of Serowe (1979–1981). *Babbler* 3: 11–17.

Hockey, P.R., Brooke, R.K., Cooper, J., Sinclair, J.C. & Tree, A.J. 1986. Rare and vagrant scolopacid waders in southern Africa. *Ostrich* 57: 37–55.

Hodgson, Malcolm 1981. Tsetse fly eradication and the birds. *Babbler* 2: 16–19.

Howard, G.W. & Aspinwall, D.R. 1984. Aerial censuses of Shoebills, Saddlebilled Storks and Wattled Cranes at the Bangweulu Swamps and Kafue Flats, Zambia. *Ostrich* 55: 207–212.

Hunter, N.D. 1984. Preliminary notes on bird distribution and rarities in south-east Botswana. *Babbler* 7: 14–20.

Hunter, N.D. 1984. Record of possible new species for Botswana—European Marsh Warbler. *Babbler* 7: 46.

Hunter, N.D. 1985. Some observations on species included in the Botswana Rarities List (1984). *Babbler* 9: 7–14.

Hunter, N.D. 1985. The Great Atlas Trek. *Babbler* 10: 8–13.

Hunter, N.D. 1985. Some interesting records from the Mafikeng area. *Babbler* 10: 37–38.

Hunter, N.D. 1986. An update on the status of White-eyes in Botswana. *Babbler* 12: 14–16.

Hunter, N.D. 1986. Withdrawal of records of Singing Bush Lark and Bluebilled Firefinch in S.E. Botswana. *Babbler* 12: 38.

Hunter, N.D. 1987. Notes on bird occurrences at a seasonal pan in southeast Botswana. *Babbler* 13: 6–10.

Hunter, N.D. 1988. Systematics of the Southern and Tropical Boubous in Botswana. *Babbler* 16: 7–10.

Hunter, Nigel 1989. Grass Owls in Botswana. *Babbler* 18: 3.

Hustler, Kit & Williamson, Craig 1985. The Rail Heron in Hwange National Park. *Honeyguide* 31: 145–147.

Hustler, K., Eriksson, M.O.G., & Skarpe, C. 1986. Status of the Whitebreasted Cormorant in the Middle Zambezi Valley. *Honeyguide* 32: 42.

Hustler, K. & Howells, W.W. 1986. Population study of Tawny Eagles in the Hwange National Park, Zimbabwe. *Ostrich* 57: 101–106.

Hustler, K. & Howells, W.W. 1988. Breeding biology of the Whiteheaded Vulture in Hwange National Park, Zimbabwe. *Ostrich* 59: 21–24.

Hustler, Kit 1983. Breeding biology of the Greater Kestrel. *Ostrich* 54: 129–140.

Hustler, K. 1986. A revised checklist of the birds of the Hwange National Park. *Honeyguide* 32: 68–87.

Hustler, K. 1988. Why are Peregrines so rare in South Africa? *Ostrich* 59: 77–78.

Hutchings, D.G., Hutton, S.M. & Jones, C.R. 1976. The geology of the Okavango Delta. *Symposium on the Okavango Delta*. Botswana Society.

Irwin, M.P.S. 1956. Notes on the drinking habits of birds in semi-desertic Bechuanaland. *Bull. Brit. Orn. Club* 76: 99–101.

Irwin, Michael P. Stuart 1981. *The birds of Zimbabwe*. Quest Publishing, Salisbury.

Irwin, M.P. Stuart 1982. On the supposed occurrence of the African Grass Owl in South West Africa/Namibia and the validity of the race *Tyto capensis damarensis* Roberts. *Honeyguide* 111/112: 12–14.

Irwin, M.P.S. 1982. The status of the Desert Cisticola in the Middle Zambezi Valley, with notes on geographical variation. *Honeyguide* 111/112: 67–68.

Irwin, M.P.S. 1984. The status of Ross's Violet Lourie in the Okavango region of northern Botswana. *Honeyguide* 30: 76.

Irwin, Michael P. Stuart, Niven, P.N.F. & Winterbottom, J.M. 1969. Some birds of the lower Chobe river area, Botswana. *Arnoldia* 4: 21 1–40.

Jackson, H.D. & Steyn, P. 1968. Bird notes from the Nata sector of the Makarikari Pan, Botswana. *Ostrich* 39: 3–8

Jackson, H.D. 1970. Further records of European Nightjar in south-eastern Africa. *Bull. Brit. Orn. Club* 90: 135.

Jackson, H.D. 1978. Nightjar distribution in Rhodesia. *Arnoldia* 8: 1–29.

Jackson, H.D. 1987. Nightjar notes from the Katambora and Kazungulu Development Areas of Zimbabwe. *Ostrich* 58: 141–143.

Johnson, D.N. 1984. The Wattled Crane, a conservation priority. *Babbler* 7: 23.

Jones, M.A. 1987. Egyptian Vulture in Hwange National Park. *Honeyguide* 33: 14–15.

Jones, P.J. 1978. Overlap of breeding and moult in the Whitebrowed Sparrowweaver in northwestern Botswana. *Ostrich* 49: 21–24.

Jones, P.J. 1979. The moult of the Little Bee-eater in northwestern Botswana. *Ostrich* 50: 183–185.

Kalikawe, Mary C. 1990. Problem birds (Pied Crow at Palapye). *Babbler* 19: 17–18.

Keast, Allen 1985. Physical geography relative to ornithogeography: Africa compared to South America and Australia. *Proc. 6th Pan-Afr. Orn. Congr.*: 347–373.

Kgasa, M.L.A. 1972. The development of seTswana. *Botswana Notes and Records* 4: 107–109.

Koen, J.H. 1988. Birds of the Eastern Caprivi. *Southern Birds* 15.

Kvist, Anders 1989. The birds of Serowe and its surroundings. *Babbler* 18: 5–33.

Lancaster, I.N. 1974. The origins and development of the Pans of the Southern Kalahari. *Botswana Notes and Records* 6: 223.

Liversedge, T.N. 1980. A study of Pel's Fishing Owl *Scotopelia peli* Bonaparte in the 'Pan Handle' region of the Okavango delta, Botswana. *Proc. 4th Pan-Afr. Orn. Congr.*: 291–299.

Liversedge, R. 1987. The sensitive Skimmer. *Babbler* 13: 31.

MacDonald, I.A.W. 1986. Range expansion in the Pied Barbet and the spread of alien tree species in southern Africa. *Ostrich* 57: 75–94.

Maclean, G.L. 1970. An analysis of the avifauna of the southern Kalahari Gemsbok National Park. *Zool. africana* 5: 249–273.

Maclean, G.L. 1973. The Sociable Weaver, Part 1: Description, distribution, dispersion and populations. *Ostrich* 44: 170–175.

Maclean, G.L. 1985. *Roberts' birds of southern Africa.* Fifth Edition. The Trustees of the John Voelcker Bird Book Fund, Cape Town.

Maclean, G.L. 1987. Seasonal changes in the birdlife in northeastern Botswana. *Bokmakierie* 39: 109–111.

Maclean, G.L. 1987. The Sociable Weaver: a cooperative bird community. *Babbler* 14: 7–17.

Main, Michael 1987. *Kalahari: life's variety in dune and delta.* Southern Book Publishers, Johannesburg.

Markus, M.B. 1963. Occurrence of Streakyheaded Canary in northeastern Bechuanaland Protectorate. *Ostrich* 34: 171.

Mathews, N.J.C. 1979. Observation of the Shoebill in the Okavango Swamps. *Ostrich* 50: 185.

McGowan, J. & McGowan, G. 1988. A suburban breed of Gabar Goshawk. *Babbler* 16: 16–17.

McGowan, J. & McGowan, G. 1988. Opportunistic breeding of Dwarf Bittern in the Palapye area. *Babbler* 16: 19–20.

Milstein, P. le S. 1975. The biology of Barberspan, with special reference to the avifauna. *Ostrich Suppl.* 11.

Moreau, R.E. 1966. *The bird faunas of Africa and its islands.* Academic Press, London.

Moreau, R.E. 1972. *The Palaearctic-African bird migration systems.* Academic Press, London.

Osborne, Timothy 1986. Notes on raptors in Botswana. *Babbler* 12: 7–8.

Paterson, Mary L. 1962. Some interesting records from Bechuanaland Protectorate. *Ostrich* 33: 21–22.

Payne, R.B. 1973. Behaviour, mimetic songs and song dialects, and relationships of the parasitic indigobirds (*Vidua*) of Africa. *Orn. Monogr.* 11: 1–333.

Payne, R.B. 1985. Song populations and dispersal in Steelblue and Purple Widowfinches. *Ostrich* 56: 135–146.

Pearson, D.J., Nikolaus, G. & Ash, J.S. 1985. The southward migration of Palaearctic passerines through northeast and east tropical Africa: a review. *Proc. 6th Pan-Afr. Orn. Congr.*: 243–261.

Penry, E.H. 1975. Terek Sandpiper *Xenus terek* in Zambia: a summary of recent records 1971–1974. *Bull. Zambian Orn. Soc.* 7: 88–90.

Penry, E.H. 1975. Palaearctic migrant ducks in Zambia. *Bull. Zambian Orn. Soc.* 7: 8–11.

Penry, E.H. 1978. The House Sparrow—a successful opportunist? *Bull. Zambian Orn. Soc.* 10: 25–27.

Penry, E.H. 1979. The Rock Pratincole at Greystone, Kitwe and a review of its migratory movements. *Bull. Zambian Orn. Soc.* 11: 20–32.

Penry, E.H. 1979. Sight records of Sooty Falcon *Falco concolor* in Zambia. *Bull. Brit. Orn. Club* 99: 63–65.

Penry, E.H. 1979. Sooty Falcon *Falco concolor* in Zambia 1977–1979. *Bull. Zambian Orn. Soc.* 11: 14–19.

Penry, E.H. 1981. Palaearctic migrants at Jwaneng. *Babbler* 2: 16–19.

Penry, E.H. 1982. The status of the Maccoa Duck in Botswana. *Honeyguide* 111/112: 60–61.

Penry, E.H. 1982. Migrations—statements, problems and lists. *Babbler* 3: 18–21.

Penry, E.H. 1982. Caspian Plovers in breeding dress in Botswana. *Babbler* 4: 23.

Penry, E.H. 1982. The Rock Thrushes of south-eastern Botswana. *Babbler* 4: 23–24.

Penry, E.H. 1983. The Botswana Bird Atlas. *Bokmakierie* 35: 88–90.

Penry, E.H. 1983. Highlights of a January visit to the Kokong area. *Babbler* 4: 14–16.

Penry, E.H. 1984. Beware of Widow Finches. *Babbler* 7: 24.

Penry, E.H. 1985. Top Twenty-two: Botswana's common birds. *Babbler* 10: 36.

Penry, E.H. 1985. An atlas trip to the northwest of Francistown and Serowe. *Babbler* 9: 15–19.

Penry, E.H. 1986. The distribution and status of the South African Cliff Swallow *Hirundo spilodera* in Botswana. *Babbler* 12: 9–13.

Penry, E.H. 1986. Threatened birds of Botswana, Part 1: the major issues. *Babbler* 11: 6–8.

Penry, E.H. 1986. Threatened birds of Botswana, Part 2: other species to consider. *Babbler* 11: 9–12.

Penry, E.H. 1986. A review of Sharptailed Glossy Starling sightings in Botswana. *Babbler* 11: 26–27.

Penry, E.H. 1986. Yellowbellied Bulbul on the upper reaches of the Motloutse river. *Babbler* 11: 31.

Penry, E.H. 1986. Highlights of a bird atlas trip to areas around Nata, Motloutse River and Shoshong. *Babbler* 12: 19–22.

Penry, E.H. 1987. Bradfield's Hornbill in Croton oil trees at Savuti. *Babbler* 14: 23–24.

Penry, E.H. 1988. A review of the status of the Fairy Flycatcher in Botswana. *Babbler* 15: 21–25.

Penry, E.H. 1988. The formation and basis of the Botswana Bird Atlas. *Proc. 6th Pan-Afr. Orn. Congr.*: 185–194.

Penry, E.H. 1988. Recent records of European Marsh Warbler in the O.F.S. *Mirafra* 5: 2.

Penry, E.H. 1988. An analysis of Palaearctic migrants in northwestern O.F.S. *Mirafra* 5: 4–9.

Penry, E.H. 1990. Short notes on the European Roller *Coracias garrulus* in Botswana. *Babbler* 19 11–13.

Penry, E.H. & Tarboton, W.R. 1990. Redwinged Pratincole breeding at Lake Ngami. *Babbler* 19: 7–10.

Pike, J.G. 1971. Rainfall over Botswana. *Botswana Notes and Records*, Special Edition 1:76.

Pollard, C.J.W. 1987. Whimbrel at Kazungula. *Honeyguide* 33: 15.

Pryce, Elaine 1986. Letter from Shakawe. *Babbler* 11: 33–34.

Pryce, Elaine 1989. A Black Cuckoo raised by Swamp Boubous. *Babbler* 18: 38.

Randall, R.D. 1987. Whitebrowed Scrub Robins *Erythropygia leucophrys* in the central Okavango delta—a distinct subspecies or race? *Babbler* 14: 22.

Randall, Richard 1988. The Natal Nightjar *Caprimulgus natalensis* in the Okavango Delta. *Babbler* 15: 30–31.

Randall, Richard 1988. The Ardeidae at the Xaxaba reed-bed. *Babbler* 15: 31–32.

Randall, Richard 1988. Kingfishers at Kasane. *Babbler* 15: 32–33.

Randall, Richard 1988. A flock of over 200 Wattled Cranes near the Boro River. *Babbler* 16: 18–19.

Randall, Richard 1989. An account of a birding trip to the Piajio area, Chief's Island, in the Okavango Delta. *Babbler* 17: 23–33.

Randall, R.D. 1990. Black Coucals around Xaxaba. *Babbler* 19: 14.

Reeves, C.V. 1972. Earthquakes in Ngamiland. *Botswana Notes and Records* 4: 257–261.

Rowan, M.K. 1983. *The doves, parrots, louries and cuckoos of southern Africa.* David Phillip, Cape Town.

Rutherford, M.C. & Westfall, R.H. 1986. Biomes of Southern Africa—an objective categorisation. *Mem. Bot. Surv. Sth. Afr.* No.54.

Scott, J.A. 1975. Observations on the breeding of the Woollynecked Stork. *Ostrich* 46: 201–207.

Siegfried, W.R. 1966. The status of the Cattle Egret in South Africa with notes on the neighbouring territories. *Ostrich* 37: 157–169.

Siegfried, W.R. 1967. The distribution and status of the Black Stork in southern Africa. *Ostrich* 38: 179–185.

Siegfried, W.R. 1968. Relative abundance of birds of prey in the Cape Province. *Ostrich* 39: 253–258.

Silitshena, R.M.K. 1978. Notes on some characteristics of population that has migrated permanently to the lands in the Kweneng District. *Botswana Notes and Records* 10: 149–157.

Skead, D.M. 1975. Ecological studies on four estrildines in the Central Transvaal. *Ostrich Suppl.* 11.

Skead, David M. 1973. Redbacked Shrikes returning to the same wintering ground. *Ostrich* 56: 278–280.

Skinner, N.J. 1985. Migration patterns for the more common Palaearctic migrants observed in south-eastern Botswana. *Babbler* 9: 20–21.

Skinner, N.J. 1986. Revised Botswana Rarities List—as at July 1 1986. *Babbler* 12: 37.

Skinner, N.J. 1988. The breeding of birds in Botswana. *Babbler* 15: 7–17.

Skinner, N.J. 1988. The Nest Record Card scheme. *Babbler* 15: 32–33.

Skinner, N.J. 1989. The breeding season of five plover species in Botswana. *Babbler* 17: 14–18.

Smith, P.A. 1976. An outline of the vegetation of the Okavango drainage system. *Symposium on the Okavango Delta.* Botswana Society, 93–112.

Smithers, R.H.N. 1961. A species new to the South African list from the Bechuanaland Protectorate. *Ostrich* 37: 144.

Smithers, R.H.N. 1964. *A checklist of the birds of the Bechuanaland Protectorate and the Caprivi Strip.* Trustees of the National Museums of Southern Rhodesia, Cambridge.

Snow, D. 1978. *An atlas of speciation in African non-passerine birds.* Trustees of the British Museum (Natural History), London.

Soroczynski, Mike 1989. The occurrence of wader species in the Jwaneng mining area. *Babbler* 17: 38–39.

Soroczynski, Mike 1990. Bird species observed at Jwaneng Golf Club dam. *Babbler* 19: 16–17.

Spawls, Stephen 1987. Sighting of a Lizard Buzzard at Moeding College. *Babbler* 14: 22–23.

Summers-Smith, D. 1983. The Great Sparrow *Passer motitensis motitensis* (A. Smith) 1836. *Babbler* 6: 9–15.

Tarboton, W.R., Clinning, C.F. & Grond, M. 1975. Whiskered Tern breeding in the Transvaal. *Ostrich* 46: 188.

Tarboton, W.R. & Allan, D.G. 1984. The status and conservation of birds of prey in the Transvaal. *Transvaal Museum Monograph* 3.

Tarboton, W.R., Kemp, M.I. & Kemp, A.C. 1987. *Birds of the Transvaal.* Transvaal Museum, Pretoria.

Tarboton, W.R. 1980. Avian populations in Transvaal savanna. *Proc. 4th Pan-Afr. Orn. Congr.*: 113–124.

Tarboton, W.R. 1982. Breeding status of Black Stork in the Transvaal. *Ostrich* 53: 151–156.

Tinley, K.L. 1973. An ecological reconnaissance of the Moremi Wildlife Reserve, northern Okavango Swamps, Botswana. *Okavango Wildlife Society.*

Tree, A.J. 1972. Mass wintering of Palaearctic waders at Lake Ngami, Botswana in 1970. *Ostrich* 43: 139.

Tree, A.J. 1972. Pectoral Sandpiper *Calidris melanotus* in Botswana. *Ostrich* 43: 184.

Tree, A.J. 1972. Ornithological comparisons between differing dry seasons at a pan in Botswana. *Ostrich* 43: 165–168.

Tree, A.J. 1979. Biology of the Greenshank in southern Africa. *Ostrich* 50: 240–251.

Tree, A.J. 1986. Redwinged Pratincole breeding on Darwendale Dam in 1986. *Honeyguide* 33: 15–17.

Tree, A.J. 1986. What is the status of the Pearlbreasted Swallow in Zimbabwe? *Honeyguide* 32: 65–67.

Tree, A.J. 1986. The European Sandmartin in Zimbabwe. *Honeyguide* 32: 5–9.

Tree, A.J. 1986. The Banded Sand Martin in Zimbabwe. *Honeyguide* 32: 10–12.

Tomlinson, D.N.S. 1979. Interspecific relations in a mixed heronry. *Ostrich* 50: 193–198.

Underhill, L.G. 1987. Waders (*Charadrii*) and other water birds at Langebaan Lagoon, South Africa, 1975–1986. *Ostrich* 58: 145–155.

UNESCO/AETFAT/UNSO 1981. *The vegetation map of Africa.* Unesco, Paris.

Urban, Emil K. 1985. Status of cranes in Africa. *Proc. 6th Pan-Afr. Orn. Congr.*: 315–329.

Urban, E.K. 1985. Monitoring cranes in Botswana. *Babbler* 10: 6–7.

Urban, Emil K., Fry, C. Hilary & Keith, Stuart 1986. *The birds of Africa*, Volume 2. Academic Press, London.

Urquhart, Ewan 1986. European Marsh Harrier at Mana Pools National Park. *Honeyguide* 32: 152.

Van der Merwe, Francois 1981. Review of the status and biology of the Black Harrier. *Ostrich* 52: 193–207.

Van Voorthuizen, E.G. 1976. The Mopane tree. *Botswana Notes and Records* 8: 223–230.

Van Voorthuizen, E.G. 1976. Preliminary utilisation studies with special reference to Western Ngwaketse. *Botswana Notes and Records* 8: 157–163.

Vernon, C.J. 1962. Passerinae at Francistown, Bechuanaland Protectorate. *Ostrich* 33: 38.

Vernon, C.J. 1962. The occurrence of the South African Cliff Swallow in southern Rhodesia. *Ostrich* 33: 53.

Vernon, C.J. 1970. Palaearctic warblers in the Transvaal. *Ostrich* 41: 218.

Vernon, C.J. 1983. Notes on the Monotonous or White-tailed Bush Lark in Zimbabwe. *Honeyguide* 113: 19–21.

Vernon, C.J. 1985. Bird populations in two woodlands near Lake Kyle, Zimbabwe. *Honeyguide* 31: 148–161.

Vernon, C.J. 1987. On the Eastern Greenbacked Honeyguide. *Honeyguide* 33: 6–12.

Von Richter, W. 1970. Wildlife and rural economy in S.W. Botswana. *Botswana Notes and Records* 2: 85–94.

Von Richter, W. 1973. Recent publications in the field of wildlife and related topics in Botswana. *Botswana Notes and Records* 5: 220–224.

Wall, H. 1992. A Cape Robin *Cossypha caffra* in Gaborone—first record for Botswana. *Babbler* 24: 26–27.

Wayland, E.J. 1981. Past climates and present groundwater supplies in the Bechuanaland Protectorate. *Botswana Notes and Records* 13: 13–18.

Weare, P.R. & Yalala, A. 1971. Provisional vegetation map for Botswana. *Botswana Notes and Records* 3: 131–147.

White, F. 1983. *The vegetation of Africa.* UNESCO/AETFAT/UNSO, Paris.

Wilson, J.R. 1981. Sightings of Pectoral Sandpiper, Knot, Redshank and Bartailed Godwit in Botswana. *Ostrich* 52: 255.

Wilson, J.R. 1981. Observation of waders at Mogobane Dam, southeast Botswana. *Babbler* 1: 8–11.

Wilson, J.R. 1981. The birds of the Pitsane grasslands. *Babbler* 2: 20–21.

Wilson, J.R. 1984. The avifauna of the Lobatse area, S.E. Botswana. *Babbler* 8: 17–43.

Wilson, R.M. 1985. Birdwatching in the Selebi-Phikwe District. *Babbler* 10: 8–13.

Winterbottom, J.M. 1967. On some birds from the Kunene river, South West Africa. *Ostrich* 38: 155.

Winterbottom, J.M. 1969. On the birds of the sandveld Kalahari of South West Africa. *Ostrich* 40: 182–239.

Woollard, E. & Woollard, J. 1989. Some records on Hadeda Ibises in Gaborone Village. *Babbler* 17: 38–39.

Index of English names

Index of scientific names